D1223741

The Office Laboratory

second edition

Lois Anne Addison, MA, MLT, (ASCP)
Office Laboratory Consultant
Dunrobin, Ontario, Canada

Formerly Laboratory Coordinator, Family Practice Center
North Carolina Memorial Hospital
Department of Family Medicine
University of North Carolina
Chapel Hill, North Carolina

Paul M. Fischer, MD
Associate Professor
Department of Family Medicine
Medical College of Georgia
Augusta, Georgia

APPLETON & LANGE
Norwalk, Connecticut

0-8385-7244-8

Notice: Our knowledge in clinical sciences is constantly changing. As new information becomes available, changes in treatment and in the use of drugs become necessary. The authors and the publisher of this volume have taken care to make certain that the doses of drugs and schedules of treatment are correct and compatible with the standards generally accepted at the time of publication. The reader is advised to consult carefully the instruction and information material included in the package insert of each drug or therapeutic agent before administration. This advice is especially important when using new or infrequently used drugs.

90 91 92 93 94 / 10 9 8 7 6 5 4 3 2 1

Prentice Hall International (UK) Limited, *London*
Prentice Hall of Australia Pty. Limited, *Sydney*
Prentice Hall Canada, Inc., *Toronto*
Prentice Hall Hispanoamericana, S.A., *Mexico*
Prentice Hall of India Private Limited, *New Delhi*
Prentice Hall of Japan, Inc., *Tokyo*
Simon & Schuster Asia Pte. Ltd., *Singapore*
Editoria Prentice Hall do Brasil Ltda., *Rio de Janeiro*
Prentice Hall, *Englewood Cliffs, New Jersey*

Addison, Lois A.
The office laboratory / Lois Anne Addison. Paul M. Fischer. — 2nd ed.
p. cm.
Rev. ed. of: The Office Laboratory / Paul M. Fischer . . .(et al.).
c1983.
Includes bibliographies and index.
ISBN 0-8385-7244-8
1. Diagnosis, Laboratory. I. Fischer, Paul M., M.D. II. Title.
(DNLM: 1. Diagnosis, Laboratory. 2. Diagnostic Tests, Routine—methods. 3. Health Facilities. 4. Laboratories—organization & administration. 5. Physician's Offices. QY 4 A225o)
RB37.A33 1990
616.07'5—dc20
DNLM/DLC
for Library of Congress

89-1751
CI

Production Editor: Eileen Lagoss Burns
Designer: Michael J. Kelly

PRINTED IN THE UNITED STATES OF AMERICA

To Dorothy and Asma

Contents

Preface

In the 5 years since we published the first edition of this book, we have seen the subject of office laboratories move from relative obscurity to national prominence. Office laboratory issues have been debated in the lay press (*Wall Street Journal* and *The New York Times*) and in the halls of Congress. We have been actively involved in this national discussion and have tried to bring our perspective to it.

Much of the national debate has centered on concerns about the quality of office testing. We are extremely optimistic that these concerns about test quality will be resolved as clinicians learn how to be more effective in their role as "laboratory directors." We hope that this book is a resource for that process.

We have learned a great deal about "quality control." Often we have found that the traditional dogma was based on assumptions that were not supported by data and that are, to a large extent, outdated by technology. We have therefore taken a fresh look at the whole issue of quality assurance, pushing to identify what can go wrong and how to check for these *real* problems. We have tried to provide reasonable and meaningful quality assurance guidelines. In doing so, our recommendations are sometimes in conflict with standard dogma. Where possible, we have taken into consideration the fact that the clinical context often provides the final quality control check on any laboratory testing. The clinician's role in spotting "laboratory error" has been ignored for too long.

In the end, it has been our work, as an office laboratory technician and as a practicing family physician, that has provided the experiences on which this book is based. It has been written to provide useful information to clinicians and laboratory workers. It is our hope that it will be used day after day until its edges are ragged and its pages are splattered with stain.

Lois Anne Addison, MA, MLT (ASCP)
Dunrobin, Ontario, Canada

Paul M. Fischer, MD
Augusta, Georgia

Acknowledgments

We wish to thank:

The residents in our departments. Their desire to learn and their expectations of reliable results have pushed us to be learners with them as well as their teachers;

Our patients, who have endured our poking and prodding and have trusted us with their health care;

The staff of our office laboratories, including Rosemarie Sally and Beverly Anderson in Chapel Hill, and Dianne Mathis, Pamela Mentrup, and Tonita Dennis in Augusta;

The staff in the North Carolina Memorial Hospital, Department of Hospital Laboratories, who have become our friends and colleagues;

Jan Bane and Velveeta Tanksley for typing the manuscript and enduring untold numbers of corrections;

Joseph Tollison, MD, Chairman, Department of Family Medicine, Medical College of Georgia, who has supported our involvement in the national debate of office laboratory issues; and

Richard Coolen, PhD, Eastman Kodak Company, who has shown great patience in teaching a family doctor to think like a clinical chemist.

CHAPTER 1

How to Use This Book

SECTION 1: Introduction

We wrote the first edition of this book because there was no "bench manual" written for the office laboratory. That remains true even now that there have been several books written about office testing. This book is a practical operating manual for the small office laboratory. It is grounded in our 10 years of experience in such laboratories. You will find that it is filled with the kind of information that is usually not written down but is passed person to person by those who actually work in the laboratory.

Sometimes "what is" is best defined by "what it is not." This is true of *The Office Laboratory*. There is little emphasis on theoretic information. We have tried to provide only the information necessary to perform and interpret laboratory tests. Also, it is not an exhaustive review of all possible techniques or methodologies for performing each test, nor are specific brands of equipment evaluated. While the actual test procedures are described in great detail, we do not mean to suggest that our way is the only way to perform these tests. We have found that these procedures work in the office setting and that if followed they will provide reliable results.

The Office Laboratory has been written with three different uses in mind: as a procedure manual, as a clinician's guide to directing an office laboratory, and as an instruction manual. Both SI and conventional units are used. The SI units are given first, the conventional units follow in parentheses.

SECTION 2: Bench Manual

Each of the procedures described in this book is written in clear and easy-to-follow steps, without jargon or technical terminology. We assumed that the reader has little, if any, previous laboratory knowledge. To provide useful information, the procedures are organized as follows:

1. Background
2. Test indication
3. Test materials
4. Specimen collection
5. Test procedure
6. Reference values/test results/interpretation
7. Quality control
8. Sources of error
9. Comments
10. References

There are occasional deviations from this format, but by and large it is followed.

A bench manual serves as the standard for how tests are to be performed in the laboratory. Having all laboratory staff perform the test in the same way helps to reduce error. If some tests that you perform are not described in this book, or you perform the test in a different manner, use the suggested format to write your own procedure for your laboratory's manual.

In addition to describing how to actually perform the test, a bench manual serves as a reference manual for a wide range of information about the test. While you may remember how to perform the test, you may forget how long the specimen is "good" (i.e., when performing a sedimentation rate, an EDTA tube of blood is good for 2 hours at room temperature, 12 hours if refrigerated), the causes of a false-positive ferning test, or the reference values for an O'Sullivan glucose tolerance test.

SECTION 3: Clinicians' Guide to Directing an Office Laboratory

The clinician fills many roles in the office laboratory, including teacher, technician, administrator, and designer. Yet few have had any formal training to prepare them for these functions.

If you are setting up a practice, you will probably be planning your new office. Chapter 5 ("Laboratory Design") offers help in designing the laboratory space. Chapter 10 ("Microscope"), Chapter 11 ("Centrifuge"), Section 3 of Chapter 15 ("Hematology"), Section 1 of Chapter 18 ("Microbiology"), and Chapter 13 ("Chemistry Testing") should serve as useful guidelines for purchasing equipment. Chapter 6 ("Personnel") provides helpful information on how to choose a laboratory worker as well as how to analyze personnel difficulties. In fact, it can serve as a reference for issues about hiring any office worker. Chapter 7 ("Reference Laboratories") lists questions to ask when deciding which reference laboratory to use.

Chapter 3 ("Choosing Office Tests") provides the kind of information that you need in order to decide which tests to perform in the office and which to send to a reference laboratory.

Finally, Chapter 8 ("Quality Assurance") offers an overall approach to quality assurance. This has become a more crucial issue now that office laboratories must be licensed.

In a small office the clinician may be responsible for performing some testing or for teaching the procedures to the nursing personnel. The following section explains how to use this book as an instruction manual.

SECTION 4: Instruction Manual

The procedure sections (chapters 10 through 21) can serve as a basic workbook for teaching how to perform common laboratory tests. The learner may be a member of the clerical staff, a medical student, resident, family nurse practitioner, physician's assistant, or laboratory technician student. The teaching may take place in a classroom, a teaching laboratory, or in a functioning office laboratory. Whatever the location or the type of learner, the approach is similar.

It is a common experience in medicine to be expected to know something that you have never been taught—ask any third-year medical student who has been asked to draw blood from a patient for the first time. This "see one, do one, teach one" approach is totally inappropriate for learning good laboratory technique.

The teachers' first responsibility is to decide what can be taught in the allotted time. There is never enough time to teach everything. Decide what are the most important procedures, and use the time to teach them well. Set down clear and specific objectives (i.e., "The student will be able to differentiate segmented neutrophils, bands, and mononuclear cells" rather than "The student will be able to examine a peripheral smear."). Ask "What are the minimal tasks that the student should be able to competently perform at the end of the training?" Explain your expectations and how you will determine if the student has met them.

This book can be easily read, and therefore teaching time does not need to be spent lecturing on its content. Instead, the time should be used for practical, hands-on learning. Look through the microscope. Practice blood drawing. Read culture plates. These activities are fun, and you will therefore be well received as a teacher.

In each procedure section some background clinical information is provided. This material is important because it helps the student to put the test into a clinical context so that the test becomes relevant. If patient specimens are presented as unknowns with some clinical history, the teaching will be more interesting and the learning more fun.

The most effective use of laboratory teaching time is to spread it out over an extended period of time. The student can then practice what is learned in the daily clinical setting. A single block of time, on the other hand, is useful only for teaching a few concepts. Remember, you are primarily teaching skills, not facts.

If you are the student and are learning a test procedure for the first time, read the entire test section in detail. For some sections it is best to also read the introduction to the chapter. If a test kit is used, read the product insert as well. Underline and note all information that you want to remember. Then lay out all of the test materials. Practice performing the test procedure until you feel comfortable with its steps. This may take some time, so be patient. In some cases you may want to get added experience in a larger reference laboratory (e.g., a hospital laboratory for the interpretation of urine sediments). After you have gained confidence, check yourself by running duplicate samples or by sending a second specimen to a reference laboratory for testing. In this way you will be assured that you are ready to perform testing on patient specimens.

If you have previous experience with hospital laboratory procedures, much of this manual will be familiar. Doing a Gram's stain is something that you can do without thinking. But there are other tests, such as a vaginal wet preparation, that may be unfamiliar to you. Furthermore, there will be differences between the procedures as done in the hospital and in the office laboratory.

The functioning office laboratory is the ideal setting to teach office laboratory procedures. Here patient specimens arrive continuously, and diagnoses are continuously made. This adds "life" to the learning experience. What better way to learn to perform a pregnancy test than with a fresh urine specimen from an excited and hopeful mother-to-be. Large-group teaching can be efficient for the teacher, but it lacks the excitement of one-on-one teaching for the student.

Anyone who is involved in laboratory teaching will want to start a collection of microscope "teaching slides." Techniques for preparing these slides are described in the Appendix. You may also want to start a reference file of journal articles and a library. The books and articles referred to in the procedure sections of this book are a good start for such a library.

Finally, we would appreciate readers' comments, techniques, or helpful hints. Write to The Office Laboratory, c/o Appleton & Lange, 25 Van Zant St., E. Norwalk, CT 06855.

CHAPTER 2

The Office Laboratory and the Changing Health Care Environment

"It is dangerous to make predictions—especially about the future."
Kerr White, MD

SECTION 1: Introduction

In the early 1980s when we were preparing the first edition of this book, there was little interest in the office laboratory. Physicians had always done a few simple tests in their office, and it appeared that they would continue to do so. There was no way to anticipate the growth and changes that have occurred in this field in the past 5 years. These changes are the result of new technology, new regulations, and a dramatic shift in the way physicians are reimbursed for their services. In this chapter some of these changes are summarized to provide the reader with a sense of how office laboratory testing fits into the broader medical environment.

SECTION 2: New Technology

The two most influential technologic innovations in office testing are the use of monoclonal antibody test systems and the development of small multichemistry analyzers. Monoclonal antibodies are homogeneous immunoglobulins directed at a single antigen. The antibody molecules are identical and can be produced continuously in large quantities from cell cultures. To make a test the monoclonal antibodies are linked to a variety of reactions so that an endpoint can be detected when the monoclonal antibody attaches to the specific test antigen in the patient specimen.

Monoclonal antibody tests are widely used for the rapid detection of group A streptococcus in patients with sore throats and to detect human chorionic gonadotropin (hCG) in urine (i.e., pregnancy tests). These two types of rapid tests have dramatically changed the care of patients seen in the office. Group A streptococcus pharyngitis can now be diagnosed while the patient waits, and the decision to use antibiotics can be made rationally. This is in contrast to several years ago when 1 to 2 days were required before a throat culture could reliably identify patients with streptococcus pharyngitis.

Likewise, the new urine pregnancy tests can rapidly diagnose pregnancy prior to the missed menses. Furthermore, they are very sensitive in identifying patients with ectopic pregnancies. This is in contrast to the first generation of slide pregnancy tests that were not reliably positive until seven weeks after conception and that were notoriously negative in patients with ectopic pregnancies.

In the new few years we expect that many new monoclonal antibody tests will be available for the office laboratory. These should include more sensitive and specific tests for blood in the stool, other rapid microbiology tests (chlamydia and gonorrhea), tests for human immunodeficiency virus (HIV), and new screening tests for cancer. In each case a scientist must first identify an antigen that is specific for a disease and then link the detection of that antigen through a monoclonal antibody system.

The second area of rapid technologic change has been the development of small chemistry analyzers for the office laboratory. Many of these instruments are based on "dry reagent" systems. Dry reagent systems were first developed for the dip-stick technology that is routinely used in urine testing. Several reactions can be linked together on different layers of a dry reagent pad. Dry reagents have eliminated the need to store, measure, and handle the wet reagents that were, until recently, the basis for most chemistry testing.

Manufacturers have linked dry reagent testing to very sophisticated, microprocessor-controlled instruments. The computers in these instruments are able to evaluate the slide, specimen, reaction temperature, and reaction time; to set the proper wavelength for reading the endpoint; and, finally, to calculate the test value from this reading. The small boxes that sit in office laboratories today can perform up to 30 different chemistry tests. Twenty years ago it would have required a large, sophisticated laboratory and a small army of technicians to do these same tests.

Many instruments now incorporate bar-code technology. A variety of information can be encoded on this bar-code system. The bar-code system can tell the instrument the type of test that is being run, whether the reagents are outdated, and even what calibration parameters to set. Some large laboratories use bar coding to track patient specimens throughout the hospital. This level of sophistication is usually not needed for most office laboratories.

Some of the currently available chemistry instruments can test whole-blood specimens. This saves time because plasma or serum does not need to be generated. The procedure also has the added advantage of being able to use fingerstick specimens, which can be helpful in pediatric patients or elderly patients with difficult veins. For whole-blood–based instruments, the initial specimen may be whole blood, but the analyzed specimen is usually plasma. A variety of filtration or centrifugation steps are involved to separate the plasma from the blood cells. We expect that many manufacturers will move in the direction of whole blood technology because of its advantages in the office setting.

A final innovation in chemistry testing is the use of kit tests to quantitatively measure various analytes without the need for an instrument. The early use of this technology has been limited to measuring blood levels of therapeutic drugs (e.g., theophylline); however, we anticipate that the uses will broaden.

SECTION 3: Reimbursement

If the changes in office laboratory technology can be described as "rapid," then the changes in the reimbursement area should be described as "supersonic." It would be impossible and foolish to try to provide timely and accurate information about the ever-changing reimbursement situation. To paraphrase what Nebraskans say about their weather, "If you don't like the reimbursement plan, wait an hour." Nevertheless, there are some themes that will likely influence how physicians are reimbursed for office testing.

The most important factor is that the payers of health care (i.e., Medicaid, Medicare, and private insurance companies) will continue to reimburse at lower and lower rates. The rationale for this is that the rates charged in the past often failed to reflect actual test costs. Not too long ago physicians or reference laboratories could more or less charge whatever they wanted and expect full reimbursement. Physicians could even add on an "interpretation charge" for work done by the reference laboratory.

Traditionally the reimbursement for a test has been based on the cost of doing the test when it was first introduced. At the time of introduction, the test might have been done in only a few laboratories with very expensive technology. As the test became cheaper to perform, there were few mechanisms to drive the reimbursement rate down to a level compatible with the actual test costs. The health care payers are now trying to pressure the system so that reimbursement rates are at their lowest possible marketplace level. In addition, there are moves to establish "appropriate" tests given a specific diagnosis. Reimbursement levels in the future will be lower and will be more strictly tied to international classification of disease (ICD) diagnosis.

In addition to lower reimbursement, most office laboratories will be competing in a world of universal reimbursement rates. This means that the small office laboratory that does manual white blood cell counts (a very costly and labor intensive procedure) will be reimbursed at the same rate as a centralized reference laboratory with automated instruments that can do 1,000 complete blood counts (CBCs) per hour. This shift to universal reimbursement rates has been brought about in part by the efforts of laboratorians who have argued that their laboratories had an unfair cost disadvantage because they were forced to meet expensive government regulations. The quality control procedures that are used in a typical hospital laboratory account for 30% of all operating costs!

It is advisable that most physicians not view the office laboratory as a major source of new revenue but rather as a service to their patients. What this means in practice is that physicians will continue to provide tests that are clinically useful as long as they do not lose money in performing the test.

Hospital diagnosis-related groups (DRGs) have had an interesting effect on office laboratory testing. They immediately forced physicians to do more intensive outpatient treatment because of early hospital discharge. This has led to an increase in the amount of testing done in office laboratories. The second effect has been a willingness for hospitals to accept the test results from office laboratories. After all, any test done on a hospitalized patient under DRGs now becomes a cost to the hospital. If the physician does the needed admission testing in his or her office laboratory and then incorporates this into the hospital record, hospitals will save money. This has led to a dramatic reversal in the way hospital laboratorians view office laboratories. It has also raised concerns about the equivalence of testing performed in the office laboratory.

The final trend that will likely affect reimbursement for office testing is the shift to prepaid, capitated, health care (i.e., health maintenance organizations [HMOs], preferred provider organizations [PPOs], etc). Office laboratories are usually more expensive in their cost-per-test rates than reference laboratories. However, the availability of office testing can lead to a marked increase in the efficiency of providing patient care because the test results are available when the patient is seen by the physician. The overall cost for care may therefore be lower.

We expect that in the long run the improved efficiency of providing care with a well-equipped office laboratory will be supported by the prepaid health care systems. This view is not now shared by many HMO administrators, who have centralized testing in an effort to reduce costs per test.

SECTION 4: Regulation

In 1967 Congress passed the first Clinical Laboratories Improvement Act (CLIA). In passing CLIA Congress mandated that the Centers for Disease Control (CDC) develop standards for quality assurance programs for hospital and reference laboratories. Office laboratories were exempted under the initial legislation. The program developed by the CDC included on-site inspection and proficiency testing and required training of personnel. Congress's intent for this legislation was to provide regulations that would ensure that clinical testing was of a uniformly high quality.

In the late 1970s and early 1980s there was new interest in the quality of testing in office laboratories. This paralleled the great increase in the volume and variety of testing that was being done at those sites. By 1987 nearly one third of all states had enacted regulations for office laboratories. However, with the exception of a few states, these regulations were never seriously enforced.

By 1988 a score of news reports on television, in magazines, and in newspapers discussed the potential threat to the public that was caused by poor laboratory testing. Largely in response to this publicity, Congress held several hearings to discuss changes in laboratory regulation. Papanicolaou (Pap) smear testing and office laboratory testing received special attention in these hearings. The result of this activity was the passage of the Clinical Laboratory Improvement Amendment of 1988. For the first time this bill established the federal licensure of all office laboratories.

At the time of this writing the specific details of the office laboratory licensure program had yet to be worked out by the Health Care Financing Administration. However, it is likely that this program will include some of the following elements:

- Licensure. Congress decided not to tie this program merely to Medicare and Medicaid reimbursement programs. Instead, it decided that all laboratories required licensure regardless of the source of patient payment.
- Limited laboratory exemption. Those laboratories that limited their testing to easy-to-use technologies would be exempted from any regulated activity except to apply for licensure.
- Proficiency testing. All nonexempted laboratories would be required to participate in an approved proficiency testing program and to submit the results of

their laboratory on proficiency testing to the federal government. This serves as the principal method by which the federal government is assured that the testing is of a high quality.

- On-site inspections. On-site laboratory inspections will be required for some laboratories. On-site inspection has been used in the past as one routine part of the licensure process for hospital and reference laboratories. However, inspection will probably not be required of all office laboratories because of the monumental task involved (10,000 hospital and reference laboratories compared to over 100,000 office laboratories).
- Deemed status. Congress agreed to give some private organizations the power to accredit laboratories. At the time of this writing, the only organization that appears to be likely to receive deemed status is the Commission on Office Laboratory Assessment (COLA). This voluntary organization was developed by the joint efforts of the American Academy of Family Physicians, American Society of Internal Medicine, College of American Pathologists, and American Medical Association.

No one likes to be regulated. This is certainly true of those clinicians who operate office laboratories. It is Congress' hope that the regulation of office laboratories will promote the improvement of the quality of testing in these laboratories without compromising the ability of physicians to test in their office.

SECTION 5: Liability

Whenever a physician extends the scope of his or her practice, additional liability occurs. For the physician who is expanding the testing within the office laboratory, a new and special increased liability exists. This added risk can be minimized if steps are taken to prevent errors and if corrective action is taken once an error is discovered. This section describes how to best manage the increased liability when directing an office laboratory.

First, it should be noted that there have been very few malpractice cases related to office laboratory testing. There have in fact been very few cases related to *any* laboratory testing. We believe that the reason for this is that a clinician usually interfaces between the patient and the test. One of the clinician's responsibilities is to see if the test result "makes sense" in light of what is known about the patient. When it does not make sense, the clinician often says that the result is a "laboratory error." Often this is true.

When the clinician functions in this capacity, he or she serves as a final quality control check for the laboratory. It is in this way that many erroneous laboratory results fail to harm the patient and therefore do not lead to malpractice cases. This may change as the public becomes more aware of the laboratory's role in their health care and more aware that laboratory staff can be sued.

One issue that is addressed in liability law is the "standard of care." The traditional standard of care has been based on the standards in the community in which the physician practices (i.e., the "locality rule"). However, the standard of care relating to laboratory testing is a nationally recognized standard of care. This is because of the well-published guidelines that laboratories are expected to follow. It is likely that a physician operating an office laboratory will therefore be subject to the national standards of laboratory practice rather than just the standards found in local office laboratories.

Since the physician serves as the director of the office laboratory, he or she is held responsible for the testing of all individuals who work in the office laboratory under the director's supervision. This is the same as for hospital-based pathologists who are held liable for any errors made by laboratory technicians in the hospital laboratory.

The most important step for the laboratory director to take in limiting liability is to ensure good testing and to record the quality of this testing by following a standard quality-control program. The details of such a program are outlined throughout this book. For liability purposes it is not enough just to have done quality control procedures. In addition, the results of these procedures *must be recorded*. It is hardly convincing to tell a judge that daily controls are run for each chemistry test unless the records are available to prove that it was done and to document how good the testing really was.

In all laboratories, even the best hospital laboratories, mistakes will occur. It has been estimated that even in good laboratories, transcriptional errors in reporting results occur at a frequency of about 1% to 2%! If a problem is identified, do not ignore it. In a malpractice case it will be important to show that when problems were identified, corrective action was taken.

Finally, it is important not to cover up an error. The clinician should immediately contact the patient, tell them that an error was made, and make appropriate plans for follow-up. In the office environment, where the patient knows the physician and the office staff, patients are usually understanding. One of us recently had to inform a middle-aged woman that she did not in fact have gonorrhea, as the laboratory had originally reported. The honesty of the clinician and the understanding of the patient can prevent court cases when such sensitive errors are made.

CHAPTER 3

Choosing Tests for the Office Laboratory

Until very recently it was easy to answer the question "What tests should be done in the office laboratory?" Most laboratory tests required expensive equipment and specially trained technicians. They were therefore performed in reference laboratories after specimen collection in the physician's office. Office laboratory testing was usually limited to hematocrits, urinalysis, and other common but simple procedures.

Today's technology has changed all of that. In an environment where almost anything is possible, the physician must make some hard decisions about which tests to perform in the office setting. This short chapter is designed to outline the issues that should be considered in making these decisions.

1. What is the level of test reimbursement?

   The current trend is toward a system of universal rates for test reimbursement. Gone are the days when a physician would be reimbursed at a higher rate for a test than the reference laboratory doing the same procedure. Also gone are the days when there were great differences among states on reimbursement levels or between the reimbursement from private insurers compared to Medicare/Medicaid. Also gone are the days when the reimbursement for a test was considerably greater than the actual cost to do the test. Health care in the United States is quickly moving toward a system of universal rates of reimbursement in which rates are set at "what the market will bear."

   While financial incentives have driven much of the discussion about office laboratory testing, we have found that most physicians choose to do office testing because of the way that it improves the efficiency and quality of patient care. Few clinicians go to the trouble of setting up a laboratory just to increase their profits. Most physicians will choose to perform a test in the laboratory if they can at least "break even."

   The crucial figure in the financial decision is not what is charged but rather what is the standard rate of reimbursement. A good starting point is therefore to obtain a list of rates of reimbursement for laboratory tests from your state Medicare provider. The standard rates of reimbursement should be compared to your costs for doing the test. There is no simple equation to calculate these costs; however, some costs are often overlooked. Be sure to factor in the following costs:

a. Reagents. This should be an easy cost to estimate, based on test volume. In general, dry chemistry reagents cost more than wet chemistry reagents, and individually packaged reagents are more expensive than those that are bulk packaged. For infrequently performed tests, your greatest reagent expense may be from unused and outdated reagents.
b. Calibration. Instruments that require frequent calibration can be very costly. For example, some of the early chemistry analyzers had to be recalibrated each week. This requires staff time, the cost of "calibration" solutions, and other reagents.
c. Controls. In large hospital and reference laboratories, it is estimated that quality control costs account for 30% of all laboratory expenses! In office laboratories the low test volume means that controls may be run almost as frequently as the tests themselves.
d. Instruments. It is not uncommon for an office chemistry analyzer or hematology cell counter to cost from $5,000 to $20,000. This makes the laboratory equipment one of the most expensive purchases for an office practice. Manufacturers offer a variety of plans, including purchase, lease, or reagent rental. In the latter, the use of the equipment is related to the purchase of a given quantity of reagents each month.
e. Maintenance contracts. Even "trouble and maintenance-free" equipment can break down. The reality of this fact is clear when you realize that the cost for the average maintenance contract is 10% to 15% of the purchase price. When signing up for this type of agreement, be sure to find out if it includes all of the routine preventive maintenance, whether "loaner" instruments are available when yours is being fixed, and what is the guaranteed response time when you have a problem.
f. Licensure. As of 1990 all office laboratories will be required to have a license. In addition to the actual licensing fee, this may mean a variety of other new costs, such as more quality control testing and participation in a proficiency testing program.
g. Labor. This is the biggest expense for any laboratory. It is difficult to calculate a "real labor cost" for the office laboratory because there is usually some mixing of responsibilities among all of the office staff. Furthermore, reducing the laboratory staff (and therefore the quantity of office testing) will *produce* other labor costs, such as nurses calling patients the next day with test results and front desk staff pulling charts and filing laboratory slips from the reference laboratory.

2. Can your office staff properly perform the test?

As described in chapter 6, the ability to do good laboratory testing is not a universal trait. Today's kits and instruments are greatly simplified, yet there is no such thing as a "foolproof" test. You should know whether your staff has adequate time, training, and skills to do the tests. If you rely on busy nursing staff to do the majority of the testing, you will probably want to limit the type and volume of testing that your laboratory performs.

3. Is the laboratory big enough?

   Look carefully at your laboratory space. Is there sufficient bench-top space for a new instrument? Is the refrigerator large enough to store new types of reagents? We have known many laboratories that thought they were buying a "typewriter-size instrument" and had forgotten about pipettes, test-tube racks, refrigerated reagents, and so on.

4. What is the test volume?

   As a general rule of thumb, the fewer tests that are done in a given period of time, the more costly each test will be. In our experience it is almost impossible to make an accurate "guess" about how many tests you order. There is a general tendency to overestimate this figure. It is therefore better to actually keep a record of the tests to be sure that the test volume will justify switching from the reference laboratory to the office laboratory for a particular procedure.

5. Can you purchase the reagents in a quantity suitable for your needs?

   The key points to consider here are the storage space required to store bulk reagents and the shelf life of the reagents. Storage space can be a problem for supplies such as urine containers. You might be able to buy them inexpensively if you order 1,000 at a time; however, you would probably have too little space to store them. Reagent shelf life is a second issue. The most common situation here is that the reagents become outdated before they can be used. This may be the critical factor in deciding to do an office test for the more uncommon procedures.

6. How much staff time is required?

   If staff are currently underused, then adding new tests may not affect the overall office personnel costs. However, if staff are already maximally working, it might require adding a new person. In this case a little bit of testing can be a very expensive proposition.

   Testing for group A streptococcus in patients with pharyngitis is a good example of how complicated the issue of personnel time can be. If a standard throat culture is done, the hands-on time to streak the plate and read the culture may be only a few minutes. But staff are then often involved in calling patients back, calling in prescriptions, pulling charts to record the culture results, and so on. Doing a rapid streptococcus test often requires 10 to 15 minutes of hands-on time by the staff. But the test is done while the patient is in the office so that no follow-up effort is required. In our experience this makes rapid streptococcus testing very competitive with throat culturing in terms of the personnel time.

7. What laboratory services are available?

   As described in Chapter 7, every office requires a reference laboratory to perform the testing that is not done in the office laboratory. If your office is located next to the hospital and you use the hospital laboratory for much of your testing, then you may be able to limit the testing done in the office laboratory. If, on the other hand, your practice is not conveniently located near a referral laboratory, you will probably depend on a commercial reference laboratory with courier or mailed specimen handling. In the latter case most results would not be available for 2 to 3 days. In such an office it would often improve patient care if more testing were done in the office laboratory.

8. Is there a good clinical reason to perform the test?

   The availability of testing is part of the accepted "standard of care." If you are doing obstetrics, it is considered essential to be able to rapidly detect proteinuria to help in the diagnosis of preeclampsia. If you are caring for a diabetic, it is the standard of care to be able to rapidly assess the level of the blood glucose. To treat women with a vaginal discharge, it is considered the standard of care to be able to perform a vaginal wet prep. We expect that as technology continues to improve, more testing will be considered "essential" as part of providing good ambulatory care.

CHAPTER 4

How Good Is the Test?

"The diagnosis of disease is often easy, often difficult and often impossible."
Peter Mere Latham (1789–1875)

SECTION 1: Introduction

The number of clinical laboratory tests performed in the United States increased in the 1980s at a rate of 9% per year. The cost for these tests is expected to reach $6 billion by 1990. Laboratory testing currently accounts for 10% of all health care expenditures.

The clinician of the 1950s had few laboratory tests available in the hospital and even fewer tests in the office. Today the clinician can order innumerable specialized laboratory procedures from hospital and reference laboratories. Furthermore, the clinician's office laboratory has grown in size, importance, and sophistication. Many office laboratories now perform procedures that were unavailable even to research laboratories 20 years ago.

Despite the rapid growth in laboratory medicine and the clinician's increased reliance on laboratory results, little has been done to formally educate clinicians about the appropriate use of laboratory test data. In many medical training programs, the accepted approach to ordering laboratory tests is to "cover all the bases." The finding of a low hemoglobin level often leads to the reflex ordering of tests for serum iron, iron-binding capacity, reticulocyte count, Coombs' test, folate, B_{12}, and hemoglobin electrophoresis. All too often a test result is viewed as the only needed criterion for the presence or absence of disease. In some instances basic clinical judgment is totally replaced by broad-spectrum testing: "The results of ten blood chemistries are all normal, which means that you must be in good health!" Such conclusions demonstrate the misunderstanding and misuse of laboratory data.

To properly use the laboratory, the clinician must understand the relationships between the test procedure, the patient, the patient population, and the disease entity. The difficulty in understanding these relationships was illustrated by a study at Harvard Medical School. Researchers asked 20 fourth-year medical students, 20 house officers, and 20 attending physicians the following question: "If a test to detect a disease whose prevalence is 1 in 1,000 has a false positive rate of 5%, what is the chance that a person found to have a positive result actually has the disease, assuming that you know nothing about the person's symptoms or signs?" The correct answer (which is, less than 2%) was given by only four students, three residents, and four attending physicians! The most frequent incorrect answer was 95%.

This chapter will examine several aspects of testing. First, the various uses for laboratory tests in clinical practice will be reviewed. Next, the statistical concepts needed to properly understand testing will be introduced, followed by the patient and procedure factors that affect test results. Finally, the issues of cost-effectiveness in the use of laboratory tests will be presented.

At first glance this chapter may appear to be just a "bunch of irrelevant statistics." Nothing could be further from the truth! The reality of clinical medicine is such that it may be impossible to apply a test result to an individual patient with any degree of quantifiable certainty, but the concepts in this chapter form the basis for many of the common clinical behaviors of test interpretation:

- Have you recently ignored a borderline elevated alkaline phosphatase in an otherwise healthy person who had no good reason to have either bone or biliary disease?
- Have you treated someone with antibiotics who had pharyngitis, even though the rapid streptococcus test was negative?
- Have you used the term "laboratory error" for an unexpected test result?

There are very good reasons for each of these clinical behaviors. By the end of this chapter, you will know why.

SECTION 2: Clinical Uses for Laboratory Testing

"Diagnosis is not the end, but the beginning of practice."
Martin Fischer (1879–1962)

Laboratory testing has a wide variety of uses in modern clinical medicine. In his book on this subject, Lundberg lists 32 different reasons, including "nothing else to do." Some of the most common reasons for ordering tests include:

- Diagnosis. The most common reason for ordering a test is to confirm a diagnosis that is suspected by the clinical history or the physical examination. The urine pregnancy test in a woman who has missed a period is an example of a procedure that is used in this way. Such a test should be *specific* for the diagnosis that is suspected.
- Screening. Some common illnesses are asymptomatic and undetectable by physical examination in the early stages. Laboratory tests can be used to screen healthy patients to detect these unsuspected illnesses. The Pap smear is used in this way to detect cervical cancer. Such screening tests should be *sensitive* to detect all possible cases of the disease.
- Disease monitoring. Certain chronic diseases can be followed over time with laboratory results. The fasting blood sugar is used in this way to monitor the course of a patient with diabetes mellitus.
- Therapy selection. Once a diagnosis has been made, the clinician must often choose from among a variety of therapies. Laboratory data can be helpful in this decision (i.e., the use of the sputum Gram's stain to choose appropriate antibiotics in a patient with pneumonia).
- To measure patient compliance. Serum drug levels allow the clinician to assess compliance with the prescribed therapy and to individualize the drug dosage. Serum aspirin and theophylline levels are used in this way in the treatment of arthritis and asthma, respectively.

- To measure therapy response. It is often important to follow a patient's response to a prescribed therapy. As an example, the hematocrit can be used to measure an anemic patient's response to oral iron therapy.
- Medicolegal Aids. The clinician is called upon to document health or disease for disability hearings, insurance purposes, and employment examinations. Laboratory testing is often relied upon in these cases to verify clinical findings. Common examples of this include the use of the VDRL as part of the premarital examination or the serum alcohol level in the patient suspected of intoxication.
- Patient education. Office laboratory testing is a tremendous resource for patient education. Patients can be shown testing methods (i.e., teaching a new diabetic the proper technique for measuring whole blood glucose). Children often enjoy looking under the microscope, and hopeful parents enjoy watching their positive pregnancy test develop.
- Research. The research laboratories found in tertiary care centers provide much of the data that has advanced modern medicine. Office laboratories can provide similar help in solving questions about common clinical problems, many of which have not been adequately studied. What better place than the office laboratory for research on the laboratory findings in vaginitis or streptococcal pharyngitis.

SECTION 3: There Is No Perfect Test

"A judicious man looks at statistics not to get knowledge but to save himself from having ignorance foisted on him."

Thomas Carlyle (1795–1881)

When faced with a patient's new complaint, a clinician's first responsibility is to "make the diagnosis." This process typically involves taking a medical history, doing a physical examination, ordering a confirmatory laboratory test, and then weighing the possibilities. Too often the clinician equates a positive result from a laboratory test as conclusive evidence of the disease and a negative result as "ruling out" the suspected illness. In a perfect world with perfect clinical testing, this would be true. That is, unfortunately, not the case.

Each test has a *false-positive* and a *false-negative* rate. The false-positive rate is the frequency of positive tests in people without the disease in question. False positives may be due to a variety of factors, including

- Patient factors, such as a falsely positive pregnancy test in a woman with a choriocarcinoma
- Specimen collection. If a urine sample is allowed to remain unrefrigerated before plating, the bacteria will multiply. This may result in a falsely positive urine culture
- Test factors. Over-reading a throat culture that has beta hemolysis due to organisms other than group A streptococci will produce a false-positive result

The false-negative rate is the frequency of negative tests in patients with the disease in question. False negatives may be due to a variety of factors, including

- Patient factors, such as a falsely negative pregnancy test because a woman has had several cups of coffee and has produced a very dilute urine
- Specimen collection, such as a gonorrhea culture that fails to grow out the organism because the swab was streaked onto a cold agar plate right out of the refrigerator
- Test factors, such as a falsely negative mononucleosis test in a person early in the course of the disease, before the test is reliably positive

For each test there will therefore be

- True positives (TP): the number of diseased patients with positive tests
- False positives (FP): the number of nondiseased patients with positive tests
- True negatives (TN): the number of nondiseased patients with negative tests
- False negatives (FN): the number of diseased patients with negative tests

The value for these four types of results allows us to determine a test's *sensitivity* and *specificity*. The *sensitivity* of a test is the rate of positive results obtained when a test is applied to patients known to have the disease (i.e., the percentage of all people with the disease who will have a positive test). Sensitivity is expressed as a percentage and is defined mathematically as:

$$\text{Sensitivity} = \frac{\text{True positives}}{\text{Total diseased patients}}$$

$$= \frac{\text{TP}}{\text{TP} + \text{FN}} \times 100$$

An example of a sensitive test is the MONO-TEST. This test has a 98% sensitivity (i.e., in 100 patients with known infectious mononucleosis, 98 will have a positive test). The remaining 2% represent false negatives.

The *specificity* of a test is the rate of negative results obtained when a test is applied to patients known to be free of the disease (i.e., the percentage of people without the disease will have a negative test). Specificity is defined mathematically as:

$$\text{Specificity} = \frac{\text{True negatives}}{\text{Total nondiseased patients}}$$

$$= \frac{\text{TN}}{\text{TN} + \text{FP}} \times 100$$

Of the two concepts (sensitivity and specificity), specificity is the one most commonly misunderstood. Specificity refers to how well a negative test correctly predicts the absence of the disease. The MONO-TEST has a 99% specificity; therefore if 100 people without infectious mononucleosis are given the test, 99 will have a negative test. The remaining 1% represents false positives.

To apply these two concepts to an individual patient's test results, the disease *prevalence* must be known. Prevalence is the frequency of diseased patients in a given population at a point in time. It is usually expressed as a ratio of diseased people to the total population (i.e., one in ten Americans is obese). Prevalence is often confused with *incidence*. Incidence is the number of new cases of a disease reported during a specified time interval (i.e., nine persons out of 100 developed flu during the 1988 to 1989 winter). Prevalence is a function of both the disease incidence and the duration of the disease. A chronic disease, such as diabetes, might have a low incidence (i.e., few new cases per year) but a high prevalence because the disease lasts for many years.

The clinical context in which the test is ordered determines the disease prevalence. This is a difficult and important concept to grasp. Consider the example of the MONO-TEST. It could be used as a part of a routine screening panel for every asymptomatic patient who comes to a clinician's office. Alternatively, it could be used as a diagnostic test for every patient who has a sore throat. Finally, it could be used in the patient population with fever, sore throat, and generalized adenopathy. These represent three populations, each with a different prevalence of infectious mononucleosis.

We know that the MONO-TEST has a 99% specificity and therefore 1% of all patients without the disease will have a positive test. If this test were used in a screening panel on every asymptomatic patient, the majority of positive test results would represent false positives in this low prevalence population. Therefore a positive test would instinctively be discarded in that situation as "laboratory error." In the patient population with fever, sore throat, and generalized adenopathy, the prevalence of infectious mononucleosis is much higher. A positive test in this situation would be viewed with more justification as confirmative proof of the disease. In these two cases the test specificity and sensitivity are the same! Yet by experience a positive test is interpreted in totally different ways.

In a patient care setting a common question is "What is the likelihood that the patient with a positive test has the disease that is being tested for?" This value is termed the *predictive value for a positive test* (+PV) and is defined as the percentage of patients with a positive test who are diseased:

$$+\text{PV} = \frac{\text{True positive tests}}{\text{Total positive tests}}$$

$$= \frac{\text{TP}}{\text{TP} + \text{FP}} \times 100$$

Alternatively, the *predictive value for a negative test* (−PV) is the percentage of patients with a negative test who are not diseased:

$$-\text{PV} = \frac{\text{True negative tests}}{\text{Total negative tests}}$$

$$= \frac{\text{TN}}{\text{TN} + \text{FN}} \times 100$$

In common usage the term "predictive value" refers to the +PV.

As has been seen, the clinician intuitively interprets laboratory results in light of the presumed likelihood that the patient is diseased. Knowing the test sensitivity, the test specificity, and the disease prevalence, the test's predictive value can be calculated. The easiest way to calculate this is with a truth table. This table is based on the premise that in any population a patient can be placed into one of four categories based on the test result and the presence or absence of the disease. These four groups are true positives, true negatives, false positives, and false negatives. See Table 4–1.

Let us analyze the problem presented to the Harvard Medical School staff. A test has a 95% specificity and a 100% sensitivity. The disease prevalence is 1 in 1,000. An arbitrary patient population size of 10,000 is chosen. (The lower the disease prevalence, the larger this chosen population size should be.) If 10,000 patients were to be tested, the number of diseased patients would be

$$10{,}000 \times 0.001 = 10$$

Since the test sensitivity is 100%, all of these patients would have a positive test.

TABLE 4–1. TRUTH TABLE

	+ Tests	− Tests	Total
Diseased	TP	FN	TP + FN
Nondiseased	FP	TN	FP + TN
Total	TP + FP	FN + TN	TP + FN + FP + TN

TP, true positive; FN, false negative; FP, false positive; TN, true negative.

The number of nondiseased patients would be 10,000 − 10 = 9,990. Since the test has a 95% specificity, it will have a 5% false-positive rate in this group of patients. The number of patients with false-positive tests is therefore

$$5\% \times 9{,}990 = 500$$

The remaining 9,490 patients will have true-negative test results. These data can be put into a truth table:

	+ tests	− tests	Total
Diseased	10	0	10
Nondiseased	500	9,490	9,990
Total	510	9,490	10,000

The positive predictive value is the total number of true positives divided by the total number of positive test results:

$$+\text{PV} = \frac{\text{TP}}{\text{TP} + \text{FP}} = \frac{10}{510} = 1.96\%$$

Less than 2% of the 510 patients with a positive test will have the disease in this patient population! If a clinician used good clinical judgment to narrow down the patient population so that the disease prevalence was 1 in 10, the truth table would show:

	+ Tests	− Tests	Total
Diseased	1,000	0	1,000
Nondiseased	450	8,550	9,000
Total	1,450	8,550	10,000

The +PV is then = 1,000/1,450 = 69%.

This greatly improved predictive value is the reason why careful clinical judgment and proper test selection are needed at all points in the clinical decision-making process.

A second method to calculate the positive predictive value is to use the 200-year-old formula developed by Bayes. This formula can be algebraically derived from the previously given definitions of specificity, sensitivity, prevalence, and predictive value. Bayes formula states that

$$+\text{PV} = \frac{(\text{Prev})\,(\text{Sens})}{[(\text{Prev})\,(\text{Sens})] + [(1-\text{Prev})\,(1-\text{Spec})]}$$

This forbidding looking formula may be quicker to use but in practice is not as conceptually satisfying as filling in the numbers in a truth table.

Following are three clinical examples of the use of predictive value in laboratory diagnosis. Both truth tables and Bayes formulations are presented.

Example 1: A throat culture is used to screen patients for group A streptococcal infection. In the office setting it has been shown to have a sensitivity of 90%, but the specificity is only 70%. The 30% false-positive rate is because of patients who are streptococcus carriers but in whom the organism is not an active pathogen. The prevalence of strep pharyngitis in children with a sore throat is reported to be 20%. What is the positive predictive value of this test in a group of symptomatic children?

An arbitrary population size of 1,000 is chosen. Since the disease prevalence is 20%, the number of diseased patients is .20 × 1,000 = 200. The nondiseased group is therefore 800. The test sensitivity is 90%; therefore, 90% of the 200 diseased patients will have a positive test:

$$\text{TP} = 0.90 \times 200 = 180$$

There will be 20 false-negative results (i.e., patients with the disease who have negative throat cultures).

The specificity is 70%; therefore 70% of nondiseased patients will have negative tests. The number of true negatives is

$$\text{TN} = 0.70 \times 800 = 560$$

TABLE 4–2. TRUTH TABLE FOR STREPTOCOCCUS CULTURE

	+ Tests	− Tests	Total
Diseased	180	20	200
Nondiseased	240	560	800
Total	420	580	1,000

Specificity, 70%; sensitivity, 90%; prevalence, 20%; population size, 1,000.

The remaining 240 patients will be false positives. The positive predictive value is the number of true positives divided by the total number of positives:

$$+\text{PV} = \frac{180}{180 + 240} = 43\%$$

Table 4–2 is a truth table for throat cultures. By Bayes formula:

$$+\text{PV} = \frac{(0.20)\,(0.90)}{[(0.20)\,(0.90)] + [(1-0.20)(1-0.70)]} = 43\%$$

This low predictive value (43%) is a result of both the low prevalence of the disease in the given population and the high rate of false-positive tests.

Example 2: The MONO-TEST has a specificity of 99% and a sensitivity of 98%. Six percent of patients with clinically suspected infectious mononucleosis actually have the disease. What is the positive predictive value of this test in the population of patients who are suspected of having the disease? (See Table 4–3.)

By Bayes Formula:

$$+\text{PV} = \frac{(0.06)\,(0.98)}{[(0.06)\,(0.98) + (1-0.06)\,(1-0.99)]} = 87\%$$

Therefore only 87% of patients with clinically suspected infectious mononucleosis and a positive Monospot actually have the disease! This surprisingly low +PV is in spite of a high sensitivity and specificity and is due to the low disease prevalence—even in that group of patients who are clinically suspected of having infectious mononucleosis.

Example 3: What would be the positive predictive value of a test with a 95% sensitivity and a 100% specificity if the disease prevalence was 1 in 1,000? (If the answer is not obviously 100% to you, go back and think about the concepts presented so far in this chapter.)

As has been explained, each test has a sensitivity and specificity. When these are applied to a particular population, a predictive value for a positive or negative test can

TABLE 4–3. TRUTH TABLE FOR MONO-TEST

	+ Tests	− Tests	Total
Diseased	59	1	60
Nondiseased	9	931	940
Total	68	932	1,000

Sensitivity, 98%; specificity, 99%; prevalence, 6%; population size, 1,000.

$$+\text{PV} = \frac{59}{68} = 87\%$$

be determined. The differences in predictive values require that the clinician use the right test for each clinical need. There are no perfect tests, but there are proper and improper clinical uses for any test.

A test with a high sensitivity will be effective in identifying all patients with a disease. Unfortunately, a high rate of false positives may be the price paid for this high sensitivity. Such a test is useful for screening large populations to identify people with an unsuspected disease. The patients with a positive test can then be further evaluated with a more specific test to separate those who are true positives from those who are falsely positive. An example of such a sensitive test is the newborn test for phenylketonuria (PKU). This is a common screening blood test. A positive test is suggestive of PKU but requires a more specific test (the serum phenylalanine) to confirm the diagnosis.

A test with a high specificity is used to diagnose disease in a situation in which the frequency of false-positive values must be minimized because treatment of a falsely positive patient would have serious and unwanted effects. In the office laboratory, the testing for gonorrhea should have a high specificity. All positive tests have serious social and medical consequences. The frequency of false positives must therefore be minimized to prevent the unnecessary treatment of nondiseased patients and their sexual partners.

At any given disease prevalence, an increasing specificity of the test will result in an increase in the +PV of the test (i.e., a positive test is more likely to indicate disease). Likewise, an increase in the sensitivity will result in an increase in the −PV of the test (i.e., a negative test is more likely to indicate the absence of the disease).

When the specificity and sensitivity are added together, the maximum summed value is 200% (100% + 100%). If flipping a coin were used as a clinical test to differentiate diseased from healthy patients, its specificity would be 50%, and its sensitivity would be 50%. The sum of these two would be 100%. Any test with a combined specificity and sensitivity equal to or less than 100% is no better than chance—and there are such tests! One do-it-yourself urine pregnancy test was studied by the Centers for Disease Control (CDC). The test claimed to detect substances found in the urine of pregnant women 2 to 4 weeks after conception. In a carefully controlled study, the CDC found that the test had a sensitivity of 50% and a specificity of 40.9%. The test was therefore no better than chance for detecting pregnancy!

In clinical practice the sensitivity and specificity of a particular test may vary from those cited in the literature. One example of this is the disparity between throat cultures done by the hospital bacteriology laboratories and cultures done in the clinician's office. Differences in techniques and differences in the level of skill of the person interpreting the culture account for the differences seen in throat culture sensitivity and specificity in these two laboratory settings.

A second problem with being able to quantitatively use predictive value calculations in practice is that the disease prevalence is usually not known. Many of the cited prevalence rates come from research that was carried out on very selected populations of patients. The prevalence rates in these settings are likely to be very different from the rates found in a primary care office. How many of your patients with headache have a brain tumor? How many of your patients with fatigue have chronic mononucleosis syndrome? How many with unexplained weight loss are found to have cancer? These are common clinical problems for which good data about disease prevalence are hard to find.

A final problem with putting predictive value into quantifiable practice is that there is often debate about the "gold standard" (i.e., the test that you use to definitively know whether a patient is diseased). Even for a condition as well studied as streptococcal pharyngitis, this is so. Is the gold standard a culture? If so, what technique do you use for culturing? Do you use ASO titers? If the titers go up, it is likely that the patient had a streptococcal infection; but many patients who are believed to be infected never have a rise in titer. A major obstacle to studying streptococcal pharyngitis is that we currently have no reliable way to identify which patients have viral pharyngitis and are carriers as opposed to those who actually have "strep pharyngitis."

SECTION 4: Ranges

"There is no royal road to diagnosis."

Robert Tuttle Morris (1857–1945)

Until now tests have been referred to as being either positive or negative. The sensitivity and specificity of a test are inherent in the actual procedure used for testing. This "all-or-nothing" type of testing is seen in many office laboratory tests: pregnancy tests, gonorrhea cultures, ferning tests, or skin scrapings for fungi. The test is either positive or negative. These are called "qualitative tests."

There is another group of tests whose results describe a range of values. These are called "quantitative tests." These tests include the serum glucose, hematocrit, colony count on a urine culture, or level of proteinuria. In each case test results will range over a variety of values, and it is the clinician's job to decide if the value represents health or disease. Fortunately, sensitivity, specificity, and predictive value are applicable to this second type of testing.

In the early 1800s the German mathematician Johann Gauss noticed that a large number of random observations produced a symmetrical bell-shaped curve around a peak that was the mean value. This "Gaussian curve" (or normal curve) has been used to describe many clinical tests. The variability of any population can be quantified by calculating a "standard deviation" (SD). The Gaussian curve is shaped such that +1 to −1 SD encompass 68% of all results, and +2 to −2 SD encompass 95% of results.

In many clinical situations the 95% limits are arbitrarily assigned as the high and low borders for the *normal range*. If a test is within the normal range, it falls within the 95% area of the curve for values obtained from a healthy population.

Unfortunately there are many difficulties in applying a statistical concept such as a normal curve to a laboratory definition of "normal," or to a clinical definition of "healthy." This is because biologic populations are not homogeneous, and test values often do not nicely fit into a Gaussian distribution. In fact, of the 12 common blood chemistry tests found on a screening panel, only albumin fits a normal curve. If a sufficient number of albumin tests are performed on a population of healthy patients, the number of values that fall within +2 to −2 SD of the mean will be 95% of all of the tests.

To solve the problem of non-Gaussian distributions, another type of statistic must be used. For tests of this sort, a large number of healthy patient values are used to construct a curve of test results. The upper and lower 2.5% of values are then arbitrarily chosen as the borders for the "reference range." Note that the word "normal" is not used. It is all too easy to confuse normal with healthy, and so the term "reference range" is more appropriate. A reference range is determined for each different laboratory for each test. The same test done by two different laboratories may have different reference ranges because of subtle differences in procedure or differences in the population used to establish the reference range. It is important to know the ranges for the particular laboratory that is doing your testing. It is unwise to use nationally published ranges without first checking on the ranges established by the laboratory performing your testing.

SECTION 5: Patient Factors That Affect Laboratory Values

"To know what kind of a person has a disease is as essential as to know what kind of disease a person has."

Francis Scott Smyth (1895–)

Since people are not a single homogeneous group (a condition that even statisticians would dread), they sometimes need to be divided into smaller, more homogeneous test populations. Age and sex are the two characteristics that are most commonly used to divide populations for developing specific laboratory test reference ranges. There are, in fact, only a few tests for which results are independent of age and sex. These include Na, K, Cl, CO_2, and pH. For most of the other common tests, the patient's age and sex determine a specific reference range. For example, the reference values for the white blood count and differential counts in children are very different than for adults. The clinician should be aware of these different ranges even if the laboratory reports only a single reference range for all population groups.

There are a wide variety of patient factors less obvious than age or sex that affect test values.

- Posture. Standing causes blood to pool in the legs, which leads to leakage of fluid from the intravascular space. Bed-bound patients who do not have these forces at work can have chemistry values out of the reference range established on healthy and active people. Albumin is an example of a test affected by this factor.
- Diet. Food intake will affect a variety of tests, including the serum glucose, uric acid, and triglycerides. This is the factor that limits the usefulness of the random glucose test in the assessment of a patient with diabetes mellitus.
- Time of day. Plasma cortisol levels vary with the body's diurnal cycles.
- Activity: Vigorous physical activity can result in the finding of protein, red blood cells, and even casts in the urinalysis of people with healthy renal function. This has been referred to as "athletic pseudonephritis."
- Pregnancy. This will affect a wide range of laboratory tests, including the hematocrit, thyroid function tests, and the erythrocyte sedimentation rate.

- Medication. Drugs will affect many tests, including such obvious ones as the prothrombin time (PT) with sodium warfarin (Coumadin) and the elevated thyroxine seen in patients taking birth control pills.

This list of patient factors that affect laboratory tests is far from complete. A comprehensive listing of these patient factors can be found in two books: *Effects of Disease on Clinical Laboratory Tests* and *Effects of Drugs on Clinical Laboratory Tests.* Each book can be ordered by writing to American Association for Clinical Chemistry, 1725 K St. NW, Suite 903, Washington, DC 20006; or call 202-835-8753. Both books were updated in 1989.

SECTION 6: Test Factors That Affect Laboratory Values

"When laboratory tests conflict with clinical judgment, don't discard the latter before repeating the laboratory test."

Paul Reznikoff (1895–)

Precision and *accuracy* are the final two concepts that must be understood about laboratory testing. Precision refers to the reproducibility of a test. Each quantitative test can be described by a level of precision (i.e., a level of reproducibility obtained when aliquots of a single sample are tested repeatedly).

An example of precision would be to run a series of hematocrits on a single specimen of blood. In common office practice, the microhematocrit has a precision of ± 2%. Serial measurements on the same sample by the same person and with the same technique should fall within 4% of each other. Other examples of precision in the office laboratory can be determined for any quantitative test, such as the serum glucose, white blood count, or sedimentation rate. Office laboratory equipment will commonly measure serum glucose to a precision of ± 10 mg/dL. More sophisticated equipment can produce a precision of ± 1 mg/dL, but this added precision is not cost-effective for routine clinical use. Test imprecision is conveniently used by clinicians. It is what is often blamed when an unexpected value is considered a laboratory error.

The accuracy of a test is a measure of how closely the test result approaches the true quantitative value. An example of an accuracy problem would be to use an inadequate centrifugation time for performing a microhematocrit. If the same shortened centrifuge time period is used with each test, the precision would be fine (i.e., all tests would be within 4% of each other); but the cells would not be adequately packed and the hematocrit values would all be higher than the true value. They would be precise but inaccurate.

SECTION 7: The More I Know of Testing, the Less of It I Do

"A misleading symptom is misleading only to one able to be misled."
Sir Heneage Ogilvie (1887–1974)

When the factors that affect a person's health are considered, 90% of the variability has been found to be due to genetics, socioeconomics, and life-style. Only about 10% of the variability is explained by access to modern medical care! There has been little effort to identify what portion of this 10% is due to diagnostic testing. It is likely, however, that diagnostic testing is responsible for a smaller improvement in public health than surgery or pharmacotherapy.

One fact that is clear is that clinicians are often not critical when ordering a test. One study has shown that only 5% of laboratory tests result in a change in a patient's therapy! It is often assumed that more testing means better care.

Laboratories have been equally at fault in contributing to this "more-is-better" philosophy. Today if a clinician wants to do a screening test for anemia, it is likely that the laboratory will report a CBC done on an automated instrument that includes 10 to 15 separate pieces of data. Some of this information will be duplicative (i.e., both hemoglobin and hematocrit), and some of it will be just plain irrelevant. An excellent example of this is the red blood cell count. This is a test reported on every CBC and yet one that has almost no clinical utility and for which most clinicians do not know how to use the data. The red blood cell count is on the CBC report not because the clinician needed it but rather because the cell counter must measure it to calculate some of the other red blood cell parameters.

A number of tests has resulted in a great deal of "noise" that interferes with the clinician's ability to "see" those laboratory results that will really make a difference. As a survival skill, clinicians have trained themselves to ignore laboratory data. They see a mildly elevated alkaline phosphatase in a patient not known to have biliary or bony disease and they ignore it. This act is often associated with a tinge of guilt and a concern about malpractice.

Ignoring the mildly elevated alkaline phosphatase is actually the clinician's intuitive way to put predictive value theory to work. He or she knows that the sensitivity for this test is low. Even in Paget's disease, 10% of patients will have a normal alkaline phosphatase! Likewise the specificity is very low. The clinician may have worked up several "healthy" patients with mildly elevated values and found no disease.

Next the clinician factors in the prevalence of the disease. Certainly any of the serious diseases associated with an elevated alkaline phosphatase are rare in the population as a whole. Even more importantly though, they are extremely rare in an asymptomatic person. From this intuitive process, the clinician assumes that the positive predictive value is low and moves on to the next laboratory value without missing a step.

It is currently in fashion to discuss predictive value in the care of patients. While we believe that the concepts outlined in this chapter are useful in more critically interpreting tests, we have found it nearly impossible to apply the equations in a quantifiable way to the care of an individual patient. The sensitivity or specificity of a test are rarely known. For many diseases there is no clear gold standard. There are almost no data about the prevalences of even the most common diseases seen in primary care. These prevalences are likely to be different in one practice from those in another. Furthermore, the natural history of diseases has been almost completely ignored by decision analysis. Tests are not either positive or negative. Rather they become sufficiently positive at a certain point in the natural history of a disease to be called "positive."

It is likely in the very near future that health care professionals will have both better tests and a better understanding of how to use test information in the care of patients. Until then the best that can be said is that there are a few instances when a test is clearly appropriate, a few when it is clearly inappropriate, and many other instances that are up to debate. Clinicians live in a sea of uncertainty.

CHAPTER 5

Laboratory Design

The planning, design, and construction of the laboratory are critical for the successful operation of a medical office. Yet such detailed planning is frequently neglected when a medical office building is being conceived. In some instances the "laboratory" is relegated to a closet or the end of a hall. Architects are often unfamiliar with the special needs of an office laboratory and will provide a design based on "kitchen" principles. Although both laboratories and kitchens have sinks and counters, important differences exist.

The planning of the laboratory must be a cooperative effort among the clinician, the architect, and the laboratory personnel. The clinician will need to decide what tests are to be performed and what volume of work is to be expected. The architect will translate these requirements into specific plans. The laboratory technician can provide information about space needs and equipment.

Large hospital or commercial laboratories must meet many specific building codes in terms of a laboratory's size, drainage pipes, counter surfaces, and air exhaust. Many of these requirements are due to the hospital laboratories' use of hazardous chemicals. In contrast, office laboratory testing does not usually require the use of caustic chemicals or dangerous biologic materials. There are therefore no specific building codes for these laboratories. Instead, office laboratories are governed by the same commercial building codes that apply to the rest of the office building.

This chapter outlines the steps needed to design a new office laboratory. It can also be used as a guide to redesign or improve an existing laboratory. The planning of a "generic" laboratory is included, but each office will have slightly different needs and resources, so each laboratory will need to be customized accordingly.

ASSESSMENT OF LABORATORY FUNCTION

The clinician must first determine the laboratory needs for the practice and clearly communicate these needs to the architect. This assessment should include the characteristics of the practice, the personnel, and which tests are to be performed.

General Practice Characteristics

The clinician must outline the type of practice (e.g., family medicine, pediatrics, internal medicine), the office location, the number of clinicians using the office, the proximity of the office to the reference laboratory, the projected number of patient visits, and the demographics of the patient population.

Laboratory Tests Offered

As a general guideline, the tests described in this book should be considered as reasonable for most primary care office laboratories. For a fuller discussion of test selection, refer to Chapter 3 ("Choosing Office Tests") and to the test procedure chapters (chapters 13 through 21).

Staffing and Organizing the Laboratory

The architect will need to know who will be performing most of the testing and what other office responsibilities that person will have. If a nurse performs the laboratory tests, the laboratory should be convenient to his or her other work areas. If a laboratory technician is employed, the laboratory will serve as both his or her primary work area and office. Modifications will be needed if the laboratory space is to be used for functions other than performing tests (e.g., storage space or the handling of laboratory-related administrative work).

Laboratory Workload and Space

The clinician should estimate the volume of laboratory work to to done. This will depend on the number of clinicians, the number of patients seen, the clinician's test-ordering practices, and the types of tests that are performed in the office. It is more difficult to expand laboratory space than it is to expand space for other office activities. It is therefore wise to initially design the laboratory to accommodate the long-term projected needs of the office. It is easy to add new tests and even new equipment into an existing laboratory, but it is difficult to increase the laboratory size.

Administrative Functions

The laboratory technician may be asked to keep records of tests performed, handle reference laboratory test results, order supplies, maintain quality control records, and report test results to patients over the telephone. If these administrative responsibilities are performed by the laboratory technician, appropriate desk and filing space should be planned into the laboratory design.

ARCHITECTURAL CONCEPTS

Once the clinician has determined the laboratory needs, the architect will translate them into a workable design.

Location Within the Office

The laboratory must occupy a central site within the office. It is usually convenient to place it near the entrance of the office so that patients coming for laboratory work can move directly from the waiting room, without passing through other areas. The laboratory should also be close to the business area so that patients' records can be easily accessed if the results are to be entered directly into the charts. If requisition forms or computers are used, proximity to the medical records' area is less important. The laboratory should be adjacent to a bathroom so that urine specimens can be collected without inconvenience to patients or staff. It is less critical for the laboratory to be near the examination rooms or the clinician's office. The architect must consider both staff and patient movement within the office before deciding on the optimum site for the laboratory (Fig. 5–1). Avoid having any of the laboratory space in the normal flow of patients (i.e., a hallway). Once the

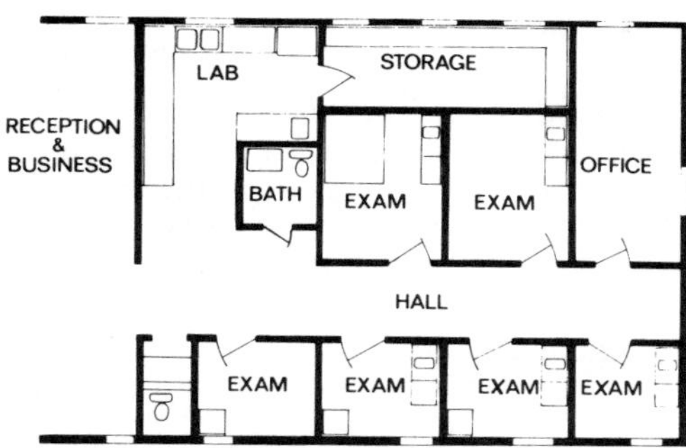

Figure 5–1. The laboratory should be placed in a central area in the office that is close to the patient waiting room, a bathroom, and the administrative area.

architect has developed a blueprint go through common patient scenarios (i.e., check in, examination, blood drawing) to see if they "work."

Laboratory Procedure Zones

The laboratory can be designed so that activities are grouped by function. This will assure the efficient use of space within the laboratory. A sample laboratory layout is shown in Figure 5-2.

Testing Area

This is the area where specimen analysis takes place. It should provide ample work surfaces for procedures and major equipment. Supplies should be readily accessible. There must be good lighting and sufficient electrical outlets. This area should be separated from the patient flow areas because of potential biologic and safety hazards.

Administrative Area

This area will include room for filing cabinets, book shelves, desks, a telephone, computer equipment, and so on.

Blood Drawing Area

The blood drawing area should provide some privacy. Female patients sometimes have to remove a tight-sleeved blouse or dress to have blood drawn. Consider having a curtain, sliding door, or screen to close off this area from the rest of the laboratory.

Bathroom

This should be directly adjacent to the laboratory and easily accessible to patients from the examination rooms. A pass-through window in the wall with doors on either side allows the transfer of urine specimens from the bathroom to the laboratory without patient embarrassment.

ARCHITECTURAL DETAILS

The architect will decide on many details about the laboratory:

Space

The laboratory size depends on the volume of laboratory tests, the desired equipment, the persons performing the testing, and the laboratory administrative functions. Ninety square feet is a minimum area for a two-clinician office with a full-time laboratory technician.

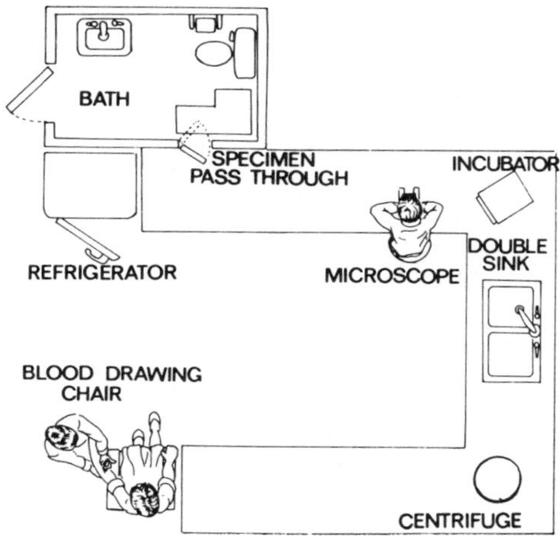

Figure 5–2. A sample small laboratory arrangement.

Shape
The floor plan will need to be individualized for each office. A U shape with the testing area at the base of the U minimizes traffic through the laboratory and maximizes available bench-top space.

Walls
The walls can be made of any material that meets local building codes. A temporary wall can be used if future expansion is planned.

Ceilings
Acoustical tile ceilings should be used to minimize the noise from vibrating laboratory equipment.

Windows
The laboratory should contain a minimum number of windows to limit fluctuations in room temperature. Temperature changes will affect many laboratory instruments. However, having at least one window helps worker morale. If there is a window, use blinds to control lighting. (It is helpful to shut out all day light when using a polarized microscope.)

Floors
The laboratory floor should not be carpeted because of the problem with chemical spills. Seamless floor is most easily maintained, but vinyl tiles are also satisfactory.

Bathroom
One- or two-patient bathrooms are needed for a two-clinician practice. One of these bathrooms should be adjacent to the laboratory but should have its entrance onto a patient-care hallway. One bathroom should be equipped with guard rails and a high toilet seat for handicapped patients. There should also be sufficient space to turn a wheelchair around in this bathroom and an extra wide door. A wheel chair-accessible bathroom requires about twice as much space as a normal office bathroom.

Air Exhaust
There is no need for specialized air venting systems or a hood in the office laboratory. However, the laboratory must have good ventilation with an exhaust to the outside. This is the same type of ventilation that will be used for the bathrooms in the office.

Electrical Outlets
For most office laboratory equipment 110 voltage is needed. Each outlet should be grounded. There should be an ample number of outlets in the laboratory, especially in the testing area. Most office laboratories will require six to ten outlets. Multiple outlets can be easily installed in the form of an outlet strip that is attached to the wall several inches above the bench top. If possible have the circuit breaker box near the laboratory so that it can be quickly reached in case of an electrical emergency.

Equipment
The laboratory equipment will vary with the type of tests performed. In general, space should be planned for a refrigerator (either full or half size), incubator, centrifuge, microscope, and a desk-top chemistry analyzer. Many test reagents require refrigeration, so a full-sized refrigerator should be considered for most office laboratories.

Plumbing
A double sink made of a nonstaining material, such as stainless steel, should be installed. Since no caustic chemicals are used in the office laboratory, there is no need for special pipes or other special plumbing.

Benches
Formica or metal bench tops should be used, since they repel water and are not easily stained. Expensive benches, such as those used in hospital laboratories, are not needed. The bench heights can be varied to provide areas for sitting (29 to 30 inches) as well as standing (36 to 37 inches). Knee

space should be provided beneath those benches where the technician will be sitting (e.g., beneath the microscope). Thirty running feet of bench space are needed for the testing function in most office laboratories.

Storage

Ample storage space can be provided by placing shelves over the benches. Adjustable shelving is the most versatile. A single shelf for holding chemical reagents should be placed over the sink. Cabinets are more expensive than shelves, but they provide storage space that can be easily locked. If space is limited, the cabinets should have sliding rather than swinging doors. Drawers are an inefficient storage system for most laboratory needs.

Telephone

A wall telephone with a long extension cord should be placed in the laboratory. It should be within easy reach of the laboratory's administrative area. Consider placing it lower on the wall than usual so that it can be answered while sitting down.

Furniture

The laboratory should be equipped with at least one height-adjustable stool with back support. The stool must be on casters to provide easy mobility within the laboratory.

Colors

All colors in the laboratory should be subdued. The laboratory is generally a small area with a great deal of activity. Loud colors will make working in the laboratory less enjoyable.

Lighting

There should be both surface lighting (under the bottom of the overhead cabinets) and ceiling lighting. Fluorescent tube lighting provides more light for the price than does incandescent lighting. Ask the architect to design enough lighting to provide 75 foot-candles of light on the countertops.

Temperature

The temperature in the laboratory should be stable, between 67 and 75 degrees Fahrenheit. If possible the laboratory should have its own thermostat control.

File Cabinet

At least one four-drawer file cabinet is needed to store test results.

Wall Space

With its many cabinets, the laboratory may end up with little wall space. Be sure to leave enough room for a wall clock (with a second hand), an eye-wash station, and a fire-blanket box.

CONCLUSION

This chapter provides an overview of planning and designing an office laboratory. It is important that the clinician work closely with the architect and the laboratory technician in this planning. Remember to allow enough space for laboratory expansion. With proper planning the long-range laboratory needs of the practice can be included in the design, however, we have found that eventually all laboratories become too small.

CHAPTER 6

Personnel

SECTION 1: Introduction

Since all laboratories use similar reagents and instruments, it is the personnel who distinguish a good laboratory from one that is bad. This is particularly true in office laboratories where much of the testing is done by individuals without formal laboratory training. Several surveys of office laboratory personnel have shown that in only about one third of all office laboratories is there a supervisory person with specific laboratory training. When one considers *all* individuals who actually perform tests, as we did in Georgia for the Centers for Disease Control (CDC), only 10% of office laboratory workers are formally trained.

This chapter describes the characteristics needed in a good office laboratory staff person, the certification for formally trained laboratory personnel, how to hire a person for your laboratory, and, finally, how to monitor and remedy personnel problems.

SECTION 2: Laboratory Personnel Characteristics

Not everyone has the same skills and talents. Some people are more technically oriented. Some have better interpersonal skills. Some have excellent hand-eye coordination while others are less adept. It is, in part, these differences that lead some physicians to become surgeons while others become psychiatrists. In the allied health fields, some go into nursing while others become medical technologists.

A conflict can arise when a person in one field is expected to be competent in a second. This has been known to happen in the office laboratory environment. For example, an office nurse who has excellent "bedside manners" and who has always been rewarded by her personal contacts with patients is suddenly asked to run a complicated chemistry analyzer. This may be the same person who avoids using electronic banking machines because she does not "like" them and does not "understand" them. Being an excellent nurse may not translate into being an excellent laboratory person.

Some individuals have "laboratory

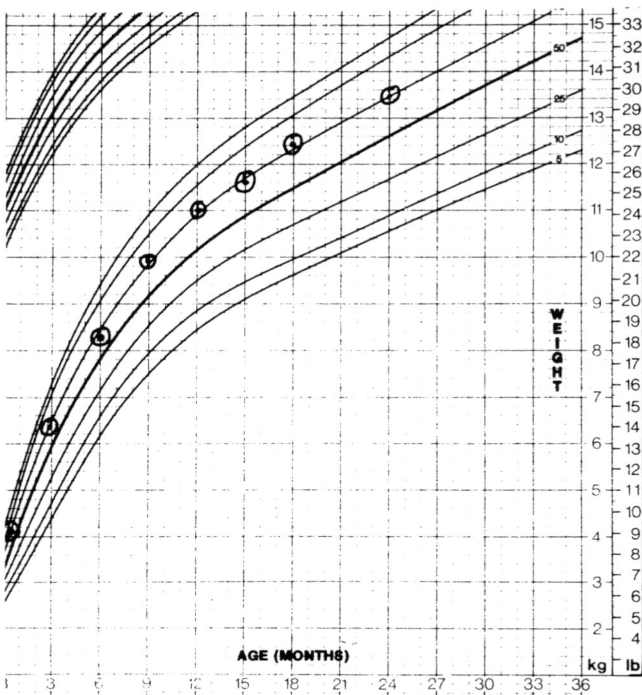

Figure 6–1. Office growth chart of a young boy. The weights were taken and plotted by a nurse who paid very close attention to the data collection.

dyslexia.'' This is an inability to function in a ''technical'' manner. Figures 6–1 and 6-2 demonstrate this concept with a common nursing function—weighing children on the office scales. Both figures show growth curves of a young boy. Figure 6–1 is done by someone who pays close attention (often subconsciously) to the details of weighing a child. All of the points sit closely to the expected curve (i.e., there is high ''precision''). Figure 6-2, on the other hand, is a growth curve completed by an office staff person who has laboratory dyslexia. The points are scattered widely above and below the line. This sort of variation is not due to changes in the child but rather to changes in how the weighing was performed.

Upon checking into how the weights were obtained for Figure 6-2, the likely finding is that the child was weighed in clothes on one occasion and naked on the next. The baby scales may not have been zeroed before taking the weight. If the child was fussy, the baby may have been weighed in the parent's arms on the adult scales and then the parent's weight subtracted. If the child was really fussy the nurse may just have asked the mother ''How much does Johnny weigh?'' While each of these is a reasonable way to collect data, the various methods will lead to great imprecision. When plotting these scattered points on the growth curve, a person with laboratory dyslexia does not ''see'' a problem with the data! (This is an actual example from the authors' practice. If you find it unbelievable, you should take a look at the growth charts in your own office.)

Office laboratory staff must be able to take a ''technical view of the world.'' Such persons enjoy figuring out how things

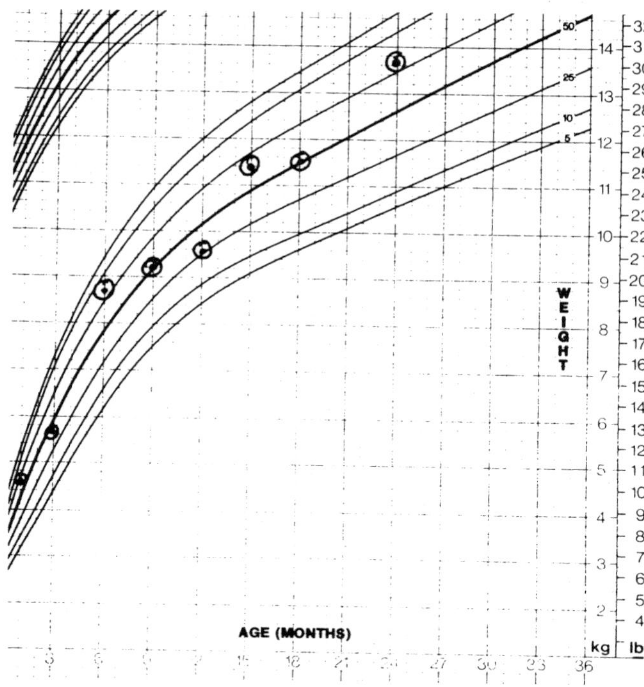

Figure 6–2. Growth chart of a young boy. The weights were taken and plotted by an office worker with laboratory dyslexia.

work. They are not frightened by instrument manuals. They are the person in the office who is relied on to fix the copy machine.

The second important characteristic is attention to detail. These individuals pay attention to the smallest irregularity. In their own lives they balance their checkbook to the penny and find this both easy and rewarding.

The next characteristic is excellent eye-hand coordination. This skill is important for pipetting, turning the microscope stage while doing a white blood count differential, and trouble shooting instrumentation. If you remember back to your high school and college chemistry classes when you did acid-base titration curves, there was always one individual in the class whose answer fell out of the range of the rest of the class. That individual probably had problems with eye-hand coordination.

The office laboratory worker must be able to deal with a great deal of stress. He or she is often required to do two things at once and will get a telephone call in the middle. This is in contrast to many hospital laboratorians, who are given protected time, a single test responsibility, and an even day-to-day work load.

The final quality that is essential in office laboratory staff is excellent interpersonal skills. This quality may be a relatively minor requirement for hospital or reference laboratory employees, but it is essential for those who work in an office laboratory. They will need to calm frightened children, discuss sensitive laboratory results with patients, and describe occasionally awkward specimen collection procedures (i.e., 24-hour urine collection). All of these require good communication skills, maturity, and tact.

SECTION 3: Role of Office Laboratorian

In a large group practice the office laboratorian's responsibilities will be very similar to the role of a technician in a hospital or reference laboratory. However, in smaller, more typical office laboratories, their role is unique in that they are directly involved with patients and are able to have a collegial relationship with the physician. These are two qualities that make the work in an office laboratory more attractive to some individuals than work in a hospital laboratory. The office laboratory staff get to know patients personally, are aware of their health problems, and are able to correlate these problems with the laboratory findings. In addition, they are often responsible for the full range of clinical tests. This is in contrast to the hospital laboratory, where a single individual rarely works with patients, has little contact with physicians, and may spend all day doing one type of test.

The office laboratorian is usually responsible for a variety of administrative tasks related to the laboratory. These include reporting of test procedures, keeping track of inventory, organizing a quality assurance program, and calculating the costs for doing testing. This broad range of administrative skills makes the office laboratorian's job a challenging one.

In smaller offices it is common for the office laboratory person to be cross trained for other office duties. These include helping with typical nursing functions, serving

TABLE 6-1. MINIMUM REQUIREMENTS FOR CERTIFICATION EXAMINATIONS BY AGENCY AND CERTIFICANT DESIGNATION FOR LABORATORY PERSONNEL

Technician		Technologist	
Designation	*Requirements*	*Designation*	*Requirements*
CLA (ASCP) Clinical laboratory assistant	One year training after secondary school	MT (ASCP) Medical technologist	BS degree with specified course work
MLT (ASCP) Medical laboratory technician	Formal training with associate degree or two years of college	CLS (NCA) Clinical laboratory scientist	Degree or college credit and experience
CLT (NCA) Clinical laboratory technician	Formal training or 4 years experience	MT (AMT) Medical technologist	3 years formal education and continuing experience
MLT (AMT) Medical laboratory technician	1 or 2 years training, often proprietary	RMT (ISCLT) Registered medical technologist	On-the-job training and experience
RLT (ISCLT) Registered laboratory technician	On-the-job training and experience	CLT (HHS) Clinical laboratory technologist	4 years experience or combination of education and experience

as an x-ray technician, doing the billing, and making appointments. Even if these are not usual responsibilities, it is not uncommon for the office laboratorian to be expected to "fill in" when someone is sick or on vacation.

SECTION 4: Technician/Technologist

As the office practice expands, there will come a time when the clinician will begin to consider hiring a full-time person for the office laboratory. This usually means somebody with formal training as a "technician" or a "technologist." There is a great deal of confusion regarding the different types of laboratory worker and their qualifications. Generally, technologists have more formal education than technicians. In traditional laboratories technologists often assume the supervisory role, while technicians are more likely to be doing the actual testing.

Unfortunately there is no standardized curriculum or certification process for technologist or technician. Table 6–1 lists the different types of designations along with the requirements for certification. Table 6–2 lists the names and addresses of those organizations that are involved in the certification of clinical laboratorians. For most office laboratories, an individual with either a technician or technologist background should be able to function as the laboratory supervisor.

The salaries of technologists are generally equivalent to those of registered nurses, while the salaries of technicians are equivalent to those of medical assistants or licensed practical nurses. A 1985 report of salaries at public health laboratories indicated that the average hourly pay for a technologist was \$7.45 to \$10.20, for a technician \$5.80 to \$7.90.

In many parts of the United States the office staff include certified medical assistants (CMA) or registered medical assistants (RMA). These personnel usually

TABLE 6–2. CERTIFICATION AGENCIES

1. ASCP—American Society of Clinical Pathologists
   Board of Registry
   PO Box 12270
   Chicago, IL 60612
   (312) 738-1336
2. AAMA—American Association of Medical Assistants
   Suite 1571
   20 N Wacker Dr
   Chicago, IL 60606
   (312) 899-1500
3. AMT—American Medical Technologists
   710 Higgins Rd
   Park Ridge, IL 60068
   (312) 823-5169
4. NCA—National Certification Agency for Medical Laboratory Personnel
   1725 DeSales St NW, Suite 403
   Washington, DC 20036
   (202) 429-0149
5. ISCLT—International Society for Laboratory Technology
   Suite 918
   818 Olive St
   St. Louis, MO 63101
   (314) 241-1445
6. HHS—United States Department of Health and Human Services
   Professional Examination Service, Inc
   475 Riverside Dr
   New York, NY 10115

receive 6 to 12 months of training that includes performing some basic laboratory tests, x-rays, and electrocardiograms; learning medical terminology; and managing billing. Their laboratory training will usually include blood-drawing skills, dip-stick tests, and hematocrits.

SECTION 5: Writing A Job Description

Long and complicated job descriptions are not the norm in the hiring practices of office personnel. It is, however, useful to sit down and decide what specific responsibilities the laboratorian will have. If you write down the responsibilities, you can give the list to applicants to review, use the list as a reference for discussions during the interviewing process, and use it to evaluate future performance. This will help to avoid situations such as, "You never told me that I was going to have to cover the front desk every Saturday morning." A good job description provides the basis for cooperative understanding among staff workers, ensures good working standards, and is a foundation for promotion or salary increases. Some typical responsibilities include:

ADMINISTRATIVE RESPONSIBILITIES

- Organize and maintain a filing system to include test reports.
- Maintain quality controls, and record the results of quality control procedures.
- Maintain a procedure manual.
- Evaluate equipment for cost-effectiveness.
- Identify local consultants and reference laboratories.
- Plan work schedules based on test requirements, patient load, and personnel activity.
- Observe safety precautions in storage and disposal of laboratory waste.
- Maintain an inventory of supplies and reagents. Interact with suppliers.
- Maintain good working relationships with the local laboratories.

TESTING RESPONSIBILITIES

- Perform a limited number of standard bacteriologic, hematologic, and chemical tests.
- Maintain equipment.
- Successfully participate in a proficiency testing program.
- Establish and maintain the laboratory's accreditation.

SUPERVISION

- Supervise other office staff who may be doing laboratory testing (i.e., a nurse who will be taught to draw blood).
- Ensure that any office staff member who works in the laboratory is adequately trained.
- Interact with nursing personnel on issues relating to patient care in the laboratory (i.e., the "patient flow" from examination rooms to the laboratory for blood drawing).

SECTION 6: Office Policies

Before hiring any new personnel, it is a good idea to write down the office's policies. It is very common for policies to evolve over time; however, they do provide a useful framework for the *do's* and *don'ts* of the office environment. The policy should include the following items:

- Dress code
- Personal conduct in the office
  - Confidentiality
  - Eating, drinking, or smoking
  - Tidiness
  - Punctuality
  - Discipline
- Telephone usage
  - Personal calls
  - Long distance calls
- Handling of reports or patient records
- Public relations
  - With patients
  - With nurses and receptionists
  - With detail men and laboratory representatives
- Supplies
  - File maintenance
  - Receipts
  - Inventory
  - Ordering
- Unusual situations
  - Managing complaints
  - Errors in reporting
  - Accidents in the laboratory
  - Termination policy
- Employee's role
  - Lines of communication and responsibilities
  - Work schedule
  - Pay scale, frequency of pay, and deductions
  - Work record
  - Periodic performance appraisal
  - Schedulde for promotion or pay increase
  - Lunch hours/coffee breaks
- Fringe benefits. The following are the minimum fringe benefits recommended by the American Society for Medical Technology.
  - Sick leave: One day of sick leave at regular pay per month
  - Holidays: A minimum of eight paid days per year
  - Vacation: Vacations shall begin to accrue with the date of employment but shall not be taken before the employee has worked for six months. This should include 10 working days for 1 to 2 years of employment, 15 days for 3 to 10 years of employment, and 20 days for over 10 years of employment.
  - Maternity leave: Pregnancy and childbirth-related leave shall be treated in the same terms as other temporary disabilities. The policy should conform to the Equal Employment Opportunity Commission's guidelines on discrimination because of sex.
  - Insurance benefits: Employees shall be eligible for major medical and disability insurance paid by the employer.
  - Pension plan: After 1 year of employment all employees shall be eligible for a pension plan.
  - Continuing education policy: At least 5 days per year should be given for continuing education. All expenses shall be paid by the employee to attend seminars, workshops, short courses, or scientific sessions.
- Surety bonding: All employees in the office who handle valuables or cash shall be covered by a special insurance bond. Surety bonding should be one of the conditions for employment and involves checking prospective employees for evidence of a criminal record.

SECTION 7: Job Advertising

All laboratories are now facing problems in finding qualified staff. This is because the number of people entering the laboratory technologist/technician field has decreased. This is primarily due to the limited advancement available in these positions as well as the opening up of many new careers to women. (Compare, for example, the earning potential of a person with a masters in business administration [MBA] and an MLT.)

In terms of attracting qualified laboratory staff, the office laboratory has some advantages over hospital or reference laboratories. It is not unusual for employees of hospital laboratories to get "burned out" from years of night shifts, limited patient contact, being tied to a single instrument, or the rigid hierarchy of authority found in many large laboratories. In contrast, office laboratories provide daytime work, close patient contact, a broad variety of responsibilities, and frequently a collegial relationship with physicians.

As is often the case, word of mouth is usually the most successful way to find a new office laboratorian. One potential source is the night staff at a local hospital laboratory. Another good source is direct contact with the community colleges that provide medical technician training. Advertising through a local newspaper or by a sign on the hospital bulletin board may also be useful.

The advertisement should request completion of an application form or submission of a full curriculum vitae. The candidate should be required to list all previous jobs in chronological order and provide references from individuals.

SECTION 8: The Interview

The applicant's interview should be scheduled with the clinician and the other people in the office who are likely to interact with the new person. Remember, if the laboratorian cannot get along with nursing staff, it does not matter whether he or she gets along with you! The interview should include a discussion of the hours of work, description of the work, salary scale, date of employment, length of probationary services, and fringe benefits. Some employers will mention the length of notice required by either side to terminate the contract as well as surety bonding as a condition of employment.

It is illegal to ask certain questions that deal with equal opportunity discrimination. The following items are considered personal and unlawful in the interviewing selection process:

- Age
- Date of birth or place of birth
- Number of children
- Marital status
- Names of relatives
- Arrangements for child care while working
- Spouse's occupation
- Ethnic background
- Criminal record
- Details of personal appearance (i.e., weight)
- Illness record
- Physical disabilities

Many applicants will provide some of this data in an effort to give the employer a better idea of their background; however, it is illegal to specifically request this information.

Following the interview, it is wise to check the references by telephone. This can uncover hidden problems and may reveal more about the personality and work habits of the applicant than can be gleaned from the interview.

It is also important to document proof of American citizenship (i.e., birth certificate) or a green card (US Immigration) and resident permission to work if the applicant is not an American citizen. Check the educational diplomas and all registrations or certifications. The societies that issue registration or certification (Table 6–2) provide yearbooks that can be checked for details on these items.

If an applicant is particularly good, invite him or her back and request that he or she demonstrate proficiency in some of the common skills. Select procedures that are frequently performed in your office, such as a differential white blood cell count, evaluation of Gram's stain slides, and blood drawing. This demonstration of skills serves the same purpose as a "typing test" that is given to an applicant for a secretarial position.

Once a decision to hire the applicant has been made, draw up an agreement, to be signed by both parties (each retaining a copy), for a one-month trial of work.

Even if a trained laboratorian has been hired, it is important during that first month to carefully review the applicant's work. Many technicians in hospital laboratories have never done a vaginal wet prep. Many may have spent the last 5 years doing chemistry tests and may have forgotten how to do a differential white blood cell count. Also pay attention to how the employee interacts with patients and other office staff.

A conference should be held between the technician and the clinician at 2 weeks and again at 4 weeks after beginning work. By the fourth week a reasonable decision regarding the permanent hiring of the laboratory worker can be made.

SECTION 9: Work Evaluation

It is far easier in the average office to get feedback about the performance of the nursing staff or the front desk staff than the laboratory staff. You will know about the front desk staff by whether or not the books balance, whether or not your patient scheduling is smooth, and by patients telling you how they were treated. Likewise it is easy to evaluate the nursing staff because patients will often be candid with you when they have been treated well or poorly by your office nurse. In contrast, unless you specifically look for performance data on the office laboratorian, you may not realize how well or poorly she performs. This evaluation should include:

- Regularly scheduled meetings between the physician and the laboratorian. This may seem like a luxury in the rush of a busy office, but it is essential and should be done at least twice a year. You will be surprised how often a problem remains quiet until this meeting. If no problems exist, it is a great opportunity to provide positive reinforcement for a "job well done."

- Review of the laboratory's quality control data. This should be reviewed and initialed each month by the clinician. In 5 to 10 minutes you will get a good idea about both the quantity of testing that is being done and the overall quality. Initial each of the quality control forms. Advise your staff to record *all* of the quality control data. If a point is "out of control," it should be recorded and then repeated to be sure that it comes into control.

- Proficiency testing. Sit down with the office laboratorian each quarter and review the results of your proficiency testing. If your proficiency testing is acceptable, it is likely that your patient testing is also acceptable.

SECTION 10: Handling Performance Problems

There is a tendency to want to blame someone when things are not running smoothly. A more useful conceptual framework is to identify whether the performance is due to:

- Skill or knowledge deficiency. In this case the person does not have the capability or the necessary training to respond appropriately.
- Incentive or motivational deficiency. In this case the person in not personally interested or appropriately rewarded for performing correctly. (A common example of this is whole blood glucose testing by nurses on hospital floors. These nurses are usually busy and feel that this testing is the "lab's job," so their performance may suffer. If it deteriorates sufficiently, then the laboratory will take over the testing again.)
- Environmental constraint. In this case the person was prevented from responding appropriately because of a problem in the environment. The constant interruption of a telephone in an office laboratory is one example.
- Physical deficiency. In this case the person is not physically capable of performing the task. This might be due to poor eyesight, limited intelligence, or bad eye-hand coordination.

Crowley and Tillman have outlined questions for each of these areas. These questions may be useful in evaluating a performance problem.

1. *Skill or Knowledge Deficiency*
   Has a person received training on this test before?
   Was the training of poor quality?
   Has the person performed this test satisfactorily before?
   Can the person perform the test satisfactorily while being observed?
   Is the test very complex, requiring independent judgment?
   Is the test performed very infrequently?
2. *Incentive/Motivational Deficiency*
   Is the person aware of the consequences of poor performance?
   Is the test considered to be socially distasteful?
   Does the person get feedback on his or her work?
   Is there any disagreement about the procedure used in the test?
   Is punishment used when performance is poor?
   Does the person have a history of motivational problems?
3. *Environmental Constraints*
   Is equipment malfunction involved?
   Are supplies adequate?
   Is the work place adequate and well organized?

Is the work area safe, pleasant, and well lighted?
Did the person have other duties scheduled at the same time?
Is there a clear understanding of who does what and when?

4. *Physical Deficiency*
Is there a problem with visual acuity?
Is there a problem with color blindness?
Is there a problem with motor dexterity?

Although these questions are only illustrative, they do suggest the types of data that need to be collected to solve a problem. This approach can be more constructive than just finding out "who is to blame."

SECTION 11: Firing

When a performance problem is identified, it is essential that you maintain documents on the problems and the dates that they occur. Give a verbal warning about the deficiencies and indicate the areas of improvement that are needed. Keep a record of this conversation in the employee's file. If no improvement occurs, give the employee a written warning, mentioning the previous problems and the verbal warning. Indicate again, in writing, the deficiencies, and state that if they are not corrected, disciplinary action will be taken. Give the technician a copy of this written warning.

Dismissal must be based on a repeated pattern of deficiencies, a failure to remedy them, and an inability to achieve adequate performance. Dismissal is painful for both the employee and the employer but not nearly as painful as continued inadequate performance, which compromises patient care.

REFERENCE

Crowley JR, Tillman M. Performance audit in the selection and management of personnel. *Primary Care Clin Office Pract.* 1986;13:617–631.

CHAPTER 7

Reference Laboratories

SECTION 1: Introduction

In chapters 13 through 21 of this book, a wide range of laboratory tests that can be effectively undertaken in the office setting have been described. The advantages of office testing include rapid diagnosis while the patient is waiting (i.e., a white blood cell count in the evaluation of a febrile patient) and the testing of specimens that are unstable and therefore cannot be sent to an outside laboratory without deterioration (i.e., the microscopic urinalysis). It is estimated that half of all tests on ambulatory patients are now done in office laboratories. The remaining tests are performed by reference laboratories using specimens collected in the clinician's office. These reference laboratories are an integral part of the diagnostic services available through any clinician's office.

SECTION 2: Choosing a Reference Laboratory

The clinician will probably choose to use more than one reference laboratory. A commercial laboratory may be used for routine serum chemistries. The local hospital laboratory could be used for stat or weekend laboratory results. The state laboratory may provide specialized services, such as well-water fluoridation levels. The Centers for Disease Control (CDC) may be sent specimens for culturing unusual organisms. The clinician should choose the laboratories that optimize the services available to his patients, provide reliable testing, and provide this testing at the most economical costs.

One of the most important decisions to be made in setting up and developing the office practice is the selection of the reference laboratory. In most communities the clinician will have several choices among commercial and hospital laboratories. Information about the services that each laboratory provides can be obtained from the respective representatives, who will, on request, often visit the clinician's office. It is advisable to explore with this representative the availability of the following services:

- Does the laboratory provide blood tubes, needles, urine tubes, Pap smear supplies, and mailing tubes? There should be no cost for these supplies.
- How often are specimens collected from the office? Most commercial laboratories

provide a pickup service one or two times a day.

- How quickly will test results be available? For the common serum chemistry tests, a 1- or 2-day turnaround time is desirable. Bacterial identification and sensitivities should be available in 3 to 4 days. Tissue pathology specimens often require 4 to 7 days.
- Are stat results available? If so, how long does it take to get test results?
- What services are available on weekends, holidays, and nights?
- How are test reports returned to the office? This will vary with each commercial laboratory. Some return reports in the mail; others return reports with the courier who picks up specimens from the office. A few commercial laboratories provide a computer terminal in the physician's office that directly reports the test results as they are available.
- What test panels are available? Many commercial laboratories will provide customized panels (i.e., SMA-12) for the individual clinician at a reduced cost compared to the individual test prices.
- What quality control procedures are followed? The common practice is to regularly run blind controls and to recheck samples that yield abnormal values. For pathology specimens the original reading may be done by a pathology technician, but all abnormal results should be checked by a pathologist. See Chapter 19–2 ("The Pap Test") for a description of what to look for in a cytopathology laboratory.
- Does the reference laboratory provide regular bulletins about new tests? These bulletins can serve as educational updates about laboratory diagnosis for the clinician.
- Does the reference laboratory provide assistance with the management of the procedures run in the physician's office laboratory? Some reference laboratories have designated laboratory technicians to provide assistance to the practicing clinician. They will train new office personnel and will visit the office to solve procedural problems that the clinician has encountered in the office laboratory.
- Does the reference laboratory provide a quality control serivce for the procedures performed in the office laboratory? Such programs involve sending the clinician biologic specimens for testing and giving feedback on the adequacy of the office test performance.
- What tests are available? Be sure that the reference laboratory provides all of the tests that you will routinely require.
- What are the fees charged for individual tests? Compare the fee schedules for each of the available reference laboratories.
- Will the reference laboratory provide a disposal service for contaminated media?
- Are important abnormal results immediately called to the office? These are often referred to as "panic values" (e.g., a potassium of 2.5).
- Is the laboratory licensed and accredited? Licensure is done by the state or federal government. Federal licensing is based on the Clinical Laboratory Improvement Act (1967, 1988). Any laboratory that bills Medicare is licensed. Accreditation is voluntary and is given by organizations such as the College of American Pathologists or the Joint Commission of Accreditation of Health Care Organizations.
- Does it provide adequate instructions to the office staff about specimen volumes, paper work, and so on? Many laboratories provide a manual with complete information on specimen collection and handling.
- Is the laboratory friendly and helpful when dealing with your office staff? (It is not uncommon for your office to have one to two telephone contacts with the reference laboratory each day. This should not be an aggravating experience!)

- Are the laboratory's reports readable and suitable for filing in your charts? Do the reports include the information you need to properly interpret the test results (e.g., reference ranges for serum chemistries, explanation of cytologic terms used in reporting Pap smears).

After you have collected the information from the reference laboratory representative, you should contact other clinicians in the area. Talk with them about the services provided and about any problems that they have encountered. The reference laboratory representative will be able to provide you with the names of clinicians in the area who use the laboratory's services.

In some areas the choice of a reference laboratory will be clear cut, but often there will be several commercial reference laboratories as well as a local hospital laboratory to choose from. Reference laboratories are a profitable enterprise, and the marketplace has become very competitive.

SECTION 3: Commercial Laboratories

Large commercial reference laboratories are very efficient organizations. A single laboratory may serve several states by courier service. This system is very cost-effective for routine tests such as chemistry panels and Pap smears. However, these laboratories may suffer from being too large to be personal and too distant to be of help for nonroutine testing. The local hospital laboratory has the advantage in these other areas. The clinician can easily be on a first-name basis with the hospital pathologist or technologist. This is helpful in getting questions answered about laboratory reports or in solving problems that you might have with your office laboratory procedures. Hospital laboratories are convenient for stat tests. The clinician can carry the specimen to the laboratory and have the results in a few hours. The local hospital laboratory can also be a source for purchasing reagents (e.g., culture plates from the microbiology laboratory). Many offices may find that good working relationships with both a large commercial and a local hospital laboratory are needed to provide the full range of reference laboratory services.

Once a decision is made to use the services of a particular reference laboratory, that laboratory's representative is again asked to come to the office. The details of the services are worked out. Be sure to discuss pickup schedule, billing, special test panels, and supplies. It is unusual and unnecessary to sign a written contract.

The clinician and the laboratory technologist should pay attention to the reference laboratory services. If there are things that you would like to have changed, the reference laboratory is often able to accommodate your needs.

SECTION 4: CDC

The central laboratory for the national public health laboratory system is located at the CDC in Atlanta. Its laboratories perform a wide variety of test procedures, many of which are not available elsewhere in the country. Specimen handling for CDC testing is done through the state laboratories. Information about the tests available through the CDC can be obtained from the state laboratories or directly from the CDC: write to US Department of Health and Human Services, Public Health Service, Centers for Disease Control, Bureau of Laboratories, Atlanta, GA 30333; or call 404-329-3286.

SECTION 5: State Laboratories

Each state and territory in the United States has a state public laboratory. This is usually a section of the state department of health. The services of the individual state laboratories vary widely and are rapidly changing. As a general policy, these laboratories provide services that are either not available through commercial laboratories (e.g., viral cultures) or that are dictated by state regulations (e.g., newborn phenylketonuria [PKU] testing). Most state laboratories can provide the clinician with a bulletin about their services, fees, and specimen handling.

The procedures performed by some state laboratories include

- Diagnostic bacteriology: bacterial and mycobacterial culture and drug sensitivities
- Mycology: fungal cultures
- Parasitology: examinations for malaria, pinworms, and fecal ova and parasites
- Virology: rabies identification and other viral cultures
- Immunology: serologies for syphilis, fungi, bacteria, rickettsia, viruses, and parasites
- Hematology: coagulation studies, blood grouping, and hemoglobin electrophoresis
- Rickettsia: identification
- Chemistry: testing for inborn errors of metabolism
- Pathology: Papanicolaou staining of cervical cytology specimens
- Environmental: water testing and fluoride levels in drinking water
- Toxicology: serum testing in alcohol, drugs, and heavy metals

The clinician is advised to consult with the state public health laboratory to identify which of these services are currently available.

Some state laboratories now function as a regulatory agency for office laboratories. As such, they may provide and grade proficiency testing services. Finally, many state laboratories have a laboratory improvement section. This section provides consultation with clinicians about office laboratory problems. They offer training courses in laboratory procedures and management.

STATE AND TERRITORIAL PUBLIC HEALTH LABORATORIES

Bureau of Clinical Laboratories
State Department of Public Health
University Dr
Montgomery, AL 36130
205-277-8660

Section of Laboratories
Alaska Division of Public Health
Pouch H-06-D
Juneau, AK 99811
907-465-3140

State Laboratory Services
Arizona Department of Health Services
1520 W Adams St
Phoenix, AZ 85007
602-542-1194

Division of Public Health Laboratories
State Department of Health
4815 W Markham St
Little Rock, AR 72201
501-661-2191

Division of Laboratories
California Department of Health
2151 Berkeley Way, Room 452
Berkeley, CA 94704
415-540-2408

Division of Laboratories
Department of Public Health
4210 E 11th Ave
Denver, CO 80220
303-331-4700

Division of Laboratories
State Department of Health
PO Box 1689
Hartford, CT 06101
203-566-5102

Office of Laboratories
Division of Public Health
PO Box 618
Dover, DE 19903
302-736-4734

Bureau of Laboratories
Department of Human Services
300 Indiana Ave NW, Room 6154
Washington, DC 20001
202-727-0557

Office of Laboratory Services
Department of Health and Rehabilitation Services
PO Box 210
Jacksonville, FL 32231
904-359-6145

Division of Laboratories
Georgia Department of Human Resources
47 Trinity Ave, Room 13–4
Atlanta, GA 30334
404-656-4850

Public Health and Social Services
PO Box 2816
Agana, Guam 96910

Laboratories Branch
State Department of Health
PO Box 3378
Honolulu, HI 96801
808-548-6324

Bureau of Laboratories
Department of Health and Welfare
2220 Old Penitentiary Rd
Boise, ID 83712
208-334-2235

Division of Public Health Laboratories
State Department of Public Health
825 North Rutledge
Springfield, IL 62794
217-782-6562

Bureau of Laboratories
State Board of Health
1330 W Michigan St
Indianapolis, IN 46206-1964
317-633-0221

State Hygienic Laboratory
University of Iowa
Iowa City, IA 52242
319-335-4500

Laboratory Services and Research
Department of Health and Environment
Forbes Bldg 740
Topeka, KS 66620
913-296-1620

Division of Laboratory Services
Department of Human Resources
275 E Main St
Frankfort, KY 40621
502-564-4446

Public Health Laboratory
Department of Health and Human Resources
325 Loyola Ave, 7th Floor
New Orleans, LA 70112
504-568-5375

Public Health Laboratory
Department of Human Services
State House- Station No. 12
Augusta, ME 04333
207-289-2727

Division of Laboratories
State Department of Health and Mental Hygiene
PO Box 2355
Baltimore, MD 21203
301-225-6100

State Laboratory Institute
Department of Public Health
305 South St
Jamaica Plain, MA 02130
617-522-3700

Laboratory and Epidemiological Services Administration
Michigan Department of Public Health
PO Box 30035
Lansing, MI 48909
517-335-8063

Public Health Laboratories
Minnesota Department of Health
717 Delaware St, SE
Minneapolis, MN 55440
612-623-5640

Public Health Laboratories
State Department of Health
PO Box 1700
Jackson, MS 39205
601-960-7582

State Public Health Laboratory
307 W McCarty
Jefferson City, MO 65101
314-751-3334

Public Health Laboratory Division
Department of Health and Environmental Sciences
Cogswell Bldg
Helena, MT 59620
406-444-2642

Division of Laboratories
State Department of Health
PO Box 2755
Lincoln, NE 68502
402-471-2123

Nevada State Health Department
Department of Human Resources
505 E King St
Carson City, NV 89710
702-885-4475

Public Health Laboratories
State Laboratory Bldg
6 Hazen Dr
Concord, NH 03301
603-271-4657

Division of Laboratories
New Jersey Public Health Department
CN 360
Trenton, NJ 08625
609-292-5605

Science Laboratory Division
Department of Health and Social Services
700 Camino de Salud NE
Albuquerque, NM 87106
505-841-2500

Division of Clinical Sciences
State Department of Health
Room E 260, Empire State Plaza
Albany, NY 12201
518-474-4170

Public Health Laboratory
Division of Health Services
PO Box 28047
Raleigh, NC 27611
919-733-7834

Consolidated Laboratories Branch
State Department of Health
1205 Avenue A West
Bismark, ND 58502-5520
701-224-2384

Division of Public Health Laboratories
State Department of Health
Box 2568
Columbus, OH 43266-0068
614-421-1078

Laboratory Services
State Department of Health
PO Box 24106
Oklahoma City, OK 73124
405-271-5070

Public Health Laboratory
PO Box 275
Portland, OR 97207
503--229-5882

Bureau of Laboratories
Pennsylvania Department of Health
PO Box 500
Exton, PA 19341-0500
215-363-8500

Institute of Health Laboratories
Department of Health
Bldg A—Call Box 70184
San Juan, PR 00936
809-764-6945

Division of Laboratories
State Department of Health
50 Orms St
Providence, RI 02904
401-274-1011

Bureau of Laboratories
Deparment of Health and Environmental Control
PO Box 2202
Columbia, SC 29202
803-737-7001

Division of Laboratories
State Department of Health
50 E Capitol
Pierre, SD 57501-5093
605-773-3368

Division of Laboratory Services
State Department of Health and Environment
630 Ben Allen Rd
Nashville, TN 37219-5402
615-262-6300

Bureau of Laboratories
Texas Department of Health
1100 W 49th St
Austin, TX 78756
512-458-7318

Utah State Health Laboratory
44 Medical Dr, Room 207
Salt Lake City, UT 84113
801-533-6131

Department of Health Laboratories
State Department of Health
195 Colchester Ave
Burlington, VT 05402-0070
802-863-7335

Division of Consolidated Laboratory Services
Commonwealth of Virginia
Box 1877
Richmond, VA 23215
804-786-5167

Bureau of Public Health Laboratories
Virgin Islands Department of Health
PO Box 8585
St. Thomas, VI 00801
809-776-8311

Office of Public Health Laboratories
1610 NE 150th St
Seattle, WA 98155-7224
206-361-2800

Office of Laboratory Services
167 11th Ave
So. Charleston, WV 25303
304-348-3530

Wisconsin State Laboratory of Hygiene
465 Henry Mall
Madison, WI 53706
608-262-1293

Public Health Laboratory Services
Division of Health and Medical Services
Hathaway Bldg, Room 517
Cheyenne, WY 82001
307-777-7431

CHAPTER 8

Reliable Results: A Quality Assurance Program

"I worry whoever thought up the term 'quality control'—thought if we didn't control it, it would get out of hand."

Jane Wagner (for Lily Tomlin). *The Search for Signs of Intelligent Life in the Universe*

SECTION 1: Introduction

In their office laboratories clinicians often approach the issue of test quality in an informal and "intuitive" manner. Their logic is "I know the patients and therefore I know when the test is wrong." There can be no doubt that the physician's knowledge of the patient is one important part in the laboratory's overall quality assurance program. Clinicians often *do* know when the test is wrong. But such an intuitive approach to laboratory quality assurance is *not* adequate in this day and age. It is too often clinically impossible to know when a laboratory has erred.

Most hospital and reference laboratories provide a high quality of testing. This, however, was not always the case. In the late 1950s several investigators examined how well these laboratories performed such common tests as blood urea nitrogen (BUN) and hemoglobins. To their surprise, they found a wide range of laboratory values when a single specimen was tested. What followed from these studies was a series of regulations and the adoption of a systematic process to ensure test quality. Unlike the clinicians' personal and intuitive quality assurance efforts, hospital laboratories developed formal programs that were uniform throughout the country. These quality assurance programs are to a large extent responsible for the high quality of testing that clinicians have now come to rely on.

It would be easy to recommend for office laboratories the same quality control standards that are used in large hospital and reference laboratories. Unfortunately we have found that these standards are often irrelevant, inappropriate, and impractical for office settings. One good example is the standard recommendation for how to "properly" quality control a chemistry instrument. The usual recommendation is that the instrument should be "controlled" for each analyte at two levels (a high and a low) every shift. These recommendations were established in the 1960s when instrument and reagent stability was poor. The early instruments would fluctuate widely within a few hours time.

Most modern chemistry systems are more stable and therefore do not require such an intensive control process. We have recently reviewed 1 year's data on "daily" chemistry controls in our laboratory. A total of 1,430 controls was done. Of these, only six were "out of control"; and of these six, five came back into control just by repeating the control testing. In only one case did the control provide any useful information (i.e., that the instrument needed recalibration). We were obviously doing a great number of unnecessary controls. A practical quality assurance program should be designed to tell you when you have a problem with your testing and to indicate what sort of problem there may be. The decision on the frequency of controls must be rooted in common sense and must be backed by data.

Standard quality assurance recommendations have often been based on "controlling" the instrument or the reagents while discounting the importance of the person doing the testing. This is a major mistake for office testing where the test operator is the most variable factor. We have seen, for example, offices in which five different individuals used a chemistry instrument but only one did the daily controls. When asked why this was so, we were told that "She does them because she can always get them in." If only one person is able to get the controls in, then it is likely that the other people are having problems when testing patient specimens.

This chapter describes one approach to quality assurance in the office laboratory. While our recommendations may be more rigorous than your current laboratory practice, they are less strict than those practiced by hospital laboratories. They are based on our best efforts to develop guidelines that are both useful and practical in the office setting.

Finally, you may find that our recommendations differ from those of a diagnostic equipment manufacturer or from your state regulations. Where such differences exist, collect data and determine what type of quality assurance is most likely to be needed in your laboratory. We have found that when the people who review laboratories for licensure are presented with data, most appreciate a reasonable attempt to control those testing factors that need to be controlled in the particular laboratory setting.

SECTION 2: Management

In any laboratory that is used by more than one clinician, a single physician should be identified as the laboratory director. In a real sense each of the involved clinicians will share in the major decisions (e.g., expenditures for equipment, hiring and firing, which tests to offer), but one person needs to be responsible for supervising the laboratory's day-to-day activities. This person should have knowledge about how each test is done, who is qualified to do the tests, the quantity of reagents ordered, the test charges, and the quality control test results.

It also makes good sense to identify a single staff person to be responsible for the majority of the laboratory's administrative functions, including setting up and maintaining the quality assurance system. This individual may be a formally trained laboratorian (i.e., medical technologist or medical laboratory technician), a nurse, or even a detail-oriented staff person. In many offices all of the staff are involved in some laboratory testing. However, the laboratory will run more efficiently if a single person is responsible for evaluating new tests, ordering reagents, interacting with instrument manufacturers, and communicating with the hospital or reference laboratory.

SECTION 3: ''Controlling'' the Test

In most practices it is common at the end of the day for one staff person to ''balance the ledger.'' The charges for the day and the income received are tallied to be sure that everything is in order. This step serves as a ''control'' of the office billing system. In the same way the office laboratory needs controls to be sure that the test systems are working properly.

''Running a control'' basically means testing a ''known'' specimen just as you would a patient specimen. Then the result is checked to make sure that you have the correct answer, or that you were ''in control.'' Since controls are tested just like a patient specimen, they check the entire test system (i.e., the person doing the testing, the reagents, and the instrument).

It is conceptually useful to separate the control process for qualitative tests from quantitative tests. Qualitative tests provide a ''yes/no'' answer. These include such things as pregnancy tests, agglutination tests for mononucleosis, sickle cell screens, and so on. On the other hand, quantitative tests give you a specific, numerical value (e.g., hematocrit or glucose).

Consider pregnancy tests. When these were done with agglutination end points, it was useful to do both positive and negative controls. These controls served as comparisons for what a positive or negative result looked like. Within the past few years a new generation of pregnancy tests has evolved based on enzyme-linked immunosorbent assay (ELISA) technology. These tests are more sensitive (i.e., they detect pregnancy earlier), are more specific (i.e., they are rarely falsely positive), and the end points are colorametric and therefore easier to identify. Many of the newer pregnancy tests include a built-in positive control on the filter paper. Therefore each time that a specimen is run, a positive control becomes visible. This ensures that the reagents are all right and that the test steps were followed properly. Because of the built-in positive control and the low rates of false-positive tests, we do not recommend that any other controls be used in quality controlling an ELISA pregnancy test.

It is useful to look at an example from industry to decide on appropriate quality controls for quantitative tests. One of the major diagnostic instrument companies was manufacturing a large chemistry analyzer for hospital laboratories. They had in the past put each analyzer through a set of control checks at the end of the assembly process. As part of a major reorganization of the manufacturing process, they attempted to ''build quality into the assembly process'' rather than ''check for it at the end.'' Each person on the assembly line was given both increased control over and responsibility for his or her part of the assembly. As quality improved, the company went from checking each instrument with controls at the end of assembly to every other, then every fourth, then every tenth, and, finally, every 20th instrument. This was done gradually, while looking at the failure rates when the instruments were placed in hospitals. The failure rates did not increase, even though the amount of ''end-of-the-line'' controls had dropped tremendously.

Such an approach can be easily taken in the office laboratory, but it requires that you regularly look at how well your laboratory is performing. Do not base such an approach on your ''guess'' of how stable and accurate your testing is!

We suggest that any time a quantitative test is added to the laboratory or a new person is expected to do testing that you do daily controls for about 2 months (i.e., about 20 to 30 control values). At the end of this time look at the numbers. Do they

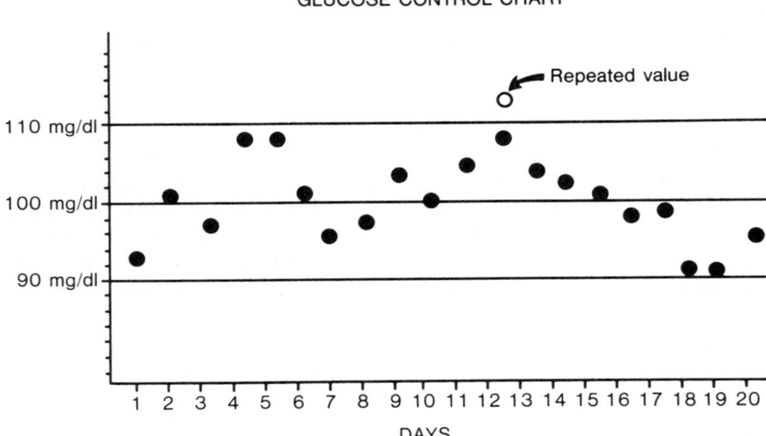

Figure 8–1. A control chart with 20 days of testing of a control with a known mean of 100 mg/dL and an acceptable performance range of 90 to 110 mg/dL.

all cluster around the mean value of the control? Is there a great deal of scatter between days? Are there great shifts in the values over time? If there appears to be good stability, then it may be appropriate to change controls to every other day. After 2 to 3 months, look again at the control values. Do they confirm that it was all right to go to every-other-day controls? Did you get out of control without knowing it? By this process of continual review, you will know how to fine tune your quality control procedures. Controls tell you a great deal more than just whether you are in control today. They also give you information about your laboratory's performance over time.

It has been tradition in large laboratories to take all of the control data from a single month and to calculate the standard deviation (SD). The SD is a measure of variability. In a "normal" distribution, 68% of values fall within 1 SD of the mean, while 95% fall within 2 SDs of the mean. We believe that for office laboratories it makes more sense to plot the data on a daily basis and to regularly "eyeball it" instead of occasionally calculating a SD.

To create such a graph, put the mean value of the control material on a midline and the acceptable control ranges above and below it. Each day that a control is performed, mark the date, the initials of the person doing the control, and then plot the value. You should be able to easily identify the following patterns:

- Normal. Figure 8–1 shows a control chart plotted from the data on a glucose control solution. The X axis represents 20 consecutive days that a control was run. The control material is known to have a mean value of 100 mg/dL, and, according to the manufacturer, its acceptable range is between 90 and 110. Most of the laboratory's values fall within this range, and they appear to be evenly distributed above and below the mean. One point was above the 110 level (i.e., the open circle on the 13th day). This was repeated, and the result was a value in control. This laboratory appears to be doing good testing.
- Bias. This will be recognized because the majority of the points fall either above or below the mean line (Fig. 8–2). The two most common causes for bias are a calibration error or a slight mistake in reconstituting a control solution.

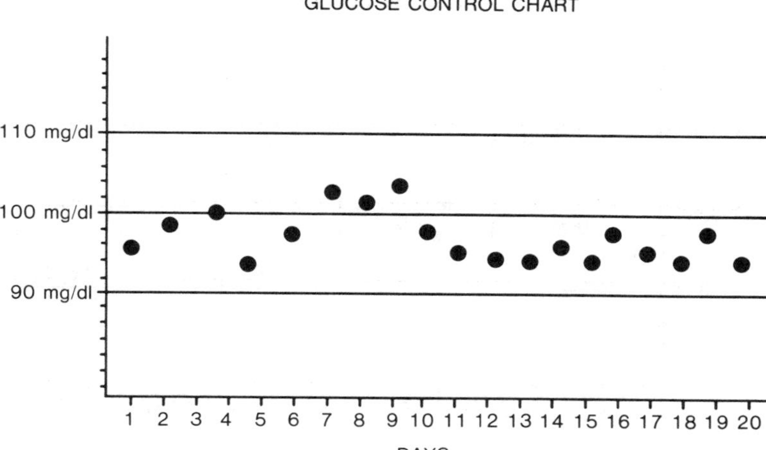

Figure 8–2. Testing bias.

- Excessive scatter. This is indicated by seeing that the points are widely scattered above or below the mean line (Fig. 8–3). Controls may frequently need to be rerun because the first value is outside of the acceptable range. The most common cause for this situation is operator variability (i.e., the person doing the test).
- Drift. Drift occurs when over time the values move away from the mean line (Fig. 8–4). This is commonly caused by reagents that are outdated or by an instrument that is losing its calibration.
- Shift. Shift occurs when the control values suddenly move from clustering around the mean line to a line above or

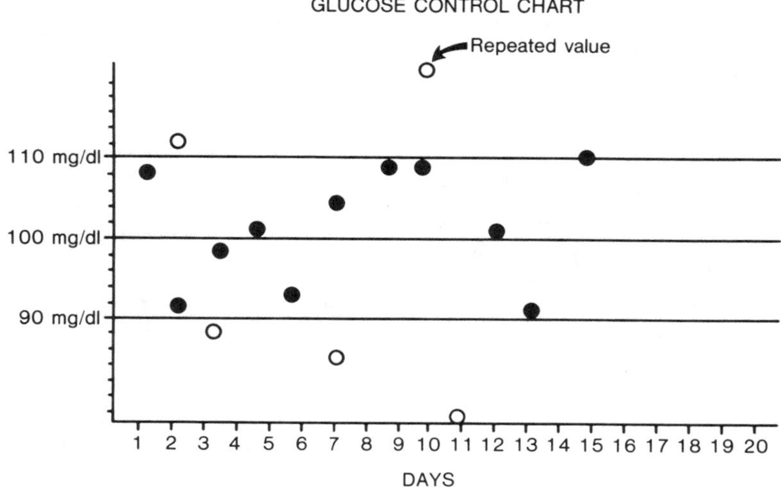

Figure 8–3. Excessive scatter (i.e., poor testing precision).

GLUCOSE CONTROL CHART

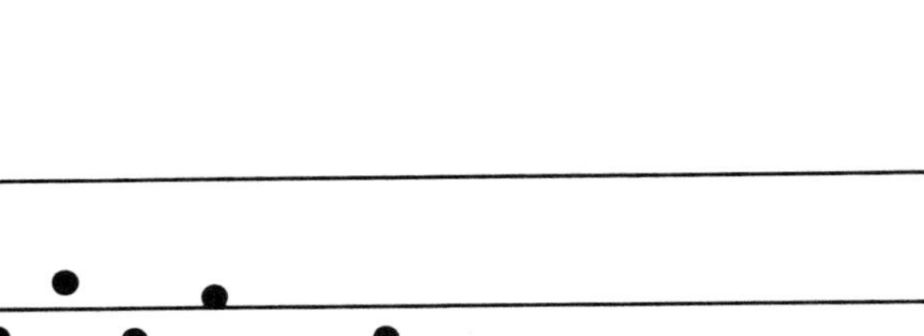

Figure 8–4. Drift in control values.

below the mean (Fig. 8–5). The most common causes for this are recalibration of the instrument, a sudden deterioration of the reagents (i.e., were they left out overnight?), or a new person doing the testing.

With time, some office laboratory staff may decide not to plot the points but rather just to include them in a control table (Fig. 8–6). This is a little faster but may also be somewhat harder to interpret.

It is important that the laboratory director review and initial the control data each month. This is far more than just a formality! It is an opportunity for the laboratory director to review the quality of the testing in the laboratory and for the director and the laboratory staff to discuss any problems.

GLUCOSE CONTROL CHART

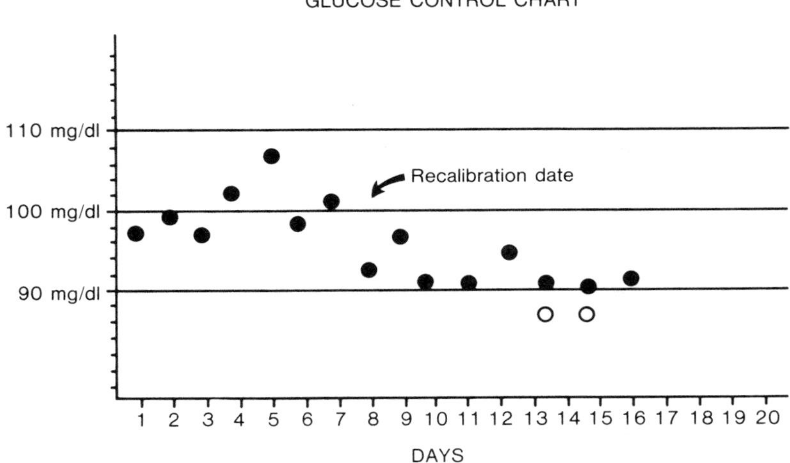

Figure 8–5. Shift in control values.

Meter: # ______________

		Chemstrip bg Strips		Lo-normal Control						Hi Control					
Date	Tech	Lot #	Exp. Date	Lot #	Exp. Date	Open Exp. Date	Actual value	Acceptable Range	Tested Value	Lot #	Exp. Date	Open Exp. Date	Actual Value	Acceptable Range	Tested Value

Figure 8–6. Accucheck Quality Control Form.

SECTION 4: Proficiency Testing (PT)

The quality controls that are run serve as an "internal" check on the quality of the testing. Proficiency testing, on the other hand, serves as an "external" check. Proficiency testing involves receiving unknown specimens on a regular basis and then reporting your laboratory's results on these unknowns. In this way you can see how well your laboratory performs compared to other similar laboratories.

Until recently it was difficult for an office laboratory to find a proficiency testing program that was specifically designed to meet office laboratory needs. This has recently changed. Proficiency testing programs are now offered through a variety of professional societies (e.g., College of American Pathologists, 5202 Old Orchard Rd, Skokie, IL 60077-1034, telephone 312-966-5700; The American Society of Internal Medicine, 1101 Vermont Ave NW, Washington, DC 20005-3457, telephone 202-289-1700; The American Academy of Family Physicians, PO Box 8723, Kansas City, MO 64114, telephone 800-274-2237), by several instrument manufacturers, as well as by some state health departments.

In addition to serving as part of the quality assurance program, proficiency testing is now used by the federal government in licensure of office laboratories. Adequate testing in a PT program is therefore the mechanism by which the government is assured that your laboratory produces good patient test results.

The biggest problem with proficiency testing is determining what the "right" answer is. Several mechanisms are used:

- Reference laboratories. A number of excellent laboratories are chosen to serve as reference laboratories. The results that these laboratories get are then grouped. If they all agree, then their answer is considered to be the correct answer.
- All-Method Comparison. For quantitative tests, the correct answer is often assumed to be the mean of all of the participant values. The range of correct answers is then determined by either 2 to 3 SD from this all-method mean.
- Single-method mean. It is not uncommon for two different types of test methods to give somewhat different answers on a single specimen. When this occurs it makes more sense to judge the PT participants on the single-method mean for their method instead of the all-method mean. (A description of how to get the most information from a PT result is given in Chapter 13, section 6E of this book.)

Participation in a proficiency testing program provides a good opportunity for the laboratory director and the staff to work together. For example, we have found it useful for the laboratory director and staff to review the slides of unknown blood smears as well as the Gram's stains of the organisms sent out as part of the gonorrhea PT challenges.

SECTION 5: Documentation

Patient's medical records are needed for medical purposes, legal purposes, and as an aid to remember important aspects of the patient's care. The documentation of the

Date	Tech	Refrig (4-8 °C)	Incubator (35-37 °C)	Room (22-25 °C)	Comment

Figure 8–7. Temperature chart.

MONO Test. * ________________
(Name of Kit)

Date	Tech	Kit #	Kit Exp Date	Pos Control #/Exp. Date	Neg Control #/Exp. Date	Comment

*Controls come with kit.

Figure 8–8. Mono-Test Quality Control Form.

laboratory's performance functions in a similar way. Remember, "If it isn't recorded, it wasn't done."

Most laboratories find it convenient to develop a variety of forms to record the important information. Samples of these types of forms have been shown (Figs. 8–6 to 8–8). In general, a quality control form should include

- the date,
- the initials of the technician,
- the reagent lot number,
- the expiration date of the reagents,

- the acceptable range of values,
- the control result, and
- comments.

Some laboratories find it convenient to keep all of the quality control forms in a single notebook. Others may find it easier to fold the quality control sheets up and keep them in the test kit's box or on a clipboard next to the instrument. This makes it easier to remember to record the result whenever a control is run. It also makes it easy to check if the controls need to be run. If the quality control forms are kept with the test kits, then a "quality control notebook" can be used to store the older forms that are filled with data.

SECTION 6: Supplies

It is essential to know how each reagent should be stored and when it becomes outdated. Even blood-drawing tubes have expiration dates! This is because over time they may lose their vacuum or the effectiveness of the additives may diminish.

Any single item may have several different expiration dates (i.e., the overall kit expiration date if the reagents are unopened, the expiration date once the reagents are opened, the expiration date after the reagents are reconstituted, an expiration date if they are refrigerated, or an expiration date if stored at room temperature). Whenever a new item comes into the laboratory, be sure to read the directions about the storage conditions. If it is a kit, then perhaps only some of the reagents need to be refrigerated. Some reagents (i.e., urine dipsticks) need to be kept away from the light and heat. Do not store such reagents in a shelf over counter lighting. The heat from the bulb can damage the reagents.

It is important to use things up before they expire. There are a variety of ways to manage this:

- For fast moving items (i.e., pregnancy tests, rapid streptococcus kits) the authors use a grease pencil and write the expiration date on the box in large letters.
- For slower moving reagents keep a card file. Use one card for each month. As reagents are brought into the laboratory you can check to see when they will outdate. Pull the card corresponding to this date, and mark the name of the item. Then once each month you can review those reagents that will be outdating and therefore will need to be reordered.
- Write clearly in permanent ink the reconstituted expiration date on all reagent bottles.
- Make a habit to check the expiration date on any item that you use just before you use it. This will become second nature with time.
- When ordering supplies, always ask for the longest possible expiration date.
- Ask about shelf life whenever a new test is being considered. In a small-volume laboratory it is not uncommon for shelf life to be *the* critical element in deciding which test system to buy.
- Use up the oldest reagents first.
- *Never* use expired supplies or reagents.

You may find it convenient to start an alphabetized file of all reagents and supplies. On each 3 × 5 in file card include storage conditions, where to order it, the cost, the quantity that is usually purchased, how long it takes to order, where you can get it in an emergency, and so on.

SECTION 7: Equipment

A file should be maintained on each piece of equipment. This file should include the instruction manual, warranty information, equipment information, and maintenance log (Fig. 8–9).

You can also keep any bills for special service or parts in this file. This file serves the same purpose as the records in your glove compartment that pertain to your car's warranty and maintenance. For instruments that require periodic preventive maintenance (i.e., some hematology analyzers), it is useful to mark the date for future maintenance on a master calendar that is kept in the laboratory.

Equipment Record

Type of Equipment Cost ______________
Brand ______________ Model ______________ Serial # ______________
Date of Purchase ______________ From ______________
Warranty Period ______________
Hotline Problem telephone number 1-800-______________
Service Contract: Name ______________ Cost ______________
Address ______________ Telephone # ______________

Recommended Preventative Maintenance
(Type in what is to be done and at what regular interval.)

Maintenance and Repair Log

Date	Problem	What Done	By Whom	Cost	Comments

Figure 8–9. Equipment information and service record.

SECTION 8: Procedure Manuals

Every laboratory must have a procedure manual. The principle uses for such a manual are to record how tests are actually being done, to train new people in the laboratory, and to remind those who regularly work in the laboratory how to do tests that are infrequently performed. It is far better for the procedure manual to be useful (i.e., have torn pages and penciled in comments) than to be nicely typed and collecting dust on the shelf.

This book is designed to serve as a procedure manual. It can be used as such for those tests that are described. For other tests the procedure manual can be the instruction pages that come with test kits, or the test sections can be prepared by the office staff.

The test descriptions in the procedure manual should include the following information:

1. Name of the test
2. Name of the kit or product used
3. Background or principals of the test
4. The required test materials
5. How to prepare and store reagents, standards, and controls
6. How to prepare equipment for use
7. Specimens
   a. Collection (e.g., In what tube, at what time, and what are the absolute minimum volumes?)
   b. Storage (e.g., Do you keep it at room temperature or put it in the refrigerator?)
   c. Specimen processing (e.g., Do you need to spin down the specimen first?)
   d. What is an unacceptable specimen (e.g., urine that has sat too long or serum that is lipemic)
8. The actual test steps
9. Any necessary calculations
10. Special requirements
11. Test interpretations
12. Quality assurance procedures, including calibration, controls, and what to do when "out of control"
13. Sources of error
14. References

In addition to describing each test procedure, include a "general policies" section. In this section describe how specimen labeling should be done, list the "panic values" (i.e., the test result level that should be immediately brought to the attention of the clinician), and specify how the laboratory results can be given out (i.e., Can a patient's gonorrhea culture be released by the office staff to the spouse of a patient?) There should also be a safety section in the procedure manual. (See Chapter 8.)

It is a good habit to review your procedure manual once a year and to initial each procedure at that time. Invariably you will need to update some procedures, drop some tests that are no longer being done, and add the new test procedures.

SECTION 9: Personnel Education

A majority of the office staff who do laboratory testing have not completed technician or technologist training programs. The same is true for many hospital laboratories, where many of the employees have only on-the-job training.

In general, those individuals who perform good laboratory testing pay attention to details, are not intimidated by technical equipment (i.e., copy machines, computers, etc.) and have good manual skills.

It is essential that the office laboratory staff have continuing education. We have found the following to be useful types of continuing education:

- Clinical correlation. Laboratory staff get tremendous satisfaction from being able to correlate what they are doing in the laboratory with the rest of the patient's care. Why was the test requested? How did the clinician use the laboratory information? What happened to the patient? Regular professional communication between the clinician and staff not only improves the satisfaction with the job but actually helps them to do better laboratory testing. (Isolation from patients is one of the major complaints of the laboratory staff in hospital and reference laboratories.)
- Journals. There is growing literature that directly relates either to office laboratory testing or to the illnesses for which the office laboratory is crucial in diagnosing (e.g., vaginitis). The clinician should make a habit of giving copies of these articles to the laboratory staff.
- Hospital laboratories. Most pathologists who run hospital laboratories are willing to have their laboratory personnel teach the office staff. It may be possible, for example, to have the office staff learn common urine microscopy by spending time in the hospital laboratory.
- Industry representatives. The diagnostic industry is beginning to assume a larger responsibility for helping the office laboratory staff. For example, most manufacturers have an 800 number that can be called to trouble shoot common problems. There remains, however, a wide range of actual support by companies or instrument dealers. This level of support is worth investigating before making a major instrument purchase.
- Workshops. With the growth in office laboratory testing, more relevant workshops for office staff are being developed. The state health department or the regional meetings of medical technology associations are two good sources.
- This book can also serve as a good source of continuing education for staff in the office laboratory.

CHAPTER 9

Office Laboratory Safety

SECTION 1: Introduction

Laboratory safety has often been an area characterized by widespread hypocrisy—with little similarity between "official guidelines" and "actual practice." This has been due to both the uncritical acceptance of unrealistic recommendations as well as the widespread laboratory worker belief that "it won't happen to me."

The concern for laboratory safety has increased because of the rising prevalence of human immunodeficiency virus (HIV), the growing understanding of these infections, and the obvious occupational risk for health care workers. Laboratory workers are at a real, but in general, small risk for catching acquired immunodeficiency syndrome (AIDS). This is because of their frequent contact with blood. The Centers for Disease Control (CDC) estimates that 0.5% of infected needle sticks will lead to HIV transmission. Since the average office laboratory worker will be exposed to several thousand patient specimens each year and since in even low-risk patient populations at least one in 20 patients will come from a high-risk group (i.e., gay males and intravenous [IV] drug abusers), there is an obvious reason for concern.

It is true that the office laboratory environment is safer than the hospital or reference laboratory. This is because the number of specimens handled tends to be fewer and the patients less sick. It is also true that the laboratory worker in the office setting often knows the patients well and may therefore know if contact with their body fluids is a risk, although there are always surprises. The question is, however, "How much risk should be taken?" There is also the issue of creating a second class of patients—those who are treated with gloves. By treating all patients the same, no one is singled out. You can then truthfully tell everyone "we wear gloves for all patients."

In this chapter we will outline the common safety concerns for the office laboratory worker. It must be remembered that safety involves not just the laboratory worker but also the cleaning staff who come in contact with the laboratory trash as well as the community, whose environment needs protection from biohazard contamination.

SECTION 2: General Safety of the Laboratory Worker

Table 9–1 lists some of the common pathogens that are considered to be potential hazards from contact with infected blood or body fluids. Of these, the virus that causes AIDS (HIV) and the viruses that cause hepatitis are potentially the most serious and are likely to be the commonest pathogens in the laboratory environment.

TABLE 9–1. DISEASES TRANSMITTED THROUGH CONTACT WITH INFECTED BLOOD OR BODY FLUIDS

AIDS
Arthropod-borne viral fevers, including dengue, yellow fever, and Rocky Mountain fever
Leptospirosis
Malaria
Syphilis: neonatal, primary, or secondary with skin or mucous membrane lesion
Viral hepatitis, type B, including HBs antigen carrier state
Viral hepatitis, type non-A and non-B

The CDC has established guidelines for a system of "universal precautions" in an effort to limit laboratory worker risks. The term universal precautions refers to the fact that *all* patients are assumed to be potentially infected. This is a dramatic change from the previous laboratory practice, which was to give special care to only those specimens from patients known to be in a high-risk group. The CDC's guidelines are based on the recognition that only some body fluids are infectious and that different types of exposures lead to different levels of risks:

1. Blood is the single most important source for HIV and hepatitis B virus (HBV) infection.
2. Semen and vaginal secretions are also risks and should be handled with universal precautions. They are known to contain the viruses and have been shown epidemiologically to cause spread of the diseases. They have, however, not been shown to cause transmission from occupational exposure.
3. In addition to blood, semen, and vaginal secretions, universal precautions should be applied whenever handling tissues, cerebral spinal fluid, synovial fluid, peritoneal fluid, or amniotic fluid. The risks of transmission to laboratory workers is unknown; however, it is considered to be real.
4. No universal precautions are required for the handling of urine, feces, nasal secretions, sputum, saliva, sweat, tears, or vomitus as long as none of these are bloody. HIV or HBV have been found in some of these fluids; however, epidemiologic evidence does not implicate these as causes for transmission of disease.

SECTION 3: Essential Aspects of Laboratory Safety

GLOVES

Gloves are the principal intervention for ensuring universal precautions. It is recognized that gloves will not prevent an injury from needle sticks; however, they will reduce the risks from cuts on the hand. Look at your own hands. It is very common to have cuts, cuticle tears, etc. All are potential entry sites for HIV, HBV, or other pathogens. Contact with breaks in the skin is believed to be a lower risk than the 0.5% risk from needle sticks. The frequency of contact, however, is much greater than for needle sticks if gloves are not used.

The CDC recommends that gloves be used whenever drawing a patient's blood. They recognize, however, that in some institutions with low cases of HIV or HBV, the requirement of gloves for all phlebotomy may be relaxed. Even in this setting, though, gloves should always be made available to any health care worker who wishes to use them. In addition, the CDC recommends that gloves be used:

- Whenever the health care worker has cuts, scratches, or other breaks in his or her skin (which is most of the time).
- In any situation in which hand contamination with blood is likely to occur.
- When performing finger or heel sticks on infants or children.
- Whenever training to do phlebotomy.

For those laboratory workers who are not used to drawing blood with gloves, universal precautions may seem like an unnecessary hassle. A "retraining period" is necessary to develop the sensitivity of touch through the gloves to permit finding difficult veins. This generally takes about 2 weeks of concentrated effort before it becomes routine. It is best to wear the smallest gloves that fit comfortably. This permits greater sensitivity. When an exceptionally difficult patient needs blood drawn, try using a pair of surgical gloves. Fresh gloves should be worn for each patient. Do not wash them for reuse. In a busy office laboratory where specimens are frequently being drawn, processed, or examined, you are likely to find yourself wearing gloves most of the time.

HANDWASHING

The use of gloves does not eliminate the need for hand washing. Hands should be washed before drawing blood, touching your own eyes or mouth (i.e., eating, drinking, or smoking), before you go home for the day, and immediately after spilling any reagent or specimen on them. It is easy to wash your hands if an antimicrobial liquid soap is kept in a plunger-top container next to the sink. To prevent severe drying of the skin, wet your hands with water before applying the soap.

AVOIDING NEEDLE STICKS

The single most important hazard to the office laboratory worker is the possibility of a needle stick with a contaminated needle. This almost always occurs when the worker goes to recap the needle after blood drawing. In addition, housekeeping staff can be punctured when they handle containers into which used needles and scalpels have been thrown. The best way to avoid both of these potential risks is to have special containers for disposing of "sharps" (i.e., needles, scalpels, etc.). Specially made boxes include holder tops that permit the needles to be detached. Such a container should be small enough to be placed in the work area; should be spill proof, tamper proof, and opaque; should have puncture proof sides; and should be autoclavable (Fig. 9–1). In the past it was recommended that needles be bent or cut to prevent their reuse by illicit drug users. This practice is no longer recommended because it increases the potential risks of the laboratory worker receiving an accidental puncture. Several companies now manufacture sharp containers (Sage Products, 1-800-323-2220; Bio-Safety Systems, 1-800-421-6556; and Eagleguard Syringe Safety, 1-800-458-0506).

REMEMBER:

Do not remove needles by hand.
Do not recap needles.
Do not bend, break, or attempt to destroy needles or scalpels.
Do not leave contaminated needles laying around.

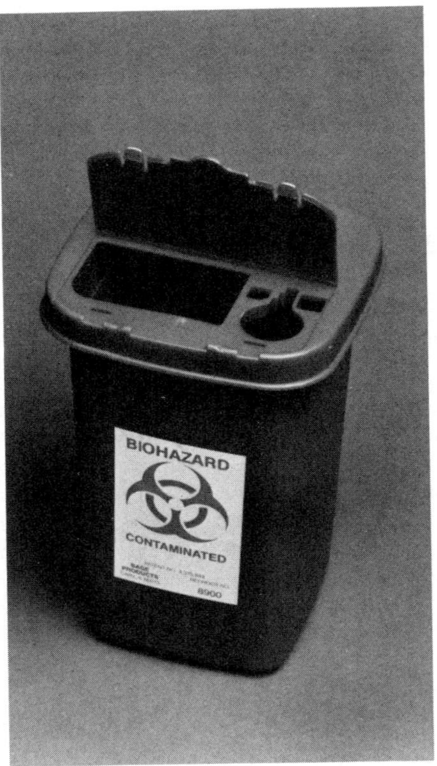

Figure 9–1. A red "sharps" disposal box. Note the slots for holding the needle when it is unscrewed from the plastic holder.

PIPETTING

In the past it was common laboratory practice to draw fluids into pipettes by mouth. This is now considered an unacceptable practice both because of the biologic as well as the chemical risks. *Never mouth pipette!*

Pipette bulbs should always be used instead of mouth pipetting. If small, inexpensive pipette bulbs are used and become contaminated (i.e., the specimen is drawn up into the bulb), throw them away. If expensive, safety pipette bulbs become contaminated, rinse out the bulb with disinfectant and set it aside to air dry. Disposable pipettes should be used whenever possible. These are relatively inexpensive and do not need to be cleaned. If nondisposable glass pipettes are required, rinse them out after use and place them in a pan filled with disinfectant to soak overnight before final washing.

REMEMBER:

Never mouth pipette.

Do not lay contaminated pipettes on the work bench.

EATING, DRINKING, AND SMOKING

Eating, drinking, and smoking have no place in the laboratory. This is because the hands can become contaminated with infectious organisms, which can then be spread to the worker's mouth. Also, do not store food, drink, or medications in the laboratory refrigerator. Post a sign on the front of the refrigerator to remind staff that it should not be used for storing lunches. Tuna fish and HIV do not go well together.

DISINFECTANT

A variety of disinfectants can be purchased. In the past a 1:10 dilution of household bleach was recommended. While this solution is cheap, it is irritating to the hands as well as the nose. There is also some evidence that the solution loses its effectiveness after the dilution is made. We now suggest a 5% solution of Staphene. This is effective against bacterial as well as viral organisms, including HIV. This disinfectant should be used to wipe down the counter tops and the inside of centrifuges at the end of each day. If a specimen spill should occur, wipe up any visible material and then clean with a disinfectant. The paper towels that are used for this cleaning should be put into a special container for infected waste. Plastic containers with lids that are marked as biologic hazards or special autoclavable plastic bags can be used for this purpose. Gloves should be worn during the cleaning and decontaminating procedures.

WASTE DISPOSAL

This is a somewhat problematic area with few widely accepted standards of practice. The U.S. Environmental Protection Agency has yet to develop regulations for the safe disposal of medical waste. States vary widely in their approach to regulation. The state laboratory will be able to tell you about restrictions in your area.

There is no evidence to suggest that hospital or laboratory waste is any more infectious than residential waste, nor is there evidence that hospital waste has caused disease in the community as a result of improper disposal. Most of the care taken, therefore, in dealing with waste disposal is to prevent causing infections during the actual handling of the material. For the laboratory setting, this usually refers to unnecessary needle sticks or cuts from broken glass.

Soft disposals (i.e., syringes without needles, bloody gauze, contaminated paper towels, gloves, etc.) also require special handling. These potentially infected wastes should not be thrown into the routine office rubbish. Instead, they should be collected in large plastic jugs (e.g., 24-hour urine collections) placed in convenient locations throughout the practice. The jugs should be marked with "contaminated" or "Biohazard" tape (Fig. 9–2).

All infected wastes should be placed in an autoclavable container and then either incinerated or autoclaved before disposal. This cannot be done in the office autoclave because of its size and the resulting smell. In some areas a specialized waste industry is available to handle these materials. Your local hospital laboratory should also be able to provide you with recommendations.

Figure 9–2. Twenty-four hour urine collection jar marked with "biohazard" stickers. This jar can be used to hold infected waste.

Blood, urine, and most reagents can be poured down the drain in the laboratory sink. Pour very carefully and avoid splashes. Be sure to use plenty of water to rinse the sink after emptying the specimen or reagent container.

Special precautions should be taken when disposing of the sodium azide reagents, which are used in some pregnancy tests. These reagents should be washed down the sink with plenty of water to prevent a possible explosion when the azide reacts with the metal in some plumbing pipes. The water should be left running for about a minute after flushing the azide solution.

Many states have developed regulations that control the disposal of culture plates. The typical office autoclave is not suitable for sterilizing culture plates or other waste materials. One option is to place contaminated plates in special contaminated containers and then to have them processed by the local hospital. A second option is to gently squirt the surface of the plate with disinfectant (i.e., a 1:10 solution of household bleach or a 5% solution of Staphene). This is done over the sink with a squirt bottle. Flood the plate with the disinfectant and then permit it to "soak in." The decontaminated culture plates can then be double bagged and discarded with the regular office garbage.

HEALTH CARE WORKER TESTING AND IMMUNIZATION

Hepatitis B is the most common life-threatening pathogen that the average health care worker will be exposed to. Fortunately the hepatitis B vaccine has been developed. It is recommended for all office staff who draw blood or process patient specimens. A major drawback is its cost (about $100 per person immunized). However, this is a small cost compared to the large costs of contracting the disease. The original hepatitis B vaccine came from pooled sera, and there was some concern about the possibility of transmission of HIV from this material. A great deal of evidence has now shown the vaccine to be safe. However, a recombinant DNA hepatitis B vaccine is now available (Recombivax). This should also be safe; however, there is some question about a lower antibody response when this vaccine is used. The best source of information about current hepatitis precautions and recommendations is the *Morbidity and Mortality Weekly Reports* (MMWR) published each week by the CDC. This gives the CDC's recommendations as the are updated. A copy of this publication appears each week in the *Journal of the American Medical Association*. MMWR can be obtained from the Massachusetts Medical Society (CSPO, Box 9120, Waltham, MA 02254-9120). It is available by first- or third-class mail.

Some experts recommend that laboratory workers should be tested for their hepatitis B status prior to receiving the vaccine. This is done by drawing the hepatitis B core antibody. If positive there is no need for the vaccine. At this time there is no agreement on the need to test laboratory workers for their HIV status. Since an HIV-infected laboratory worker is unlikely to spread the disease to patients, such testing is not considered to be medically indicated.

MANAGING NEEDLE STICKS

The clinician should be contacted whenever a health care worker has a needle stick, a mucous membrane exposure (i.e., splash to the eye or mouth), or exposure through contact with an open area on the skin. The patient should be informed of the needle stick and should be tested for HIV after consent. If the patient has AIDS, is HIV antibody positive, or refuses to be tested, the health care worker should be serologically tested for HIV. If that test is negative, then the test should be repeated at intervals recommended by the CDC.

SECTION 4: Other Safety Factors

CLUTTER

Since most office laboratories are small, clutter can be a major problem. If there is a spill on a counter top, clutter can make cleaning very difficult. If necessary, build small wall shelves over the laboratory bench to store small bottles. Bottles of corrosive reagents (which are rare in office laboratories) or large bottles can usually be kept in the cabinet under the sink. Be sure that all chemical containers are tightly capped before storing.

LABORATORY EQUIPMENT

The surfaces of all equipment should be kept clean. Spills should be cleaned up as soon as they occur. There is no evidence that chemistry analyzers or blood cell counters require more than regular cleaning, even when processing specimens from patients with hepatitis or AIDS. For offices that do manual white blood cell counts, it is advisable to soak the hemocytometer and cover slip in a disinfectant solution between use.

EYEWASH

Every office laboratory should have equipment readily available to wash a worker's eyes as soon as possible after an eye accident. A very convenient wall-mounted unit with a mirror, a 32-ounce squeeze bottle, and a special fitting eyecup top is available for about $20.00 from scientific supply companies (i.e., Fisher Scientific).

FIRE EXTINGUISHER

A small, multiple-use (i.e., ordinary combustible, flammable liquids, and electrical equipment) fire extinguisher should be mounted on one wall in the laboratory. All workers should be instructed in its proper use.

FIRE BLANKET

An open flame from either an alcohol lamp or a bunsen burner is a common hazard in office laboratories. A fire blanket can be used to smother flames should a person's clothing catch fire.

ELECTRICAL PRECAUTIONS

Office laboratories should have an adequate number of grounded electrical outlets. An outlet board mounted on the wall is a convenient way to increase the number of outlets. Be sure that all laboratory equipment has safe cords and plugs and has been approved by Underwriters Laboratory (UL) or a similar group. Avoid using regular extension cords. Do not remove the grounding prong from any instrument's three-prong plugs. Do not handle electrical equipment with wet hands. Immediately disconnect any piece of equipment that produces a "tingle" when it is touched. Be sure that such equipment is repaired before using. Each laboratory and office worker should be shown the location of the fuse or breaker box in the practice. In the case of an electrical shock, be sure to turn off the power at its source before attempting to touch the person, otherwise you too will become a victim.

HANDLING SPUTUM

If there is the slightest clinical suspicion that a patient has tuberculosis (TB), the laboratory staff need to be alerted. *Mycobacterium tuberculosis* is easily airborne, and even a few of this bacteria are potentially infectious. Office laboratories do not have safety hoods, so special care must be used to prevent the creation of aerosols. Routine caution, such as wearing gloves and hand washing, is advised while handling sputum. To allow for immediate recapping of the cup, we recommend that sputum be collected in the screw-topped urine containers.

HANDLING TUBE TOPS

To avoid spilling or splashing blood when removing the rubber top from a tube of blood, a 2 × 2-inch gauze should be used (in addition to gloves) to grasp the top. Do not "pop" it off by pulling straight up; rather carefully twist the top until it comes loose. Should blood splash on you or the table top, clean it up immediately by washing the splashed area with a disinfectant. If you are sending an opened tube to the reference laboratory or need to centrifuge it, tape the top on after you have reinserted it. This prevents it from popping off.

CENTRIFUGE CARE

Centrifuges can cause severe injury if one attemps to slow them down by hand. Never put your hand into a moving centrifuge, and never open a centrifuge while it is spinning. Some centrifuges come with safety devices to prevent this type of misuse. Occasionally a specimen tube will break in the centrifuge. This will cause a lot of noise from obvious centrifuge imbalance. The centrifuge should be stopped immediately. The laboratory worker should then put on gloves and remove the tube holders (trunnions). These should be soaked in a small container filled with disinfectant. The broken glass should be carefully removed and put into a puncture proof container. Finally, the inside of the centrifuge should be cleaned with paper towels and disinfectant. Microhematocrit tubes sometimes leak, leaving a ring of blood around the inside of the centrifuge. It is therefore a good idea to routinely wipe the centrifuge out with disinfectant at the end of each day.

LABORATORY COATS

Most laboratory staff will choose to wear a knee-length labortatory coat. The primary function of the coat is to protect the worker's clothing. It is not really a protection from infectious organisms. The coats can be taken home and washed in hot water with bleach.

SAFETY MANUAL

A safety manual should be included as part of the office laboratory's procedure manual. This should include a list of the names and phone numbers of people to contact in an emergency, procedures for reporting accidents, and an accident log in which the person's name, type of accident, and date are included. If the accident is a needle stick, include the patient's name, number, and any relevant medical diagnoses.

REFERENCES

Bond W. Viral hepatitis B: Safety in the immuno chemistry laboratory. *Ligan* 1982;5(1):34-39.

Rose SL. Clinical Laboratory Safety. Philadelphia, Pa: JB Lippincott; 1984.

Update on Hepatitis B prevention. *MMWR*. 1987;36(23):353-366.

Centers for Disease Control. Update: Acquired immunodeficiency syndrome and HIV infection among health-care workers. MMWR. 1988;37(15):229-239.

Centers for Disease Control. Recommendations for prevention of HIV transmission in health-care settings. *MMWR*. 1987;36(25):15-185.

Centers for Disease Control. Update: Universal precautions for prevention of transmission of human immunodeficiency virus, hepatitis B virus, and other blood borne pathogens in health care settings. *MMWR*. 1988;37(24):377-388.

Protection of Laboratory Workers from Infectious Disease Transmitted by Blood and Tissue. Villanova, Pa: National Committee for Clinical Laboratory Standards, NCCLS document N29-T, 1989.

Some of the material for this chapter was prepared by the authors as part of CDC Cooperative Agreement No. 050/CCU00987-01 and BHP Grant 1D32PE14025.

CHAPTER 10

The Microscope

SECTION 1: Introduction

The brightfield microscope is an indispensable instrument for the office laboratory. Unlike other instruments, it will, if taken care of, last the lifetime of the practice. Despite its potential usefulness, few clinicians have been taught to use the microscope properly. This lack of instruction puts them at a disadvantage when using it themselves or when attempting to teach a new and untrained laboratory worker.

An initial and important question is what type of microscope to buy? To help you answer this question, Section 2 below provides a description of the parts and characteristics of a microscope. Section 3 offers an alphabetized glossary. These are offered not to make you an expert on microscopes but rather to familiarize you with the terminology so that you can use the appropriate language when talking with a salesperson. Section 4 offers a purchase check list and a list of manufacturers.

Once you buy your microscope you will want to take proper care of it. Section 5 describes how to clean and perform routine maintenance. Section 6 covers the troubleshooting of common problems. These two sections should keep repairs to a minimum and guarantee many years of trouble-free use.

Sections 7 through 9 serve as a detailed instruction manual to train someone who has not previously used a microscope. Section 7 describes how to use the microscope to examine "wet" specimens such as urine sediments or vaginal/cervical secretions. Section 8 reviews the use of the microscope to examine specimens under the oil immersion objective such as Gram's stains or Wright's-stained peripheral blood smears. Section 10 covers how to polarize your microscope to examine joint fluid for crystals.

All of the material has been written from a very practical point of view, with minimal discussions of optical physics. Should you find this aspect of microscopy interesting, we include some suggested readings at the end of Section 11.

The microscope should be placed in the laboratory on a counter that is free from vibrations. This means that the microscope should not be on the same counter as a centrifuge. Have a chair with good back support in front of the scope. Place the microscope so that the stage is closest to you. This arrangement makes it easier to put specimens on the stage. If you have a dual-headed scope, the stage should face the most frequent user.

If you wear eyeglasses, take them off when using the microscope, since plastic lenses can be easily scratched. However, if you are astigmatic, you will need to keep your glasses on because while the focusing mechanisms allow you to focus each eye, there is no way to adjust for astigmatism. Bifocals and trifocals are difficult to wear when using the microscope, even if you are astigmatic, because of the lines that cut across the visual field. People wearing these glasses, therefore, usually take them off when using the microscope.

When moving the microscope, do not carry it by the arm alone. Place one hand under the base and hold the arm with the other hand. This will avoid a costly accident.

SECTION 2: Description of the Microscope

To use the microscope properly, you need to be familiar with its various parts and functions. Like typewriters or personal computers, different brands share similarities but are not identical. There is, however, enough similarity so that Figures 10-1 and 10-2 can serve as general references. However, it is advisable for you to look at the manual that goes with your particular model. The microscope that is used in an office laboratory is called a "brightfield microscope."

In this section the microscope is divided into its different functional systems: illumination, support, magnification, and focusing. The essential elements associated with each function are then described. Features that you should be sure to have on your office scope are also noted.

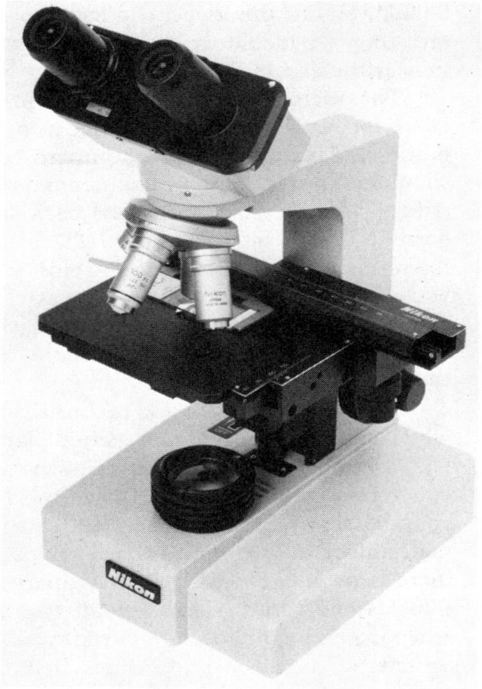

Figure 10–1. Clinical microscope suitable for office use. (Courtesy of Nikon Inc, Instrument Division.)

ILLUMINATION

The routine clinical microscope is called a compound brightfield microscope. The term brightfield is used to distinguish this *type* of compound microscope from other *types*, such as darkfield, phase, or fluorescent. With brightfield illumination, the specimen appears dark against a bright background. *This is the type of microscope you will want to purchase.*

With darkfield illumination, the specimen is light against a dark background. This type of microscope is used when examining a specimen for the organism that causes syphilis. It is not needed in routine office laboratory.

With phase contrast illumination, higher contrast is produced. This is used for looking at things that are difficult to distinguish with the regular bright-light background (e.g., platelets). This type of illumination helps in seeing living (as opposed to stained) organisms; however, it is not needed in an office laboratory.

With fluorescent illumination, special filters and stains are used that "fluoresce."

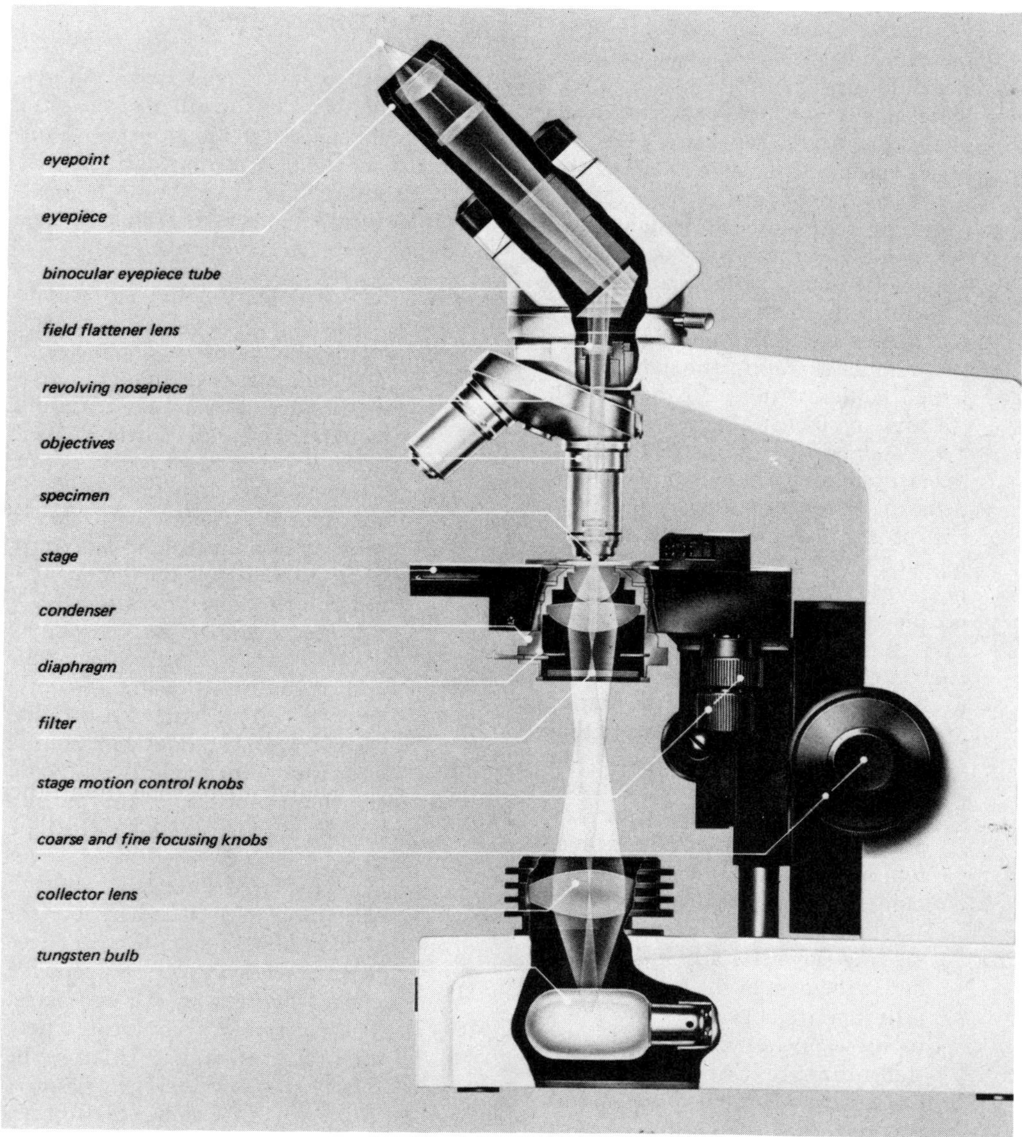

Figure 10–2. Details of the parts of a microscope. (Courtesy of Nikon Inc, Instrument Division.)

The use of this type of microscope has been greatly expanded with the advent of fluorescent monoclonal antibody test systems. At this point, however, it is not appropriate for most office laboratories.

The following terms are used in describing brightfield microscopes:

1. Light source: While older scopes used sunlight, candles, and mirrors to pro-

vide light, modern microscopes now have incandescent, tungsten, or halogen bulb illumination.

2. Field diaphragm: The exact location of this diaphragm will vary with the brand and model of microscope. It is in the lower light path, either right on top of the light source or, in new halogen scopes, in the lower body of the scope. The function of this diaphragm is to limit the field of view.
3. Condenser assembly: The light from the bulb passes up to the condenser, which collects (i.e., "condenses") the rays and focuses them on a point. The condenser system is made up of at least two different lenses. The condenser is under the stage, secured into the substage assembly. It is raised and lowered by a knob located on the side of the microscope. This knob will be on the side opposite from the knobs that control the slide movement on the stage.
   a. Aperture diaphragm: At the base of the condenser is the aperture diaphragm. It controls the angle of light entering the condenser and affects the resolution, contrast, and definition of the specimen. If too much light illuminates the field, contrast will decrease. The aperture diaphragm is adjusted by moving the little arm that sticks out at the base of the condenser. A fully open diaphragm allows you to use a lower voltage from the transformer to light the bulb. This can mean longer bulb life.
   b. Blue filter: At the bottom of most substage assemblies is a slot that holds a blue filter. The purpose of the blue filter is to "whiten" the light from the tungsten bulb. (Some scopes may have a rotating system of filters attached at some other point. Consult your product manual.)

THE SUPPORT SYSTEM

The support system is the structural system that holds all of the other parts of the microscope together and in proper position. It consists of the base, arm, stage, and substage assembly. The base should be obvious. The arm connects to the base and holds the eyepiece tube in place.

1. Stage: The *stage* is the surface on which the microscope slide is placed. There are three types of stages:
   a. The *glide stage* is a manual movement stage. You "glide" it along by hand. The slide is held in place by two pressure clips. It is the cheapest stage, but it is not adequate for most office use. This is because it is difficult to smoothly control the slide's movement. *Do not buy this type of stage.*
   b. The *mechanical stage* is moved by turning stage control knobs. This allows you to move the slide in a smooth, slow, and controlled manner. One knob moves the slide from left to right (y axis), and the other knob moves the slide up and down (x axis). *This is the type of stage that you will want.*
   c. The *graduated stage* is a mechanical stage with numbers on the slide-holding assembly. These numbers allow you to specify a given field by a set of coordinates. This permits you to relocate a field for re-examination. This is the most expensive type of stage. While it is nice to be able to relocate fields, this type of stage does not merit the extra cost for most office laboratory work.
2. Substage assembly: The substage assembly holds the condenser and has an adjustment knob that raises and lowers the condenser. The substage adjustment knob and the mechanical

stage control knobs are on opposite sides of the microscope. In most microscopes the blue filter is in the substage assembly, below the condenser.

MAGNIFICATION AND FOCUSING SYSTEM

The compound brightfield microscope magnifies the specimen by using two separate lens systems, the eyepiece or ocular lenses, and the objective lens system. The total magnification of the specimen is calculated by multiplying the eyepiece magnification times the objective magnification. The usual eyepiece magnification is 10x; therefore:

10x eyepiece × 10x low power objective = 100x total magnification

10x eyepiece × 40x high power objective = 400x total magnification

10x eyepiece × 100x oil objective = 1,000x total magnification

1. The eyepiece (or ocular system) is the part of the microscope that you look into. A monocular scope has one eyepiece, a binocular scope has two. *If you are buying a scope, get one with binocular eyepieces.* There are two types of eyepieces in common use:
   a. The huygenian eyepiece provides a small field and has a low "eyepoint" (i.e., the height above the eyepiece at which you best see the image). This means that it may be more difficult for people with glasses to use.
   b. The wide-field eyepiece provides a 25% larger field and has a higher eyepoint. It is therefore better for people who wear glasses. *If you are buying a microscope, get wide-field eyepieces.*
2. The objective lens system is the lens system nearest the specimen. The objectives are of various magnifications and are attached to a revolving nosepiece that permits easy switching. Each individual objective has a series of internal lenses. It is possible to get a nosepiece with four objectives; however, you really only need three in a physician's office laboratory. These three include:
   a. Low power (10x-dry): The lower the magnification, the shorter the objective. On a three-objective scope, this is the shortest objective. It is used to *scan* specimens before switching to higher magnification.
   b. High power (40-45x dry): This is the middle length objective. It is used to examine "wet" specimens, such as vaginal secretions and urine sediments.
   c. Oil immersion (95-100x): This is the longest objective. It is used to examine stained slides such as Gram's stains and peripheral blood smears. You must use a drop of *immersion oil* with this objective. The oil helps to capture the light rays. Avoid oil that contains polychlorinated biphenyls (PCB) or other health hazards. (Check the label.) *Do not use mineral or cedarwood oil.* Each has a low viscosity and can seep into the objective lens system. Most manufacturers recommend a standard oil such as Cargille type A immersion oil.
   d. Some microscopes have a fourth, very low-power objective (either a 2.5x or 4x scanning objective). These require the use of an auxiliary swing-in condenser to fill the entire field with light. *You do not need this extra objective in a physician's office laboratory.*
3. Focusing
   The specimen is brought into focus

by using the coarse focus knob and the fine focus knob. In addition, the ribbed band around one of the eyepieces permits focusing of an individual eyepiece, thus accommodating the differences between the visual acuity of your two eyes.

4. Concepts of Magnification and Focusing
In addition to the "physical" parts of the microscope described above, there are a series of terms used to refer to characteristics of the magnifying/focusing systems. You might wonder why microscopes do not have even greater magnification ability (1,000x). There is a limit to the amount of *useful* magnification that can be produced. "Empty magnification" involves an increase in size without any increase in specimen detail.
   a. Resolving power (RP), or resolution, is the ability to reveal fine detail. The resolution power of a microscope is the smallest distance between two points that can still be seen as two separate entities. The resolution power is a function of the wavelength of light.
   b. Numerical aperture (NA) is a mathematical concept that describes light delivered by the condenser and picked up by the objective. The higher the NA the greater the resolution power of the objective. The NA of the condenser must be greater than or equal to the NA of the objective. The usual NA of clinical scopes is 1.25. Clinical microscopes are usually sold with objectives and condenser matched to each other.
   c. Working distance is the distance between the bottom of the objective and the top of the coverglass. When you visualize the length of the different objectives (e.g., the 100x being the longest objective), it becomes obvious that the greater the magnification, the smaller the working distance. When using the standard coverslip, there is less than 0.005 cm between the oil objective and the slide when the specimen is focused. If your microscope does not have a "focus stop" preset at the factory that *prevents* the objective from touching the slide, you must exercise great care when focusing under oil to avoid damaging the lens or breaking the slide. *Choose a microscope that has a factory-set focus stop.*
   d. Depth of focus refers to the "thickness" of the specimen that can be brought into focus at one time. The greater the magnification the thinner the layer of material that can be focused. There is a wide depth of focus when using the low power (10x) objective. In contrast, with the oil objective (100x) a slight change in fine focus will put the specimen out of focus.
   e. Chromatic aberration is due to a failure to bring into focus all of the colored components of white light. Objects under the microscope will appear to have colored fringes. "Acromatic" lenses are those that have been corrected so that the different rays of light are focused, and therefore there are no color fringes to the objects.
   f. There is another type of distortion that occurs when only the center of the field is in focus. This is a special problem when examining specimens under oil. "Planachromatic" objectives correct not only for chromatic aberration but also provide a "flat" field (i.e., uniform level of focus). There is a $50 to $70 difference in cost between achromatic and planachromatic ob-

jectives. *Buy a microscope with planachromatic objectives.*

g. Parfocal means that when you are in focus at one magnification, you will only have to fine tune the focus at a different magnification. This saves a great deal of time. *Choose a parfocal microscope.*

h. Parcentered means that if you center something in the field under one magnification, it will be approximately in the center of the same field when you change magnification. This also saves time and frustration when you are looking at specific cells or areas of a field. *Choose a parcentered microscope.*

SECTION 3: Alphabetized Glossary

This is a listing of the terms commonly used when microscopes are described. It is included as a convenience so that you can look up terms quickly should you need to.

Achromatic. White light is made up of different colors of light. If there is some aberration, the component colors are not brought together, and there could be color fringes to an image. Achromatic lenses have been corrected to eliminate these color fringes.

Aperture diaphragm (or iris diaphragm). A mechanism directly beneath the condenser that controls the angle of the cone of light shining on the specimen and entering the objective. It affects the resolution, contrast, and definition. It does not control the intensity of light.

Blue filter. A filter that whitens the light by removing red from tungsten light.

Brightfield illumination. A type of illumination used for the clinical microscope. The specimen appears dark against a bright background.

Compound microscope. A "compound" microscope has two separate lens systems: objectives and eyepieces. The amount of magnification the scope provides is calculated by multiplying the strength (e.g., 10x) of eyepiece, or ocular, by the strength (10x, 40x, or 100x) of the objective. Therefore when you examine a specimen under low power, the specimen is magnified (10×4) 40 times. Under high dry the specimen is magnified 400 times (10×40), and under oil, 1,000 times (10×100).

Condenser. A system of lenses located under the stage. The function of the condenser is to collect the light rays from the light source and converge them. It has a numerical aperture (NA) rating, as do the objectives. The NA of the condenser must be greater than or equal to the NA of the objective. However, when you buy a microscope, the condenser and objectives are usually matched by the manufacturer.

Contrast. The density difference between the darkest and brightest parts of the specimen. The aperture diaphragm (located beneath the condenser) controls the contrast. Too little contrast leads to a lack of definition. Too much contrast leads to a lower resolution.

Chromatic aberration. A distortion that results in color fringes around objects. It is due to a failure to bring all of the colored components of white light into one focus.

Coverglasses. Coverslips are made of optically flat glass or plastic. The glass slips are preferred.

Darkfield illumination. A type of illumination using a different condenser or a special stop in a regular condenser so that the specimen appears bright against a dark background. This type of illumination is used for locating the *Treponema pallidum* organisms of syphilis.

Depth of focus. The thickness of a specimen that can be seen "in focus" at one time. As you increase the amount of magnification, you produce a thinner depth of focus (i.e., only a thin plane of the specimen will be in focus at one time).

Empty magnification. An increase in size without any increase in specimen detail.

Eyepiece (or ocular). The upper optical system on a microscope. If there are two eyepieces it is a binocular scope; if there is only one, it is a monocular scope.

Eyepoint. The distance above the eyepiece where you can best see the image.

Field diaphragm. A diaphragm that is built into the lower light path and limits the field of view. On some microscopes it is located over the base light source; with others it is in the lower body of the scope.

Filters. The use of filters can increase stained detail. A blue filter is the most common filter used with a tungsten-bulb light source. Tungsten gives off yellow light. The blue filter, by absorbing excess red, delivers whiter light to the specimen.

Fluorescent illumination. Special filters used to see specimens stained with "fluorescent" dyes.

Focus stop (preset). A lock in the focusing mechanism that prevents the objectives from coming into contact with the slide. Most microscopes are set at the factory and allow you to focus while looking through the microscope without fear of damaging the objective or the slide.

Glide stage. A manual movement stage. It is the cheapest kind of stage, but it is not acceptable for office use. Slight hand movements with a glide stage result in enormous shifts when looking at a specimen through the microscope.

Graduated stage. A mechanical stage with numbers on the slide-holding assembly. You can specify a particular field by giving the coordinates. When you want to relocate that field, you set the stage at those coordinates.

Immersion oil. A special oil used to deliver more light to the oil immersion objective. Do not use mineral oil or cedarwood oil. Instead use a standard oil such as Cargille type A immersion oil.

Koehler illumination. This term describes a complicated procedure that is sometimes recommended to obtain the proper illumination of a specimen for microscopic examination. The procedure is rarely used by those who spend much of their time performing microscopy. Instead it is easiest to turn down the condenser to increase the contrast when using the low-power objective. We can detect no practical advantage to using the Koehler procedure.

Magnification. The amount of magnification provided is determined by multiplying the power of the eyepiece times the magnification of the objective. Example: 10x eyepiece with a 40x objective = 400x.

Mechanical stage. A stage controlled mechanically with knobs and gears. One knob moves the slide up and down (x axis), the other right and left (y axis).

Numerical aperture (NA). A mathematical notation describing the solid cone of light coming out of the condenser that is focused by the objective. The higher the NA of the objective the greater the resolution ability of the objective.

Objectives. The lenses that are located directly above the stage. There are various magnifications of lenses:
2.5x-4x (very low): This is used for scanning and is not usually required in primary care laboratories. This objective requires an additional swing-in condenser.
10x (low, dry): This is an essential objective and is used for scanning the slide.
40x (high, dry): An essential objective used for examining wet specimens (e.g., urine sediment, vaginal specimens).
100x (oil immersion): An essential objective used to examine stained specimens (e.g., Gram's stains, Wright's-stained blood smears). This is the highest magnification available.

Phase contrast illumination. A type of illumination that uses a special condenser to increase the contrast between the specimen and background. It is useful in distinguishing elements that are difficult to see against the brightlight background, especially living organisms.

Parcentered. A focusing feature that maintains an "object" approximately in the center of the field, switching from one objective to another.

Parfocal. A focusing feature that allows for focused viewing despite switching from one objective to another. Once you have focused the specimen with one objective (e.g., low power) you will only need to make minor adjustments with the fine-focus knob to bring the specimen into focus when switching to a different objective.

Planachromatic. An objective that corrects for color fringes and shows a flat field of view (i.e., the entire field of view is in focus).

Resolution. The ability of the microscope to show detail. It refers to the smallest distance between which two points can be separately identified.

Stage. The flat surface on which the slide is placed. There are three types: glide, mechanical, and graduated.

Substage assembly. A part of the microscope that holds the condenser, the blue filter, and an adjustment knob to raise and lower the condenser.

Support system. The skeleton of the microscope that holds the various parts in proper position. It consists of the base, arm, stage, and substage assembly.

Working distance. The distance between the bottom of the objective and the top of the coverglass or slide.

SECTION 4: Purchasing Information and Checklist

Contact several microscope manufacturers (see list below) and request information and price lists. Indicate that you are interested in purchasing a microscope for your office laboratory. If possible, talk to the manufacturer's representatives rather than to general medical suppliers. The manufacturers will tell you who sells their microscopes in your area. Ask for a demonstration before making a purchase so that you and your staff can see and use a variety of microscopes. During the demonstration, look at slides with the oil objective (i.e., Gram's stains or Wright's-stained peripheral smears) and a "wet" specimen under high dry (i.e., a urine sediment).

QUESTIONS TO ASK

- Is it possible to "upgrade" the microscope (e.g., to make it a teaching scope)?
- What is the warranty?
- Is technical service available for repairs?
- Is there an 800 hot line for problems?
- Does someone come to set up the microscope properly after you purchase it?
- Is an instruction manual included?
- Does the supplier have microscope repair or servicing capabilities?

COMMENTS

In setting up a practice or in buying new equipment, physicians quickly reach a point at which they are shocked by the expense and want to cut corners. *DO NOT SCRIMP ON A MICROSCOPE.* You can expect to pay between $2,000 and $3,000 for a new, high-quality scope. While this represents a significant financial investment, this piece of equipment, properly cared for, should last the life of your practice. It will not become outdated. When deciding which microscope to purchase, consider the needs of your office laboratory and the convenience and comfort of those who will be using the microscope.

WHAT FEATURES TO ORDER

We strongly recommend that your microscope include the following features:

- Binocular eyepieces
- Mechanical stage
- Wide-field eyepieces
- Objectives
  - Oil immersion (100x)—planachromatic
  - High dry (40x to 45x)—achromatic
  - Low power (10x)—achromatic
- Parfocal
- Parcentered
- Preset focus stop

LIST OF CLINICAL MICROSCOPE MANUFACTURERS

Reichert Scientific Instruments, PO Box 123, Buffalo NY 14240; 1-800-828-1200

American Scientific Products, American Hospital Supply Corp, 1430 Waukegan Rd, McGraw Park, IL 60085; 315-689-8410

Bausch & Lomb, Scientific Optical Products, Optics Center, 1400 N Goodman St, Rochester, NY 14602; 716-338-6000

Bristoline Inc, 248 Buffalo Ave, Freeport, NY 11520; 516-623-1000

Carolina Biological Supply Co, 2700 York Rd, Burlington, NC 27215; 1-800-632-1231

E. Leitz, Inc, Link Dr, Rockleigh, NJ 07647; 201-767-1100

Edmund Scientific Co, 101 E Gloucester Pike, Barrington, NJ 08007; 1-800-257-6173

Fisher Scientific Co, 711 Forbes Ave, Pittsburgh, PA 15219; 412-562-8300

H & R Optical Systems, Inc, 6565 Odell Place, Boulder, CO 80301; 303-530-2217

Labtron Scientific Corp, 91 Cabot Ct, Hauppauge, NY 11788; 1-800-645-9066

Nikon Inc, Instrument Division, 623 Stewart Ave, Garden City, NY 11530; 516-222-0200

Olympus Corporation of American, Precision Instrument Division, 4 Nevada Dr, New Hyde Park, NY 11042; 516-488-3880

Polysciences, Inc, 400 Valley Rd, Warrington, PA 18976; 1-800-523-2575

Propper Manufacturing Co, 36-40 Skillman Ave, Long Island City, NY 11101; 718-392-6650

Seiler Instrument and Mfg Co, 170 E Kirkham, St. Louis, MO 63119; 314-968-2282

Swift Instruments, Inc, PO Box 562, San Jose, CA 95106; 408-293-2380

Tasco Scales, Inc, 7600 NW 26 St, Miami, FL 33122; 305-591-3670

Unitron Instruments, Inc, 175 Express St, Plainview, NY 11803; 516-822-4601

Vickers Instruments, Inc, Joyce Loebl Division, 300 Commercial St, Malden, MA 02148; 1-800-VICKERS

Western Scientific Co, Inc, 1577 Colorado Blvd, Los Angeles, CA 90041; 213-257-0832

Carl Zeiss, Inc, One Zeiss Dr, Thornwood, NY 10594; 914-747-1800

CHOOSING A USED MICROSCOPE

- It is best to choose name brands.
- Be clear about all features.
  (The word "planachromatic" will be inscribed on such objectives.)
- Contact the manufacturer to inquire about service and technical assistance.
- Make sure that the instruction manual is included in the purchase, or order it from the manufacturer if it has been lost.
- Have someone very familiar with the microscope check it out for you.
- Was it placed in a high-use or misuse setting (i.e., the house office laboratory in a busy teaching hospital)?
- Do all of the gears work smoothly? Are they too "tight" or too "loose"?
- Is the image with each objective focused and clean?
- Is the light source adequate for examining routine specimens under the 100x objective?
- Examine the service records. Hospital laboratories have their microscopes routinely serviced at least once a year.

ACCESSORIES

Locks

If security is an issue, Design Instruments (404 N Bouser, Bldg B, Richardson, TX 75080) offers excellent microscope locks for under $40. Write for their catalog.

Eyeshields

Soft rubber eyeshields are also available from Design Instruments. They protect the laboratory worker's eyeglasses from being scratched from contact with the microscope eyepieces.

Polarizing Filters

The polarizing systems built into the microscope usually cost between $200 and $400. However, a simple polarizing kit (costing approximately $15) is available from J and G Manufacturing Co, PO Box 42, Midlothian, VA 23113. It comes with simple instructions. (See Owen DSA. A cheap and useful compensated polarizing microscope. *N Engl J Med* 1971;285:1152.)

SECTION 5: How to Clean the Microscope and Perform Routine Preventive Maintenance

If you take good, regular care of your microscope, it will last the lifetime of your practice and will rarely, if ever, require professional cleaning or repair. However, if you and your staff are careless, you may end up spending hundreds of dollars on new objectives and reconditioning efforts.

It is important that you have lens paper readily available. This can be ordered from any scientific supply company or from a camera shop. Take the larger sheets and cut them into smaller 2 × 2-inch pieces. Put these in the top half of an empty slide box, and leave the box top next to the microscope where the sheets will be handy for wiping immersion oil from the 100x objective.

Dust can collect on various parts of the microscope and will produce dirty eyepieces, condensers, or light sources. Also, over time dust can make the mechanical parts move with difficulty. Microscopes usually come with a dust cover, or one can be ordered. Put the dust cover over the microscope every evening when the laboratory is closed.

If you wear glasses, the "dirt" you notice in the microscope field may be on your glasses (i.e., dirt or scratches). Jiggle your glasses as you look through the microscope. If the "dirt" moves, clean your glasses and check them for scratches.

It is similarly easy to determine if the eyepieces are responsible for the "dirt" that you see. Rotate each eyepiece separately. If there is "dirt," you will see it move as you rotate an eyepiece. If the dirt moves but does not go away with simple cleaning, then the dirt is probably on the inner surface of the eyepiece. This requires a thorough eyepiece cleaning.

It is more difficult to evaluate the objectives. Usually you will not see "dirt" the way you do with dirty glasses or eyepieces. Instead, dirty objectives produce a blurry, unsharp image, no matter how much you focus.

ROUTINE DAILY CLEANING

- *Eyepieces:* Wipe the outside lens surface with lens paper as needed.
- *Objectives:* Wipe immersion oil off the lens surface immediately after each use. If you drag the high-dry objective through the oil or if urine or other fluids get onto any objectives, clean it right away.
- *Stage:* Wipe off the stage with an alcohol wipe or lens paper whenever needed.
- *Condenser:* Wipe off the lens surface with a piece of lens paper dampened with alcohol or lens cleaner.
- *Light source:* Wipe off with a piece of lens paper.

MONTHLY THOROUGH CLEANING

1. *Eyepieces:* If there is dirt on one of the eyepieces that cannot be removed by simple cleaning, perform the following procedures:
   a. Place a clean cloth on the bench next to the microscope.
   b. Remove one eyepiece from the body of the microscope.
   c. Poke a full-sized sheet of lens paper into the body opening to prevent dust from entering the inside of the scope.
   d. Put the eyepiece on the cloth with the outer surface facing you.

e. Dampen a piece of lens paper with alcohol or lens cleaner and use this to wipe the outer surface.
f. Use a clean piece of lens paper to dry the lens.
g. Turn the eyepiece so that the inner surface faces you.
h. Dampen a piece of lens paper with alcohol or lens cleaner and use a cotton-tipped swab to move it over the entire inner surface of the eyepiece lens.
i. Use a second clean piece of lens paper to dry the lens—again using a cotton-tipped swab to move it all over the inner lens.
j. Remove the lens paper from the eyepiece opening and insert the eyepiece into the tube.
k. Look through the eyepiece to make sure that you have removed any dirt. If not, repeat those steps until it is clean.
l. Perform the same steps on the other eyepiece.

2. *Objectives:*
   a. Spread a clean cloth on the bench next to the microscope. Make sure that you have lots of lens paper and cotton-tipped swabs available.
   b. Unscrew one of the objectives from the nosepiece. Place a piece of lens paper in the nosepiece hole so that dust does not get into the microscope.
   c. Place the objective (screw end down) on the cloth.
   d. Wet a cotton swab with alcohol or lens cleaner and wipe the outer surface of the objective lens. DO NOT CLEAN THE INSIDE SURFACE.
   e. Repeat this process with a clean swab.
   f. Dry the lens with a clean dry swab.
   g. Remove the lens paper from the nosepiece hole and screw the objective back into place.
   h. Repeat this process for all objectives. You may want to use one or two extra alcohol-wet cotton swabs on the oil objective.
   i. If the image is blurred when using the oil objective, repeat the cleaning process again. If this does not correct the problem, it is very likely that oil has seeped into the internal lens system of the objective. It will most likely need to be replaced. At this point you will need help from a microscope repair technician.

MICROSCOPE REPAIR/SERVICE CONTRACT

If your microscope needs repair, consult a microscope service specialist. You can contact the manufacturer for the name of an authorized service center. (This information may be in the warranty book.) Another source of information is your local hospital. Call the laboratory and find out who performs its regular servicing.

If you take good *regular* care of your microscope, you will rarely, if ever, need a microscope technician. Because of this there is really no need to purchase a service contract.

EQUIPMENT RECORD

Whenever you purchase a piece of laboratory equipment, you should create a file with all of the important information. This file should include the instruction manual, the warranty information, the sales invoice, and a repair/maintenance record (Fig. 10–3).

Equipment: ______________________________

Manufacturer's Name: ______________________________

Address: ______________________________

____

Telephone: ______________________________

Model Number: ______________ Serial Number: ______________

Date of Purchase: ______________ Purchase Price: ______________

Supplier's Name: ______________________________

Address: ______________________________

____

Telephone: ______________________________

Warranty Period: ______________________________

Repair Person's Name: ______________________________

Address: ______________________________

____

Telephone: ______________________________

Problem Hotline (if available): ______________________________

Date	Problem	How Fixed	By Whom	Cost

Figure 10–3. REPAIR/MAINTENANCE RECORD

SECTION 6: Troubleshooting Problems

Table 10–1 is a chart of common difficulties that you might have with your microscope. Before you call in outside help, consult this table to determine if you can correct the problem yourself. This approach will save both time and money.

TABLE 10–1. COMMON PROBLEMS WITH THE MICROSCOPE

Problem	Cause	Corrective Action
	Magnification System	
Incomplete binocular vision	Eyepiece distance incorrect	Adjust distance
Field of view cut off	Objective out of position	Rotate nosepiece until objective "clicks" into place
Image not sharp	No oil used with 100x objective	Place a drop of oil on the slide
	Urine or oil on the 40x objective	Remove the objective and clean it properly
	Slide is upside down (can focus on low, not under oil)	Turn slide over
	Oil has dried on oil objective	Remove objective and clean it properly
	Urine has dried on 40x objective	Remove objective and clean it properly
	Oil may be inside objective	Remove objective; hold up to the light with the screw end near your eye. If you see an irregularly shaped circle that changes shape as you rotate objective, there is oil inside. Replace objective
Dirt seen in field	Dirty eyepiece (rotate and see if dirt moves)	Clean
	Dirty eyeglasses (move them to see if dirt moves)	Clean
	Dirt on condenser or objective	Clean

TABLE 10–1. Continued

Problem	Cause	Corrective Action
	Illumination System	
Excessive image contrast	Condenser too low	Raise
	Diaphragm too closed	Open
Light intensity too low	Light set too low	Increase light power
	Condenser too low	Raise
	Closed diaphragm	Open
Light flickers	Bulb needs replacement	Replace
	Loose connection	Consult manufacturer
Bulb does not light	Bulb is burned out	Replace
	Fuse is burned	Replace
	Cord is unplugged	Plug in
	Poor electrical connection	Consult manufacturer
	Focusing System	
Unable to fully focus	Slide upside down	Turn slide over
	Fine adjustment is fully turned in one direction	Center fine adjustment knob, and refocus with coarse, then fine, focus knobs
Focus knob is too tight	Tension ring is too tight	Consult manual and loosen
	Gears need oiling	Have system serviced
Stage drops	Tension ring is too loose	Consult manual for tightening instructions

SECTION 7: How to Examine "Wet" Specimens Under the Microscope (e.g., Urine Sediment, Vaginal/Cervical Secretions)

1. Rotate the objective nosepiece until the low-power (10x) objective is securely in place. If the microscope has three objectives, low power is the shortest. If it has four objectives, the 10x objective is next to the shortest. If the nosepiece is not properly seated, you will see a partially darkened field—not a full, round circle of light. The objective usually makes a slight "click" when it is properly engaged.

2. Look at the microscope from the side, and turn the coarse focus knob until the objective and stage are the maximum distance apart.

3. Place a cover-slipped specimen onto the stage and secure it between the fingers of the slide holder assembly. The cover slip protects the lens on the objective and produces a uniform specimen thickness.
4. Turn on the light source. When looking at wet specimens, use low light. If there is too much light you will not be able to see the cells in the specimen. If there is a natural density filter, switch it into place.
5. Lower the condenser to its lowest position.
6. Make sure that the field and aperture diaphragms are now open.
7. Adjust the binocular eyepieces so that when you look through them you see a single, round field. They can be adjusted by pulling them together or by pushing them apart. Some scopes have a small knob in between the eyepieces to use for this adjustment. If you can see through only one eyepiece or if there is a black area of interference, then the eyepieces are not properly spaced. If you have a monocular scope, keep both eyes open while looking through the microscope. This will be difficult at first but does prevent eye strain.
8. Use the coarse focus knob to bring the objective as close to the slide as possible. If your microscope has a preset focus stop, the objective *cannot* touch the slide. You can therefore turn the coarse focus knob without watching from the side to see how close it gets. If your microscope *does not have the preset focus stop feature,* you should always *watch* from the side while bringing the objective close to the slide. (Consult the reference manual for your microscope to find out whether it has a preset stop feature.) *Remember the direction (clockwise/counterclockwise)* in which you turned the coarse focus knob to bring the slide and objective close together. You will want to turn it in the opposite direction in the next step.
9. Look at the two eyepieces. One is plain and one is ribbed. The ribbed eyepiece permits you to separately focus the two eyepieces to adjust for the acuity differences between your eyes. Close the eye that looks through the ribbed eyepiece. While looking at the specimen through the plain eyepiece, slowly turn the coarse focus knob until the specimen comes into focus. Finish focusing the plain eyepiece by turning the fine focus knob. You may have to play with the fine focus knob a bit, turning first one way, then another, to get things into sharp focus.
10. Open the eye that will be looking through the ribbed eye piece. Close the other eye. Turn the *ribbed band* (not the whole eyepiece) first one way and then the other until the specimen comes into sharp focus. *Do not use the coarse or fine-focus knobs* when focusing this second eye. Once you have adjusted the scope to both your eyes, the coarse/fine focus knobs should bring things into focus for *both* eyes. You should not have to refocus each eye separately every time you try to focus on a specimen. However, if someone else has used the microscope since you last set it for your eyes, you will have to go through the separate focusing steps again.
11. If you need more light, adjust the light source. But remember, with wet specimens you need very subdued light.
12. Use the low-power (10x) objective to scan the specimen. To look at different fields, move the slide by turning the knobs that control the stage/slide holder assembly. One knob moves the specimen up and down (x axis), the other moves it from left to right (y axis). Practive moving these controls.

After a while you will not have to think about which knob does what.

13. As you move from field to field you will have to keep focusing with the fine-focus knob. Wet specimens have depth (i.e., all the cells are not in the same focal plane); therefore you have to constantly move the fine-focus control.
14. When you have finished scanning, you will want to look at the specimen under greater magnification. Find an area of the slide that you want to examine more closely, and center this in the low-power field. Bring it into fine focus.
15. Switch to the 40x (high-dry) objective by rotating the nosepiece.
16. Look through the eyepieces. You should be able to bring the specimen into sharp focus by slightly turning the fine-focus knob. If your scope requires more than fine turning to bring the specimen into focus, you will probably have to go back through the coarse focusing steps again.
17. You may need to increase the light at this point. If you have used a neutral-density filter when scanning, switch it out of the light path now. This may be all the additional light that is required, or you may need to increase the transformer voltage.
18. As you examine several fields you will need to continually readjust the fine focus. Examine at least ten different fields.
19. When you are finished examining the slide, remove it from the stage. If any of the specimen has spilled onto the stage, wipe it up immediately with gauze or an alcohol wipe.
20. Turn off the light.

PRACTICE EXERCISE

1. Take a urine or vaginal specimen that has lots of white cells. Follow steps 1 through 16 to examine the slide. Then, with either the 10x or 40x objective in place, find a field full of white cells. Now move the condenser all the way up and increase the light source to its maximum. What happens? Watch the specimen as you lower the condenser and lower the amount of light.
2. Take a specimen and put it into focus. Then turn the fine focus a little one way, then the other. Wet specimens have a certain thickness. Not all of the cells in the specimen are at the same depth, so that not all of the specimen is "in focus" at the same time.

COMMENTS

1. Always coverslip wet specimens. This protects the objectives and gives better specimen detail. Glass coverslips are preferred. 22 x 30 mm No. 1 1/2 glass coverslips are preferred.
2. Always use subdued light when looking at wet specimens. Keep the condenser down and the light at the low setting.

SECTION 8: How to Use the Microscope to Examine Stained Slides Under Oil (e.g., Gram's Stains and Peripheral Blood Smears)

1. Rotate the objective nosepiece until the low-power (10x) objective is securely in place. If the microscope has three objectives, the low power is the shortest. If it has four objectives, the 10x objective is next to the shortest. If the nosepiece is not properly sealted, you will see a partially darkened field—not a full, round circle of light. The objective usually makes a slight "click" when it is properly in place.
2. Look at the microscope from the side and turn the coarse focus knob until the bottom of the objective and the stage are the maximum distance apart.
3. Place the slide with the stained specimen, specimen side up, onto the stage and secure it between the fingers of the slide holder assembly.
4. Turn on the light source at the lowest setting.
5. Raise the condenser to its highest position.
6. Make sure that all of the diaphragms are open.
7. Adjust the binocular eyepieces so that you see a single, round lit field when looking through them. They can be adjusted by pulling them together or by pushing them apart. Some scopes have a small knob in between the eyepieces to use for adjustment. If you can see through only one eyepiece or if there is a black area of interference, then the eyepieces are not properly spaced. If you have a monocular scope, keep both eyes open while looking through the microscope. While this may be difficult at first, it prevents eye strain.
8. Use the coarse focus knob to bring the objective as close to the slide as possible. If your microscope has a preset focus stop, the objective *cannot* touch or go through the slide, and you can turn the coarse-focus knob without watching from the side to see how close it gets. If your microscope does not have the preset focus stop feature, you should always *watch* from the side while bringing the objective close to the slide. (Consult the reference manual for your microscope to ascertain if it has the preset stop feature.) *Do not touch the slide with the objective.* Remember the direction (clockwise/counterclockwise) in which you turned the coarse-focus knob to bring the slide and objective close together. You will want to turn it in the opposite direction in the next step.
9. Look at the two eyepieces. One is plain and one is ribbed. Close the eye that looks through the ribbed eyepiece. While looking at the specimen through the plain eyepiece, slowly turn the coarse focus knob until the specimen comes into focus. Finish focusing the plain eyepiece by turning the fine-focus knob. You may have to play with the fine-focus knob a bit, turning first one way and then the other, to get things into sharp focus.
10. Open the eye that will be looking through the ribbed eyepiece. Close the other eye. Turn the *ribbed band*

(not the whole eyepiece) first one way and then the other until the specimen comes into sharp focus. *Do not use the coarse- or fine-focus knobs* when focusing this second eye. Once you have adjusted the scope to both your eyes, the coarse-/fine-focus knobs should bring things into focus for *both* eyes. You should not have to refocus each eye separately every time you try to focus on a specimen. However, if someone else has used the scope since you last set it for both your eyes, you will have to go through the separate focusing steps again.

11. If the field seems too dark, turn the light source up. You do not want the light to be so bright that you feel "blinded," but neither do you want to look through a glass darkly.
12. Using the lower power (10x) objective, scan the specimen by moving the slide using the knobs that control the stage/slide-holder assembly. One knob moves the specimen up and down (x axis), the other moves it from left to right. Practice moving these controls. After awhile you will not have to think about which knob does what.
13. As you move from field to field you will need to keep focusing with the fine-focus knob.
14. Find an area of the slide that you want to examine more closely. If you are examining a sputum Gram's stain, find a clump of white blood cells. Center it in the low-power field. Adjust the fine focus.
15. Turn the objective nosepiece to a position midway between the 10x and the 100x objectives.
16. Put a small drop of immersion oil on the slide where the light from the condenser shines through the specimen. Use Cargille type A immersion oil. Do not use cedarwood or mineral oil.
17. Turn the objective nosepiece to the 100x objective. *Do not drag the 40x objective through the oil. If that should happen, clean it off immediately* by wiping the objective with lens paper.
18. The specimen should almost be in focus. Bring it into sharp focus by further adjusting the fine-focus knob. If it is not possible to bring the slide into sharp focus:
    a. The slide may be upside down (i.e., specimen side is facing down).
    b. The drop of oil may be too small.
    c. The oil objective may be bumping against the slide holder assembly. Locate a field away from the edge of the slide and refocus.
    d. Your microscope may require more than just fine tuning when you change objectives. If this is true, bring the 100x objective as close as possible to the slide. Looking through the eyepiece, *slowly* turn the coarse focus knob until the specimen comes into crude focus. You should not have to focus each eye separately at this point.
19. You may need to increase the light when you switch to the oil objective.
20. Examine at least ten different fields using the 100x oil objective.
21. When you have finished examining the slide, rotate the nosepiece so that the 100x objective is not in the "view" position. Wipe the tip of the objective with lens paper to remove the oil. *Never leave oil on the objective tip, and never leave the 100x objective sitting in a drop of oil.* This can ruin the objective, and it will cost $200 to $300 to replace.
22. Remove the slide. If you want to file it, remove the oil by dipping the slide into some xylene and gently wiping it with a soft cloth. The specimen will not rub off.
23. Turn off the light.

PRACTICE EXERCISE

Take a specimen slide and put it on the stage upside down. Focus under low power. Add the drop of oil and try to focus under 100x. What happens?

SECTION 9: Abbreviated Procedure for Examining Specimens Under the Microscope

1. Rotate the nosepiece until the low-power objective is securely in place.
2. Look at the microscope from the side and turn the coarse-focus knob until the bottom of the objective and the stage are the maximum distance apart.
3. Place a specimen slide on the stage and secure it between the clips.
4. Set the light to its lowest setting.
5. Set the condenser to its lowest setting.
6. Make sure the diaphragms are all open.
7. Adjust the eyepieces.
8. Use the coarse focus to bring the objective as close to the slide as possible without touching it.
9. Bring the specimen into focus using one eye (plain eyepiece). Adjust the coarse and fine focus. Finishing focusing the other eye by rotating the ribbed eyepiece.
10. Adjust the light as necessary.
11. Scan the specimen under low power.

WET SPECIMEN (High Dry)	***OIL IMMERSION (100x)***
12. Adjust the fine focus to study cells at different levels in wet specimens.	**12.** Raise the condenser.
13. Pinpoint an area of interest and move it to the center of the field.	**13.** Adjust the fine focus as needed.

WET SPECIMEN (High Dry)	***OIL IMMERSION (100x)***
14. Switch to the high dry.	**14.** Pinpoint an area of interest and move it to the center of the field.
15. Adjust the fine focus.	**15.** Switch the nosepiece until halfway between 10x and 100x.
16. If necessary, increase the light.	**16.** Place a drop of immersion oil on the spot where the light shines through the slide.
17. Examine at least ten fields, continually adjusting the fine focus.	**17.** Switch to the oil immersion objective (100x).
18. When finished, remove the slide	**18.** Adjust the focus.
19. Turn off the light	**19.** Examine at least ten fields.
	20. Rotate the objective out of view and wipe off any oil with lens paper.
	21. Remove the slide.
	22. Turn off the light.

SECTION 10: How to Polarize Your Office Microscope

Frequently the clinician wants to know if the patient with a painful joint has gout, pseudogout, or an infection. The ability to polarize the office microscope and to examine an aspirated specimen immediately facilitates both the quality and timeliness of the patient's care.

Polarizing systems that are available from manufacturers (i.e., that are built "into" the microscope) are relatively expensive, ranging from $200 to $400. These systems include a polarized analyzer that fits into the body of the microscope, a second polarizer that fits over the light source, and a full-wave compensation plate that fits over the light-source polarizer.

Another very simple polarizing kit that costs only $15 is available from J & G Manufacturing Company (PO Box 42, Midlothian, VA 23113). The kit includes a round piece of polarized glass that fits into the barrel of the microscope, a square piece of polarized glass, which is placed over the light source, and a glass slide full-wave compensator that fits over the polarized square. (Fig. 10–4). See Chapter 21 for a discussion of how to prepare a joint fluid specimen for a microscope crystal examination.

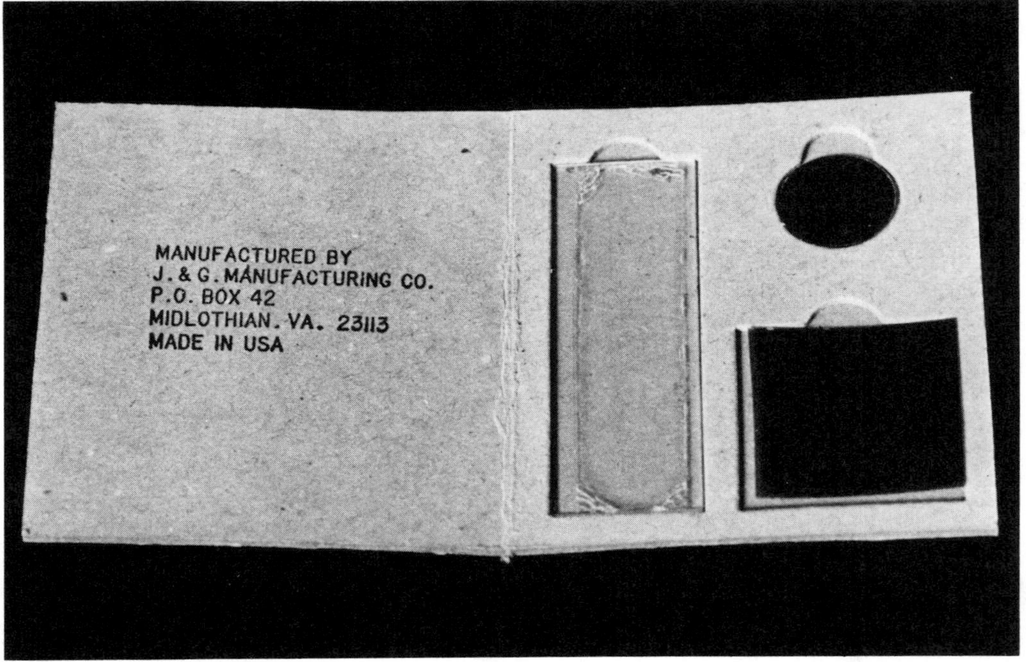

Figure 10–4. An inexpensive microscope polarizing kit (J & G Manufacturing Co.)

POLARIZING YOUR SCOPE USING THE J & G KIT

1. Look at your microscope. It should be possible to remove the part of the scope that houses the eyepieces (i.e., the eyepiece tube) from the rest of the scope. Usually there is a small screw knob. Turn it to loosen and remove the eyepiece tube. Lay it carefully on the bench (Fig. 10–5).
2. Look down into the scope. You should see a glass lens.
3. Place the round piece of polarized glass over this lens (Fig. 10–6). Make sure that it is clean (i.e., no fingerprints, etc.)
4. Replace the top part of the microscope and tighten the screw.
5. If your scope has a blue filter in a slot at the bottom of the condenser assembly, remove it (Fig. 10–7). (The blue filter is there to make the light from the tungsten bulb whiter.) If you forget and do not remove it, the background of the polarized fields will be excessively dark and blue. However, although more difficult to see, the crystals are still visible and remain the correct color.)
6. Place the square piece of polarized glass over the base light source (Fig. 10–8).
7. Turn on the light source. You will probably need the maximum light.
8. With the high-power objective in place, look through the eyepieces. You should see a light green field, unless the polarizer is by chance in just the right position.
9. Slowly rotate the base-light polarizer. At some point the field will slowly become very dark. Leave the base polarizer in this position. Your scope is now "polarized."
10. It is probably unlikely that you have a gout reference slide the first time that you go through this set-up procedure. To check the polarization, make a cornstarch test slide. Using ordinary cornstarch, place a very small amount on a glass slide. Add a

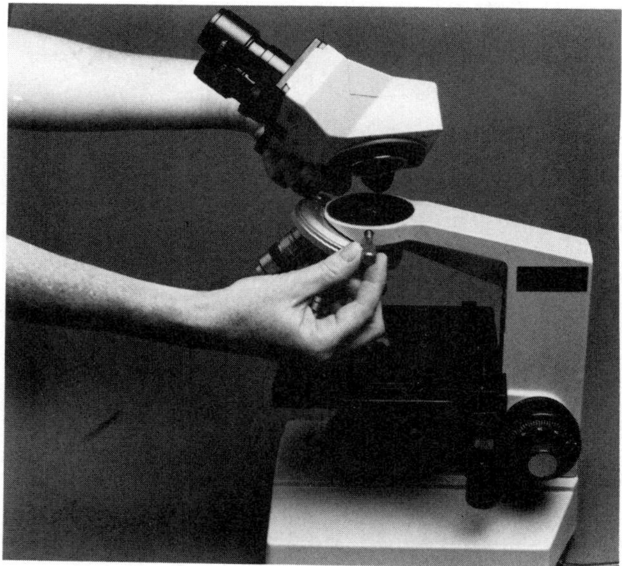

Figure 10–5. Remove the eyepiece tube from the microscope.

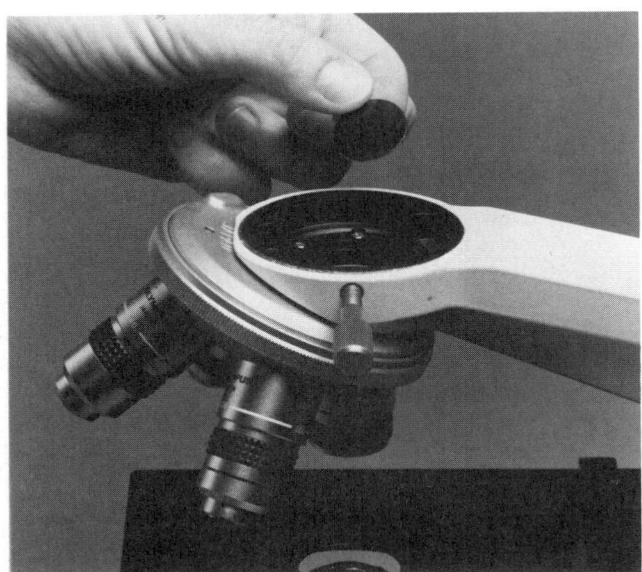

Figure 10–6. Place the round polarized glass over the lens.

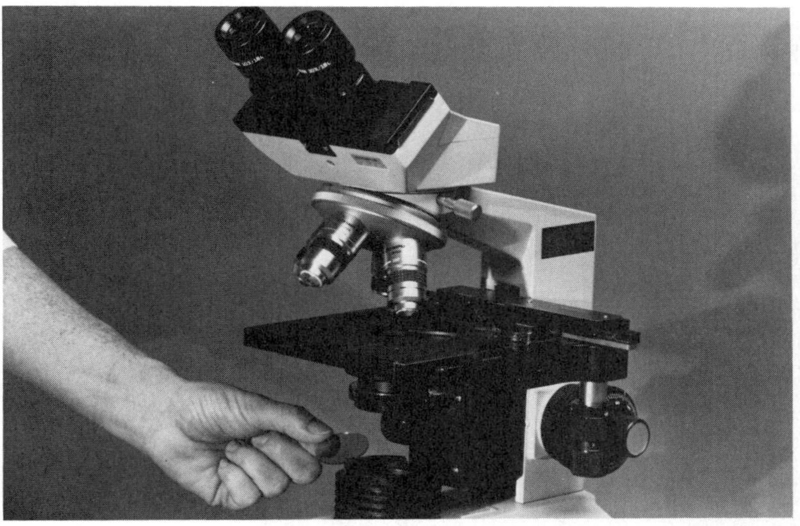

Figure 10–7. Remove the blue filter from the bottom of the condenser assembly.

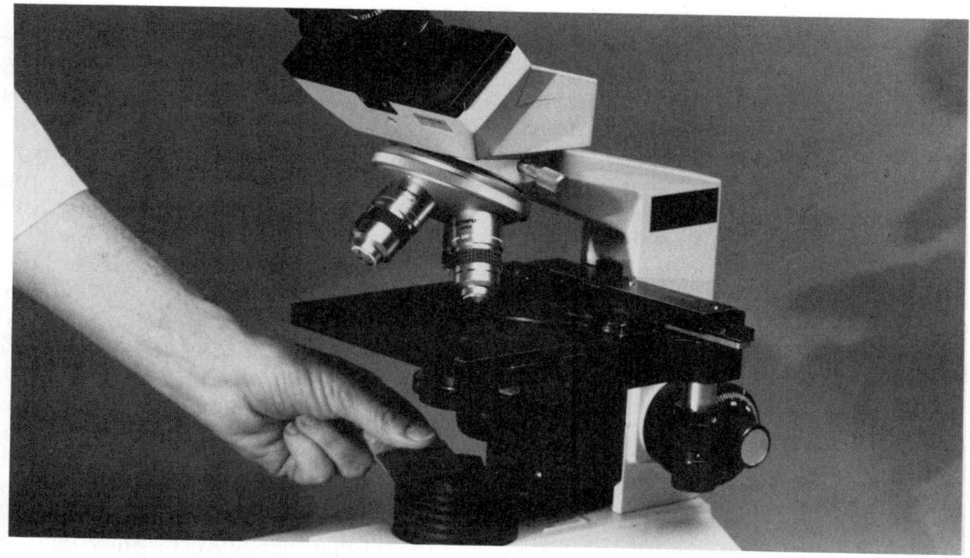

Figure 10–8. Place the square polarized glass over the base light source.

few drops of normal saline, mix with a wooden applicator stick, and put a coverglass over the specimen.

11. Place the slide on the stage between the fingers of the slide-holding assembly.
12. Switch to the low-power objective and focus. You should see small, white, maltese crosses against a dark background (Fig. 10–9). The background will probably not be as dark as it appeared using the high-dry objective in steps 8 and 9.
13. Switch to the high-dry objective and refocus. You may have to add more light. At this point you have polar-

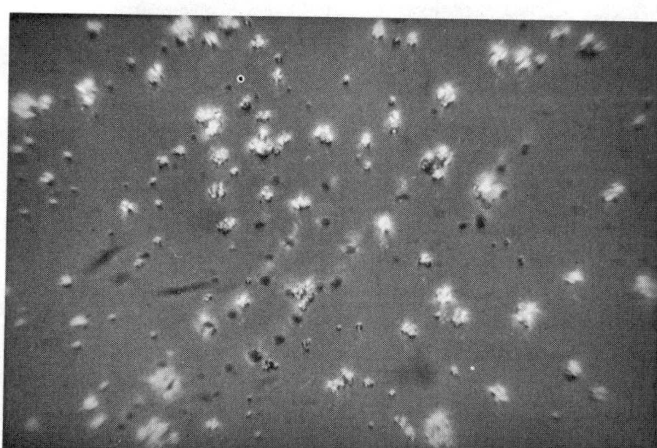

Figure 10–9. Polarized cornstarch crystals. Viewed under low power objective.

ized your scope and have demonstrated a double refractile or "birefringent" substance.

When a first-order red compensator is now added, you are able to determine "positive" or "negative" birefringence:

(a) positive birefringence:
The crystal appears blue when parallel to the axis of the first-ordered compensator. This is seen with the calcium pyrophosphate crystals of pseudogout.

(b) negative birefringence:
The crystals appear yellow when parallel to the axis of the first-ordered compensator. This is seen with the uric acid crystals of gout.

14. Take the compensator that comes with the kit and lay it across the base polarizer (Fig. 10-10). The light should still be on.
15. Rotate the compensator until the background becomes reddish or fuschia. You will notice that the starch granules are now yellow and blue. Look at one crystal. Note which part is blue and which part is yellow. Rotate the compensator 90 degrees. Notice how the colors have switched. Because of the round shape of starch crystals, you will not be able to determine their relationship to the long axis.
16. Gout crystals are needle shaped, while calcium pyrophosphate crystals are small rectangles or rhomboid plates. Because of their shape it is possible to determine if they are parallel to or perpendicular to the long axis and therefore whether they are positively or negatively birefringent (see Chapter 21, Section 7).
17. When you have finished the polarized examination, remove the compensator slide and the base polarizer. For convenience you can leave the body polarizer in place. This may tint the field green and may require a little more light than usual, but it

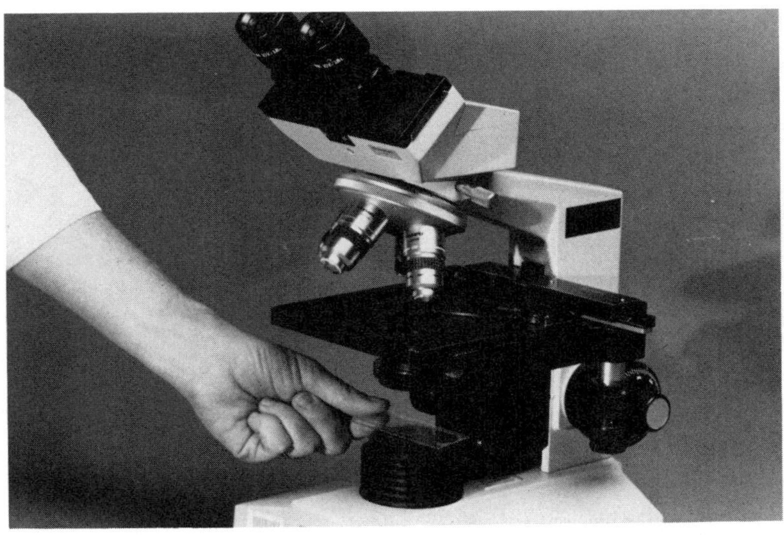

Figure 10–10. Place the compensator across the base polarizer.

is more convenient than having to take the microscope apart every time you want polarization.

18. Replace the filter.
19. Turn off the light source.

POLARIZING YOUR MICROSCOPE WHEN IT HAS BUILT-IN POLARIZATION

1. From your manual, determine what setting introduces the body polarizing filter into the line of light and adjust the setting.
2. If your scope has a blue filter in a slot at the bottom of the condenser assembly, remove it (Fig. 10–7). (The blue filter is there to make the light from the tungsten bulb whiter. If you forget and do not remove it, the background of the polarized field will be excessively dark and blue. (Although slightly more difficult to see, the crystals are still visible and remain the correct color.)
3. Place the base-light polarizer over the light source. If there is a first-order red compensator in the base-polarizing unit, swing it out of the light path.
4. Turn on the light source. You will need a great deal of light.
5. With the high-power objective in place, look through the eyepieces. You should see a light green field, unless the polarizer is by chance in just the right position.
6. Slowly rotate the base-light polarizer. At some point the field will become dark. Leave the base polarizer in this position. Your scope is now "polarized."
7. It is probably unlikely that you have a gout reference slide the first time that you go through this set-up procedure. To check the polarization, make a cornstarch test slide. Using ordinary cornstarch, place a *very* small amount on a glass slide. Add a few drops of normal saline, mix with wooden applicator stick, and put a coverglass over the specimen.
8. Place the slide on the stage between the fingers of the slide-holding assembly.
9. Switch to the low-power objective and focus. You should see small, white, maltese crosses against a dark background. (The background will probably not be as dark as it appeared using the high-dry objective in steps 5 and 6, Fig. 10–9).
10. Switch to the high-dry objective and refocus. You may have to add more light. At this point you have polarized your scope and have demonstrated a double-refractile or "birefringent" substance. When a first-order red compensator is added, you are able to determine "positive" or "negative" birefringence:
    (a) positive birefringence:
    The crystals appear blue when parallel to the axis of the first-ordered compensator. This is seen with the calcium pyrophosphate crystals of pseudogout.
    (b) negative birefringence:
    The crystals appear yellow when parallel to the axis of the first-ordered compensator. This is seen with the uric acid crystals of gout.
11. If your base polarizer comes with a compensator, swing it into the light path. It should be marked in some way so that you can determine the long axis. It is designed so that you can change the axis. You should see an orange or fuschia background. (Go to step 15.)
12. If there is no compensator with your base polarizer, take a plain glass microscope slide. Using old-fashioned (Scotch Red Plaid) cellophane tape, put a strip of tape lengthwise on the

slide. Press down on the tape with a piece of gauze to eliminate air bubbles. Add another layer of tape. Once again, eliminate air bubbles. You now have a first-ordered red compensator (Fig. 10–10).

13. Place the compensator across the base polarizer. The light should still be on.
14. Rotate the handmade compensator until the background becomes reddish or fuschia.
15. You will notice that the starch granules are now yellow and blue. Look at one crystal. Note which part is blue and which part is yellow. Rotate the compensator 90 degrees. Notice how the colors have switched. Because of the round nature of the crystals, you cannot determine the relationship to the long axis.
16. Gout crystals are needle shaped, while calcium pyrophosphate crystals are small rectangles or rhomboid plates. Because of their shape, it is possible to determine if they are parallel to or perpendicular to the long axis and therefore whether they are positively or negatively birefringent (see Chapter 21, Section 7).
17. When you have finished with a polarized examination, remove the compensator slide and the base polarizer. For convenience you can leave the body polarizer in place. This can tint the field green and may require a little more light than usual but is more convenient than having to take the microscope apart every time that you want polarization.
18. Replace the blue filter.
19. Turn off the light source.

SECTION 11: References/Additional Readings

The following books and journal article are recommended if you have an interest in learning about the more technical aspects of microscopy.

- Locquin M, Langeron M; Hillman H, trans-ed. *Handbook of Microscopy*. London, England: Butterworths; 1983.
- Needham GH. *The Practical Use of the Microscope Including Photomicrography*. Springfield, Ill: Charles C Thomas; 1958.
- Owen DSA. A cheap and useful compensated polarizing microscope. *N Engl J Med*. 1971; 285:1152.
- Rochow TG, Rochow EG. *An Introduction to Microscopy by Means of Light, Electrons, X-Rays, or Ultrasound*. New York: Plenum; 1978.

Use and Care of the Microscope

- AO Reichert Scientific Instruments, Customer Service, PO Box 123, Buffalo, NY 14240. This nice manual can be obtained free of charge by writing to AO Reichert.
- Wilson MB. *The Science and Art of Basic Microscopy*. Bellaire, Texas: American Society for Medical Technology, 1976.

The material in this chapter has been based, in part, on training manuals written by the authors as part of CDC Cooperative Agreement No. 050/CCU400987-01 and BHS Grant 1 D32 PE.14025.

CHAPTER 11

Centrifugation

SECTION 1: Introduction

Many patient specimens require some centrifugation before the sample is ready for testing. Centrifugation is nothing more than the separation of different elements within a specimen based on their density. In the office setting a centrifuge is used to generate serum or plasma from blood, to concentrate the sediment in a urine specimen, and to perform a microhematocrit. Table 11–1 lists some centrifuge manufacturers.

SECTION 2: Types of Centrifuges

There are now four basic types of centrifuges found in office laboratories:

1. Microhematocrit centrifuge: This centrifuge is used to perform a microhematocrit and is designed to hold standard 75-mm hematocrit tubes and to spin them at a speed of 8,000 to 11,000 RPM. This results in adequate cell packing to produce a reliable hematocrit in 3 to 5 minutes. A centrifuge for this purpose should hold about six to eight tubes, which allows duplicate samples to run on several patients at one time. A typical centrifuge costs about $500. Larger models that hold up to 24 microhematocrit tubes are more expensive and are not needed in the office laboratory.
2. Urine/blood-tube centrifuge: These centrifuges are used for spinning urine samples and blood tubes. They generally have a maximum speed of 5,000 rpm and hold from six to twelve 15-mL tubes. Some can also be adapted to spin 4- or 7-mL tubes by changing the centrifuge head. Most of these centrifuges include a timer to automatically shut off the centrifuge. Many also have a locked cover that prevents opening while the centrifuge is spinning.
3. Combination centrifuge: This centrifuge can spin both urine/blood tubes and microhematocrits. Its purchase price (about $1,000) is less than the combined cost to buy both a microhematocrit and a blood/urine tube centrifuge. An example of such a centrifuge is the TRIAC (Clay Adams). The TRIAC has three push-button settings (Fig. 11–1): (1) Microhematocrit: 12 capillary tubes for three minutes at 10,300 rpm; (2) Urine: Eight standard 12-mL tubes for 6 minutes at 2,300 rpm; (3) Blood: Eight tubes (15, 10, or 7 mL) for 5 minutes at 3,000 rpm (Fig. 11–2).

   This instrument also has a push-button brake to slow down the spinning centrifuge. This brake can be activated

TABLE 11–1. MANUFACTURERS OF OFFICE LABORATORY CENTRIFUGES

American Scientific Products
1430 Waukegon Rd
McGraw Park, IL 60085
312-689-8410

Bio Dynamics
PO Box 50100
Indianapolis, IN 46250
1-800-428-5074

Clay Adams
299 Webro Rd
Parsippany, NJ 07054
1-800-638-1532

Damon IEC Division
300 Second Ave
Needham Heights, MA 02194
1-800-225-8856

Fisher Scientific
711 Forbes Ave
Pittsburgh, PA 15219
1-800-282-0578

StatSpin Technologies
588 Pleasant St
Norwood, MA 02062
617-551-0100

only if the top is down. You can save valuable bench space in a small laboratory by having only one centrifuge for multiple functions.

4. StatSpin: This instrument is the first major improvement in centrifugation in many years and is ideally suited to the office laboratory (Fig. 11–3). It is designed to spin small volumes at very high speeds. This speed considerably shortens the time that it has traditionally taken to process specimens. The StatSpin will not, however, spin regular-sized (7 mL) blood collection tubes. You will need to use either specially designed tubes or the microtubes ("bullets") used to collect pediatric samples (Fig. 11–4). Laboratories with a StatSpin will also need a second centrifuge to spin down the blood tubes sent to a reference laboratory for further testing.

A StatSpin will complete a microhematocrit in 2 minutes, urine for microscopy in 45 seconds, and plasma for chemistry testing in only 30 seconds. The findings for urine microscopy (e.g., number of RBC/HPF) are similar when using the small urine tubes and a Stat-

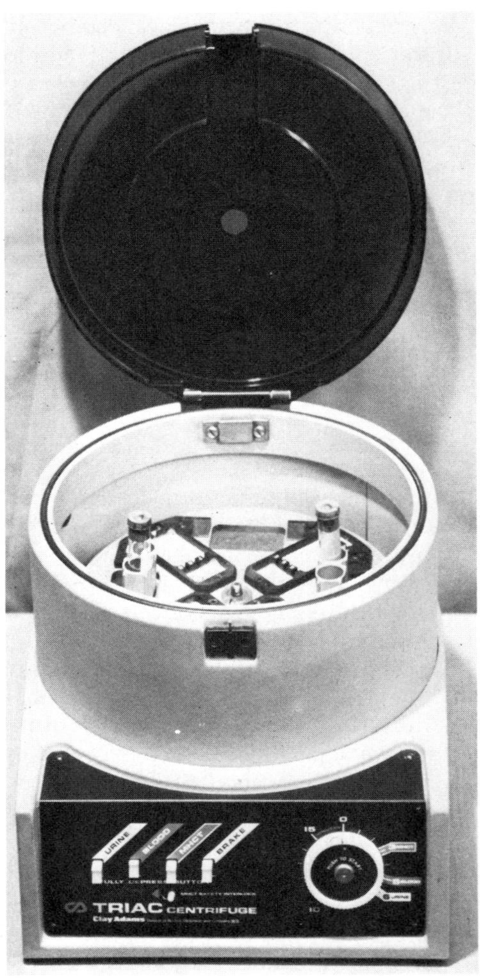

Figure 11–1. TRIAC multipurpose centrifuge.

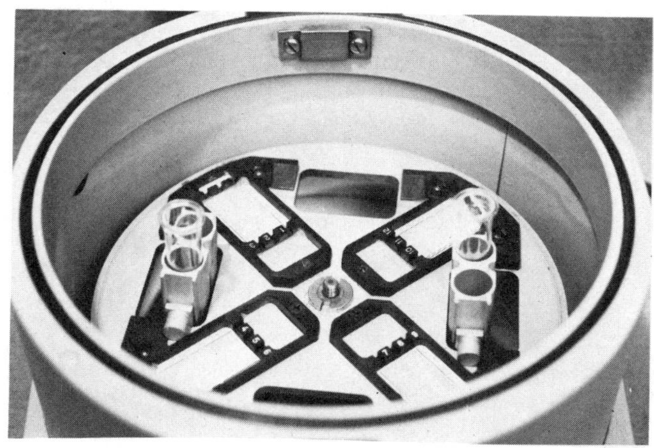

Figure 11–2. Head of a TRIAC centrifuge. Note the urine tubes in the trunnions and the numbered slots for microhematocrit capillary tubes.

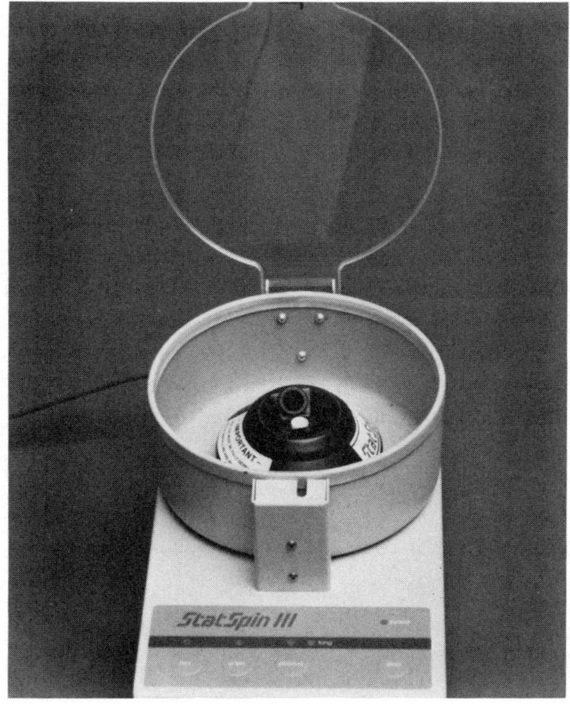

Figure 11–3. StatSpin centrifuge.

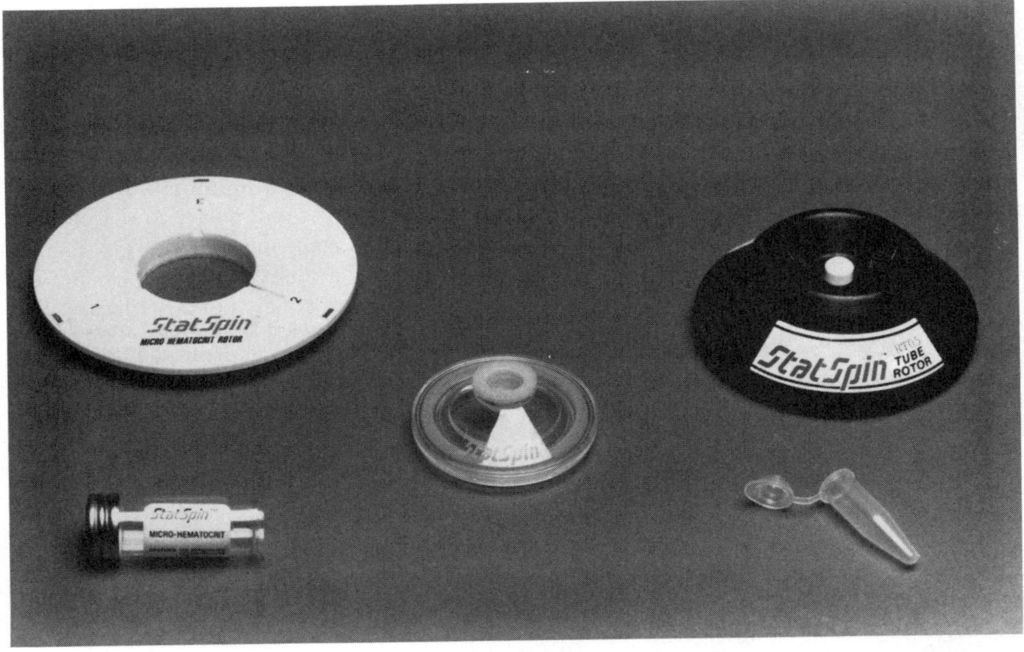

Figure 11–4. StatSpin rotors and specimen collection devices.

Spin, as would be found if the urine sediment had been processed in the traditional manner.

Other advantages to the StatSpin are its small size and its very quiet operation. A variety of models are available. The approximate cost is $900.

SECTION 3: Centrifuge Use

1. The specific assembly and maintenance instructions for each centrifuge are provided by the manufacturer. An equipment maintenance sheet such as shown in the quality assurance chapter should be kept up to date on each instrument (Fig. 8–9).
2. All loads in the centrifuge should be bal-

anced! It is critical that you read the instrument manual to find out how to properly balance the loads in *your* centrifuge. Keep different-sized tubes filled with water in a rack near the centrifuge to use as a counter-balance when a single tube of blood or urine is to be spun. If the centrifuge is not balanced, it will make an awful noise and will ''walk'' across the bench. Immediately stop the centrifuge to avoid instrument damage. Sometimes even a ''properly'' balanced centrifuge will do this. The commonest cause for this problem is a cracked tube that has leaked fluid.

3. The centrifuge should be placed on a stable, balanced countertop. It is best to place it on a separate surface not connected to the bench holding the microscope or other sensitive instrumentation, as the centrifuge's vibration may interfere with other testing.
4. Always keep the lid closed until a specimen has stopped spinning. Many centrifuges will shut off power to the motor if the lid is opened.
5. Do not slow down the centrifuge with your hands, pencils, or other items! This can injure both the equipment and yourself.
6. Clean the inside of the centrifuge bowl each day with a bactericidal solution such as Staphene. Also clean after any spills.

SECTION 4: Quality Assurance

Hospital laboratories are required to regularly measure their centrifuge's speed with a tachometer and its timer with a stopwatch. These laboratories must sometimes separate very similar specimen fractions (i.e., white blood cells from red blood cells from platelets). This process requires very specific timing and centrifuge speeds. Because of this a great deal of attention has been paid to the measurement of the centrifuge's performance.

This same level of concern is excessive for office laboratories because centrifuge problems are obvious to the laboratory worker. When spinning blood it is apparent whether plasma or serum have been separated from the red blood cells. For urine specimens, using the same volume of urine and the same centrifugation time is sufficient to ensure a clinically useful level of reproducibility of the sediment findings. For microhematocrits, the end point is ''maximal packing'' of the red blood cells. As explained in the hematology chapter, this can be easily quality controlled by respinning a microhematocrit once a month to be sure that no further packing occurs.

For most office laboratory needs the ''correct'' centrifugation is ''enough'' (i.e., sufficient centrifugation to generate the specimen of interest). The specific rpm and timing are relatively noncritical. As the National Committee for Clinical Laboratory Standards states, ''Documentation to support the recommendations on centrifuge time and speed was not found in the scientific literature. Current laboratory practice appears to be. . .based on a time honored practice of empirical observation).''

REFERENCES

Procedures for the Handling and Processing of Blood Specimens. National Committee for Clinical Laboratory Standards. DOC. H18-T, p. 224, 1984.

SECTION 5: Centrifuge Maintenance and Troubleshooting

A properly cared for centrifuge should rarely require professional repair. Table 11–2 lists some common centrifuge problems and the appropriate corrective action.

TABLE 11–2. COMMON PROBLEMS WITH THE CENTRIFUGE

Problem	Cause	Corrective Action
Centrifuge fails to start	Cord unplugged	Plug in cord
	Cover not closed	Fully close cover
	Mechanical defect	Check instruction manual
Excessive vibration	Unbalanced load	Balance specimens; check for fluid in trunnions
	Rubber feet worn	Replace worn feet
	Specimen broke in trunnion	Clean out centrifuge
	Motor brushes worn	Check manual for brush replacements
Blood empties from capillary tubes	Check capillary tube gasket	Replace gasket

CHAPTER 12

Blood Collection

SECTION 1: Introduction

Blood drawing is one of the most commonly used and least frequently taught clinical skills. When properly performed it can be nontraumatic for any patient. However, all too often it is done by people who are inadequately trained and therefore nervous about the procedure. This can result in a bad experience for the patient, who may even refuse to return to the office for future medical care. Health care workers have won few friends with their needles and tubes of blood.

This chapter describes several of the common outpatient techniques for blood collection: infant heel sticks, finger sticks, and venous punctures. These techniques are used for collecting specimens for hematology, coagulation, and chemistry testing. The tests may be performed either in the office laboratory or at an outside reference laboratory. The techniques for blood culture collection, arterial ''sticks,'' and venous punctures on newborns are not described, since these procedures are usually confined to the hospital setting.

Drawing blood is a skill that is mastered only with practice. Part of this skill involves the actual manual technique. A second, often overlooked, aspect is how to approach the patient who is having his or her blood drawn. Attention to both considerations is essential if you are to become a ''good'' blood drawer.

Your beginning attempts at blood drawing will be awkward and difficult. But after some practice and experience, you will hear those unbelievable words from a patient, ''Why I didn't even feel it. That was the best blood drawing that I have ever had.'' Becoming an excellent blood drawer is totally a learned experience.

SECTION 2: Approach to the Patient

Having blood drawn, even a simple finger stick, is a universally painful experience. The best that you can do is to make blood drawing a bearable experience. One way is to provide distractions (i.e., things that patients can focus on to avoid concentrating on what is happening to them). In our laboratory, animal pictures hang directly across from the blood-drawing chair. Both children and adults look at these pictures when their blood is being drawn. Some patients so love the ''animal wall'' that they have brought in pictures of their favorite pets. Answering questions such as ''Who's the cat lover here?'' or ''What kind of dogs are they?'' has gotten both patients and laboratory staff through many difficult venipunctures.

When you are first learning to draw blood, you will focus solely on the technical aspects of obtaining the specimen. As your confidence grows, you will find that you can perform the procedure and talk to the patient at the same time. This talk, or "chatter," helps distract the patient from the pain of the needle stick. It is important to develop "chatter" skills.

Some patients who are nervous will become talkative themselves. They may tell you important clinical information that they have forgotten to tell the clinician. Ask these patients if they have talked to their doctor about this. If they have not, tell these patients to remember to bring this up at the next office visit, or tell the clinician yourself.

Some patients will tell you that they feel faint just before or after the blood drawing. Others will warn you that they always faint when their blood is drawn. Take these patients to an examination room and draw their blood while they are lying down. Do not discount these patient warnings!

Everyone desires privacy when their blood is drawn. The blood drawing area of the office laboratory should therefore permit some privacy.

Before drawing the blood it is necessary to identify the patient and the test that is to be done. In an office setting many patients are known to the laboratory worker by sight, but others will be strangers. Instead of asking "Are you Mr. Jones?," ask "What is your name?" A nervous patient may answer "yes" to the first question, even if it is not his name. Some offices will include the patient's birth date on the laboratory slip to avoid confusion with other patients who have the same name. As with the name, ask the patient for his or her birth date rather than giving a date and asking for verification. In an attempt to make light of it all, you can jokingly tell the patient who has given the "correct" answer that he or she has won a prize—a stick in the arm. This usually produces a chuckle.

Always wash your hands before and after you collect blood, even though you will be wearing gloves. Washing your hands in front of the patient is psychologically comforting. This is especially important if patients see you handling other laboratory specimens, such as urine samples.

An individual approach must be taken for patients of different ages:

Infants

It can be distressing for the blood drawer to do a heel stick on a sleeping baby! We feel that it is best for the baby to be held by the parent during the procedure. If the parent refuses, we lay the baby on his or her stomach on an examination table and collect the heel-stick specimen. Most parents are not used to seeing the heel stick done. It is therefore important to describe the procedure to them. We tell the parent that the initial pain from the stick is very short lived. The crying that comes later is an expression of anger. Babies do not like to be restrained or to have their legs held. Frequently, releasing the foot will be enough to stop the infant's crying. We talk to the baby to calm it throughout the procedure. For our own benefit, we like to spend a few minutes stroking or holding the baby after the blood collection is finished.

Preverbal Children

No matter what the age, talk to the child. Children understand speech long before they can speak. In very young children, just the sound of the voice can be comforting. If a finger stick is to be done, explain the procedure to the child and do a mock demonstration. If a venous stick is required, you will need assistance from a parent or another office worker. Children are often much stronger than is anticipated. A short period with good restraint is less traumatic than "fighting with a kid" for a long time.

Verbal Children

The absolute, cardinal, unbreakable rule when dealing with children is to never lie about the pain. *Do not tell them that it will not hurt.* Often the parent has told the child that it will be painless, but you must tell the child that it will hurt some. One of the authors was hysterical for years about having blood drawn because she had been lied to about the pain as a small child.

The first step is to explain the procedure to the child and to answer all questions. Children will ask how long it takes to fill the blood tube (15 seconds for a pediatric tube and 30 seconds for a regular-sized tube) or how much it will hurt (less than skinning a knee). If it is a finger stick for a hematocrit, tell them that the capillary tube is like a magic straw that will fill up without anyone sucking on it or that it is similar to a thermometer and that they can watch the blood rise as the mercury does. If it is a venous stick, explain what the tourniquet is for and have the child touch the vein. Children are fascinated by the needle. They may be worried that the entire needle is going to be put in their arm. Show them how far it will be inserted.

If at all possible, make a contract with the child. Most children will remain still if they have an incentive. Explain that screaming and yelling are acceptable but that the child must not move because if there is movement, you might miss the vein and he or she would have to be stuck again. Children can also be "bribed." Tell them that you will show them their peripheral blood smear under the microscope, if such a test was ordered. We keep a variety of specimens to show children, such as a black widow spider in formalin and a flea mounted on a microscope slide. Even very small children are able to see the flea through the microscope. Inexpensive gifts, such as "stickies" (colorful stick-on patches) can also be given so that the child will hold still. As a last resort, you can demonstrate a finger stick on yourself or have someone draw your blood to show them that it is not the worst thing in the world. Children are usually impressed with your courage and often settle down and allow the procedure.

Continuously talk with the child while the blood is being drawn. Most children can talk for hours about their favorite television show. Talking distracts them from what you are doing to them. If all else fails, ask for assistance to hold the child so that the specimen can be drawn.

Adults

Most adults know that they cannot fight or scream, but they may nevertheless be very nervous. Acknowledge their fears. Any basic relaxation technique can be used to calm an adult. Take your time. Speak in a soft, calming voice. Distract the patient by talking about a common interest. Some patients may want to imagine that they are someplace else, such as at the beach. Others can be instructed to concentrate on breathing slowly and moving the air in and out of their lungs.

Because we have been so successful at calming people, we know that it can be done. It does take patience, empathy, and time, but this is always time and effort well spent. One good experience can change a person's entire attitude toward having his or her blood drawn.

SECTION 3: Blood-Drawing Equipment and Supplies

A section of the laboratory should be specially arranged for drawing blood. This area should permit patient privacy during the procedure. Special blood-drawing chairs can be purchased for patient seating; however, they cost about $400. All that is really needed is a chair with flat arms across which a board is placed for supporting the patient's arm. A plastic-covered board can be easily wiped clean if blood is spilled on it.

A cabinet can be placed next to the chair so that tubes and needles are within easy reach. The cabinet should be placed on the side of the chair that is most convenient for the laboratory worker. This will depend on whether he or she is right or left handed. Tubes can be easily stored on top of this cabinet in the square plastic containers that are sold for food storage. Plastic urine specimen cups can be used to store needles and lancets. It is also convenient to have a blood-drawing tray that can be carried to the examination rooms. A plastic tray with a handle is best. All routine blood collection supplies should be kept on the tray. Be sure that all the tubes on the tray are kept in date (Fig. 12–1).

Most of the supplies required for blood drawing will be available from your reference laboratory. The reference laboratory will also provide you with a manual that describes the tests that are available, the blood tube that should be used for the test specimen, the quantity of blood required, and any special handling needs. Ask the reference laboratory for a listing of the *absolute minimum volume* of blood required for each test. This information is especially important when drawing blood from small children, where small volumes are common. A copy of the reference laboratory manual should be kept in the cabinet next to the blood-drawing chair for easy reference by the laboratory worker.

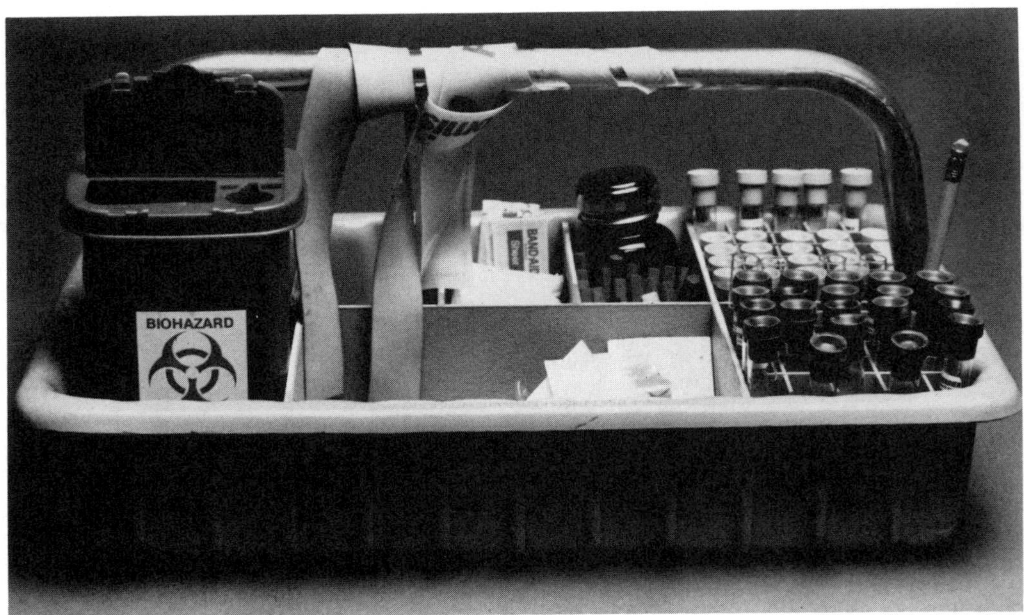

Figure 12–1. A well-stocked blood drawing tray can be carried to the patient examination room for specimen collection.

Supplies

Vinyl Gloves: Wear the smallest gloves that are comfortable. Buy nonsterile gloves, 100 per box. You may need to evaluate several brands, since some tear easily. Use latex surgical gloves for patients with difficult veins, since they permit better "feel."

Blood Collection Vacuum Tubes: There are many types of tubes available. Each has a different additive and is used for a different type of test. Some tubes are very specialized. For example, special tubes are needed for serum zinc levels because the stoppers in regular tubes contain zinc. All tubes are prepared with a certain amount of negative pressure so that they fill to a specified volume. Familiarize yourself with the level that marks a "filled" tube. Tubes come in regular and pediatric sizes. Pediatric tubes are smaller and have less negative pressure. These smaller tubes are useful for both children and for adult patients whose veins tend to collapse with the higher negative pressure of a standard-sized tube. Tubes should not be used after the expiration date because they may lose their vacuum and therefore may fail to draw up the proper volume of blood. Tubes are traditionally identified by the color of their stopper (e.g., a "red top" tube). The commonly used tubes include:

- *Red-top (plain) tubes:* These tubes contain no anticoagulant. They are commonly referred to as "clot" tubes. They can be used for almost any test requiring serum. This includes many chemistry and blood bank tests.
- *Red-/gray-top tubes with gel at the bottom:* These tubes are also "clot" tubes. They contain no anticoagulant. However, these tubes contain an inert barrier material that has a specific gravity between that of serum and clotted red blood cells. When the tube is spun after the blood is clotted, the gel moves up to seal off the serum from the cells. This prevents contamination of the serum from the chemicals released by hemolyzing red cells. These tubes are used for most serum chemistry tests. They cannot be used for blood bank tests because the gel coats the cells, giving a false positive Coombs' tests.
- *Royal blue-top tubes:* These are also "clot" tubes containing no anticoagulants but are cleaner than red-top tubes. They are used by some laboratories for drug testing (e.g., theophylline, digoxin).
- *Purple (lavender)-top tubes:* These tubes contain the anticoagulant EDTA in either liquid or powder form. They are used for routine hematology tests.
- *Light blue-top tubes:* This regular-sized tube has 0.5 mL of the anticoagulant sodium citrate. The tube is used for routine coagulation studies, such as prothrombin time (PT) and partial thromboplastin time (PTT). The tube must be fully filled to ensure the proper ratio of blood to citrate. If you have a patient with problem veins, draw a pediatric blue-top tube rather than risking incomplete filling of the standard-sized tube. Some laboratories may advise you to remove some of the citrate from the blue-top tube if a patient has polycythemia. This is because the volume of citrate is based on the plasma volume, and these patients have less plasma per volume of whole blood.
- *Grey-top tubes:* These tubes contain potassium oxalate and sodium fluoride, which inhibit glycolysis. These tubes are used by some laboratories for plasma glucose testing. Blood drawn in these tubes cannot be used for testing with whole-blood glucose analzyers.
- *Green-top tubes:* These tubes contain heparin. They are not used for many routine tests by reference laboratories. However, if you want to perform "stat" chemistries, they can be used to generate

plasma specimens. This is a time-saving advantage, since you immediately spin the specimen instead of waiting for the blood to clot. Check with your chemistry instrument manufacturer to find out which of the chemistry tests can be run on a heparinized plasma specimen.

Microtubes: A variety of very small plastic tubes are available for collecting specimens from infants using a heel-stick technique or from adults using a finger-stick technique. Each tube comes with a special capillary blood collector that channels free-flowing capillary blood into the plastic microtube or "bullet" (Fig. 12–2). We have not found these capillary collector tops very easy to use. Instead, we use plain Natelson capillary tubes to collect the blood and to then transfer it to the microtubes.

There are a variety of brands of microtubes available. The volume of blood contained and the markings on the side of the microtubes vary with the manufacturer. Read the insert for specific information. Also check with your reference laboratory for the minimum amount of specimen necessary for specific tests.

- *Red-top microtubes:* These small tubes have a small quantity of serum separator gel. When they are filled the capillary collector top is pulled off and is replaced with a tightly fitting red top.
- *Purple-top microtubes:* Like the regular-sized purple-top tube, these microtubes contain the anticoagulant EDTA. After the specimen is collected, a purple-top is placed onto the tube and the tube is then gently inverted ten times to ensure adequate anticoagulation. The specimen can be used for white blood cell and red blood cell counts, red cell indices, hemoglobin, and hematocrit (within 4 hours of the specimen collection), as well as for platelet count, reticulocyte count, and peripheral smear (within 2 hours of the collection).
- *Pink-top microtubes:* These are used for blood-bank tests as well as for other tests that require serum without the separation gel.

Microhematocrit capillary tubes: These are very small, thin, glass, capillary tubes that are used to collect blood for the microhematocrit test. You may need to try several brands before finding one that does not break in the centrifuge (Fig. 12–3).

Natalson capillary tubes: These are long, glass, capillary tubes that are used to

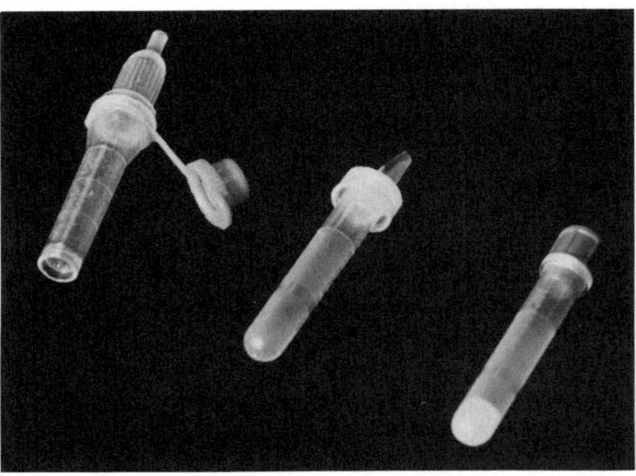

Figure 12–2. Small plastic microtubes can be used to draw capillary specimens. These come with plastic tops to direct the blood into the tube.

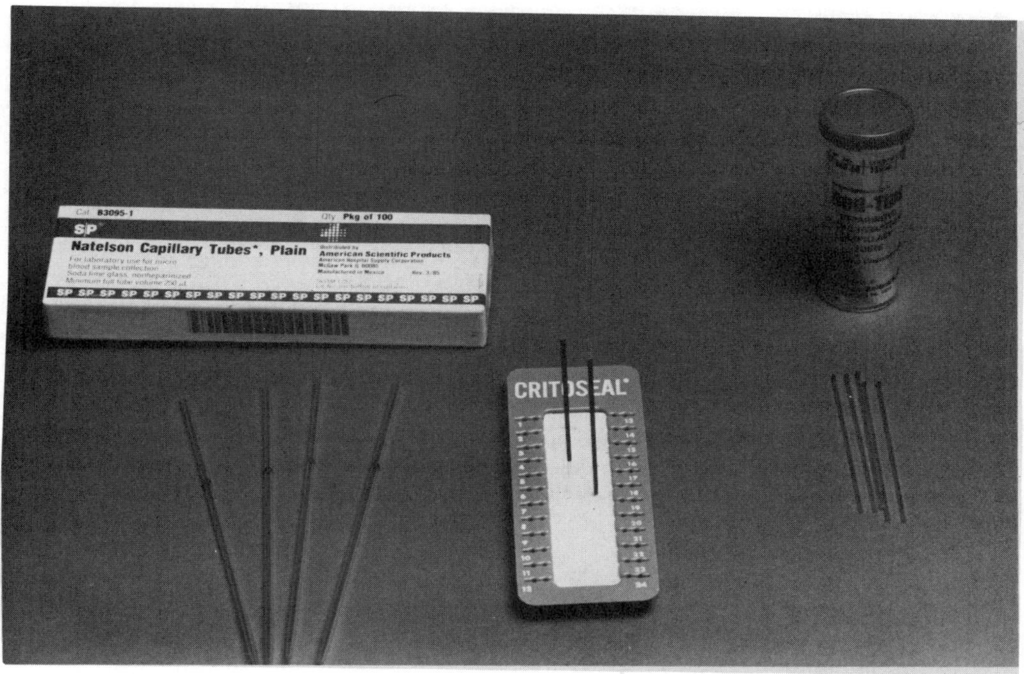

Figure 12–3. Microhematocrit capillary tubes are used for the hematocrit test. The end of the tube is plugged with sealant. Larger Natalson tubes are used to collect specimens from an infant by heel stick.

collect blood specimens from an infant by a heel stick. They can be purchased either heparinized or nonheparinized. Heparin interferes with a variety of tests, including the red blood cell indices, peripheral smear, and one method of phenylketonuria (PKU) testing. The nonheparinized tubes are therefore best for most office laboratory use. After the blood has been collected into the Natalson tube, it can be transferred to a microtube for transport or specimen processing (Fig. 12–3).

Needles: Several types of needles are needed for blood drawing:

- Multidraw needles: These needles are used when more than one tube of blood is required. They have a rubber sleeve over the end of the needle that punctures the tube stopper. The sleeve moves back and forth over the tip of the stopper-puncturing needle and prevents blood from escaping into the plastic holder when the tubes are being changed. The needle is screwed into a plastic holder when blood is drawn. A 21-gauge, thin-walled, 1.0- or 1.5-inch needle is most commonly used (Fig. 12–4).
- Single-draw needles: These needles are used when only one tube of blood is drawn with the evacuated blood system. They have no rubber sleeve over the end of the stopper-puncturing needle. The needle is not as strong as a multidraw needle, so it may bend if you try to fill more than one tube. A 22-gauge, thin-walled, 1.0- or 1.5-inch needle is most frequently used (Fig. 12–4).

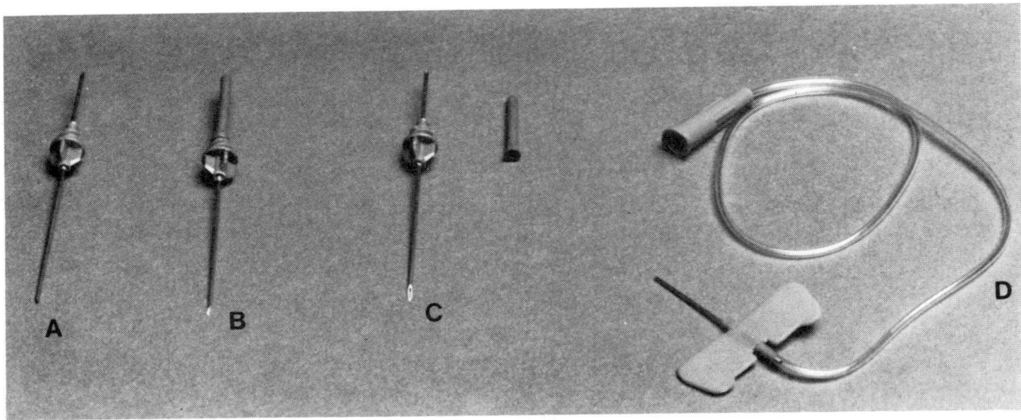

Figure 12–4. A variety of needles are used to draw venous specimens including **(A)** single-draw needles, **(B)** multi-draw needles, **(C)** multi-draw needles with sleeve removed, and **(D)** butterfly needles.

- Butterfly (scalp-vein) needles: These needles are held in a soft, plastic, butterfly-shaped handle and are attached to tubing that is fitted with a Luer-end (Fig. 12–4). A syringe can be directly attached to the Luer fitting, or you can use a Luer-adapter, which allows a blood tube to be drawn directly from the butterfly needle. A butterfly needle provides much easier control than a regular needle and syringe. This is especially helpful when drawing blood from young children or patients with very difficult veins. A 23-gauge butterfly needle is commonly used. We suggest that you use pediatric-sized tubes and a small plastic needle holder (Fig. 12–12) when using the Luer adapter to draw directly into tubes. (See p. 133 for detailed instructions.)
- Syringe needles: Regular 21- to 24-gauge needles with Luer ends can be used to draw blood with a syringe.

Plastic Tube/Needle Holders: Plastic holders are used for drawing blood with evacuated tubes. The needle is screwed into one end of the holder and blood tubes are pushed onto the other end. These holders come in separate sizes for regular and pediatric tubes. When they get dirty, the holders can be cleaned with water and a solution of 5% Staphene (Fig. 12–5).

Syringes: Sterile plastic disposable syringes of 3, 5, 10, and 20 mL sizes can be used to draw blood. Syringes provide the "ultimate" in blood-drawing control, but they are problematic when drawing blood for tests that require anticoagulated blood. Therefore be very careful that the blood in the syringe does not clot before transferring it to anticoagulant tubes.

Tourniquets: Tourniquets are used to distend the veins for a venous puncture. Pieces of Penrose latex tubing can be used. For an adult tourniquet, use five eighth-inch tubing cut to an 18-inch length. For pediatric tourniquets, use five sixteenth-inch tubing cut to a 12-inch length. Both widths come in boxes of 36-inch length tubing. These tourniquets can be wiped clean with an alcohol wipe when soiled, but they should be replaced frequently. Some laboratory workers prefer to use specially made gum-rubber straps with self-adhesive Vel-

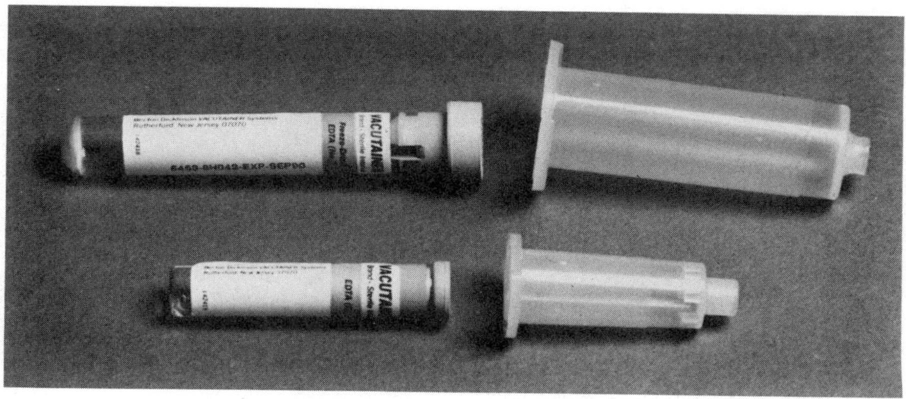

Figure 12–5. Plastic tube/needle holders come in both regular (top) and pediatric (bottom) tube sizes.

cro closures. They can be purchased in pediatric and adult sizes. Pharmaceutical companies also give these tourniquets away as promotional aids.

Spring-loaded device for finger sticks: There are a variety of spring-loaded lancet devices available (Fig. 12–6). Some of the manufacturers include:

- Autolance: Becton Dickinson Consumer Products 1-800-237-4554 (9 AM to 5 PM EST)
- Autoclix: Bio-Dynamics (Boehringer Mannheim Co) 1-800-428-4674 (8:15 AM to 5 PM EST) in Indiana 1-800-382-5200
- Autolet Ames Division, Miles Laboratories, Inc Customer Service, 1-800-348-8100
- Auto-Lancet Palco Laboratories, 1595 Soquel Dr, Santa Cruz, CA 95065
- Monojector lancet device: Monoject, Division of Sherwood Medical St. Louis, MO 63103
- Glucolet Ames Division, Miles Laboratories, Inc Customer Service, 1-800-348-8100

The advantage of these devices is that they produce far less pain than the standard lancet technique. Patients who refused finger sticks in the past accept the procedure after one stick with a spring-loaded device. The manufacturer will indicate which lancets are used with their model.

Regular lancets: These lancets are used for finger and heel sticks. They can be purchased with a regular point (blue package) that penetrates to 2.4 mm or with a long point (red package) that allows a 5-mm penetration.

Sharps disposal box: These containers need to be made of opaque, puncture-proof, autoclavable plastic. They should have a top that permits you to unscrew the needle without recapping. A variety of containers are now available (Sage Products, 1-800-323-2220; Bio-Safety Systems, 1-800-421-6556; and Eagleguard Syringe Safety, 1-800-458-0506). They should be small enough to

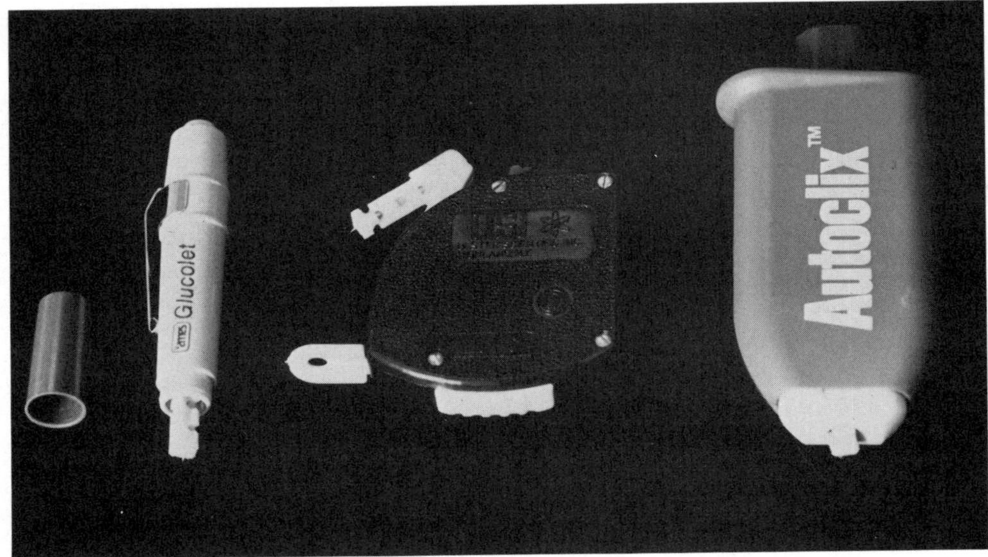

Figure 12–6. A variety of spring loaded lancets are available for drawing capillary specimens.

be placed in the work area (Fig. 9–1). Arrangements should be made with the reference laboratory or the local hospital for disposal of these containers.

Contaminated softs diposal box: This container is used to dispose of bloody gauze and partially filled tubes. It is convenient to use large plastic jugs. They should be disposed of similarly to the sharps box. They should be marked with "Contaminated" "Biohazard" tape (Fig. 9–2).

"Contaminated" "Biohazard" label tape: Rolls of yellow tape with the word "contaminated" in red letters can be purchased. Containers with used blood-drawing supplies should be labeled with this tape.

"Special precautions" labels: In the past, special precautionary labels (i.e., "hepatitis," "blood, body fluid precautions") were attached to tubes from patients with suspected or known blood-borne infections. The Centers for Disease Control (CDC) has recommended that universal blood and body-fluid precautions be implemented, thus eliminating the need for special warning labels. Check with your reference laboratory to determine if they are using the universal precautions system.

Adhesive labels: Rolls of 1.0 × 2.5-inch adhesive labels are convenient for labeling blood tubes with the patient's name and the date that the sample was drawn.

Alcohol wipes: Individually sealed sterile packets contain pads soaked in 70% isopropyl alcohol. They are usually supplied in boxes of 100.

Povidone-iodine wipes: These individually sealed packets contain pads soaked in povidone-iodine antiseptic. They are used in the collection of blood culture specimens.

Nonsterile 2 × 2-inch gauze squares: These squares can be folded in quarters and taped to the puncture site to control bleeding after the stick. This is less expensive and provides more pressure on the wound than do regular bandaids.

Sterile 3 × 3-inch gauze squares: These squares are supplied two to a package. They are used when performing infant heel sticks.

Ammonia inhalant capsules: These capsules can be crushed between the fingers and passed under the nose of a patient who feels faint.

Plastic wash bottle: A small squeeze bottle should be filled with a 5% solution of Staphene to wipe up small, spilled blood specimens.

Disinfectant: A disinfectant should be used to clean the counter tops in the laboratory each day. A cheap and effective general germicidal solution is 5% Staphene. It is more stable and less irritating than bleach solution. Wear gloves to protect your hands when using the disinfectant solution. Wear a laboratory coat to protect your clothes.

SECTION 4: Capillary Specimens

Capillary blood is obtained from heel sticks in infants and finger sticks in children and adults. Capillary blood is a mixture of blood from arterioles, venules, and capillaries. The higher pressure in the arterial side results in a larger portion of the specimen coming from the arterioles. Capillary blood specimens are therefore considered to be arterial blood as opposed to venous blood. The specimen may also contain interstitial and intracellular fluids. Excessive milking of the skin during the collection procedure will result in an increase in the level of these other fluids.

When performing a skin puncture it is important to make sure that the puncture site is not edematous. Edema fluid will produce error in some tests. Cold or cyanotic skin is also a source of error for capillary specimens. Such skin will need to be warmed before blood is drawn. This can be done by holding the hand under the warm tap water or by wrapping the foot or hand in a washcloth that has been wet with very warm water.

HEEL PUNCTURE

The heel puncture is used to collect blood specimens from children younger than 3 months of age. This technique is most frequently used to collect the PKU screening test specimen and to follow the bilirubin values in newborns with neonatal jaundice.

Recent studies have shown that the heel stick can result in some injury to infants. Osteochondritis and calcified nodules have been documented after heel sticks. Current criteria for a safe heel stick include:

- The puncture should be no deeper than 2.4 mm (i.e., the depth of a regular-point lancet). A surgical blade or a long-point lancet are not acceptable instruments.
- Do not repuncture the sites of previous heel sticks.
- Do not stab the posterior curvature of the heel.
- Puncture only the most lateral and medial positions of the bottom of the heel (Fig. 12–7).

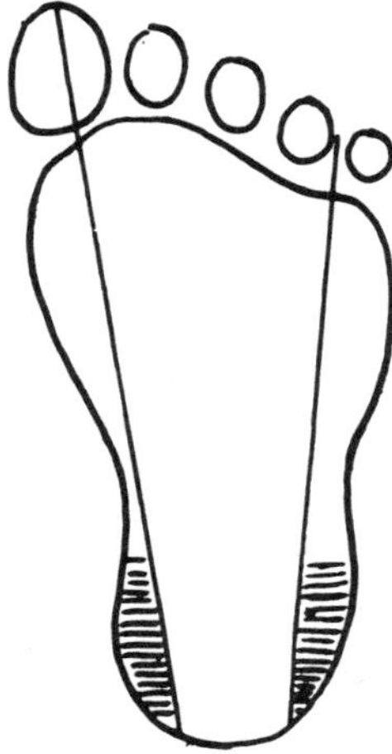

Figure 12–7. Heel sticks should be performed only on the lateral and medial curves of the heel (the etched areas in the figure).

Procedure:

1. Gather together all of the needed materials. This will include an alcohol or povidone-iodine wipe, sterile gauze, regular lancet, and the test supplies. For a bilirubin or other reference laboratory tests, the blood is collected in a capillary blood microtube. The specimens for PKU and other neonatal screening tests are often collected on a special filter paper supplied by the state laboratory. Follow the collection instructions on the PKU form.

   Some laboratory workers prefer drawing blood into a Natalson tube and then transferring it to a microtainer. A heparinized tube can be used for a bilirubin test, but the nonheparinized Natalson tube must be used for hematology tests and PKU testing done by the Guthrie method.
2. If there is another adult who can hold the infant, drape the person's lap with a towel to protect his or her clothing from dropped blood.
3. Remove the baby's blanket and clothes. Babies have strong legs. They often pull their feet out of your hand and spread blood around as they kick.
4. The infant should be held so that the feet are free and hanging down. If there is no one to hold the child, lay him or her stomach down on the center of an examination table.
5. Put on gloves.
6. Examine the infant's feet. Note any previous sticks and avoid these areas.
7. Wet a clean washcloth with warm water. Do not use hot water, which could burn the infant's skin. Wring out the cloth and then wrap it around the foot for several minutes. The heat increases the blood flow to the foot. Do this several times. The time spent warming the foot is well worth it.
8. Unwrap the cloth and grab the foot around the heel.
9. Clean the bottom of the heel with an alcohol or povidone-iodine wipe. Do not use ether.
10. Take a sterile 3 × 3-inch gauze and wipe the area vigorously to dry the alcohol and to increase the blood flow to the skin. If the alcohol has not dried, it can cause hemolysis of the specimen during the collection.
11. Open a blue, regular-sized (2.4 mm), sterile lancet package.
12. Hold the foot tightly and quickly stab the chosen area. Withdraw the lancet and immediately make a second stab perpendicular to the first stab, producing a "+" shaped wound. Lancets are not wide enough to produce a good heel stick with a single puncture.
13. Release the foot for a few seconds and wipe away the first drop of blood.
14. To collect the blood, hold the foot firmly, but avoid excessive squeezing. If a capillary tube is used, hold it so that the blood flows down. This

position enhances tube filling. Release the foot from time to time and let the infant kick away. This helps calm the baby and also increases blood flow to the foot.

15. When the collection is complete, hold the sterile gauze over the puncture site until bleeding stops. Do not use tape. A baby's skin is sensitive to the tape's adhesive and can become irritated.

Comments:

- If both a PKU and bilirubin are needed, it is often necessary to do two heel sticks to collect sufficient blood.
- If blood is dropped onto the baby blanket or clothes, advise the parent that it can be removed by using hydrogen peroxide. Do not hang clothing soaked in hydrogen peroxide in the sunlight, as it will turn bright yellow.
- Do not squeeze the infant's heel excessively. If the blood stops flowing, release the foot and hold the leg down for a few minutes to see if the blood will start flowing freely again. If not, perform another heel stick.

FINGER STICK

The finger is the most frequently used skin puncture site for children over 3 months of age and for adults. In small children it is best to use the thumb, since it is large and easy to hold onto if the child pulls away. In older children and adults we routinely ask the patient if he or she is right or left handed and then use the fourth finger of the nondominant hand. This finger is sufficiently large and is usually less calloused than the others. The cutaneous nerves in the fingers are concentrated in the inner part of the finger pad, so it is best to use a site halfway between the center of the finger pad and the outer edge of the finger (Fig. 12–8).

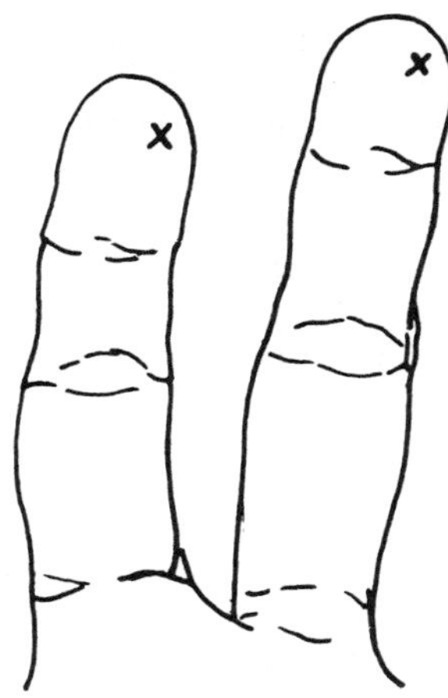

Figure 12–8. Sites for the finger stick.

Patients have a tendency to tighten the muscles in their hands during this procedure. You will notice that this turns the hand pale and reduces the blood supply to the skin. Take the hand and shake it until it is loose, and caution the patient against tightening it. We tell children to make their hands "loose like cooked spaghetti."

There are two methods for collecting the finger-stick specimen. The traditional method involves the use of a lancet. A less painful technique uses a spring-loaded instrument containing a lancet. Never use a surgical scalpel blade for this procedure.

Spring-loaded Puncture Instrument:

1. Collect the necessary supplies: a spring-loaded puncture instrument, lancet, gauze, alcohol wipe, and collection capillary tube.
2. Put on gloves.
3. Following the manufacturer's instructions, insert the lancet into the device and set the spring for use.
4. Select the finger to be stuck (i.e., the thumb in small children, the fourth

finger on the nondominant hand in adults and older children).

5. Have the patient relax his or her hand and dangle it down across the board on the blood-drawing chair.
6. Wipe the finger site vigorously with an alcohol wipe. You can help ensure a good blood flow by massaging the finger, starting at the base and moving upward.
7. Exert some pressure on the finger tip by holding it between your thumb and index finger—"pumping it up" (Fig. 12–9). This helps to ensure a satisfactory flow of blood.
8. Place the spring-loaded device against the side of the finger tip—between the center pad and outside rim of the nail (not dead center; Fig. 12–9) and release according to manufacturer's instruction.
9. Release your hold from the finger and wait a few seconds for a blood drop to begin to form.
10. Wipe away the first drop of blood. Collect the platelet-count specimen first, then blood for a peripheral smear, and then blood for any other test. Always wipe away any blood that accumulates between the collection of specimens for different tests. Platelets tend to collect at the site of the puncture. Wiping away the drops of blood ensures that the site will continue to bleed.
11. A good blood flow can be maintained by holding the patient's hand down and gently massaging the finger. However, do not squeeze the finger at the puncture site. This will introduce interstitial and intracellular fluid into the specimen and lead to inaccurate results. Moderate pressure at least 1 cm behind the puncture site is acceptable.
12. After the collection is complete, give the patient a piece of gauze to apply pressure to the wound.
13. Remove the used lancet and put it into the "contaminated" container. Make sure that the spring-loaded arm of the lancet instrument is not set. Storing the device with a relaxed spring ensures longer life.

Figure 12–9. Use of an Autolet for a fingerstick blood collection.

Regular Lancet Procedure

A regular lancet is used for the puncture if there is no spring-loaded puncture instrument available or if the patient has very tough skin. Thick skin is often seen in older patients, manual laborers, or diabetics.

1. Collect the necessary supplies: a long-point lancet, gauze, alcohol wipe, and collection pipette.
2. Put on gloves.
3. Select the finger to be stuck (i.e., the thumb in a small child, the fourth finger on the nondominant hand in older children and adults).
4. Wipe the finger vigorously with an alcohol wipe.
5. Remove the lancet from its package.

Hold it between your thumb and index finger.

6. Make a quick, firm, deep puncture; as you withdraw the lancet, twist the lancet slightly. This may sound cruel and painful, but it is necessary to ensure an adequate blood flow. Beginners will frequently be too gentle and will not get an adequate blood flow; as a result the patient must be stuck a second time.
7. Wipe away the first drop of blood and then collect the specimen. Be sure to wipe away the blood that forms between the collection of specimens for different tests.
8. After the collection is complete, give the patient a piece of gauze to apply pressure to the wound.

SECTION 5: Venous Specimens

The actual puncturing of the vein is the least time-consuming aspect of a good venipuncture. First, it is important to find the patient's "best" vein. This may take some time. A patient is made to suffer unnecessarily if the blood drawer is unwilling to spend sufficient time to find a good vein to stick. A patient's arms are different, so if there are no good veins on one arm, look at the other arm. If you have difficulty finding a vein, ask the patient where people have had the greatest luck drawing blood in the past. Some patients will volunteer this information. Always listen to the patient, for even if you should miss the vein, the patient is less likely to be upset with you if you have followed his or her advice.

TYING A TOURNIQUET (FIG 12–10 A through D)

Penrose Drain Tourniquet

If you have not done this before, you will want to practice before you start drawing blood. You will not want to convey your inexperience to the patient by fumbling with the tourniquet.

1. Put the tourniquet around the patient's arm. Hold an end in each hand and stretch the tubing by pulling both ends toward yourself. The patient must brace his or her arm to produce tension in the tubing.
2. Cross the right end over the left end. Your hands should be almost on top of each other.
3. While maintaining the tension in the tourniquet, grab both ends with the right hand. There will be a V-shaped space between the patient's arm and the tourniquet.
4. Reach through the V-space from the top with your left hand and grab the end of the tourniquet that now points toward the left. This is the top end of the tourniquet. Pull it up halfway through the V. Make sure that the very end of it hangs down below the level of the tourniquet on the arm.
5. Release the tourniquet against the patient's arm. It should be tight enough to be just uncomfortable. At first you will have a tendency to tie it much too loosely.
6. To release the tourniquet after the blood is drawn, merely pull the end that was left hanging.

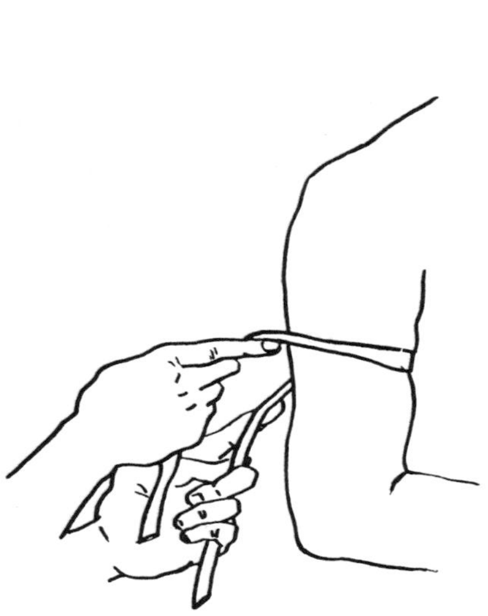

Figure 12–10A. Tying a tourniquet. Hold an end in each hand, and stretch the tubing by pulling the ends toward yourself. Cross the right hand over the left hand. They should be almost on top of each other.

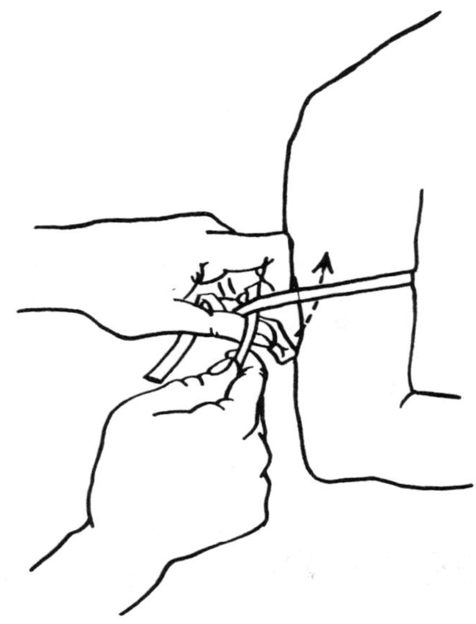

Figure 12–10B. While maintaining the tension in the tourniquet, grab both ends with the right hand. Reach through the V space from the top with your left hand and grab the end of the tourniquet that now points toward the left.

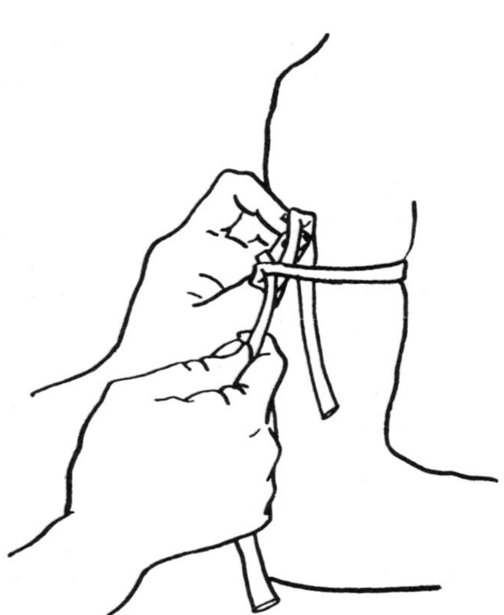

Figure 12–10C. Pull the end of the tourniquet about halfway through the V space. The very end should hang down below the level of the tourniquet on the arm.

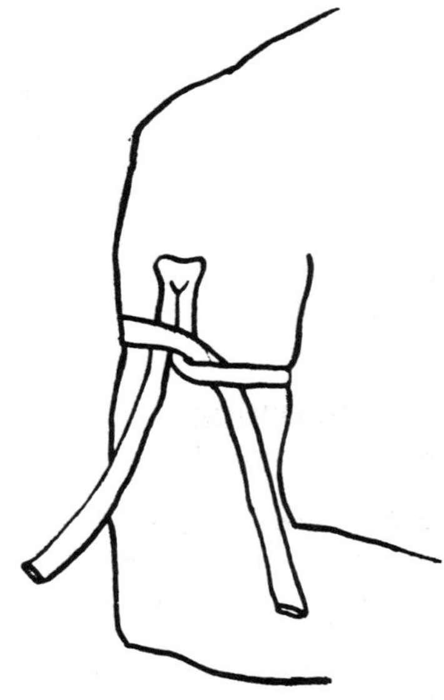

Figure 12–10D. Release the tourniquet against the patient's arm. To release the tourniquet, pull the end that was left hanging (on this drawing the end on the left).

Self-adhesive Tourniquet (Velcro)

1. Take an end of the tourniquet in each hand and put the tourniquet around the patient's arm.
2. Press the left end against the patient's arm with your left hand and pull the right end with your right hand to create tension in the tourniquet.
3. Press the Velcro® ends together. This will hold the tourniquet in place.
4. To release, merely pull the Velcro ends apart from each other.

AREAS TO AVOID WHEN DRAWING BLOOD

- *Scarred areas*: Healed burns have deep scar tissue that makes drawing blood difficult.
- *Thrombosed veins*: These veins feel thick, like a cord. They lack resilience and will roll. They may look easy, but it is usually difficult to obtain blood from them.
- *Bruised areas*: Bruising can lead to variable blood-drawing results. If you cannot find another puncture site, collect the specimen from a site that is farthest away from the trunk.
- *Mastectomy*: This surgery results in lymphostasis, which makes drawing blood difficult. If only one breast was removed, draw the specimen from the opposite arm. Mastectomy patients are usually very good about volunteering this information.
- *Dialysis patient*: Do not collect blood from the arm with the arteriovenous (A-V) shunt.
- *Intravenous catheter*: Collect the specimen from the arm without the intravenous (IV) line. If there is an IV in each arm, stop the IV for 2 minutes and take the specimen from 3 inches distal to the IV site. Always consult with the clinician before stopping the IV.

ARM-VEIN ANATOMY

The veins at the elbow (i.e., the antecubital fossa) are most commonly used for drawing venous blood. The three major veins in this area are the cephalic, median basilic, and the median cephalic (Fig. 12–11). The median cephalic is located in the middle of the antecubital fossa and is sought out first. This is the choice vein because it is better anchored in the tissue and is therefore less likely to roll than are the other veins. If the patient does not have a good median cephalic vein, examine the other veins in the area. The outer vein is the cephalic vein. It is less likely to roll than the median basilic vein. Be careful to feel for the artery in the same area when drawing from the median basilic vein. If you cannot find a suitable site at the elbow, check the rest of the arm. The vein in the center of the arm just above the antecubital fossa or the veins that travel up the forearm can be used. Veins on the wrist and on the back of the hand are last resorts because these areas have higher numbers of cutaneous nerves, which result in more pain when the skin is punctured.

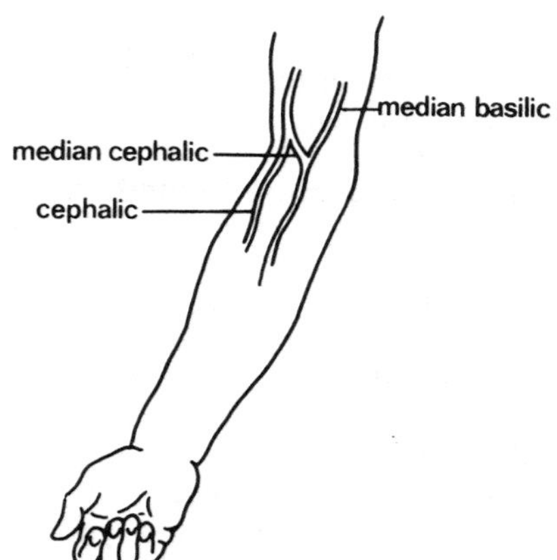

Figure 12–11. Veins of the antecubital fossa.

Techniques for Locating a Vein

1. It is crucial for a good blood drawer to develop a sensitive finger to feel for veins. The tightest gloves you can wear allow the most sensitivity to touch. Use the same "finder" finger all of the time. This is usually the index or third finger on the nondominant hand. When first learning to draw blood, it is a good idea to practice finding veins. Trace the veins in your own arms without a tourniquet to learn the location and feel of the arm veins. Once your finger becomes sensitive, it will be able to feel veins that you would miss with your other fingers.
2. Slapping or tapping the skin over a vein will sometimes cause the vein to dilate and make it easier to draw blood from it.
3. Massage the arm upward from the wrist to the elbow to force blood into the veins.
4. Ask the patient to dangle his/her arm for several minutes to distend the veins.
5. Apply heat to the arm with a towel soaked in hot water.
6. In dark-skinned patients the vein is sometimes easier to see if the skin is freshly wiped with alcohol.
7. Use a blood pressure cuff inflated between the diastolic and systolic blood pressures to distend the veins.

Routine Venipuncture Procedure

1. Collect the required materials. Check to see that the needed number and types of tubes are available. Additional tubes should be within reach in case the vacuum in a tube is lost. If the tubes contain an additive, tap the tube to be sure that the additive is in the bottom of the tube. If a peripheral smear is required, set aside two frosted microscope slides with the frosted sides up. Also set aside a plastic tube/needle holder, sterile needle, 2 × 2-inch gauze, alcohol wipe, tape, and clean tourniquet.
2. Wash your hands and put on gloves. *Always wear gloves.* If you are just learning to draw blood, it will be easy to get used to them. For people who learned to draw blood ungloved, it usually takes several weeks to get used to wearing gloves and to be able to feel difficult veins. Tight-fitting gloves are better than loose-fitting ones.
3. Ask the patient if the veins in one arm are easier to draw from than those in the other.
4. Have the patient roll up his or her sleeve and prop the arm on the crossboard of the blood-drawing chair. Some people prefer to have the arm hyperextended, but we have found it better to have the arm slightly bent at the elbow.
5. Apply the tourniquet several inches above the puncture site. It should be tight enough to be slightly uncomfortable to the patient.
6. Ask the patient to make a loose fist. Give him or her an empty tube to hold. If the patient makes too tight a fist, it will be difficult to find a good vein. Do not have the patient pump the fist, as this can affect test results. (e.g., the serum potassium may increase by 10% to 20% due to pumping).
7. Select a good vein for venipuncture. If this takes a long time, loosen the tourniquet to allow blood to flow freely for a minute and then retie the tourniquet.
8. Open the needle package, but do not remove the needle cap. (If you are going to remove the sleeve on the needle, do so now. See the section on "Special Comments" for an explanation of this technique.)
9. Screw the needle into the plastic tube/needle holder.

10. Insert the blood tube into the plastic tube/needle holder. The tube can be partially pushed onto the inside needle without losing the tube's vacuum.

11. Clean the puncture area with an alcohol wipe. If you need your sensitive finger to relocate the vein, wipe the tip of that finger with the alcohol wipe. Always tell the patient that you are cleaning your finger so that you can touch the cleaned area. Remind the patient not to move during the puncture.

12. **a.** Put the thumb of your nondominant hand just below the puncture site and pull the skin tight, or
b. Place your thumb below the puncture site and your index finger just above the puncture site and pull the skin tight, or
c. Put your sensitive finger right over the vein just proximal to the puncture site.

13. Hold the tube/needle assembly in your dominant hand. Remove the protective cap from the needle. Hold the needle with the bevel up.

14. Hold the assembly close to the skin, with the needle in line with the vein and at about a 15-degree angle to the skin. Puncture the skin and enter the vein with a very quick, one-step motion. Speed is the secret to minimizing the pain. (People who are used to starting IVs first enter the skin, then find the vein. It is better for the venipuncturist to try to enter the vein in the same motion as you puncture the skin.)

15. Once the needle is in the vein, anchor the plastic needle holder. Some technicians switch hands at this point. Stabilize the plastic holder by pressing your index finger (which is resting against the base of the neck of the holder) against the patient's arm. By stabilizing the plastic holder, you can safely change the collection tubes without moving the needle out of the vein.

16. Push the tube forward onto the needle in the end of the holder. This will puncture the stopper of the tube. If the needle is in the vein, blood will flow into the tube. If this does not happen, back the needle up a bit. This will cause the blood to flow. Hold the plastic holder so that the stopper is uppermost. This prevents the flow of the tube additives into the patient. Try to avoid having the tube's contents touch the stopper as it fills. The tourniquet can be released at this point or after the blood collection is complete.

17. When the blood tube is completely filled, the blood will stop flowing into the tube. If more than one tube is to be filled, remove the first tube and push the next tube onto the holder. Push the tube forward. An extra push is needed to push the tube's stopper through the needle. If no blood flows into the new tube, back out a bit. Be sure to stabilize the holder against the patient's arm when this is done. Otherwise the needle will move in the vein.

18. Draw blood culture or sterile tubes first, then tubes without additives, then coagulation tubes, and finally other tubes with additives. If you use a tube with an anticoagulant, gently invert this tube several times to mix the additive with the blood as soon as the tube is removed from the plastic holder. Do not shake these tubes. If you are drawing only a coagulation (i.e., blue top) tube, either use a single-draw needle and allow the first drops of blood to flow into the holder before using the tube or use a multidraw needle and collect the first drops of blood in another tube before using the blue tube. Discard the first tube. This prevents the tissue fluids from the puncture site from contaminating the specimen. Such contamination can affect coagulation studies.

19. Remove the tourniquet if you did not do so earlier.
20. Take a 2 × 2-inch gauze in the hand used for changing the blood tubes and fold it into fourths. Hold it just above the puncture site. As you quickly withdraw the needle, press the gauze against the puncture site. It is less painful if the needle is removed quickly.
21. Ask the patient to press the gauze against the puncture site, keeping the arm straight and above the level of the heart.
22. While the patient is holding the gauze, make blood smears if they are requested from the drops of blood at the tip of the needle (see Chapter 15). *Do not recap the needle*. Lock the needle hub in the appropriate hole in your sharps disposal box and unscrew the needle. It will fall into the box.
23. If the plastic holder was soiled with blood during the procedure, take it to the sink and wash it out. You can soak it in a 5% Staphene solution. Then rinse it out and set it aside to dry. Alternately, you can dispose of it in the softs contaminated box.
24. Check to make sure that the patient feels well. Check to see that the bleeding has stopped at the puncture site. If the patient is taking an anticoagulant (e.g., warfarin sodium [Coumadin]), add another folded gauze and tape both tightly to the skin to form a pressure dressing. This can be removed in about an hour.
25. Properly label, date, and time all specimens. Never leave an unmarked tube lying around because there will be no way to identify it.

SECTION 6: Special Comments

ROLLING VEINS

Veins frequently seem to have a mind of their own and will move or disappear in the presence of a needle. The veins of the antecubital fossa are the least likely to roll, but they can. The other veins in the hand and arm will often roll. To prevent this, put your thumb and third finger along the vein and spread your fingers apart. This stretches the vein and minimizes rolling. If you want to see how the vein will roll, anchor it between your thumb and third finger, then touch it with the tip of the capped needle. If you expect the vein to roll when you puncture it, insert the needle at a greater angle to the wall of the vessel.

PROBING

Probing is what you do when you do not enter the vein with the initial stick. Much of the pain associated with the venipuncture is due to entering the skin. It is therefore best not to remove the needle and start all over if you first miss the vein. Probing can be almost painless when it is properly done. Tell the patient that you are not yet in the vein and to hold still for a minute. Also tell the patient to inform you if it is too painful. Then take your sensitive finger and feel for the vein. Veins are often tough and are pushed aside by the needle. After relocating the vein with your finger, back the needle out until it is almost out of the arm. It is important not to totally back it out through the skin because the vacuum in the tube will be lost. Move the needle back toward the vein at a slightly steeper angle. Do not move the needle from side to side while it is deep in the arm because this is painful and can injure the tissues. If the patient complains of pain while you are probing, release the tourniquet, pull the needle out, and try again at a different site. Use a new, unused needle for the second stick. If you fail after the second or third attempt, have someone else try. Your anxiety level will be so high that you will often be doomed to failure after three unsuccessful attempts. The patient will be happier with someone else.

SYRINGE

Some people prefer to use a syringe to draw blood from a patient with difficult veins. The main disadvantage to this technique is that if anticoagulated blood is needed, the blood may clot in the syringe before it can be placed in the anticoagulant tubes. The actual skin puncture is the same as for the evacuated blood-tube system. Gentle back pressure is applied on the syringe plunger after the skin has been entered. Blood flows into the syringe as soon as the needle enters the vein. Do not apply too much back pressure, as this can cause the vein to collapse. Put the blood rapidly into the specimen tubes after it is collected. Take the needle off the syringe and the tops off the tubes. Do not squirt the collected blood through the needle because this may result in red blood cell hemolysis. Put the blood into the anticoagulant tubes first. Cap and mix each of these tubes. When you are finished, place a piece of tape around each stopper and tube to prevent the stoppers from popping off in transit. It is useful to have assistance during this procedure so that the blood is transferred before it clots.

MULTIDRAW NEEDLE WITH RUBBER SLEEVE

Whenever we suspect that some difficulty might be encountered in drawing the blood or that we might miss the vein, we employ a technique that is a little more messy but that combines the advantages of both the syringe and the evacuated tube systems. The rubber sleeve on the multidraw needles prevents blood from flowing into the holder when the tube is removed. When you have to probe for a vein, you will often back the needle out of the skin and lose the vacuum in the tube. If you remove the sleeve from the needle that pierces the stopper, you can probe with the tube out of the holder. When you are in the vein, blood will flow into the holder, and the tubes can be popped into the holder for the specimen collection. The tubes can get bloody on the outside with this technique, so lay them on a paper towel after they are filled. The tubes will need to be wiped off with a paper towel squirted with a 5% Staphene solution. We also wrap a piece of 2 × 2-inch gauze around the tube end of the holder to catch blood that might spill out when the plastic holder is tipped. Hold the unit with the needle pointing down to prevent spilling. You can dispose of the entire unit in your sharps disposal box, or after unscrewing the needle, wash out the holder with water and a 5% Staphene solution.

BUTTERFLY/LUER ADAPTER

This is a trick that reduces the possibility of blood clotting as well as the chances that the vein will collapse. Screw a Luer adapter into the pediatric-sized plastic needle holder. The outer fitting of this adapter is then hooked up to the Luer end of a butterfly needle and tubing. The venipuncture is performed with the butterfly needle. When

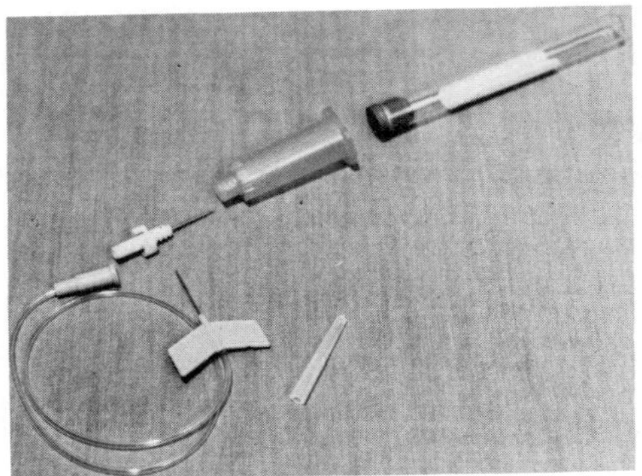

Figure 12-12. Butterfly needle, Luer adapter, plastic tube holder, and pediatric tube.

you penetrate the vein, blood will be seen in the butterfly tubing. As the blood approaches the Luer adapter, a small pediatric blood tube is pushed onto the holder. The tube is pushed forward and the blood is collected (Fig. 12-12).

REFERENCES

Blumenfield T, Turi GK, Blanc WA. Recommended sites and depth of newborn heel skin punctures based on anatomic measurements and histopathology. *Lancet*. 1979;1:213.

Hammond KB. Blood specimen collection from infants by skin puncture. *Lab Med*. 1980;11(1):9.

Product Insert: Directions for Drawing Multiple Blood Samples. Becton Dickinson, Rutherford, NJ 07070

Product Insert: Microtainer Capillary Blood Collector, Becton Dickinson Company, Rutherford, NJ 07070

Slockbower J. Venipuncture procedures. *Lab Med*. 1979;10:747.

CHAPTER 13

Chemistry Testing

SECTION 1: Introduction

In the 1983 edition of this book our recommendation was for office laboratories to limit their chemistry testing to glucose analysis. This was based on the fact that at that time most of the available chemistry instruments used slow, wet-reagent methods that required a great deal of precise handling by office laboratory staff. Since then a revolution has occurred in the technology for office chemistry testing. This revolution has been made possible because of new reagent systems (particularly monoclonal antibodies and dry-chemistry reagents), microprocessor-controlled instruments, and improved, stable electronics. We anticipate that most office laboratories will expand the range of their chemistry tests as this technology continues to improve.

The instrumentation for office chemistry testing is continuously changing. It is therefore impossible to describe in detail the operation of even the common instruments that are available. We will focus instead on the issues that should be considered when purchasing a chemistry analyzer and the steps that must be done to ensure that the instrument is operating properly.

SECTION 2: Types of Specimens

Chemistry testing can be done on three types of specimens: whole blood, serum, and plasma. All too often the differences among these three types of specimens are ignored (i.e., "A glucose is a glucose is a glucose."). This attitude can lead to problems because there are differences in how specimens must be processed, in which tests can be run on which specimens, and even in the range of normal values from one type of specimen to another.

1. Serum. Serum is the fluid that is left when all of the clotting factors have been used up. Its principal disadvantage in the office setting is the time that it takes for the blood to clot.
   a. Collect a venous specimen in a red-topped tube or a serum separator tube.
   b. Place the tube in a test-tube rack so that it sits vertically. This position speeds up the clot formation, reduces hemolysis, and minimizes clot agitation when the specimen is spun.
   c. It will take 5 to 15 minutes for the clot to completely form in the tube. The time will be longer for patients on anticoagulants. It is es-

sential that the blood be completely clotted. If you spin the tube prior to complete clot formation, then fibrin strands will appear in the serum. These strands can interfere with the test results. The easiest way to check for complete clotting is to gently tilt the tube to see if the blood still flows. Clotted blood does not flow.

d. Spin the clotted tube in a centrifuge for 5 minutes.
e. If a red-topped tube has been used and the specimen is to be sent to a reference laboratory, carefully pipet the separated serum from the clot. Transfer the serum to a second red-topped tube for shipment to the reference laboratory. Work quickly to minimize the exposure of the specimen to the air. Air exposure can lead to a loss of carbon dioxide, which may alter some test results. Tape the top of the red-topped tube that contains the serum. (Otherwise the top may pop off in transport.) As a general rule it is best to separate the serum from the clot within 2 hours of specimen collection. Some analytes (e.g., potassium) can become falsely elevated if the serum remains in contact with the clot.

2. Plasma. When blood is drawn into a tube that contains an anticoagulant (e.g., EDTA or heparin), no clot will form. The red blood cells, however, can still be separated by centrifugation. When this is done the fluid on top is "plasma." Plasma still contains the unactivated clotting factors. Plasma is a very good specimen for office chemistry testing, since the blood can be centrifuged immediately after its collection. When plasma specimens are used, chemistry results can be easily available within 10 minutes of the time that the specimen is collected.

The biggest drawback to plasma specimens is that the anticoagulants may interfere with some chemistry tests. It is therefore essential that you request information from the chemistry analyzer manufacturer about each of the tests that you are doing in the office laboratory. It is likely that many tests will be suitable for blood collection in a common purple-top (i.e., EDTA) tube; however, some tests may require collection in less common tubes (i.e., a green-top, heparinized tube).

a. Collect a venous or capillary (finger stick) specimen in an appropriate anticoagulant tube or microtainer. Check to be sure that the tube is suitable for the test that is requested.
b. The blood can be immediately centrifuged. If collected in a standard blood tube, it should be spun for 5 minutes at 3,000 rpm. If whole blood was collected in a microtainer, then it can be spun down in a STAT SPIN for 1 minute.
c. If the testing is to be done immediately, carefully pipet the desired plasma volume from the centrifuged tube or microtainer.
d. If the testing is to be done more than 2 hours later, carefully pipet off all of the plasma and transfer it to a red-top tube. Cap the tube to avoid air exposure.

3. Whole blood. There has been a recent trend toward the introduction of "whole-blood analyzers." These chemistry instruments usually include a filtration or centrifugation step that is automatically performed by the instrument. The actual test is therefore carried out on plasma. Whole blood analyzers offer at least three major advantages. First, they are quicker than serum-based tests, since the blood clotting time is elimi-

nated. Second, they require less hands-on time by staff, since the specimen requires almost no processing before testing. Finally, finger-stick (capillary) specimens can be used for almost all tests. This is helpful for both pediatric patients as well as older patients with difficult veins.

SECTION 3: Instrument Factors

None of the currently available instruments will replace your reference laboratory! Most of these instruments do single tests in a few minutes and at a reagent cost of $1 to $3 per test. This high reagent cost per test usually prohibits doing more than three tests per patient at one visit. If more than three tests are needed on a single patient, it is usually more economic to send a specimen to your reference laboratory for a chemistry panel. The office instruments can perform from 6 to 30 different types of tests. However, a reference laboratory will still be needed for the less common chemistries.

Office chemistry instruments are best suited to provide one or two clinically relevant tests while the patient waits in the office. This rapid test-turnaround time can dramatically improve patient care. Patients will no longer need to wait several days to be told about their test results. Clinicians can make rapid changes in treatment and can avoid the time spent calling patients back with laboratory results. It has been estimated that it requires 14 minutes of clinician time to follow up on a single test. This time includes reviewing the results, examining the patient chart, contacting the patient by telephone, writing a note in the chart about the telephone call, and calling the pharmacy for a prescription change. One of the principal driving forces for office chemistry testing is the fact that the test result is available while the patient is still in the office.

The ideal chemistry test for an office instrument is one that is diagnostically important and that leads to an on-the-spot change in the patient's care. Some examples of popular office chemistry tests include the measurement of glucose in the care of diabetics, the measurement of theophylline in the care of asthmatics, and the measurement of potassium in patients on diuretics. In each of these cases office testing permits the clinician to immediately assess the patient and if necessary change the treatment.

Several factors should be considered in choosing a chemistry system:

- *Test menu.* The starting point in comparing instruments is to examine the list of tests that are available. A word of caution: many manufacturers will describe tests that are not yet available but are "in development." These may be years in coming, so it is usually best to evaluate an instrument based only on the tests that are currently available.

  It is easy to be impressed by the large number of tests an instrument can perform. Here you must remember that it does not matter how many different tests are available unless the specific tests you need are included. An instrument that does only six tests might be better than an instrument that does 30 tests if the six happen to be the ones that you need.

  You may decide not to do some of the tests that can be done on your office analyzer. A common example are the hepatic enzymes. These enzymes are not usually considered to be useful for screening healthy patients. Furthermore, when hepatic disease is suspected, a patient often requires many tests (i.e., bilirubin, AST [SGOT], ALT [SGPT], LDH, alkaline phosphatase, etc.) In such a situation it is generally better to send a patient's serum out for testing to a reference laboratory. The availability of SGOT testing on an office instrument may therefore not be an important advantage!

- *Operator skills*. The next consideration for the physician should be who will be performing the tests. It is important to consider not only who is in charge of the laboratory but also what other staff persons will be expected to use the instrument. This list often includes nurses, medical assistants, and front desk staff.

  Office chemistry instruments vary tremendously in terms of the skills required of the operator. We have observed that some office staff may have "laboratory dyslexia." These staff are unable to reliably produce good chemistry results with even the most simple instrumentation. The two areas that seem to be most problematic to office staff are quantitative pipetting and instrument calibration. Pipetting can be a problem even if the instrument includes an "automated" pipetter. This is because small amounts of the sample can drop from the pipet, air can be sucked in, or fluid may stick to the outside of the tip. Since most instruments require very small sample sizes, even tiny problems with pipet technique can lead to unreliable results.
- *Test time*. The next consideration in choosing the instrument should be the speed of testing. In most offices you can expect the instrument to be used for five to ten individual tests per physician per day. This is in contrast to reference laboratories that may batch hundreds of tests together. The most critical time period is how long it takes from drawing the specimen until the result is available to the physician. This should generally be less than 30 minutes. As described above, the efficiency of office laboratory testing is dependent on doing the testing while the patient is in the office. Few patients will want to wait 30 minutes for a test result. Furthermore, few offices have sufficient space either in the waiting room or in examination rooms to permit long patient waiting times.

  Another crucial time factor is the hands-on time required to produce a test result. Most of the current office analyzers use microprocessors to automate the test steps. In comparing instruments it is important to look at the level of automation and to consider how operating the instrument might affect busy office staff.

- *Cost*. The next factor to evaluate should be the instrument cost and the reagent costs to perform a single test. The analyzers that are being sold to physicians' offices range in initial cost from $4,000 to $20,000. This will make the analyzer one of the most expensive pieces of equipment in many office practices. But the purchase price is just one of the costs in doing testing. Other operating costs include test reagents, calibrating solutions, control reagents, and staff time. The reagent costs for office chemistry instruments are high (about $1 to $3 per test). It is often advisable to purchase an instrument service contract. This will add an additional 10% per year to the purchase price.

- *Calibration*. All instruments have a "calibration period." This is the time period during which the instrument can reliably "remember" how to calculate a chemistry value from the reaction measurement. Instruments need to be periodically recalibrated. This process requires reagents with known concentrations of the analyte. For example, when you calibrate the chemistry instrument for glucose, you "set" the instrument to report results based on the calibration points of the standard solutions. The instrument uses the calibration values to then calculate the value of the patient's glucose test.

Instrument calibration can become very costly if it needs to be done frequently (i.e., once a week). Some of the currently available office chemistry instruments hold their calibration for up to 3 months. The frequency of recalibration is an important consideration in terms of the total costs for testing.

- *Performance.* The final consideration when comparing office chemistry instruments is "analytic performance" (i.e., How good is the testing?). This includes the instrument's analytic range (i.e., What is the range over which the analyte can be tested?), interfering substances (i.e., Does lipemia interfere with the test?), accuracy (i.e., How close to the "true" value will the answer on the instrument be?), and the test precision (i.e., How reproducible are the results when a single specimen is retested?).

All of the available instruments have passed Food and Drug Administration guidelines for analytic performance. This means that at least under ideal circumstances the test systems perform in a manner that provides clinically useful data. However, it is important to realize that this may have little to do with how well the instrument functions in your laboratory. The quality of the chemistry testing in your laboratory is primarily dependent on the skill of the office laboratory staff. There is unfortunately very little information available on how sensitive an instrument will be to the subtle indiscretions of an inexperienced operator. The published data do indicate, however, that there can be unexpected testing problems in "real world" settings. This fact forces the clinician to be sure that the instrument is properly operating in the office laboratory (see Section 5).

SECTION 4: Choosing the Right Chemistry Analyzer

Table 13–1 lists the manufacturers of some of the common chemistry analyzers designed for office laboratory use. A good starting point in choosing a chemistry instrument is to call each of the toll-free numbers listed for these manufacturers and to request information about their office chemistry systems.

From the brochures you may be able to narrow down your selection based on the factors listed in Section 3. It is likely that each of the instruments you read about will have both pros and cons. Getting further reliable information may be difficult.

An additional source of information is a colleague in the community who has had experience with a chemistry analyzer. Talk to that clinician about how the instrument has performed, what the "real" costs have been and the quality of the service by both the instrument dealer and the manufacturer. It is a good idea to have your staff visit an office with the analyzer so that they can get some hands-on experience. Do not be surprised if the clinician and the staff have very different opinions about a chemistry analyzer. It may be that the physician is quite happy with the results but that staff have been frustrated with trying to keep the instrument running.

Medical journals are unlikely to be of great help in choosing one instrument over

TABLE 13–1. SOME MANUFACTURERS OF OFFICE CHEMISTRY SYSTEMS

Abbot Laboratories Diagnostic Division Routes 43 and 137 Abbott Park, IL 60064 800-342-5228
Baker Instruments Diagnostic Division 100 Cascade Dr PO Box 2168 Allentown, PA 18001 800-345-3127
Becton-Dickinson Johnson Laboratories 383 Hillen Rd Towson, MD 21204 800-638-8656
Boehringer Mannheim Bio-Dynamics Division 9115 Hague Rd Indianapolis, IN 46250 800-428-5074
Dupont Company Biomedical Products Department Market St Wilmington, DE 19898 800-262-5978
Eastman Kodak Clinical Products Division 343 State St Rochester, NY 14650 800-445-6325
Electro-Nucleonics, Inc 350 Passaic Ave PO Box 2803 Fairfield, NJ 07007-2803 800-346-4364
Miles Laboratories Ames Division PO Box 70 Elkhart, IN 46515 800-348-8100
Syntex Medical Diagnostic 330 Hillview Avenue PO Box 10058 Palo Alto, CA 94303-0847 800-227-9948

another. This is because there have been very few published instrument comparisons. In addition, instruments are frequently changed, so published evaluations may not provide data that accurately reflects the current instrument model. When reviewing the medical literature, be sure to identify whether the data are from "ideal" or "real-world" settings.

Office laboratory consultation is a growing business. Medical technologists, clinical chemists, pathologists, and reference laboratories have all begun to provide such consultative services. You may find that a consultation is helpful in making a decision about a particular analyzer. Two warnings, however, are in order. First, some consultants may have hospital laboratory experience but only a limited knowledge of office laboratories. Remember, office laboratories are not merely small hospital laboratories! Second, some consultants may have agreements with instrument manufacturers. In this case their recommendations may not reflect the best analyzer for your needs.

At some point you will need to contact a dealer or a manufacturer's representative. In general, the more expensive instruments are sold by a direct sales force employed by the manufacturer. Less expensive instruments are usually sold by medical equipment dealers. The industry representatives may be more knowledgable about a single product line, while the local dealer usually has a broader range of instruments to choose from.

Ask the salesperson if it would be possible to have an in-office demonstration of the analyzer. Schedule this at a time when both you and your staff will be free. This is an expensive purchase. Do not try to evaluate an instrument between seeing patients! For more expensive analyzers, it may be possible to arrange to keep the instrument for several days on an evaluation basis. This will give your staff some hands-on experience in operating the analyzer.

Choosing an office chemistry analyzer can be a scary experience. It is not uncommon for an office to first purchase a lower-priced instrument that may be slower and have more limited testing. After 1 or 2 years many offices come to depend on the advantages of office chemistry testing. At that time it is not uncommon to invest in a more expensive instrument that provides a broader range of tests and higher quality test results.

SECTION 5: How Good Is Your Testing?

It is essential that some evaluation be performed whenever a new instrument is purchased, whenever a new test is added to an existing instrument, and, most importantly, whenever a new operator is trained. You may find that your new chemistry analyzer does not work even though the old one did or that the new potassium test does not work even though the instrument has been measuring glucose well, or that your new nurse is unable to provide good test results even though the previous nurse was able to. In your evaluation you must show that the instrument, the reagent system, and the instrument operator—all working together as a system—are able to provide high-quality testing.

In addition to telling you that the system is working, an evaluation serves two other useful purposes. First, it provides feedback to the office staff during their "learning curve." Every person, no matter how experienced, improves their testing with an instrument over an initial period of time. This is referred to as the learning curve. An evaluation period gives feedback to your staff that facilitates their learning good technique. Without this sort of formal feedback, the same technical errors can be repeated for long periods of time without either you or your staff realizing it.

The second important reason for doing an evaluation is that it gives the clinician an idea of how good the test really is. For example, you may find that the test is not as accurate at either the high or low range. Or you may see that the results are not reproducible. The results may be different than those you obtained from your reference laboratory. Occasionally you may find that the testing is not good enough to meet your clinical needs.

The essence of any test evaluation is to run a series of samples that have known values and to compare these values to the results obtained in your laboratory. An evaluation should look at the test performance at both the high and low ranges of the test. In addition, you should repeat the test often enough to develop an idea of how reproducible the testing will be. We suggest that at a minimum the evaluation should include running a high and low control once a day for 5 days. By eyeballing the five values at the two levels, you will get an idea of how well the test system is operating. It should then be fairly obvious whether you have problems that require further evaluation.

It is a good idea to use control solutions rather than patient samples for the evaluation. Controls are usually available from the manufacturer with values assigned for your particular instrument. Be sure to reconstitute the controls with the exact volume of diluent. Too much diluent will lead to low test results. Too little will lead to high test results.

It is a common practice to do "split-sample testing" as an ongoing evaluation in a laboratory. To do this sort of evaluation a sample is drawn. Part of the sample is tested, and the remainder of the specimen is sent to the reference laboratory. Split-sample testing should be done when the office laboratory result does not "make sense'' to the clinician from what is known about the patient. Be careful in interpreting the split-sample results. When the office laboratory and reference values do not agree, there is no way to know which laboratory is correct. Reference laboratories do not always come up with the correct results!

A final way to measure the quality of your testing is to participate in a proficiency testing program, as explained in section 6. Proficiency testing is basically just split-sample testing done simultaneously by many different laboratories. It permits you to evaluate your own laboratory, compared to other laboratories with the same analyzer, and to compare your specific type of analyzer with the testing done on other types of instruments.

SECTION 6: Ongoing Quality Assurance

After you are certain that the instrument/reagent/operator system is working, there must be an ongoing program to ensure that the testing continues to be accurate. Instruments malfunction. Reagents deteriorate. Operators become sloppy. An ongoing quality assurance program will tell the clinician that the test results are sufficiently reliable for making patient care decisions. Such a system has the additional benefit of providing continuous feedback to the operators to give them an idea of their technical performance. A quality assurance program should include each of the following components:

- *Reagent handling.* All reagents should be stored in the manner described by the manufacturer. They should not be used after the expiration date. This date is usually stamped on the reagent bottle or is included in the printed materials shipped with the reagents.
- *Instrument maintenance.* Today's chemistry instruments are amazingly free of routine maintenance. The details of the maintenance are described in the manufacturer's information.
- *Calibration.* Most instruments require the use of either two or three calibration points across the test's analytic range. The calibration data is held inside the electronics of the instrument for periods of time ranging from days to months. When the calibration is out of date, the instrument requires recalibration! Unfortunately most instruments will continue to function even when the calibration is long out of date.

  Each instrument system is approved by the FDA for a calibration period based on the stability of the instrument and the reagent system. In general, instruments should be recalibrated whenever the calibration is out of date and whenever a new reagent lot is used. It is important that excellent technique be used during the calibration procedure because all of the patient values obtained after the calibration will be based on that calibration result. A small error in pipetting during the reconstitution of the calibrator solution can lead to abnormally high or low patient values for months after the calibration is done.

  Some of the currently available instruments use automated calibration. Each lot of reagents is assigned a specific calibrator value that is read by the instrument in a bar-coded message on the reagent strip. In essence these instruments are recalibrated each time a test is run. This is a somewhat unorthodox approach but does avoid problems with calibration of instruments by inexperienced office staff.

- *Controls.* Control solutions are reagents with a known value that are periodically tested in the same manner as a patient specimen. The "acceptable" range of the control solution is given by the manufacturer. By running the controls you know whether the analytic system is operating properly. In most traditional laboratory settings "good quality control" has meant, at a minimum, running a high- and a low-control solution each shift. This rigid view of quality control is probably not appropriate for all types of technology. Some tests are appropriate to control frequently (i.e., daily). Others may need to be controlled only once a week. The frequency of controls should be based on the stability of the test technology. The frequency stated in the manufacturer's guidelines is usually a conservative assessment of this stability.

The most useful thing to do with the control data is to plot it on graph paper as illustrated in Chapter 8 (Figs. 8–1 to 8–5). The range of acceptable results are indicated by two lines. The control values should fall within these lines. By eyeballing this graph it is easy to see if there is a trend, an unexpected imprecision, or a sudden shift in values. Regular eyeballing of the data is a much more useful habit than occasionally calculating a standard deviation. For most office laboratory uses eyeballing is a sufficient "statistical" test of the quality control data.

The range for most available control solutions is set using 2 standard deviations (SD) of the mean. What this means is that if 20 tests are done using the coltrol solution, 19 of the 20 will fall within the reported range. One out of 20, on a statistical ground alone, will fall outside this range.

When a control test falls within the acceptable range, it is referred to as being "in control." If it falls out of the acceptable range, the test system is said to be "out of control." If the system is out of control, then no further patient testing should be done until the system is again in control. Table 13–2 lists some of the common reasons for being out of control.

There are several steps that should be followed when a control value falls out of the acceptable range. The first is to immediately reanalyze the same control. If the reanalyzed control falls within the acceptable limits, then put both of these values on the graph and proceed with patient testing. If two consecutive results fall outside the acceptable limit, it is very unlikely that this is due to chance. The next thing to do at that point is to repeat the control test using a freshly prepared control solution. If this falls within the acceptable limits, then record only this result on the graph. (The first control material was probably unsatisfactory for analysis.) If the result is still outside the acceptable limits, then the instrument should be recalibrated according to the manufacturer's directions. If the test system is still out of control after recalibration, then the manufacturer should be contacted.

When looking at a month's quality

TABLE 13–2. SOME COMMON INSTRUMENT PROBLEMS

Failure to regularly calibrate
Error in reconstituting calibration fluid
Pipetting error in running calibration
Error in reconstituting control fluid
Pipetting error in running control
Use of outdated controls
Use of outdated reagents
Reagent deterioration from improper storage
Failure to keep the instrument's optical window clean
Instrument instability
Clerical errors

control data, most of the values should center around the mean value for the control material (Fig. 8–1). If they center above or below this line, then this indicates a "positive bias" or a "negative bias" (Fig. 8–2). If there are 20 to 30 data points on this graph, then on statistical grounds alone, about one or two points should fall outside the acceptable (i.e., 2 SD) range.

We recommend that the office laboratory director initial the quality control graph on a monthly basis. This provides an opportunity for the office staff and the clinician to regularly examine the performance of the office chemistry system.

- *Proficiency testing.* A final check to be certain that the office testing is accurate is to subscribe to a proficiency testing program. But until recently it was difficult to find a proficiency testing program that was specifically designed to meet office laboratory needs. This has now changed. Proficiency testing programs are offered through a variety of professional societies (e.g., College of American Pathologists, 5202 Old Orchard Rd, Skokie, IL 60077-1034. 312-966-5700; The American Society of Internal Medicine, 1101 Vermont Ave NW, Washington, DC 20005-3457. 202-289-1700; The American Academy of Family Physicians, PO Box 8723, Kansas City, MO 64114. 800-274-2237), by several instrument manufacturers, as well as by some state health departments.

In a typical program an unknown specimen is sent to the laboratory every 3 months. You are asked to run this specimen in the same manner as you would a patient specimen and to perform all chemistry tests that are routinely done in the laboratory. When all of these results are received by the proficiency testing program, the numbers are grouped and analyzed and reported back to the participating laboratories. These results show the "all-methods mean" as well as the "single-method mean." The all-method mean is the average of all test results using all of the methods used by all of the laboratories subscribing to that proficiency testing program. The single-method mean is the same data broken down by the different instruments that are used for testing.

In addition to the mean, the reports usually give the range (i.e., the lowest and highest values reported), the number of laboratories reporting, and the SD or the coefficient of variation (CV). The SD or CV is a measure of the variation in the test results. The CV is calculated by dividing the SD by the mean. For most of the commonly available chemistry tests, good laboratories are able to produce CVs of 5% to 10% or less. This means that 68% of the time the tests will fall within 10% of the mean and that 95% of the time the results will fall within 20% of the mean.

It may be initially confusing to analyze the data from a proficiency testing program. The first thing to look at is the all-method mean. This is a good approximation of the "real" value for the specimen. Next look at the mean for the method that your office laboratory uses. If it is very different from the all-method mean, this might indicate that your method has a "bias." Then look at the CV for your method and compare it to the CVs found for other instruments. This will give you an idea of the reproducibility of your method compared to other available instruments. Finally, look at the results of your laboratory compared to other laboratories using the same instrumentation. Good test performance is indicated if most of your laboratory's results fall within 2 SD of the single-method mean.

As an example, look at Table 13–3. These are data from a recent proficiency testing challenge. Nearly 50 different methods for measuring glucose were reported by the proficiency testing program , but the data from three of the

TABLE 13–3. GLUCOSE

Method/Instrument	No. Laboratories	Mean (mg/dl)	SD (mg/dl)	CV %	Low Value (mg/dl)	High Value (mg/dl)
All Method	1752	288.6	15.0	5.2	240	338
Instrument A	227	296.8	8.2	2.7	271	321
Instrument B	202	292.9	19.4	6.6	238	354
Instrument C	31	288.5	16.9	5.8	253	320

commoner office chemistry analyzers are included. From this table you can see that 1,752 laboratories participated in this particular glucose challenge. A large number of laboratories used instrument A or B. Only 31 laboratories used instrument C. If less than 20 laboratories use a particular instrument, then the standard practice is to indicate the low value and high value but not the other statistical results.

The mean for all methods was 288.6 mg/dL. Even though many instruments were used in coming up with the data, the CV is only 5.2%. This indicates that in general laboratories are good at glucose measurements. Instrument A appears to have a bias. Its mean value is somewhat higher than that reported by the all-method mean. Instrument A also appears to have excellent precision. The CV for the 227 reporting laboratories is only 2.7%. This instrument bias may just occur on the specially prepared material sent out by the proficiency testing program and not on fresh patient specimens. If this is so, then the bias would be called a ''matrix affect''.

The mean for instrument B is much closer to the all-method mean. You can see though that this instrument's precision is not nearly as good. Its CV is 6.6%. Instrument C is quite accurate (i.e., its single-method mean is nearly identical to the all-method mean), but it also shows some problems with precision.

After a while this approach to interpreting proficiency testing results becomes fairly intuitive. It can then give you a great deal of information about both your own laboratory and also the performance of the other available instruments that are used for chemistry testing.

REFERENCES

Burke JJ, Fischer PM. A clinician's guide to the office measurement of cholesterol. *JAMA.* 1988;259:3444-3448.

CHAPTER 14

Glucose Testing

SECTION 1: Introduction

Diabetes mellitus is a common chronic disease that affects 5% of the people in the United States. It can be life threatening on its own, as in diabetic ketoacidosis. It is also a major factor in the development of cardiac, eye, cerebrovascular, renal, skin and neurologic illnesses. In pregnant patients diabetes is a cause for morbidity to both the mother and the fetus.

Blood glucose testing is important in the initial diagnosis of diabetes and in following the course of the illness. It is one of the most frequently requested blood chemistry tests in outpatient medicine. Furthermore, the results are often rapidly needed so that appropriate changes in treatment can be made. Because of these factors, most primary care laboratories should have blood glucose testing available.

The measurement of blood glucose was revolutionized in the 1970s by the introduction of small, whole-blood glucose instruments. These instruments are now commonplace in many clinical settings and are also widely used by patients in their homes. While these instruments are adequate for some screening purposes and for following the course of a single patient's illness, they are often inadequate for the initial diagnosis of diabetes. In contrast, most of the office chemistry analyzers that use serum or plasma are sufficiently accurate for all primary care needs.

We have found that there is a great deal of misunderstanding about glucose testing. Most of this relates to what is the appropriate test in a given clinical situation. All too often the distinction between a ''screening'' test and a ''diagnostic'' test is confused.

What we present is an orderly testing process based on the established protocols of the American Diabetes Association. While it is tempting to offer simple tables of values for diagnosing diabetes, the ADA criteria are too complicated for such a simple-minded approach. We therefore encourage you to completely read the entire chapter.

The National Institutes of Health (NIH) and the ADA have established a classification for the various disorders of glucose metabolism. This classification reflects the present knowledge about the pathophysiology of diabetes:

- Type 1 diabetes mellitus: This is also referred to as insulin-dependent diabetes mellitus and is primarily due to a deficiency in insulin production. Previous names included juvenile-onset, ketosis-prone, or unstable diabetes. These patients require treatment with insulin. Without insulin they are likely to go into ketoacidosis. This condition usually first occurs in the young; however, it may begin at any age. This type of diabetes accounts for 10% of all cases of diabetes mellitus.
- Type 2 diabetes mellitus: This is also referred to as noninsulin-dependent diabetes mellitus. It was previously called adult-onset diabetes; however, it may occur at any age. These patients are insulin independent (i.e., they may be treated

with insulin to avoid symptoms, but they do not become ketotic if insulin is withheld). Many of these patients are obese. Diet is therefore the first form of therapy. These patients have normal or even increased levels of insulin. Their problem is a resistance to the peripheral action of the insulin hormone rather than a deficiency in insulin production.

- Other types of diabetes mellitus: This is the smallest group of diabetic patients. In these patients the diabetes is caused by other conditions, such as pancreatic disease, Cushing's syndrome, acromegaly, insulin receptor abnormalities, genetic conditions (e.g., Huntington's chorea, hyperlipidemia), malnutrition, or drugs (e.g., steroids, estrogen, or thiazide diuretics).
- Impaired glucose tolerance (IGT): This was formerly called "chemical diabetes." These patients have higher-than-normal plasma glucose values but lower values than those that are diagnostic for diabetes. They have none of the symptoms of diabetes but are probably at an increased risk for developing diabetes. These patients should not be given the diagnosis of "diabetes" or "prediabetes" because of the serious implications that these diagnoses have for health risks, employment status, and health insurance. These patients should be followed with a fasting blood glucose once or twice a year.
- Gestational diabetes mellitus (GDM): This class includes women who develop hyperglycemia during pregnancy but who have a normal glucose tolerance when not pregnant. Any pregnant patient with a family history of diabetes, a history of unexplained fetal death, a previously large baby, or who is found to have glycosuria on routine prenatal testing should be evaluated for gestational diabetes (see Section 4). Many experts now recommend that all pregnant patients be screened for this condition.
- Previous abnormality of glucose tolerance: This class includes patients who have previously had hyperglycemia when under stress, when obese, or when pregnant but who now have a normal glucose tolerance. These patients may develop a decreased glucose tolerance later in life and should therefore be tested with a fasting glucose on a yearly basis.
- Potential abnormality of glucose tolerance: This group includes patients with close relatives who are diabetic or women who have given birth to babies weighing over 10 pounds.

SECTION 2: Nonpregnant Hyperglycemia

SCREENING TESTS

Who should be screened and what is the appropriate test for screening? The ADA recommends that only high-risk individuals be routinely screened. These include:

- Patients with a strong family history of diabetes
- Markedly obese patients
- Women who have had babies weighing greater than 9 pounds at birth or who have had a morbid obstetric history
- Patients who have recurrent infections (e.g., urinary tract infections (UTIs), skin, or genital)

The appropriate initial screening test in these individuals is a fasting glucose. "Fasting" is defined as no food or drink other than water for 10 to 14 hours.

POSITIVE SCREENING TEST VALUES

- NONPREGNANT ADULTS:
  Venous plasma or serum: greater than or equal to 6.38 mmol/L (115 mg/dL)
  Venous whole blood: greater than or equal to 5.55 mmol/L (100 mg/dL)
  Capillary whole blood: greater than or equal to 6.11 mmol/L (110 mg/dL)
- CHILDREN:
  Venous plasma or serum: greater than or equal to 7.22 mmol/L (130 mg/dL)
  Venous whole blood: greater than or equal to 6.11 mmol/L (110 mg/dL)
  Capillary whole blood: greater than or equal to 6.38 mmol/L (115 mg/dL)

Note that different values are given for each of the different types of specimens. It is therefore essential that you know what type of specimen you are dealing with. The basis for these differences is that capillary blood is considered arterial and arterial blood has a higher glucose than venous blood. This is because glucose is extracted by the tissues as the blood passes through the capillary bed. In addition, there are differences between whole blood and serum or plasma. Putting it altogether:

- Venous plasma or serum is 15% higher than venous whole blood.
- Capillary whole blood is 10% higher than venous whole blood.
- Venous plasma or serum is 5% to 7% higher than capillary whole blood.

DIAGNOSTIC TESTS

"Diagnostic" testing for diabetes is one of the most frequently misordered laboratory procedures. This is partly due to confusion about the appropriate test and partly due to the insistence of patients who request glucose tolerance tests for a variety of vague symptoms.

A diagnostic test for diabetes is indicated for:

- Patients with a positive screening test
- Patients with obvious signs and symptoms of diabetes (e.g., polydipsia, polyuria, polyphagia, and weight loss)
- Patients with an uncertain clinical picture (e.g., someone with glycosuria who has a borderline elevated random blood glucose level)

No single diagnostic test is appropriate for all clinical indications. Instead, a random glucose, a fasting glucose, or a 2-hour glucose tolerance test is indicated in different clinical settings. When done as a diagnostic test, each of these should be performed on venous serum or plasma.

Random glucose

This is used *only* for patients with *classical* signs and symptoms of diabetes (adults and children). The advantage of this test is that it can be done while the patient is seeing the clinician and requires no special preparation.

Fasting glucose

The patient needs to fast for 10 to 14 hours. Only water can be consumed.

Two-hour glucose tolerance test

The patient needs to eat a regular carbohydrate diet (> 1,500 calories or > 150 grams of carbohydrates) for at least 3 days. The details for administering a glucose tolerance test are given in Section 5 of this chapter.

Adults: This test is for the patient who has indications for the *diagnostic* testing (not screening) *and* who already had a fasting plasma/serum glucose value of < 7.77 mmol/L (140 mg/dL).

Children: This test is required in addition to fasting testing. A child must have two fasting values ≥ 7.77 mmol/L (140 mg/dL) to proceed to the tolerance test.

TEST INTERPRETATION

1. Normal plasma/serum glucose values

a. Nonpregnant adults

Fasting plasma/serum glucose	< 6.38 mmol/L (115 mg/dL)
Plasma glucose values after 75-gram oral glucose dose	**30 minutes <** 11.10 mmol/L (200 mg/dL) **60 minutes <** 11.10 mmol/L (200 mg/dL) **90 minutes <** 11.10 mmol/L (200 mg/dL) **120 minutes <** 7.77 mmol/L (140 mg/dL)

b. Children

Fasting plasma/serum glucose	< 7.22 mmol/L (130 mg/dL)
Plasma glucose value after glucose dose of 1.75 grams/kilogram ideal body weight up to maximum of 75 grams	**120 minutes <** 7.77 mmol/L (140 mg/dL)

2. Diagnostic criteria for impaired glucose tolerance

a. Nonpregnant adult (ALL of the following three results are required.):

- Fasting plasma/serum glucose of less than 7.77 mmol/L (140 mg/dL); *and*
- Two-hour oral glucose tolerance test plasma glucose level between 7.77 mmol/L (140 mg/dL) and 11.10 mmol/L (200 mg/dL); *and*
- One other oral glucose tolerance test plasma glucose level of 11.10 mmol/L (200 mg/dL) or greater (i.e., 0.50 hour, 1.0 hour, or 1.50 hour)

b. Children (Both of the following are required.):

- Fasting plasma/serum glucose concentrations of less than 7.77 mmol/L (140 mg/dL); *and*
- Two-hour oral glucose tolerance test plasma glucose level of greater than 7.77 mmol/L (140 mg/dL)

3. Diagnostic criteria for diabetes mellitus

a. Nonpregnant adults

- Random plasma/serum glucose level of 11.10 mmol/L (200 mg/dL) or greater plus classic signs and symptoms of diabetes mellitus, including polydipsia, polyuria, polyphagia, and weight loss; *or*
- Fasting plasma/serum glucose level of 7.77 mmol/L (140 mg/dL) or greater on at least two occasions; *or*
- Fasting plasma/serum glucose level less than 7.77 mmol/L (140 mg/dL) plus sustained elevated plasma/serum glucose levels during at least two oral glucose tolerance tests. The 2-hour sample and at least one other between 0.50 and 2.0 hours after 75-gram glucose dose should be 11.10 mmol/L (200

mg/dL) or greater on both tests. Oral glucose tolerance testing is not necessary if the patient has a fasting plasma/serum glucose level of 7.77 mmol/L (140 mg/dL) or greater. If the 2-hour value on the first glucose tolerance test is < 11.10 mmol/L (200 mg/dL), there is no need to perform the second test, since *both* 2-hour values must be greater than or equal to 11.10 mmol/L (200 mg/dL).

---

b. Children

- Random plasma/serum glucose level of 11.10 mmol/L (200 mg/dL) or greater plus classic signs and symptoms of diabetes mellitus, including polyuria, polydipsia, ketonuria, and rapid weight loss; or
- Fasting plasma/serum glucose level of 7.77 mmol/L (140 mg/dL) or greater on at least two occasions *and* sustained elevated plasma/serum glucose levels during at least two oral glucose tolerance tests. Both the 2-hour plasma/serum glucose and at least one other value between 0 and 2 hours (after a glucose dose of 1.75 g/kg ideal body weight up to 75 grams) should be 11.10 mmol/L (200 mg/dL) or greater.

SOURCES OF ERROR

1. MAO inhibitors, alcohol, and large quantities of salicylates all lower blood glucose levels
2. The following factors raise blood glucose levels:
   a. Drugs (i.e., thiazide diuretics, beta blockers, nicotinic acid, steroids, estrogens, phenylephrine [in cough medicines], catecholamines, dilantin)
   b. Acute medical or surgical stress (wait several months after the event before testing)
   c. Marked restriction of carbohydrate intake (i.e., dieting or malnutrition
   d. Prolonged physical inactivity (i.e., people confined to bed for 3 days or more)

COMMENTS

The conversion factor for glucose to change a value in mg/dL to mmol/L is .05551. The mmol/L values are those used in the new SI system.

REFERENCES

Lebovitz H, ed: *Physicians Guide to Non-Insulin Dependent (Type II) Diabetes.* 2nd ed. American Diabetes Association, 1988.

SECTION 3: Testing for Gestational Diabetes

BACKGROUND

Gestational diabetes mellitus (GDM) is a condition of abnormal carbohydrate metabolism that develops during pregnancy. While women with GDM may develop nongestational diabetes sometime later in life, the principal reason for identifying the condition is that such pregnancies are associated with prematurity, congenital defects, macrosomia, neonatal hypoglycemia, respiratory distress syndrome, and neonatal hyperbilirubinemia.

The original work in this area was done by O'Sullivan; therefore the testing protocols for GDM are referred to as the O'Sullivan Screen and the O'Sullivan 3-hour glucose tolerance test. The amount of glucose given as a challenge for these tests differs from the amount used in the routine glucose tolerance test for nonpregnant patients.

The O'Sullivan tests should be performed on venous serum or plasma. There is insufficient data at this point to rely on capillary specimens or whole blood glucose analyzers. This is a test with serious implications, placing a pregnancy at high risk. It should be performed on reliable and accurate equipment.

TEST INDICATIONS

All *non-high-risk patients* should be tested between 24 and 28 weeks' gestation. *High-risk* patients should be screened as early in pregnancy as possible. If the initial screen is negative, they should be rescreened between 24 and 28 weeks. Women defined as high risk include those with:

- A previous history of gestational diabetes
- A previous "large-for-gestational-age" infant
- Prior polyhydramnios
- A suspected large-for-gestational-age fetus
- Glycosuria values of 1+ or greater on two or more occasions or 2+ or greater on one occasion
- A history of increased thirst or urination
- Recurrent vaginal and urinary tract infections

O'SULLIVAN SCREENING TEST

1. No patient preparation is required. THE PATIENT DOES NOT HAVE TO FAST.
2. Give the patient 50 grams of glucose in the form of a glucose drink (i.e., Glucola or similar solutions). Tell her to take about 10 minutes to drink it. In addition, explain that if she feels sick to her stomach, she should notify you immediately. Take her to an examination room so that she can lie down. If she vomits, the test is invalid.
3. Tell the patient to wait quietly in the waiting area. She should not eat or smoke until her blood is drawn.
4. After 1 hour, draw a venous specimen for glucose testing. Use an appropriate tube. Some laboratories will use a gray-top tube, others a red top, others a green top (plasma).
5. If you have a precise and accurate glucose instrument that tests serum or plasma, you can run the test in your office. If you only have a small whole-blood glucose analyzer, send the specimen to your reference laboratory.

Interpretation

If the test value is less than 7.77 mmol/L (140 mg/dL), the screen is considered normal. If the result is 7.77 mmol/L (140 mg/dL) or higher, the screen is considered "positive," and a diagnostic O'Sullivan 3-hour glucose tolerance test should be scheduled. (Note: The 7.77 mmol/L (140 mg/dL) value is lower than the 8.33 mmol/L (150 mg/dL) value cited in O'Sullivan's original paper. The value has been lowered to produce a more sensitive screening test.)

O'SULLIVAN 3-HOUR GLUCOSE TOLERANCE TEST

Preparation

The patient should be instructed to eat an unrestricted diet (more than 150 grams carbohydrate) and is permitted her usual level of physical activity. She should then fast for at least 8 hours but not for more than 14 hours before being tested.

Test

1. Draw a fasting venous sample in the appropriate tube.
2. Give the patient a 100-gram load of glucose. (Note that this is different from the regular 75-gram challenge used when nonpregnant patients are given a 2-hour glucose tolerance test.)
3. Tell the patient to sit quietly for the next 3 hours. She may have small drinks of water to wash the taste of the glucose preparation out of her mouth, but should not smoke or eat.
4. Draw venous specimens in the appropriate tubes at 1-, 2-, and 3-hour intervals. Be sure to label each tube with the correct time. If you have a precise and accurate serum/plasma glucose instrument, it can be run in your laboratory; otherwise the specimen should be sent to your reference laboratory.

Interpretation

1. The diagnosis of gestational diabetes is established if two or more of the venous serum/plasma glucose values are equal to or greater than:

Fasting:	≥ 5.79 mmol/L (105 mg/dL)
1 hour:	≥ 10.55 mmol/L (190 mg/dL)
2 hour:	≥ 9.16 mmol/L (165 mg/dL)
3 hour:	≥ 8.05 mmol/L (145 mg/dL)

2. The test is considered normal if none or only one of the values meet or exceed the designated limit.

Comments

- We do not consider the use of capillary blood and small whole-blood glucose analyzers accurate enough for this important screening or diagnostic test.
- Whole-blood testing may, however, be used to monitor therapy of GDM after the diagnosis is made.
- The glucose conversion factor from mg/dL to mmol/L is .05551. The mmol/L value is the preferred one in the SI system of laboratory test reporting.

REFERENCES

Public health guidelines for enhancing diabetes control through maternal- and child-health programs. *MMWR.* 1986; 35(13):201-204.

O'Sullivan JB, Charles D, Mahan CM, Dendrow RV. Gestational diabetes and perinatal mortality rate. *Am J Obstet Gynecol.* 1973;116:901.

Swinker M. Routine screening for gestational diabetes mellitus in a family practice center. *J Fam Pract.* 1983;17(4):611-614.

SECTION 4: Hypoglycemia

BACKGROUND

It is much more difficult to establish a diagnosis of hypoglycemia than hyperglycemia. This is, in part, due to the difficulty in establishing criteria for a lower normal limit for blood glucose. This is especially true for women, many of whom have plasma glucose levels below 1.10 mmol/L (20 mg/dL) and remain asymptomatic. The generally accepted lower limits of normal for plasma glucose after 24 hours of fasting are 3.05 mmol/L (55 mg/dL) for men and 2.22 mmol/L (40 mg/dL) for women.

Hypoglycemia is not itself a disease. *Abnormal hypoglycemia* is diagnosed only when there is a documented low-glucose value at the same time that a patient is symptomatic and when the symptoms are relieved by ingesting glucose.

A detailed history should be taken on any patient in whom the diagnosis of hypoglycemia is considered. It is important to check for the specific symptoms, when they occur, how they are related to eating or fasting, how the symptoms are relieved, and any history of diabetes or stomach surgery.

There are two types of hypoglycemia: postprandial and fasting. Both are relatively uncommon disorders but have attracted the attention of the lay press because the symptoms of hypoglycemia are similar to those of the very common problems of anxiety and depression.

POSTPRANDIAL HYPOGLYCEMIA

This is also referred to as "reactive hypoglycemia." It is most commonly seen in patients with a history of stomach surgery and, therefore, very fast gastric emptying after a meal. The syptoms are rapid in onset and occur several hours after eating. The symptoms include sweating, palpitations, weakness, anxiety, tremor, hunger, and irritability. It is uncommon to find neuroglycopenic symptoms, such as confusion, amnesia, or seizures. To establish the diagnosis of postprandial hypoglycemia, the patient must have these symptoms at the same time as a low plasma glucose. In addition, the symptoms must be relieved by the administration of glucose.

FASTING HYPOGLYCEMIA

This disorder is seen primarily in diabetics and in alcoholics. The symptoms are more gradual in onset and more persistent than those of postprandial hypoglycemia. They occur after a long period of not eating. Neuroglycopenic symptoms, such as mental confusion, bizarre behavior, focal neurologic signs, amnesia, seizures, or unconsciousness may be seen. The diagnosis of fasting hypoglycemia is usually made by checking glucose and insulin levels during a 72-hour fast. This testing should obviously be done only in hospitalized patients who can be monitored throughout the entire test.

In the past a variety of strategies have been used to diagnose hypoglycemia. At one time a 5-hour glucose tolerance test was given. This test is now outdated. Other approaches included testing the patient after a large meal or whenever the patient was symptomatic. With these approaches it was often common to miss the symptomatic period when the patient was in the office to be tested. With the availability of reliable whole-blood glucose instrumentation, a different approach is now possible.* Using this method the patient actually has control over the glucose measurements and can then see for himself or herself whether they are caused by hypoglycemia.

This procedure requires a reliable whole-blood glucose instrument with a memory. We recommend the Ames Glucometer M. It records the date and time as well as the test result. It holds up to 350 readings and has a system of "markers" that allow you to differentiate some tests (or testers) from others. It also permits you to download the results onto a personal computer that can then graph the results.

TEST PROCEDURE

1. Take a detailed history to find out the patient's symptoms, how they are related to meals, and who told the patient

*We are indebted to Paul Dunn, MSN, FNP, Department of Family Medicine, University of North Carolina at Chapel Hill, for this protocol.

that the symptoms might be due to hypoglycemia.

2. Make a clear contract with the patient to work with you in the evaluation.
3. Ask the patient to keep a detailed 3-day diary of the following information:

Date	Time	Symptoms	Food eaten

4. When the patient returns, teach him or her to use a whole-blood glucose analyzer. Be sure that the patient is able to get a good capillary specimen and can reliably measure its glucose content.
5. Have the patient take the glucose analyzer home for 1 week. During that week he or she should measure capillary glucose before each meal and also whenever he or she has symptoms.
6. In the follow-up office visit, the clinician should review the patient's diet, the symptom record, and the glucose readings. A truly hypoglycemic patient will have symptoms at the same time that the glucose values are less than 2.78 mmol/L (50 mg/dL). Most patients will test negative and will have realized this from their own recording of the test data. At that time you can explore other reasons (e.g., stressful event, erratic eating, or overindulgence in "quick simple sugars") for his or her symptoms.

REFERENCES

Bornstein R. Meaningful screening test for reactive hypoglycemia. *Diabetes Care*. 1987; 10 (6): 792-793.

SECTION 5: Administering Glucose Tolerance Tests

The 2- and 3-hour glucose tests are similar except for the amount of oral glucose that is given.

When the patient is scheduled for testing, verify that the criteria for diagnostic testing have been met and that the patient is properly prepared for the glucose tolerance test. The test should be administered only during the morning. Diurnal variations of hormones, such as cortisol, can alter the results if done later in the day.

1. Make sure the patient has fasted for at least 10 hours.
2. Make sure that the patient ate a regular diet (> 150 grams of carbohydrates) for at least 3 days prior to testing.
3. Collect a fasting glucose sample.
4. Give the patient the appropriate amount of glucose:
   a. Nonpregnant adult: 75 grams
   b. Child: 1.75 grams/kilogram ideal body weight up to 75 grams
   c. Pregnant patient: 100 grams (O'Sullivan 3-hour)
5. Tell the patient to take 5 to 10 minutes to drink the solution. He or she may experience stomach cramping or nausea if the solution is swallowed too quickly. Tell the patient to call you if he or she feels sick. If this happens, have him or her lie down on an examination table.
6. Tell the patient that he or she can have

nothing to eat or drink during the test except for a little water. Smoking is not allowed because it can increase glucose levels. The patient should remain relatively quiet.

7. Collect additional plasma/serum glucose specimens at 30, 60, 90, and 120 minutes for a 2-hour test; or at 60, 120, and 180 minutes for a 3-hour test.
8. Send these to a reference laboratory, or test them on an accurate serum/plasma office instrument.

SECTION 6: Patient Instruction Sheet for a Glucose Tolerance Test

You are to undergo a glucose tolerance test. To achieve the most reliable test result, please:

1. Eat a regular diet (at least 150 grams of carbohydrates or 1,500 calories) for at least 3 days prior to the test.
2. Fast (i.e., do not eat or drink anything except small amounts of water for at least 10 but not for more than 14 hours before the test.) *Do not drink black coffee the morning of the test.*
3. Bring something to read or to do for the 2 or 3 hours. You will have to remain relatively quiet during the test.
4. If during the test you feel sick, please notify the laboratory staff immediately.

CHAPTER 15

Hematology

SECTION 1: Introduction

Office laboratories can be easily equipped to perform most routine hematology testing. These tests provide the clinician with critical information about hematologic disease and, just as importantly, give clues about systemic illnesses or diseases in other organ systems. For example, the presence of atypical lymphocytes on a peripheral smear may explain a patient's splenic enlargement. Likewise, the discovery of a low hematocrit might be the first clue to an occult gastrointestinal malignancy.

TESTS RECOMMENDED FOR THE OFFICE LABORATORY

Microhematocrit and Hemoglobin

Each of these tests measures the red blood cell (RBC) fraction of blood. There are only slight differences between these two tests, and an office laboratory may therefore choose to perform only one of the two tests.

White Blood Cell Count

This test is most commonly used in the assessment of patients with acute infections. It is often needed very rapidly and must therefore be done in the office, unless a reference laboratory is available for stat testing. The test is also helpful in following some chronic diseases (e.g., leukemia) or drug treatments (e.g., gold injections). The white blood cell (WBC) count can be performed either manually in a counting chamber or with an automated cell counter. Such automated equipment is now being designed specifically for office use.

Peripheral Smear and Differential Count

This is one of the most difficult tests for the laboratory worker to learn. Supervised at-the-microscope training is required. We created and described a simplified WBC differential count that is useful for people who have not had previous laboratory training. The various morphologic features of RBCs are also described.

Reticulocyte Count

This test is easy to perform in the office laboratory and can provide helpful information about the cause of an anemia.

Platelet Count

This test is best done in the office laboratory rather than the reference laboratory. Platelets are not easily preserved. A specimen sent to an outside reference laboratory may therefore result in a falsely low platelet value. Platelets are counted in a chamber just as white cells are counted. A platelet estimate can be easily done from the peripheral smear if the quantified platelet count is not required.

Erythrocyte Sedimentation Rate (ESR)

This is a nonspecific test for inflammatory diseases. It is best performed in the office laboratory because the test must be completed within a short time after the specimen collection.

Sickle Cell Test

Several office tests for sickle cell hemoglobin are available. This test should be considered for any practice that has a large number of black patients.

COLLECTION OF BLOOD

All venous blood used for office hematology testing is collected in a purple-top (EDTA) tube. Before any test is run, the contents of the tube must be thoroughly mixed. This can be done by gently inverting the tube by hand 60 times, but it is more convenient to rock the blood on an aliquot mixer. There are a variety of mixers available costing between $150 and $250 (Fig. 15–1). The tubes can be added to or taken from the mixer without stopping it. Fifteen small or ten large tubes can be rocked at one time. An aliquot mixer is well worth the money spent in terms of saved technician time.

After thorough mixing the tube must be opened and checked for clots. This is done by stirring through the specimen with a long wooden applicator stick. If clots are found, a new specimen should be drawn. *A blood clot in a purple-top tube invalidates all hematology tests.* See Chapter 12 for details of blood specimen collection.

TESTS NOT RECOMMENDED FOR THE OFFICE LABORATORY

Blood Typing

The primary use of blood typing in the outpatient setting is for prenatal blood studies in pregnant women. This testing is best done by a certified blood-bank laboratory.

Manual RBC Count

This test, when performed manually in a counting chamber, results in a precision of only ± 20%. Electronic cell counters count many more cells than can be done manually and have an increased precision of ± 2%. The error of the manual count is too high for the test to be useful for most clinical purposes. The test should therefore be performed by an electronic cell counter. Calculated cell indices that are based on the RBC count are also not recommended unless the laboratory has an automated hematology instrument.

REFERENCES

Brown B. *Hematology: Principles and Practice.* Philadelphia, Pa: Lea & Febiger; 1976.

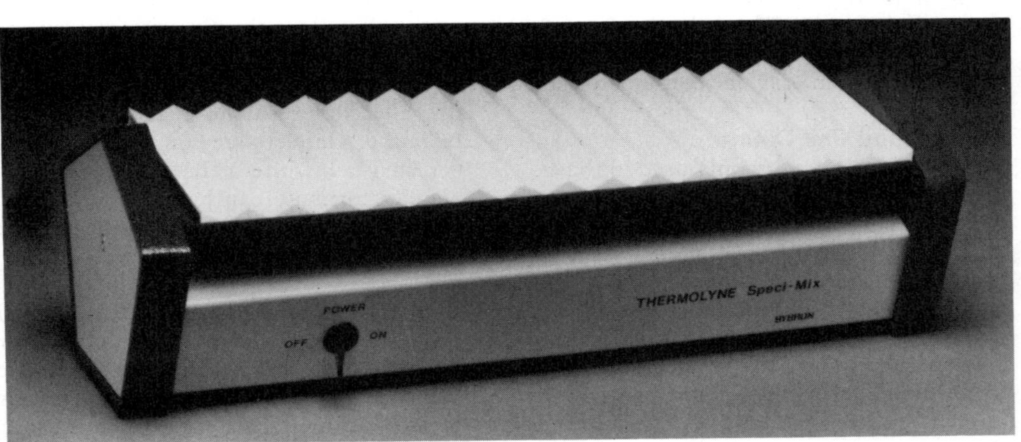

Figure 15–1. Specimen mixer (Courtesy of Barnstead/Themolyne).

Henry JB, ed. *Clinical Diagnosis and Management of Laboratory Methods.* Philadelphia, Pa: Saunders, 1979.

Miale JB. *Laboratory Medicine, Hematology.* 5th ed. St. Louis, Mo: Mosby; 1977.

Undeitz E. *Sandoz Atlas of Haematology.* 2nd ed. Basle, Switzerland: Sandoz; 1973.

Wintrobe M. *Clinical Hematology.* 7th ed. Philadelphia, Pa: Lea & Febiger; 1974.

SECTION 2: Safety

The office laboratory worker must follow certain safety procedures whenever he or she is performing hematology testing. In this section some of the most important safety "habits" relevant to hematology testing are reviewed. See chapters 9 ("Safety") and 12 ("Blood Collection") for more specific and detailed information. The routine safety procedures should include:

- Hand washing: A disinfectant soap should be used before and after drawing blood from each patient. This will prevent the spread of most infections to both the patient and the laboratory professional.
- Gloves: Vinyl gloves should be worn when preparing the specimen for testing and during the actual testing procedure.
- Glassware: Any disposable glassware that has come in contact with blood samples should be placed in the "contaminated" disposal jug.
- Universal precautions: These Centers for Disease Control (CDC) recommended precautions eliminate the need for individualized patient precautions. It is no longer necessary to identify patients who do or may have hepatitis or HIV infections. All samples are assumed to be infectious.
- Daily cleaning: The laboratory equipment, bench tops, and arms of the blood-drawing chair should be wiped once each day with a disinfectant: 5% Staphene is a good general purpose disinfectant. Discard the used towels in a container that will be autoclaved before disposal.
- Blood tubes: Tubes containing blood should not be thrown out in with the regular trash. Put them in the biohazard container or the "Contaminated" softs disposal jug (Fig. 9–2) and have them autoclaved before disposal. The reference laboratory can be asked to autoclave these discarded blood tubes. (See Chapter 9 for details.)

SECTION 3: Automated Hematology Instruments

Until very recently most office laboratories restricted their hematology testing to the use of manual methods. As is described in the rest of this chapter, manual methods can quickly provide the clinician with a great deal of valuable information. Limiting an office laboratory to manual hematology methods does not, in any major way, compromise the clinical information that is available from the office laboratory.

A variety of hematology instruments has recently become available for office laboratory use. These instruments are small, relatively easy to use, and are much less expensive than the large hematology instruments used by hospital laboratories.

Any clinician who is considering buying an automated hematology instrument should carefully consider four testing issues:

1. Who will operate the instrument? These instruments require occasional troubleshooting. Apertures become dirty. Tubing gets clogged. Diluters malfunction. Troubleshooting for these problems requires that the office worker have a "technical mentality." Most individuals with an MT or MLT degree will feel quite comfortable operating these instruments; however, those individuals without this type of formal training may not. Do not purchase an expensive laboratory instrument if the office staff all have "instrument phobia."
2. How much test accuracy is required? The automated hematology instruments are quite accurate in cell counting. For example, in a typical proficiency testing program the laboratories that use automated hematology instruments report WBC counts with a precision of less than 5%. In comparison, manual WBCs using a hemocytometer produce a precision of about 15%. The better 5% precision may be useful when testing a patient who is severely neutropenic; however, for most clinical uses the manual WBC is adequate. The automated instruments produce hemoglobins or hematocrits with the precision of 2% to 3%. The microhematocrit (section 4) is nearly as precise in the hands of a skilled office laboratory worker. Automated instruments are more accurate, but is this increased accuracy needed in your setting?
3. Test volume: The turnaround time for hematology test results is similar by either manual or automated methods. However, manual cell counting is time consuming in terms of hands-on time by staff. Therefore one of the major reasons for buying an automated instrument is that the office staff are spending large amounts of time doing manual tests. For most laboratories, the purchase of an automated hematology instrument is unnecessary unless that laboratory is doing at least five to ten WBCs per day.
4. Are RBC indices needed? The mean corpuscular volume (MCV), mean corpuscular hemoglobin (MCH), and mean corpuscular hemoglobin concentration (MCHC) cannot be easily done by manual methods. These red cell indices are sometimes clinically useful in diagnosing the cause of anemia. Having an automated hematology instrument in the office laboratory permits an on-site evaluation of these complete blood count (CBC) parameters.

OPERATING PRINCIPLES

Most automated hematology instruments have two basic functions: They count cells and they spectrophotometrically measure hemoglobin. The cell counting is based on the "impedance principle," discovered by Wallace Coulter. Cell counting is done using two electrodes between which current passes. When a cell moves into the counting chamber, the flow of current is interrupted and a cell is "counted." The degree of disturbance (i.e., impedance) of the current is proportional to the number and size of the cells. Using impedance cell counting, thousands of cells can be counted each second. In one specimen the RBCs are counted. In a second specimen the RBCs are chemically lysed and then the nuclei of the WBCs are counted. The hemoglobin that is released from the lysed red cells is chemically converted to cyanmethemoglobin, which is measured spectrophotometrically. Finally, the hematocrit, MCH, and MCHC are calculated from the measured RBC, hemoglobin, and MCV.

Within the past several years a centrifugal hematology analyzer has been marketed (QBC, Clay Adams, Inc.). In this system whole blood is spun in a reagent-coated hematocrit tube. A special microscope is then used to examine the buffy coat. By examining the sizes of the various blood layers in the tube, this instrument reports out a hematocrit, WBC, platelet count, total granulocyte count, and total nongranulocyte count. It must be recognized that since this instrument uses very different technology than the usual cell counting methods of examination of peripheral smear, the results may not always be the same as those obtained by traditional methods. The instrument is expensive but easy to use. The information that it provides may be suitable for some routine clinical purposes.

INSTRUMENTATION

Table 15–1 lists some of the manufacturers that produce automated office hematology instruments. In general, the greater the number of hematology indices, the higher the instrument purchase price. For example, a variety of instruments are available that do only hemoglobins and WBCs. They typically cost between $5,000 and $7,000. The next group of instruments do hemoglobins, WBCs, hematocrits, RBCs, and MCVs. They typically cost between $7,000 and $10,000. For an instrument that also does an automated platelet count and provides an MCH or an MCHC, expect to spend somewhere near $15,000.

One of the major differences between instruments is the way in which the specimen dilutions are made. In the lower priced instruments, the diluting is done manually using a Unopette-type system. Since the reagents are individually packaged for each test, they tend to be expensive if many tests are to be done. The next approach to diluting is to have a stand-alone electronic diluter. In such a system a large container of diluent is hooked up by tubing to a suction device. The diluter draws a specific volume of blood from a purple-top tube and mixes it with the diluent. In the most expensive instruments, the diluter is built right into the hematology analyzer. The diluter system can be a major source of test inaccuracy and technical malfunctions. If the automatic diluter systematically draws up

TABLE 15–1. MANUFACTURERS OF AUTOMATED OFFICE HEMATOLOGY INSTRUMENTS

Company
American Scientific Products 1430 Waukegan Rd McGraw Park, IL 60085 312-689-8410
Baker Diagnostics 100 Cascade, PO Box 2168 Allentown, PA 18001 800-345-3127
Bio-Dynamics Division, BMC PO Box 50100 Indianapolis, IN 46250 800-428-5074
Clay Adams 299 Webro Rd Parsippany, NJ 07054 800-638-1532
Coulter Electronics PO Box 2145 Hialeah, FL 33010 800-526-6932
Sequoia Turner 755 Ravendale Dr Mountain View, CA 94043 415-969-5533
Smith Kline Beckman 485 Potrero Ave Sunnyvale, CA 94086 800-645-5467

either too much or too little blood or diluent, then all test results will be in error. For this reason offices with less technically oriented staff should consider using prepackaged (i.e., Unopette) diluting systems.

QUALITY ASSURANCE

- Calibration. *Instrument calibration must be done meticulously.* Many of the hematology instruments that are currently available require calibration only twice a year. Any error that occurs in calibrating will therefore be reproduced with all testing for up to 6 months. Instrument manufacturers should provide assistance in calibrating if there are any problems.
- Controls. Most manufacturers recommend that controls be done on a daily basis and that they include high values, normal values, and low values for the hemoglobin and WBC. The process of controlling an automated hematology instrument can be very expensive. The controls themselves are small bottles of whole blood that have very short outdate periods. Like any other test system, the frequency of the controls should be based on the stability of the instrumentation. It is easy to examine the instrument stability by looking at the results from daily high, low, and normal control testing. If all appears well, then you may consider changing to one level of control per day and alternating, the high, low, and normal controls.
- Proficiency testing. Most proficiency testing programs include hematology. The specimens are usually received on a quarterly basis and should be processed just as a patient specimen would be done. Since most office hematology analyzers "count cells" just like the large instruments in hospital and reference laboratories, there tends to be less variation in the hematology results than is seen for the chemistry testing data.

TEST PROBLEMS

Even the simplest of these instruments requires occasional tinkering by somebody who is not afraid to "look under the hood." The commonest problems include:

- Replacement of the hemoglobin lamp. This lamp is part of the spectrophotometer that is used to measure the hemoglobin. This lamp needs to be replaced whenever the light coming from the lamp is reduced below a certain limit or when the bulb burns out.
- Tubing change. Many of the instruments pass the diluent and the specimen through a variety of plastic tubes. Eventually protein material from the specimen builds up on the inside of the tubing and produces erroneous test results. Many instruments recommend preventive maintenance for this problem by either periodically changing the tubing or

TABLE 15–2. POTENTIAL CONFOUNDERS FOR AUTOMATED CELL COUNTERS

Abnormal Samples
Hemoglobin S and C
Nucleated RBCs
Heinz and Howell-Jolly bodies
Malarial parasites
Red cell fragments
Atypical lymphocytes
Immature WBCs
Giant platelets
Interferents
Hemolysed sample
Microclots
Cold agglutinins
Hyperbilirubinemia
Hyperlipidemia
Uremia
Nonketotic hyperosmolarity
Hyperglobulinemia

flushing the system with a bleach solution.

- Pipettor variability. Automatic pipettors are tricky mechanical devices that are prone to both getting clogged and behaving in a finicky manner.

TESTING ERRORS

Testing problems may be associated with the specimen characteristics. The cell counting part of the instrument can, for example, count nucleated RBCs as leukocytes. Likewise, the spectrophotometer that measures the hemoglobin may be interfered with from a variety of substances in the sample. Table 15–2 lists the common interferences for automated cell counters.

REFERENCES

Nixon GA, Mentrup P. Automated office hematology instruments. In: Fischer PM, Addison LA, eds. *Primary Care: Office Laboratory Testing.* Philadelphia, Pa: Saunders, 1986:727-741.

SECTION 4: Microhematocrit

BACKGROUND

The hematocrit is the percentage volume of packed RBCs in a volume of whole blood. It is neither a measure of red cell mass nor of total blood volume, but it is affected by both of these factors (as in pregnancy or dehydration). A microhematocrit is a very simple and reproducible test.

In clinical practice the hematocrit is most commonly used to diagnose and follow the course of anemia. Anemia may be caused by acute blood loss, chronic blood loss, nutritional deficiencies, hereditary disorders, and a variety of chronic illnesses. The most common anemia is due to iron deficiency. This anemia is classified as microcytic (small cell) and hypochromic (pale cell). However, these red cell characteristics may not appear until the hematocrit has dropped to 30%. The peripheral smear is therefore less sensitive than the hematocrit in the early diagnosis of this anemia. The hematocrit can also be used to follow the response to therapy for anemia. The rise in the hematocrit for iron deficiency anemia usually appears about 1 week after initiating iron therapy. Treatment for iron deficiency anemia should continue for 6 months after the return to a normal hematocrit to ensure the complete restoration of bone marrow iron stores.

TEST INDICATIONS

- Screening test as part of a health maintenance examination, especially for infants, menstruating females, the elderly or those patients with a chronic disease.
- Screening test for patients taking aspirin, other anti-inflammatory drugs, Coumadin, or other drugs that can cause occult blood loss.
- To follow the response to therapy for anemia.
- To follow the course of a chronic anemia, such as thalassemia.
- To assess the degree of chronic blood loss in a patient with occult blood loss due to

gastrointestinal tumors, polyps, or ulcers.

- To assess the degree of blood loss following acute bleeding. (It may take 4 to 8 hours for fluid shifts to occur within the body after an acute blood loss. A hematocrit can be relied on only after these fluid shifts have reached equilibrium. A measurement for orthostatic blood pressure drop is a more reliable test to assess blood volume loss in an emergency situation.)
- To diagnose and follow the course of polycythemia.

TEST MATERIALS

Microhematocrit Centrifuge

A high rpm centrifuge with a special head is required to perform the microhematocrit. A specialized centrifuge can be purchased for just this purpose, or a multipurpose centrifuge can be obtained for spinning microhematocrits, urine specimens, and other blood specimens (Chapter 11, Figs. 11–1 and 11–2).

Microhematocrit Capillary Tubes

Inexpensive capillary tubes have a high incidence of breakage during the centrifuge process. We therefore use a more expensive tube (Pre-Cal Capillary Tubes, Clay-Adams, Inc). These tubes have a calibration line to which the tubes can be filled; however, exact filling is unnecessary if a reader card is used (Fig. 15–2).

Hematocrit Tube Sealant

This is a putty that is used to seal one end of the capillary tube. Several brands are available, and each works well. If the sealant gets dirty with blood, it can be wiped clean with an alcohol wipe. It is best to discard the sealant when the surface is filled with holes. Trying to repack the sealant to extend its life inevitably results in inadequate seals for the capillary tube. A sure sign that you need a fresh slab of sealant is when blood leaks out of the capillary tubes. If the material becomes brittle, it should be replaced (Fig. 15–2).

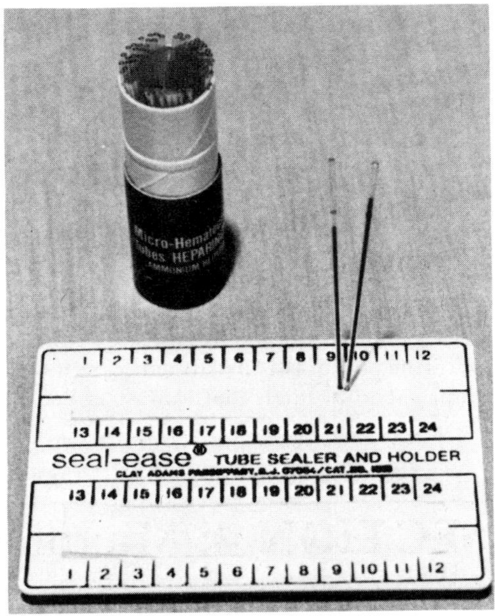

Figure 15–2. Microhematocrit capillary tubes and sealant.

Microhematocrit Reader Plastic Card

These inexpensive cards are available from scientific supply companies. We find them to be easier to use than the various machines that are sold for reading the microhematocrit tubes. The plastic card can be taped to the bench surface in front of the centrifuge for convenient use (Fig. 15–3).

Hand Magnifying Glass

A magnifying glass can be used to more accurately read the top of the red cell column in the capillary tube.

SPECIMEN COLLECTION

Both capillary and venous blood may be used for the microhematocrit. (See Chapter

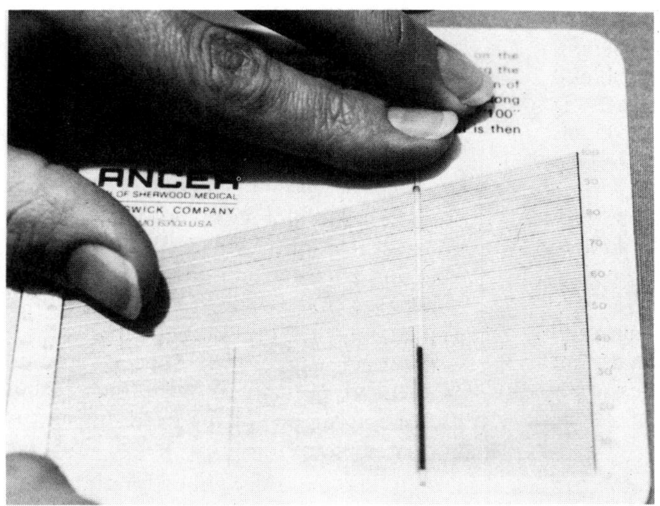

Figure 15–3. Reading the spun capillary tube on the reader card. Place the top of the clay sealant (bottom of the red-cell column) at the 0 line. Place the top of the plasma line at the 100% line. Find the point where the top of the red-cell column crosses the line on the reader card. Do not include the gray buffy coat. The identified line corresponds to the microhematocrit value.

12 for details of the collection techniques.) Two capillary tubes should be obtained for each patient who is tested. This allows an internal test control (i.e., the two tubes should give an equivalent hematocrit value) and also allows for the occasional tube that breaks during the centrifuge spinning.

Finger Stick

See Chapter 12 for a detailed description of this technique.

1. Put on gloves.
2. Wipe the finger with alcohol and allow it to dry—or wipe it vigorously with a 2 × 2-inch gauze.
3. Make a puncture wound with a sterile lancet.
4. Check that the patient's hand is relaxed. Some patients will hold their hand open, but it will be very tense. This will prevent blood flow to the small capillaries and will prevent sample collection. It may be necessary to shake the patient's hand to make it loose.
5. Do not squeeze the finger excessively. This radically alters the relationship between the plasma and the RBCs and will therefore change the hematocrit.
6. Wipe away the first drop of blood with a gauze pad.
7. Hold the capillary tube to the puncture site, tilting it in a downward direction. This will allow the tubes to fill much faster. Some varieties of capillary tubes may have a thin membrane over the end of the tube. They may not start filling immediately; however, this membrane will dissolve in several seconds.
8. Fill the capillary tube at least three-quarters full, and insert the end of the tube into the clay sealant. Either end of the tube can be put into the sealant, but using the clean end will keep the sealant surface clean. If the other end is used, it should be wiped with a gauze to clean off excess blood from the outer surface of the tube.
9. Repeat the above steps with a second capillary tube.

Heel Stick

See Chapter 12 for a description of the infant heel-stick technique.

EDTA-Anticoagulated Blood Tube

An accurate microhematocrit can be obtained using EDTA-anticoagulated blood for up to 24 hours after its collection as long as the sample has been refrigerated. If left at room temperature it is good for only 6 hours. The best practice is to put a tube into the refrigerator if it is not to be processed immediately. Then bring the tube to room temperature before processing.

1. Rock the tube on an aliquot mixer for 2 minutes. Alternately, gently invert the purple-top tube by hand 60 times to ensure adequate mixing.
2. Put on gloves.
3. Hold a 2 × 2-inch gauze around the rubber stopper and gently remove the stopper from the tube. Try to avoid splashing blood.
4. Take a wooden applicator stick and stir the blood. Pull the stick almost all the way out to see if there are any clots. If the blood has clotted, the test cannot be accurately performed.
5. Put the microhematocrit capillary tubes into the purple-top tube and collect the blood sample. Hold the tube as close to horizontal as possible without spilling the blood. Blood should rise into the tube by capillary action. If this does not happen, tap the top of the capillary tubes several times to speed up the process.
6. Wipe the blood off the outside of the capillary tube with a 2 × 2-inch gauze.
7. Place the capillary tubes into the sealant after they are at least three-fourths filled with blood. The more blood in the tube the easier the result is to read and the more accurate is the result.
8. Put the stopper back on the tube. If you are sending the tube to a reference laboratory for additional tests, secure the stopper to the tube with tape. Once a stopper has been taken off, it does not stay on tightly. The tape helps prevent spillage during transport.

TEST PROCEDURE

1. Place both capillary tubes in the centrifuge across from each other. The clay-sealed end should be on the outside, as the centrifugal forces press the cells toward the outside of the centrifuge.
2. Record the slot numbers used on the test result form. This will prevent confusion if more than one patient's samples are centrifuged at the same time.
3. Place the lid on the centrifuge head. Most centrifuges have a special spin-on head to prevent the capillary tubes from flying out of the head during the centrifugation.
4. Spin the capillary tubes for the specified time. A standard time is 3 minutes at a minimum of 10,000 rpm, but the time must be individualized for each instrument (see Chapter 11).
5. Wait until the centrifuge has come to a complete stop. It is safe to use a break if the machine has such an attachment, but it is unwise to stop the centrifuge from spinning by hand. This can do injury to both the technician and the centrifuge.
6. Remove the capillary tubes from the centrifuge and quickly read them on the hematocrit card reader. If the capillary tubes are allowed to sit for a long period of time before they are read, the results will be inaccurate.
7. To read the microhematocrit, place the tube at the right edge of the reader card with the top of the plasma line at the 100% mark. Slide the tube to the left until the top of the clay sealant is at the bottom line (0). Check that the top of the plasma is at the 100% line, the bottom of the red cell column is at the 0 line, and the capillary tube is parallel to the edge of the card (Fig. 15–3).
8. Carefully find the point where the top of the red cell column crosses the line on the reader card. Do not include the

gray buffy coat just above the red cell column as part of the reading. The identified line corresponds to the microhematocrit value. It may be necessary to use the magnifying glass to accurately identify the top of the red cell column.

9. Repeat the reading process for the second capillary tube. The results should agree within 2 percentage points. The reported value is the average of those values. If the two values differ by greater than 3 points, the entire procedure should be repeated.
10. When you are reading the microhematocrit values, be sure to note any discoloration of the plasma. Yellow plasma may be an indication of an elevated bilirubin level, and this should be brought to the attention of the clinician. Pink or red plasma suggests either hemoglobinemia or a poor collection technique with hemolysis of the specimen.

TEST INTERPRETATION

Reference values for the microhematocrit vary with sex and age (Table 15–3). The value increases as the elevation above sea level increases. People from mountainous areas have an increased hematocrit to increase their oxygen-carrying capacity in the thin mountain air.

QUALITY ASSURANCE

The microhematocrit is defined as the percentage volume of RBCs in whole blood when there is maximum packing of the RBCs. The particulars of the centrifuge speed or the length of time of centrifugation are not critical as long as maximum packing is achieved. This should be checked each month by running a routine patient sample and reading the results, then placing the capillary tube back in the centrifuge and spinning for one more minute. The second microhematocrit reading should agree with the first. If the second value is less than the first, then either the centrifugation time or the rpm speed needs to be increased to ensure maximal red cell packing. Record this quality control test in the quality control book each month.

A second method to evaluate the accuracy of your results is to do a split-sample comparison. Once a month, when you are sending a CBC to your reference laboratory, fill two microhematocrit tubes from a well-mixed purple-top tube. Run a hematocrit on this specimen and send the purple-top tube to the reference laboratory. Be sure to record the patient's name so that you can check the results when they are back. Automated hematology instruments calculate the hematocrit from the red cell size and the red cell number. This calculated hematocrit value may be about 3% less than the measured microhematocrit. This is due to the fact that there is plasma trapping in even the best microhematocrit technique.

A final quality assurance method is to participate in the hematology section of a proficiency testing program. You will be sent an unknown specimen to be used for a microhematocrit determination. The test results are then reported to the proficiency testing program, and your laboratory's performance will be evaluated.

SOURCES OF ERROR

Falsely Low Values

- Improper reading of the hematocrit value from the reader chart
- Excessive tissue juice from squeezing the finger too hard
- Inadequate mixing of an EDTA-anticoagulated tube and drawing off the specimen from the plasma-rich area at the top of the tube

TABLE 15–3. REFERENCE MICROHEMATOCRIT VALUES

Children	Average Normal %	Minimal Normal %
At birth	56.6	51.0
First day	56.1	50.5
End of first week	52.7	47.5
End of second week	49.6	44.7
End of third week	46.6	42.0
End of fourth week	44.6	40.0
End of second month	38.9	35.1
End of fourth month	36.5	32.9
End of sixth month	36.2	32.6
End of eighth month	35.8	32.3
End of tenth month	35.5	32.0
End of first year	35.2	31.7
End of second year	35.5	32.0
End of fourth year	37.1	33.4
End of sixth year	37.9	34.2
End of eighth year	38.9	35.1
End of twelfth year	39.6	35.7
Men	%	%
End of fourteenth year	44	39.6
End of eighteenth year	47	42.3
18-50 years	47	42.3
50-60 years	45	40.5
60-70 years	43	38.7
70-80 years	40	36.0
Nonpregnant women	%	%
14-50 years	42	36
50-80 years	40	36
Pregnant women	%	%
End of fourth month	42	30
End of fifth month	40	30
End of sixth month	37	30
End of seventh month	37	30
End of eighth month	39	30
End of ninth month	40	30

From Miale JB. *Laboratory Medicine, Hematology.* St Louis, Mo: Mosby; 1977:426. Reprinted with permission.

- Inadequately filling an EDTA tube with blood so that the anticoagulant overdilutes the blood sample.

Falsely High Values

- Incorrect reading of the hematocrit from the reader chart
- Inclusion of the buffy coat in the reading of the red cell column
- Inadequate mixing of an EDTA-anticoagulated tube and drawing off the specimen from the cell-rich area at the bottom of the tube
- Inadequate centrifuge speed or time, resulting in less than maximal packing of the RBCs
- Increased plasma trapping due to abnormally shaped cells (e.g., macrocytes, sickle cells, poikilocytosis). In some cases this can result in an increase as great as 20% in the test value (i.e., sickle cell patients)
- The use of EDTA-anticoagulated blood that has been stored for greater than 24 hours

COMMENTS

- A consistent technique for the microhematocrit should produce a test-to-test variation of only ± 2%. The largest source of error is the variation between readers in the reading step of the procedure.
- Air bubbles that occur when filling the microhematocrit tubes will not affect the final value because the bubbles will be expelled during the centrifugation process. This is not the case for other tests that require filling tubes with blood (e.g., the sedimentation rate).
- Remember that the office laboratory microhematocrit may be 3% higher than the calculated hematocrit obtained from a reference laboratory. This is because of the small volume of plasma that is trapped between the red cells during the packing procedure.
- A good relationship to keep in mind is that the micrahematocrit is usually three times the value of the hemoglobin value. Another way to describe this relationship is that 1 hematocrit point is equivalent to 0.34 g Hb/dL.
- The microhematocrit is greatly affected by abnormally shaped cells because such cells trap plasma and artificially elevate the value of the microhematocrit. Patients with distorted cells such as in sickle cell disease should therefore be followed by hemoglobin levels rather than microhematocrit values.

SECTION 5: Hemoglobin

BACKGROUND

Hemoglobin is the oxygen-carrying protein in RBCs. Its concentration can be measured in whole blood and is reported as grams/dL or, in the SI system, as grams/L. The hemoglobin value is used much like that of the hematocrit in clinical practice. It is commonly used to diagnose and follow patients with anemia. In patients whose anemia is accompanied by variations in cell shape (e.g., sickle cell anemia), the hemoglobin test is superior to the hematocrit. This is because abnormally shaped cells trap plasma when the hematocrit is spun down, which can result in falsely elevated values. The measurement of hemoglobin is not affected by cell size or shape.

An office laboratory may decide to perform either a microhematocrit or a hemoglobin or both. In small, limited-function laboratories, the microhematocrit is usually chosen. In larger laboratories that have access to a chemistry analyzer or automated hematology analyzer, the hemoglobin is the test of choice. Most of the chemistry analyzers that are currently used in office laboratories include hemoglobin as one of their tests. Having such an instrument may therefore preclude the need to purchase a specialized microhematocrit centrifuge.

In the first edition of this book, we described the use of the Spencer hemoglobinometer. At the time this was one of the only methods available for hemoglobin testing in the office laboratory. The Spencer method is much less accurate than the hemoglobin testing on any of the currently available chemistry analyzers. We therefore cannot recommend its continued use.

TEST INDICATIONS

See test indications for microhematocrit.

TEST MATERIALS

Refer to the instruction manual that accompanies your chemistry analyzer or automated hematology instrument.

TEST INTERPRETATION

Reference values for hemoglobin vary with the patient's age and sex (Table 15–4). The values also vary with the elevation above sea level. Hemoglobin increases as the elevation increases to compensate for a lower partial pressure of oxygen in the air.

TABLE 15–4. REFERENCE HEMOGLOBIN VALUES

Subject	g/L[a]
Adult men	139-163
Adult women	120-150
Boys	
Birth	185-215
1 month	155-185
3 months	135-165
6 months	130-160
9 months	120-140
1 year	100-140
2 years	105-142
4 years	112-143
8 years	120-148
14 years	125-150
Girls	
Birth	180-210
1 month	158-189
3 months	133-164
6 months	128-148
9 months	117-139
1 year	100-140
2 years	105-142
4 years	113-142
8 years	115-145
14 years	116-148

[a]Hb is expressed as g/L in the SI system. The previously used system reported results in g/dL. The SI results are convertible by a conversion factor of 0.1.

SOURCES OF ERROR

Falsely Low Results

Inadequate mixing of an EDTA tube and drawing off a plasma-rich specimen from the top of the tube.

Falsely High Results

Inadequate mixing of an EDTA tube and drawing off a cell-rich fraction from the bottom of the tube.

REFERENCES

Miale JB *Laboratory Medicine, Hematology*. 5th ed. St. Louis, Mo: Mosby; 1977.

SECTION 6: WBC Count

BACKGROUND

Changes in the WBC count are seen with many infectious, hematologic, inflammatory, and neoplastic diseases. This variety of diseases makes the WBC a nonspecific test. It is, however, a sensitive indicator of disease in particular clinical situations. Its degree of elevation or depression from reference values often correlates with the severity of the disease process. Furthermore, following the changes in the WBC count over time can provide information about the course of an illness.

Performing the WBC count involves diluting a volume of whole blood in a specified volume of a solution that lyses RBCs. The WBCs are not lysed and can therefore be counted, and their concentration in the whole blood can be determined. WBC values are generally expressed as cells times $10^3/\mu L$ or cells times $10^9/L$.

In the past, the diluting step was done by manually pipetting the blood and the lysing solution. This required considerable skill and resulted in wide variations in test values with only slight differences in technique. A precision of only ± 20% was possible with these cumbersome procedures. For most office laboratories the speed, accuracy, and precision of the dilution step can be greatly improved by using the Unopette System (Becton Dickinson, Rutherford, NJ 07070). Each Unopette contains a sealed plastic reservoir with a premeasured volume of reagent and a self-filling capillary pipette to draw up the blood sample. This system has been a great improvement for the office laboratory.

After the RBCs have been lysed, the WBCs in a known volume of solution are counted. This counting is done under the microscope using a hemocytometer and a special coverslip. The hemocytometer is a glass instrument with a pattern finely etched onto its surface. A specified volume of solution is contained under the etched area between the hemocytometer and its special coverslip. The WBCs in this area are counted, and from this value the WBC count can be calculated.

Even where automated counters are used, the manual method should be available as a back-up system when the automated counter is not working.

TEST INDICATIONS

- Diagnose diseases associated with a leukocytosis (e.g., appendicitis)
- Diagnose diseases associated with a leukopenia (e.g., alcoholism)
- Follow the toxic effect of some drug therapy (e.g., methotrexate)
- Follow the course of some hematologic diseases (e.g., aplastic anemia)

TEST MATERIALS

Unopette System

Unopettes are made for a variety of manual or automatic hematology procedures, including the leukocyte count, platelet count, eosinophil count, RBC count, reticulocyte count, and hemoglobin determination. Each is a single sample unit with a reagent reservoir and a capillary pipette (Fig. 15–4). We recommend the use of test 5855, which allows both a WBC and a platelet count from the same sample. Test 5805 can be used for the WBC count only, and test 5856 is designed for low leukocyte counts. Directions for these other tests are available from the manufacturer. The procedure described here is for test 5855. The reagent-containing reservoirs should be kept at room temperature and should be protected from sunlight. It is best to keep

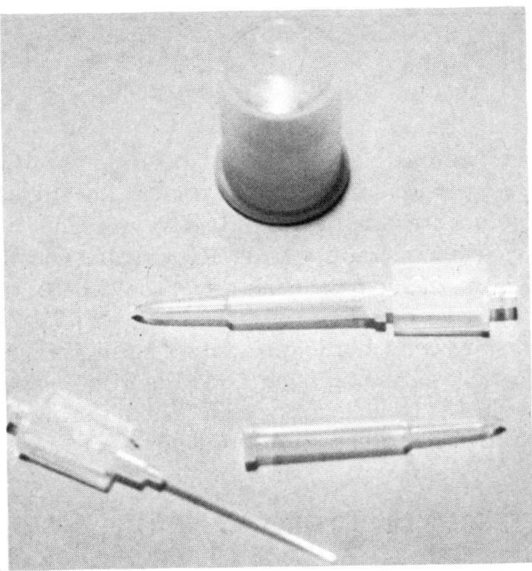

Figure 15–4. Unopette test system: reservoir and pipettes (covered and uncovered by the shield).

them in their original containers. They should not be used after the expiration date nor if the reagent solution has become cloudy.

Hemocytometer

This is a surprisingly expensive and delicate piece of equipment. It can be purchased from scientific supply companies for about $40 (Figs. 15–10 to 15–12). Hemocytometers should be handled with care to prevent scratching the counting chamber surface. They should not be used after being scratched. It is recommended that at least one extra hemocytometer be kept in the laboratory in case the original is damaged. Each hemocytometer comes with two special coverglasses. These can be easily broken. A package of spare coverglasses should therefore be kept on hand.

Petri Dish

This is used to hold the hemocytometer while the cells are settling before the counting process. Place a piece of gauze or filter paper in the bottom of the dish (Fig. 15–10).

Counter

A handheld counter is needed to keep track of the number of white cells that are counted. An inexpensive counter can be purchased from an athletic supply company. (These counters are used for counting laps for runners.)

Microscope

(See Chapter 10).

Drying Towel

A very soft cloth is needed to dry the hemocytometer and its coverslip. A soft cloth diaper is excellent.

Hemocytometer Storage

Place one plastic urine collection cup for each hemocytometer and coverslip on the back of the sink. Fill each cup with a 5% Staphene solution. Label the cup "hemocytometer/coverslip." After completing the WBC count, rinse the hemocytometer and coverslip using tap water, and then place them in the cup. Do not put more than one hemocytometer in a single cup, as they are easily scratched.

When a hemocytometer is to be used, put on gloves and take the hemocytometer and coverslip out of the Staphene solution. Rinse them with tap water and dry with the very soft cloth.

SPECIMEN COLLECTION

Capillary blood from a finger or heel stick can be collected directly into the Unopette capillary pipette. It should then be immediately placed into the reagent reservoir and mixed with the cell-lysing solution. EDTA-anticoagulated blood can be used for up to 24 hours after collection if it is refrigerated or stored at room temperature. The EDTA tube should be rocked for 2 minutes before pipetting the sample to ensure adequate blood mixing. Once the blood sample has been mixed with the diluent, it must be counted within 3 hours.

TEST PROCEDURES

Unopette Preparation

1. Place the Unopette reservoir on a flat surface. Label it with the patient's name and the time that the specimen is added to the diluent.
2. Take the Unopette shield/pipette assembly in one hand and hold the reservoir with the other hand.
3. Push the pointed tip of the shield into the reservoir neck through the plastic diaphragm that seals the reservoir. Ream out a large hole so that the pipette can be easily placed in the reservoir later in the procedure (Fig. 15–5).
4. Remove the shield/pipette assembly from the neck and the reservoir and then remove the protective shield from the pipette by twisting it off.
5. Put on gloves.
6. a. Finger/heel stick: (See Chapter 12 for a detailed description of this specimen collection technique.)
   (1) Wipe off the first drop of blood from the skin. Do not squeeze the finger.
   (2) Hold the pipette horizontally, and when a new drop of blood forms, touch the pipette tip to the drop. The pipette will automatically fill to the proper level. Do not dip the pipette in and out of the drop of blood, as this can produce air bubbles in the pipette. Such bubbles are unacceptable, and a new pipette would need to be used (Fig. 15–6).

   b. EDTA tube of blood:
   (1) After mixing hold a 2 × 2-inch gauze around the stopper and gently remove the stopper from the tube. Try to avoid splashes.

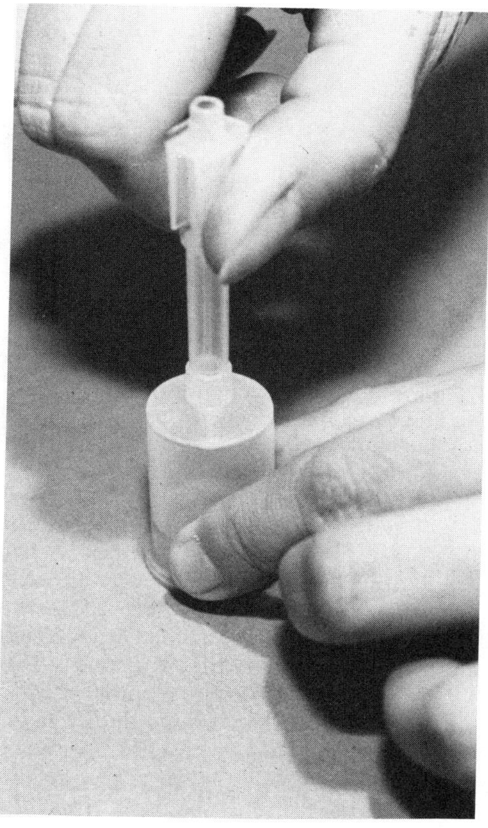

Figure 15–5. Puncturing the Unopette reservoir diaphragm.

   (2) Take a wooden applicator stick and stir the blood. Pull the stick almost all the way out to see if there are any clots. If the blood has clotted, discard the specimen and do not run the test.
   (3) Hold the blood tube as close as possible to the horizontal position without spilling any blood.
   (4) Put the pipette tip into the blood and collect the sample.
7. If the pipette does not fill rapidly, tap the open end of the pipette gently

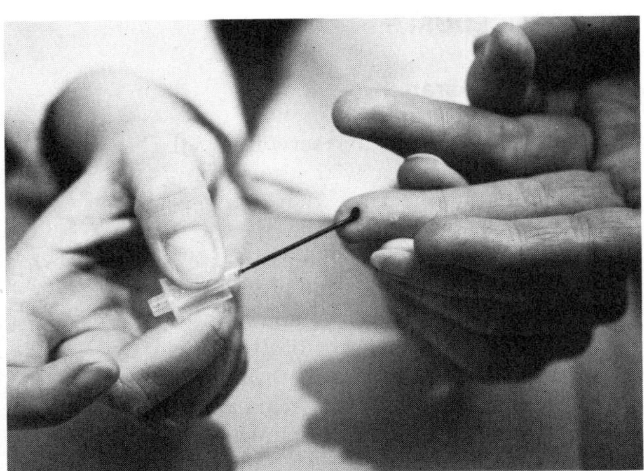

Figure 15–6. Collection of a capillary blood specimen from a finger puncture.

with your finger. It should fill to the hub of the pipette assembly when filled. Gently place your index finger over the hub end of the pipette. Make sure that no blood is forced out of the pipette when you do this. It is critical to keep the blood at the top of the hub. If the blood should go below this level, you will need to add more to bring it to the proper height.

8. Wipe off the excess blood from the outside of the capillary pipette by quickly passing a piece of gauze from the pipette hub to its tip. Be sure not to touch the gauze to the open tip, as this may result in some of the sample being drawn out.
9. You now have the filled pipette (with index finger covering the hub end) in one hand. Take the reservoir with the reamed-out diaphragm in the other hand. Squeeze the reservoir between your fingers to force out some air (Fig. 15–7). Do not squeeze too hard or you will lose some diluent. If this happens, you must begin again with a new reservoir. Put the pipette into the reservoir as far as it will go, with your index finger still covering the overflow chamber. The pipette will fit snugly into the neck of the reservoir. Relax the fingers that hold the reservoir and remove the index finger from the top of the pipette at the same time. This will pull the blood from the pipette and into the reservoir.
10. Very gently squeeze the reservoir two or three times to force the diluent into the capillary pipette to rinse out the blood (Fig. 15–8). Do not squeeze too hard or you will lose liquid out of the top of the pipette. Watch the column of fluid inside the pipette to avoid losing any of the dil-

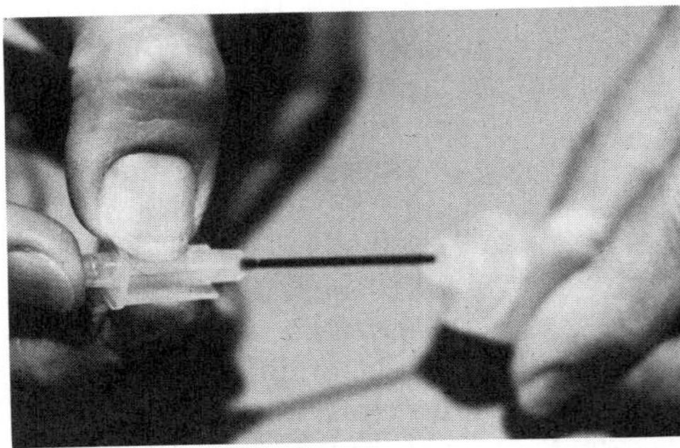

Figure 15–7. Inserting a filled pipette into the Unopette reservoir. (Ink has been used for better contrast.)

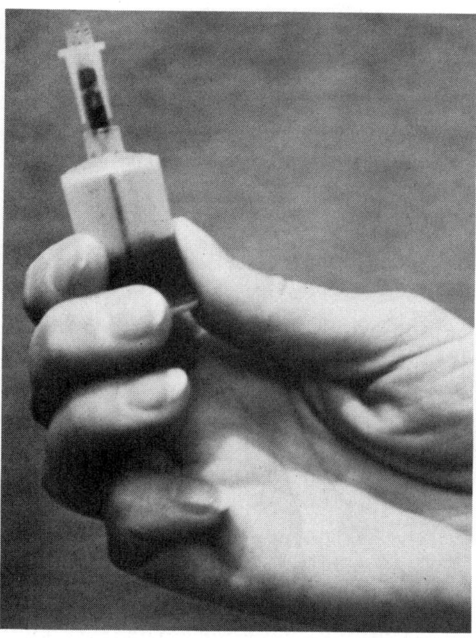

Figure 15–8. Mixing the blood specimen with the Unopette reagent. (Ink has been used for better contrast.

uent. If you do lose some of the diluent, it will be necessary to start over with a fresh sample and a new Unopette because you will have altered the dilution ratio, and therefore the count will be inaccurate.

11. Again squeeze the blood-diluent solution high into the pipette (but not overflowing). Put your index finger over the open hub end of the pipette and remove it from the reservoir.
12. Once again, squeeze the reservoir and hold it. Reseat the pipette into the reservoir neck; relax the fingers holding the reservoir and remove your index finger at the same time. The solution will be sucked into the reservoir. This guarantees that you have washed the blood sample out of the pipette.
13. Place your index finger over the top of the pipette's overflow chamber and invert the whole assembly several times to ensure complete mixing.
14. Repeat this sequence with a second reservoir and pipette so that you have two samples.
15. Place the pipette shield on the top of the pipette and set the Unopette aside for at least 5 minutes to allow complete lysing of the RBCs. If you do not wait long enough, the RBCs will not be lysed and it will be impossible to perform the count because the numerous RBCs will obscure the WBCs. Wait no more than 3 hours before counting the sample. Even the WBCs will lyse after a long period of time.

Hemocytometer Chamber Preparation

1. Use tap water to lightly dampen a piece of gauze or filter paper in the petri dish that is to hold the hemocytometer. This provides a moist atmosphere so that the white cell diluent does not dry out in the counting chamber during the test.
2. Put on gloves, take the hemocytometer and its special coverslip out of the Staphene solution, and rinse them with tap water. Dry each with a soft cloth. Be sure that there is no lint on the coverslip or the counting chambers of the hemocytometer. Do not touch these areas with your fingers, since the oil from your skin will make it difficult to properly fill the counting chamber.
3. Put the hemocytometer on top of the dampened gauze in the petri dish. Place the coverslip over the counting chamber areas of the hemocytometer.
4. Mix the solution in the Unopette by inverting the assembly several times. This will resuspend the cells in the solution. Remove the shield.

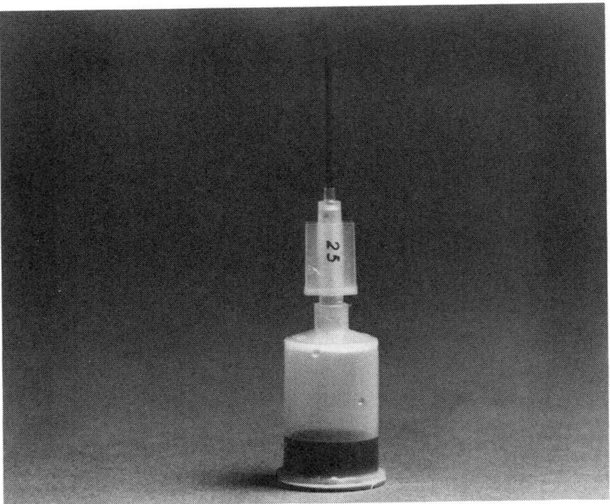

Figure 15–9. The Unopette converted to a dropper bottle.

5. Again squeeze the blood-diluent solution high into the pipette (but not overflowing). Put your index finger over the open hub end of the pipette and remove it from reservoir.
6. Once again, squeeze the reservoir and hold it. Reseat the pipette in the reservoir neck; relax the fingers holding the reservoir and remove your index finger at the same time. This guarantees that you have washed the blood sample out of the pipette.
7. Hold your index finger over the hub end of the pipette and invert the Unopette a few times. You now have a well-mixed specimen.
8. Convert the Unopette into a dropper bottle by removing the capillary pipette, turning it upside down, and then reinserting it back into the neck of the reservoir. Be sure that the pipette is securely positioned into the reservoir neck (Fig. 15–9).
9. Again invert the assembly several times; then, holding it upside down, gently squeeze the reservoir. Let the first few drops of solution drip onto a piece of gauze.
10. Put the tip of the pipette into one of the V-shaped slots on the hemocytometer and continue to gently squeeze the reservoir. The solution should flow into the counting chamber and completely fill the area between the coverslip and the counting chamber (Fig. 15–10).

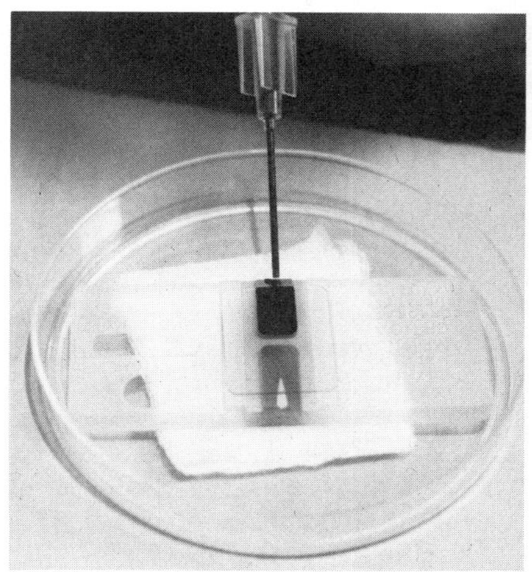

Figure 15–10. A properly filled hemocytometer counting chamber. The second counting chamber must also be properly filled before performing the cell count. (Ink has been used for better contrast.)

This will require only one or two drops of the solution. If the solution overflows the area of the counting chamber (Fig. 15–11) or if an air bubble is trapped beneath the coverslip (Fig. 15–12), you must begin again by cleaning the hemocytometer and refilling it with the cell solution. A count performed when *either* of the two counting chambers has overflowed or has an air bubble is not acceptable. If you have repeated problems with filling the hemocytometer, wipe it and the coverslip with alcohol to remove any organic film that may accumulate.

11. Repeat step 10 for the other counting chamber of the hemocytometer. Both chambers must be filled to perform a WBC count.
12. After the hemocytometer has been correctly filled, put the top on the petri dish and allow the unit to sit for at least 5 minutes. This time allows the cells to settle so that they are not moving during the counting process.

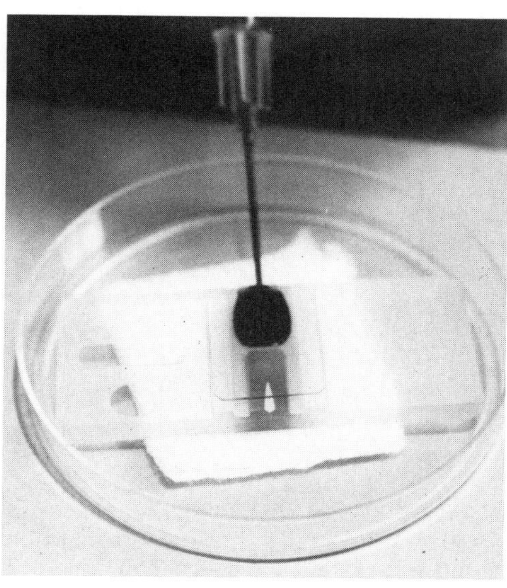

Figure 15–11. Hemocytometer counting chamber that has been overfilled (top chamber).

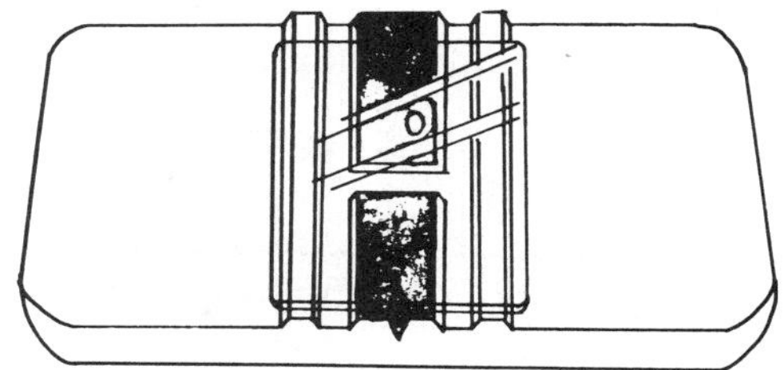

Figure 15–12. Line drawing of an air bubble under the coverglass in the top counting chamber.

Cell Counting

1. A Neubauer hemocytometer is used for the cell counting in this procedure. The etched surface of the counting chamber of this hemocytometer is composed of a square grid that is divided into nine large squares. Each of these nine large squares is subdivided further by a series of fine lines (Fig. 15–13). For the Unopette test 5885, all of the WBCs in all nine of the large squares are counted. It is best to count the cells in a systematic way so that a count can be easily resumed if there is an interruption. We

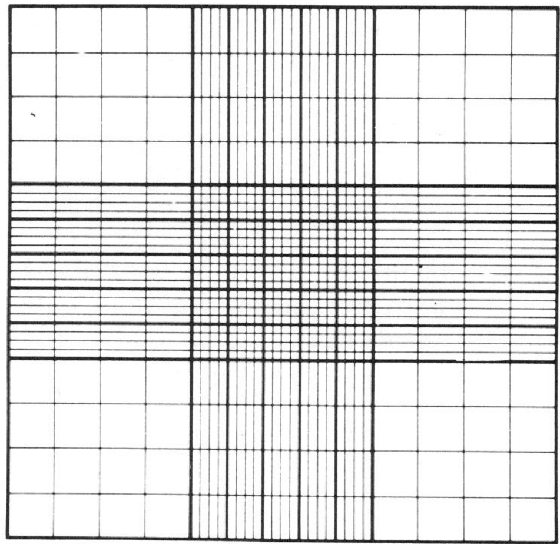

Figure 15–13. Complete Neubauer hemocytometer grid (line drawing).

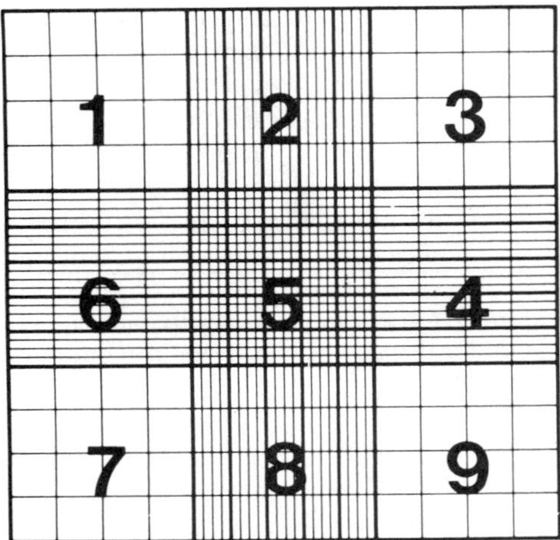

Figure 15–14. Neubauer grid reading sequence (line drawing).

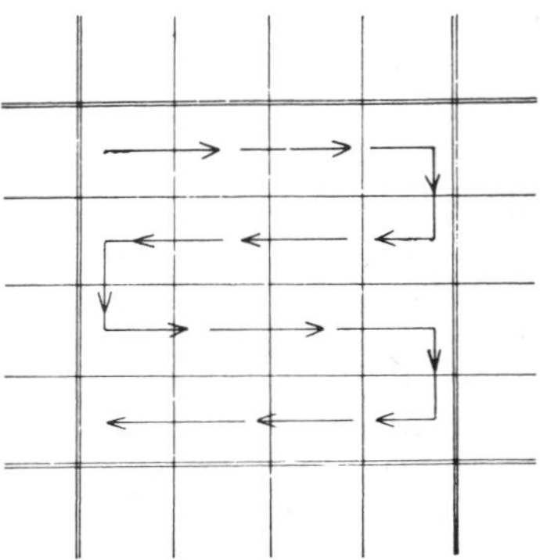

Figure 15–15. Eye movement pattern within one of the nine large squares on the Neubauer grid.

recommend that the count start with the large square in the upper left hand corner of the counting chamber and proceed to each of the remaining eight large squares, as outlined in Figure 15–14. One large square can therefore be counted without moving the stage. The stage will need to be repositioned for each of the nine large squares. The counting process within each of the nine large squares should follow the same general eye movement pattern. The small square in the upper left corner is first counted, and then the eyes move in the direction indicated in Figure 15–15.

2. During the counting process some cells will be seen right on the lines, rather than inside one of the squares. It is important that a cell be counted only once, so you must have a way of ensuring that cells sitting on lines are not counted twice. We recommend that the cells be counted as part of a square if they sit on the bottom line or on the left hand line that forms the square (Fig. 15–16).

3. Put the low power objective in place on the microscope and rotate the coarse focus knob so that the objective is as far away as possible from the stage. This will prevent hitting the objective with the hemocytometer.

4. Lower the condenser and turn the light to its lowest position. If the light on the

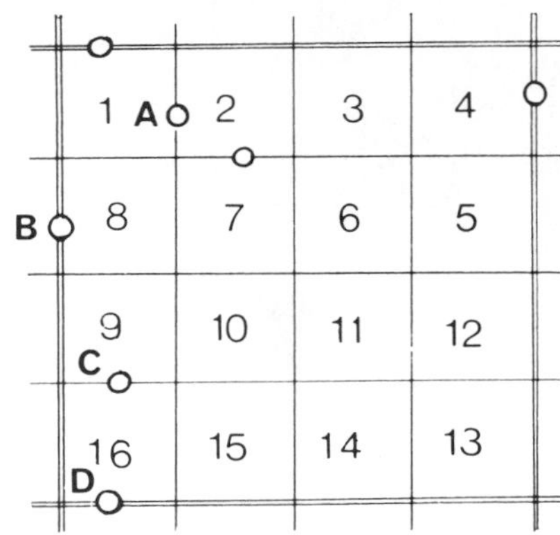

Figure 15–16. Cells are counted as part of a square if they fall on that square's bottom or left edge. Cell A is counted in square 2, cell B is counted in square 8, cell C is counted in square 9, cell D is counted in square 16.

microscope is very high, the cells cannot be easily seen.

5. Place the hemocytometer on the stage of the microscope and adjust the stage so that the light shines through the V-shaped slot of the hemocytometer. Look through the microscope and focus on the surface of the V-shaped area. To find the counting chamber, use the V as an arrow to direct you to the counting grid. The grid will sit just beyond the point of the V-shaped area.
6. Move the hemocytometer so that the upper left square of the Neubauer counting grid is centered in the microscope field. Focus sharply. The WBCs appear as round, slightly refractile cells (Fig. 15–17). If there is any question about an object, flip the microscope to the high dry objective to more closely examine it and determine if it is a WBC (Fig. 15–18). To continue counting, return to the low power objective.
7. Set the cell counter to zero. Count all of the cells in the upper left large square using the counting pattern as described. When the first large square is counted, move the stage to the second largest square and continue the cell counting. Count all nine large squares in this manner. Record the total number of cells after the ninth large square has been counted. Reset the counter to zero.
8. Repeat the counting procedure with the second counting chamber on the hemocytometer. Both chambers should be counted quickly to avoid the evaporation of the solution in the hemocytometer.
9. Compare the two results of the counting procedure and check that they are within the allowable tolerance range (Table 15–5). For example, if the first count was 51 and the second count was 65, this would be an acceptable variation. But if the second count was 66 or more, the variation between the two counts would be unacceptable. If the variation is too great, repeat the test until the two counts are within the allowable range. (It is uncommon to have to repeat the counting because of large differences in the count.)
10. Average the two counts, add 10%, then divide by 10. This number is then reported as the WBC count. (For the example above, the two counts were 51 and 65. The average is 58. Add 10% (i.e., 5.8) and the resulting number is 63.8. Finally, divide by 10 to give a WBC count of $6.38 \times 10^9/L$.)
11. You are then finished unless you want to do a platelet count on the same sample. Rinse the coverglass and the hemocytometer with tap water and place them in the appropriate plastic cup.

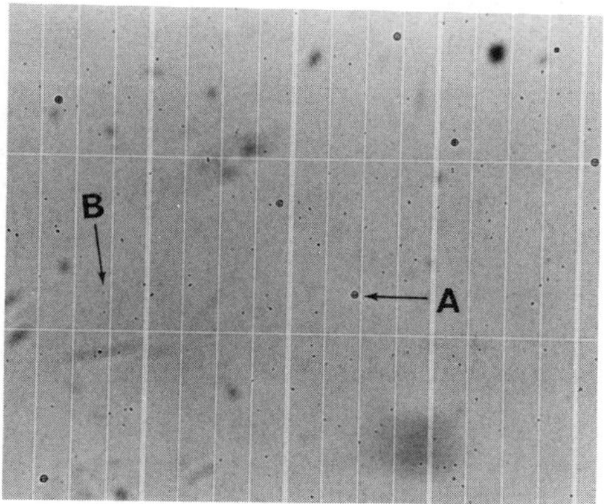

Figure 15–17. A. WBCs on part of a Neubauer grid. They are the small dark circles. **B.** The small flecks are platelets. Viewed under the low-power objective.

TEST INTERPRETATION

For adults the reference range for the WBC count is independent of age and is standardly given as 5.0 to 10.0×10^9 /L. Table 15–6 gives reference values for pediatric patients.

TABLE 15–5. STANDARD DEVIATION TOLERANCE FOR DUPLICATE HEMOCYTOMETER COUNTS

Cells Counted	Allowable Difference	Cells Counted	Allowable Difference
1-2	2	324-342	36
3	3	343-360	37
4-6	4	361-380	38
7-8	5	381-399	39
9-12	6	400-421	40
13-15	7	422-440	41
16-20	8	441-462	42
21-24	9	463-483	43
25-30	10	484-506	44
31-35	11	507-528	45
36-42	12	529-552	46
43-48	13	553-575	47
49-56	14	576-600	48
57-63	15	601-624	49
64-72	16	625-650	50
73-80	17	651-675	51
81-90	18	676-702	52
91-99	19	703-729	53
100-110	20	730-755	54
111-120	21	756-783	55
121-132	22	784-812	56
133-143	23	813-839	57
144-156	24	840-869	58
157-168	25	870-899	59
169-182	26	900-929	60
183-195	27	930-959	61
196-210	28	960-989	62
211-224	29	990-1022	63
225-240	30	1023-1053	64
241-255	31	1054-1088	65
256-272	32	1089-1121	66
273-288	33	1122-1153	67
289-306	34	1154-1189	68
307-323	35	1190-1224	69
		1225-1260	70

TABLE 15-6. PEDIATRIC REFERENCE VALUES FOR LEUKOCYTE COUNT[a]

Age	Average	95% Range
Birth	18.1	9.0–30.0
12 hours	22.8	13.0–38.0
24 hours	18.9	9.4–34.0
1 week	12.2	5.0–21.0
2 weeks	11.4	5.0–20.0
4 weeks	10.8	5.0–19.5
2 months	11.0	5.5–18.0
4 months	11.5	6.0–17.5
6 months	11.9	6.0–17.5
8 months	12.2	6.0–17.5
10 months	12.0	6.0–17.5
12 months	11.4	6.0–17.5
2 years	10.6	6.0–17.0
4 years	9.1	5.5–15.5
6 years	8.5	5.0–14.5
8 years	8.3	4.5–13.5
10 years	8.1	4.5–13.5
12 years	8.0	4.5–13.5
14 years	7.9	4.5–13.0
16 years	7.8	4.5–13.0
18 years	7.7	4.5–12.5
20 years	7.5	4.5–11.5

[a]The WBC count is reported in SI units as cells $\times$ 10^9/L.

QUALITY CONTROL

- The easiest quality assurance procedure is to routinely make two dilutions, filling one side of the chamber with the first and the other side with the second. Count both chambers of the hemocytometer and check that the two counts are within an acceptable range of each other.

  All office staff who are learning the procedure should use this two-sample method. After some experience the laboratory worker will become proficient at the pipetting steps in the procedure. You will know this because the two separate counts will usually be very similar and will always be within the acceptable range. When this occurs you may choose to use only one Unopette; however, you will want to continue to count both sides of the hemocytometer. Use the sample from the single Unopette to fill both sides. Continue to check to see that both counts are within an acceptable range. This process will control for variation in cell counting; however, it does not control for pipetting errors. It is, however, faster and obviously less costly than the two-Unopette procedure.
- A second way to check your results is to split a patient's CBC specimen once a month. Send the specimen to your refer-

ence laboratory for hematology testing. Be sure to record the patient's name so that you can check the results when they come back. Perform duplicate testing on the sample in your laboratory. Results for the WBC count should agree within 10%. Record the results of the reference laboratory testing and your office laboratory testing in the quality control book.

- Subscribe to the hematology section of a proficiency testing program. They will send samples to your laboratory for WBC testing.
- Some hematology standards recommend that a "background count" be done to check for contaminants in the diluent. We have not found this to be necessary when using the Unopette system and doing manual counts.
- Check out all new laboratory personnel on this procedure before they do testing on patient samples.

SOURCES OF ERROR

Falsely Decreased WBC Count

- The patient's finger is excessively squeezed to obtain a sample from a finger stick. This will result in more plasma than blood cells in the sample.
- Taking the sample from the top of an inadequately mixed purple-top tube.
- Inadequate filling of the capillary pipette.
- Underfilling of the counting chamber.
- Bubbles in the counting chamber.

Falsely Elevated WBC Count

- A screaming or agitated child will have an elevated WBC because the child's epinephrine causes white cells that marginate along the vessel walls to enter the circulating blood.
- Taking the sample from the bottom of an inadequately mixed purple-top tube.
- Counting a chamber that has been filled to a point where the solution overflows the counting area.

COMMENTS

- It takes some skill to fill the counting chambers of the hemocytometer. A person not used to performing this procedure should practice it with a Unopette.
- With the old acetic acid diluent method it was possible to tell the difference between a neutrophil and mononuclear cell while doing the WBC count, thus obtaining a quick differential test. The reagents used for the technique described here do not allow this.
- A book describing the use of the Unopette can be obtained free of charge by writing to Becton Dickinson Company, Rutherford, NJ 07070.
- There is a shift from reporting the relative count (i.e., the percentages of each type of WBC) to the absolute count (i.e., the total number of cells). The absolute count is calculated by multiplying the percentage (from the differential) times the WBC. For example, if the differential showed 40% lymphocytes and the WBC count was 8.0, the absolute lympyhocyte count would be $.40 \times 8.0 = 3.2 \times 10^9/L$. Reporting CBC reports in absolute counts is occasionally helpful. For example, many patients on chemotherapy will become neutropenic. The absolute neutrophil count is easiest to characterize by the degree of neutropenia. This reporting also helps to prevent the common misconception that a patient with a WBC of 4,000 and 80% lymphocytes has a lymphocytosis. In fact, this patient has a normal number of lymphocytes but has an absolute neutropenia.

REFERENCES

Albritton EC *Standard Values in Blood.* Philadelphia, Pa: Saunders; 1952.
Laboratory Procedures Using Unopette Brand System, 8th ed. Becton Dickinson Company, Rutherford, NJ 07070, 1977.
Product Information Insert: Unopette Test 5855, Becton Dickinson Company, Rutherford, NJ 07070, 1977.

SECTION 7: Platelet Count

BACKGROUND

The normal adult platelet count ranges from 140 to 400 × 10^9/L. A patient with a platelet count below 100 × 10^9/L is considered to be thrombocytopenic, but it is not until the count drops to 40 × 10^9/L that there is a problem with bleeding during surgery and not until 20 × 10^9/L that spontaneous bleeding is seen. A platelet count below 10 × 10^9/L is often associated with severe, spontaneous hemorrhage. A low platelet count can be caused by either suppressing the bone marrow's production of platelets or increasing their destruction in the blood.

Platelet counts that are elevated above 400 × 10^9/L (i.e., thrombocytosis) are seen following splenectomy, hemorrhage, surgery or with iron deficiency anemia. When the platelet count is elevated above 600 × 10^9/L it is usually due to a myeloproliferative disorder. Levels above 1,000 × 10^9/L are associated with both spontaneous thrombosis and bleeding.

As estimate of the platelet count can be easily obtained in the office by averaging the number of platelets seen on an oil-power field of a routine peripheral smear and then multiplying this number by 20. A platelet estimate is adequate for most routine blood work. If thrombocytosis or thrombocytopenia is suspected and knowing the actual count is clinically important, then a formal platelet count should be done. This is, however, not a common clinical need. Usually the clinician wants to know if there are "enough" platelets, and this can be determined from a peripheral smear.

The platelet count is best performed in the office latoratory because platelets do not store well. A platelet count sent out to a reference laboratory may therefore be falsely depressed.

TEST INDICATIONS

- Evaluation of a patient with a purpural rash, splenomegaly, or a history of abnormal bleeding
- Evaluation of platelet production in a patient with a marrow-suppressvie or marrow-infiltrative disease (e.g., cancer, folate deficiency)
- Evaluation of a patient with a disease that increases the destruction rate of platelets (e.g., lupus, prosthetic heart valve)
- Evaluation of platelet production in a patient taking a myelosuppressive drug (e.g., cytosine arabinoside, ethanol, estrogens, or thiazide diuretics)
- Evaluation of platelet production in a patient taking a drug that can cause immunologic destruction of platelets (e.g., antibiotics, methyldopa, gold salts, and aspirin)

TEST MATERIALS

All of the materials used for the WBC count are used for the platelet count. This includes a Unopette test 5855 , a hemocytometer, special glass coverslips, a petri dish, 2 × 2-inch gauze, hand-held counter, microscope, and a drying towel. Please refer to the WBC count section (Section 4) for a more detailed description of the materials that are needed.

SPECIMEN COLLECTION

Whole blood is used for a platelet count. The blood sample can be from a finger or heel stick if you are careful not to use the first drop of blood, since platelets are often clumped in the first drop. If platelet clumping is seen during the hemocytometer-counting procedure, another specimen should be drawn.

EDTA-anticoagulated blood can be used for up to 5 hours after its collection if it is kept at room temperature or for up to 24 hours if it is refrigerated.

TEST PROCEDURE

Whenever a platelet count is performed, also make a blood smear from the sample, and Wright's stain it to compare the actual platelet count with a platelet estimate.

Unopette Preparation

1. Place the Unopette reservoir on a flat surface. Label it with the patient's name and the time that the specimen is added to the diluent.
2. Take the Unopette shield/pipette assembly in one hand and hold the reservoir with the other hand.
3. Push the pointed tip of the shield into the reservoir neck through the plastic diaphragm that seals the reservoir. Ream out a large hole so that the pipette can be easily placed in the reservoir later in the procedure (Fig. 15–5).
4. Remove the shield/pipette assembly from the neck and the reservoir and then remove the protective shield from the pipette by twisting it off.
5. Put on gloves.
6. a. Finger/heel stick: (See Chapter 12 for a detailed description of this specimen collection technique.)
   (1) Wipe off the first drop of blood from the skin. Do not squeeze the finger.
   (2) Hold the pipette horizontally, and when a new drop of blood forms, touch the pipette tip to the drop. The pipette will automatically fill to the proper level. Do not dip the pipette in and out of the drop of blood, as this can produce air bubbles in the pipette. Such bubbles are unacceptable, and a new pipette would need to be used (Fig. 15–6).

   b. EDTA tube of blood:
   (1) After mixing hold a 2 × 2-inch gauze around the stopper and gently remove the stopper from the tube. Try to avoid splashes.
   (2) Take a wooden applicator stick and stir the blood. Pull the stick almost all the way out to see if there are any clots. If the blood has clotted, discard the specimen and do not run the test.
   (3) Hold the blood tube as close as possible to the horizontal position without spilling any blood.
   (4) Put the pipette tip into the blood and collect the sample.
7. If the pipette does not fill rapidly, tap

the open end of the pipette gently with your finger. It should fill to the hub of the pipette assembly when filled. Gently place your index finger over the hub end of the pipette. Make sure that no blood is forced out of the pipette when you do this. It is critical to keep the blood at the top of the hub. If the blood should go below this level, you will need to add more blood to bring it to the proper height.

8. Wipe the excess blood from the outside of the capillary pipette by quickly passing a piece of gauze from the pipette hub to its tip. Be sure not to touch the gauze to the open tip, as this may result in some of the sample being drawn out.
9. You now have the filled pipette (with index finger covering the hub end) in one hand. Take the reservoir with the reamed-out diaphragm in the other hand. Squeeze the reservoir between your fingers to force out some air (Fig. 15–7). Do not squeeze too hard or you will lose some diluent. If this happens you must begin again with a new reservoir. Put the pipette into the reservoir as far as it will go, with your index finger still covering the overflow chamber. The pipette will fit snugly into the neck of the reservoir. Relax the finger that holds the reservoir and remove the index finger from the top of the pipette at the same time. This will pull the blood from the pipette and into the reservoir.
10. Very gently squeeze the reservoir two or three times to force the diluent into the capillary pipette and rinse out the blood (Fig. 15–8). Do not squeeze too hard or you will lose liquid out of the top of the pipette. Watch the column of fluid inside the pipette to avoid losing any of the diluent. If you do lose some of the diluent, it will be necessary to start over with a fresh sample and a new Unopette because you will have altered the dilution ratio; therefore the count will be inaccurate.
11. Again squeeze the blood-diluent solution high into the pipette (but not overflowing). Put your index finger over the open hub end of the pipette and remove it from the reservoir.
12. Once again, squeeze the reservoir and hold it. Reseat the pipette into the reservoir neck; relax the fingers holding the reservoir and remove your index finger at the same time. The solution will be sucked into the reservoir. This guarantees that you have washed the blood sample out of the pipette.
13. Place your index finger over the top of the pipette's overflow chamber and invert the whole assembly several times to ensure complete mixing.
14. Repeat this sequence with a second reservoir and pipette so that you have two samples.
15. Place the pipette shield on the top of the pipette and set the Unopette aside for at least 5 minutes to allow complete lysing of the RBCs. If you do not wait long enough, the RBCs will not be lysed and it will be impossible to perform the count because the numerous RBCs will obscure the platelets. Wait no more than 3 hours before counting the sample. Even the WBCs will lyse after a long period of time.

Hemocytometer Chamber Preparation

1. Use tap water to lightly dampen a piece of gauze or filter paper in the petri dish that is to hold the hemocytometer. This provides a moist atmosphere so that the platelet diluent will not dry out in the counting chamber.

2. Put on gloves, take the hemocytometer and its special coverslip out of the Staphene solution, and rinse them with tap water. Dry each with a soft cloth. Be sure that there is no lint on the coverslip or the counting chambers of the hemocytometer. Do not touch these areas with your fingers, since the oil from your skin will make it difficult to properly fill the counting chamber.
3. Put the hemocytometer on top of the dampened gauze in the petri dish. Place the coverslip over the counting chamber areas of the hemocytometer.
4. Mix the solution in the Unopette by inverting the assembly several times. This will resuspend the platelets in the solution. Remove the shield.
5. Again squeeze the blood-diluent solution high into the pipette (but not overflowing). Put your index finger over the open hub end of the pipette and remove it from the reservoir.
6. Once again squeeze the reservoir and hold it. Reseat the pipette in the reservoir neck; relax the fingers holding the reservoir and remove your index finger at the same time. This guarantees that you have washed the blood sample out of the pipette.
7. Hold your index finger over the hub end of the pipette and invert the Unopette a few times. You now have a well-mixed specimen.
8. Convert the assembly into a dropper bottle by removing the capillary pipette, turning it upside down, and then reinserting it back into the neck of the reservoir. Be sure that the pipette is securely positioned into the reservoir neck (Fig. 15–9).
9. Again invert the assembly several times; then, holding it upside down, gently squeeze the reservoir. Let the first few drops of solution drip onto a piece of gauze.
10. Put the tip of the pipette into one of the V-shaped slots on the hemocytometer and continue to gently squeeze the reservoir. The solution should flow into the counting chamber and completely fill the area between the coverslip and the counting chamber (Fig. 15–10). This will require only one or two drops of the solution. If the solution overflows the area of the counting chamber (Fig. 15–11) or if an air bubble is trapped beneath the coverslip (Fig. 15–12), you must begin again by cleaning the hemocytometer and refilling it with the cell solution. A count is unacceptable if performed when *either* one of the two counting chambers has overflowed or has an air bubble. If you have repeated problems with filling the hemocytometer, wipe it and the coverslip with alcohol to remove any organic film that sometimes accumulates.
11. Repeat this sequence for the other counting chamber of the hemocytometer, using the second sample. Both chambers must be filled to perform a platelet count.
12. After the hemocytometer has been correctly filled, put the top of the petri dish and allow the unit to sit for at least 5 minutes. This is to allow the platelets to settle so that they are not moving during the counting process. This keeps them all in one focal plane under the microscope.

PLATELET COUNTING

1. Platelets are counted only in the large central square of the Neubauer grid on the hemocytometer (i.e., the square marked "5" in Fig. 15–14). This large square is subdivided into 25 smaller squares, and each of these is subdivided into 16 even smaller squares. The same counting pattern is used for the platelet count, as is described in the WBC count section (Fig. 15–15); i.e., starting in the

upper left small square and proceeding to the right).

2. Move the microscope stage until the large central square is centered in the field. Focus sharply using low power, then change to the high-dry objective. Reposition the hemocytometer so that the upper left-hand small square is centered in the field. Refocus the microscope and lower the light. The small square, with its 16 smaller subdivisions, should just fit into the entire field with a high-dry objective.
3. Platelets are not as easy to recognize as are WBCs in the WBC count. They appear as small, dark objects that vary in size and shape (Fig. 15–18). It is necessary to focus up and down during the entire counting procedure to see all of the platelets at different focal levels in the counting chamber. This can be minimized by allowing the chamber to sit in the petri dish for 5 minutes. Small pieces of debris may also be seen, but they are distinguishable from platelets because debris is usually refractile.

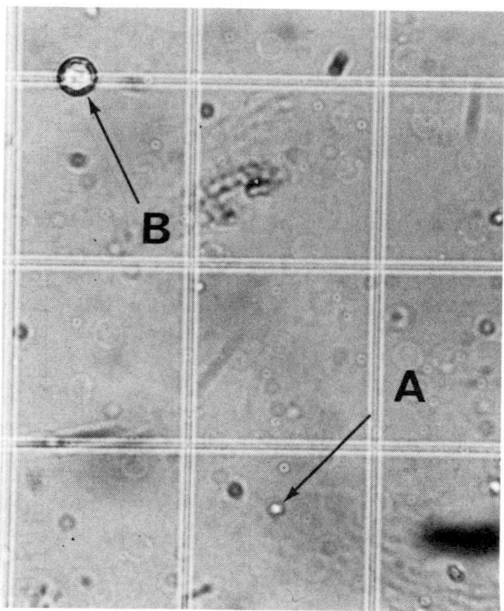

Figure 15–18. View of part of a Neubauer grid. **A** Platelet. **B** WBC. Viewed under the high-dry objective.

4. Set the hand-held counter to zero, and count the 16 squares in the upper left-hand square. Then move to the next square. Continue counting until all of the 25 squares have been counted. Record this number.
5. Reset the counter to zero, and count the platelets in the 25 small squares that make up the central large square of the counting chamber on the other side of the hemocytometer.
6. Check that the two counts are within an acceptable range by comparing the values to Table 15–5. If the difference is not of an acceptable range, repeat the count by refilling the hemocytometer with more solution from the Unopette reservoir. Mix the solution well before doing this.
7. Average the two counts (e.g., if the first count was 220 platelets and the second count was 248 platelets, the average is 234, and the reported platelet count is $234 \times 10^9/L$.
8. Report this number as the platelet count.

TEST INTERPRETATION

1. Reference Values

   First week of life: $84 \times 10^9/L$ to $478 \times 10^9/L$

   After first week of life to adult: $140 \times 10^9/L$ to $400 \times 10^9/L$
2. Clinically Important Values,

   $<10 \times 10^9/L$

   Severe thrombocytopenia—associated with spontaneous bleeding

   $<40 \times 10^9/L$

   Moderate thrombocytopenia—associated with bleeding during minor surgery

$<100 \times 10^9/L$
Mild thrombocytopenia—associated with bleeding during major surgery

$140 \times 10^9/L$ to $400 \times 10^9/L$
Normal range

$>400 \times 10^9/L$
Thrombocytosis

$>600 \times 10^9/L$
Moderate thrombocytosis—often due to a myeloproliferative disorder

$>1{,}000 \times 10^9/L$
Severe thrombocytosis—associated with both spontaneous thrombosis and bleeding

QUALITY ASSURANCE

- The easiest quality assurance procedure is to collect duplicate specimens and to fill one chamber with one specimen and the other chamber with the other specimen. Then count both chambers of the hemocytometer and check that the two counts are within an acceptable range of each other. (Table 15–5). All office staff who are learning the procedure should use this two-sample method. After some experience the laboratory worker will become proficient at the pipetting step in the procedure. You will know this because the two separate counts will usually be very similar and will always be within the acceptable range. When this occurs you may choose to use only one Unopette; however, you will want to continue to count boths sides of the hemocytometer. Use the sample from the single Unopette to fill both sides. Continue to check to see that both counts are within an acceptable range. This process will control for variation in cell counting; however, it does not control for pipetting errors. It is, however, faster and obviously less costly than the two-Unopette procedure.
- Do a peripheral smear for every patient who has a platelet count done. Stain the smear with Wright's stain (see sections 8 and 9, this chapter) and examine the slide for a platelet estimate (section 10, this chapter). This will give you a check on any gross differences between the smear and the count. It will also help you to perform a better platelet estimate on the routine peripheral smear.
- Another external check is to once a month split a CBC specimen sent to your reference laboratory for hematology testing. Note the patient's name so that you can check the results when they come back. Perform duplicate testing on the sample in your laboratory. Results for the platelet count should agree within 10%. Record the results of the reference laboratory testing and your office testing in the quality control book. Use this specimen for other hematology controls.
- Some hematology standards recommend doing a "diluent background count" (i.e., doing a count without any blood specimen). This checks for diluent contaminants. We have not found this to be necessary when using the prepackaged Unopette diluent system and manual counting.

SOURCES OF ERROR

The same technical problems encountered with the WBC count will result in inaccurate platelet counts. The Unopette manufacturer claims a precision of $\pm$ 15 $\times$ 10^9/L platelets for the described method.

COMMENTS

- This is a difficult procedure to explain in written form. We suggest that if you have not previously done platelet counts, do the procedure several times as described. Then arrange to spend some time with the hematology technician in your local hospital laboratory to master the procedure.
- It is important to allow the "settling" time so that the platelets have settled to the same focal plane.

REFERENCES

Laboratory Procedures Using Unopette Brand Systems. 8th ed. Becton Dickinson Company, 1977.

A book describing the use of the Unopette can be obtained free of charge by writing to Becton Dickinson Company, Rutherford, NJ 07070.

Product Inserts: Unopette Test 5855, Becton Dickinson Company, Rutherford, NJ 07070, 1977.

SECTION 8: Making Blood Smear Slides

The WBC differential, platelet estimate, and examination of RBC morphology all require that a blood smear slide be prepared and properly stained. Making an adequate blood-smear slide is a skill that can only be learned and mastered with time and practice.

SPECIMEN COLLECTION

Smears can be prepared from blood that has been collected in a variety of ways. Frosted glass slides that have been wiped free of dust should be used. Two slides should be prepared for any test that requires a smear to be made. The slides should be labeled and dated with the frosted side up.

Finger/Heel Stick

(See Chapter12 for a detailed description of this technique.)

1. Wipe the first drop of blood away from the lancet puncture site. This will remove platelets that immediately aggregate and that would otherwise prevent a good blood flow.
2. Touch the labeled side of the frosted slide to the drop of blood so that the blood sample is placed near to the frosted edge of the slide. Try to avoid touching the patient's finger or heel with the slide, as this will smear the blood drop.
3. Prepare the smeared slide as described below. Then collect a second specimen on a second slide. Do not wait until both specimens have been collected to smear the first slide because the blood will quickly clot. Waiting will also result in platelet clumping in the specimen.

Venous Sample

If a venous specimen is obtained for other laboratory testing, the blood smear can be prepared from the drop of blood that remains on the tip of the blood-drawing needle at the end of the specimen collection.

1. Leave the tube pushed forward in the plastic holder when you withdraw the needle from the patient's arm.
2. If the tube has been completely filled, there will be no more vacuum pressure, and a drop of blood will form at the tip of the needle. If the drop does not form, sharply shake it until a drop forms at the needle tip. Touch the drop at the tip of the needle to the glass slide.
3. Place a drop of blood on each of two slides near the frosted tip of the slides. Quickly streak the slides as described below.
4. If still no blood is obtained, a sample can be obtained from a purple-top tube.

EDTA (Purple-Top) Vacutainer Tube

Blood smear slides can be prepared for up to 2 hours after blood has been collected in an EDTA tube. It may, however, be difficult to assess RBC morphology from a smear made with EDTA-anticoagulated blood, but a

WBC differential and a platelet estimate can be performed.

1. Rock the EDTA blood tube to fully mix the blood sample.
2. Put on gloves if you do not already have them on.
3. Hold a 2 × 2-inch gauze around the rubber stopper and gently remove the stopper from the tube. Try to avoid splashing blood.
4. Taken a wooden applicator and stir the blood. Pull the stick almost all the way out to see if there are clots. If the blood has clotted, discard the specimen. It is not acceptable.
5. Dip the stick in again. Remove it and touch it near to the frosted end of the slide, leaving a small drop. Alternately, you can use a plastic transfer pipette.
6. Repeat for a second slide.
7. Prepare the slides immediately as described below.

PROCEDURE

1. Place the slide with the drop of blood on a table top with the frosted end to the right.
2. Hold the nonfrosted end of the slide with the thumb and third finger on the left hand (FIg. 15–19).

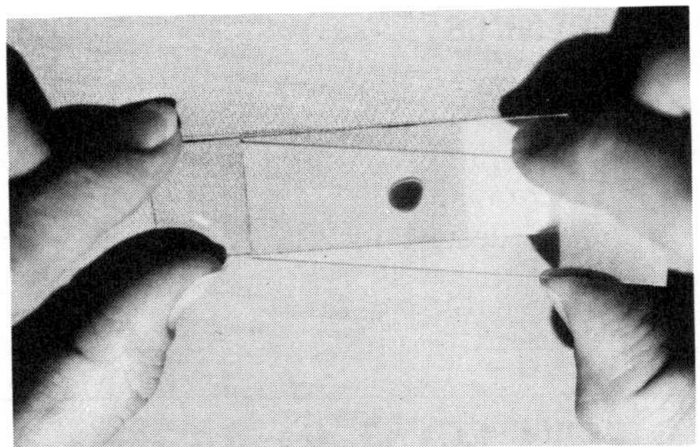

Figure 15–19. Making a peripheral smear: placing second slide at 30-degree angle to first slide.

3. Pick up a second slide with the right hand.
4. Place the short edge of this second slide on the midportion of the first slide so that the two slides form about a 30-degree angle. *The smaller the drop of blood the smaller the angle between the two slides should be. The larger the drop the larger the angle* (Fig. 15–19).
5. Pull the right-hand slide along the flat surface of the left-hand slide toward the drop of blood.
6. When the slide backs into the drop, the blood will spread along the edge of the right-hand slide (Fig. 15–20).
7. Push the right-hand slide quickly and smoothly away from the frosted end of the specimen slide, producing an even

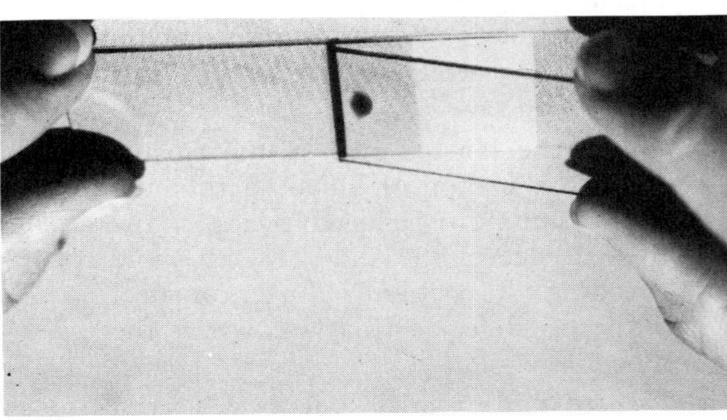

Figure 15–20. Making a peripheral smear: blood spreading along edge of second slide.

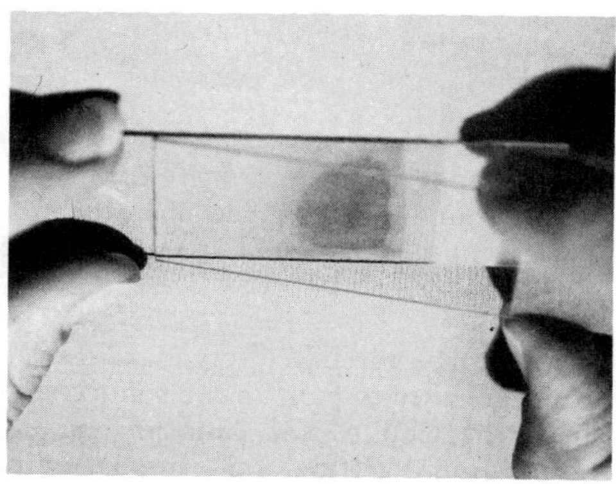

Figure 15–21. Making a peripheral smear: even spreading of blood sample.

spreading of the blood sample (Fig. 15–21).

8. Lift the right-hand slide off the specimen slide about two thirds of the way down the specimen slide.
9. Smear the second specimen slide in the same manner.
10. Pick up each slide by the nonfrosted end and wave it gently in the air to quickly dry. If the slides are allowed to slowly air dry on the bench top, there will be considerable changes in the RBC morphology. Rapid drying is especially important in humid weather.
11. To check that you have an acceptable slide, hold the dried, unstained slide up to the light. The reflected light from the slide should produce a rainbowlike effect (i.e., the "feathered edge") at the nonfrosted end of the slide.
12. Label each slide with the patient's name and the date.
13. The smear is now ready for Wright's staining. This should be done as soon as possible and within 2 hours.

COMMENTS

- Preparing blood smears is a learned skill that requires practice, patience, and a box of slides to practice with. We suggest that beginners use a tube of EDTA-anticoagulated blood to practice the technique. A plastic transfer pipette, or a glass Pasteur pipette, or a wooden applicator stick can be used to put the drop of blood on the slide. It is a good idea to vary the size of the drops of blood so that the proper angle for each size of drop can be learned.
- Delay in smearing the blood after the drop has been placed on the slide will produce an abnormal distribution of cells, with the larger cells at the thinner edges of the smear. Also, platelets will aggregate when they touch the glass. Artifactual red cell rouleaux formation and blood clotting can also be seen.
- Poor quality slides may be the result of dirty slides, too large or too small a drop of blood, pushing down too hard on the spreader slide, or lifting the spreader slide too early from the specimen slide.
- If you want to concentrate the larger cells in one area of the slide to allow for easier examination, you can lift the spreader slide off before the blood drop is completely spread.

SECTION 9: Wright's Stain

BACKGROUND

Wright's stain is the standard hematology stain. It is to hematology what Gram's stain is to microbiology. There are numerous other special hematologic stains, but aside from the reticulocyte stain, they are not needed in the office laboratory.

Historically, Wright's staining involved the mixing of dry reagents, the use of special buffers, and the careful monitoring of the reagent pH with a pH meter. This process has been greatly simplified by the availability of fully prepared, modified Wright's stains. The procedure described below is based on the use of the Camco Quik Stain, which is one such modified Wright's stain. It is easy to use, gives fine staining results, and is stable in office conditions.

TEST MATERIALS

Camco Quick Stain II

This is a modified Wright's stain for easy office use. It contains methyl alcohol, Wright's stain, alcohol-soluble buffers, and a stabilizing agent. It is manufacturered by Cambridge Chemical Products Inc, (6880 NW 17th Ave, Fort Lauderdale, FL 33309) and is distributed by Scientific Products (product number B4129).

Coplin Jars

These are special staining jars with slots to hold the slides upright during the staining procedure. The jars have screw-on lids so that the reagents do not evaporate over time (Fig. 15–22). It is most convenient to have three Coplin jars for the staining procedure.

Distilled Water

Gallon jugs of distilled water are available from most pharmacies or super markets. The water should be kept near the sink.

Filter Paper

Small, round, regular-grade laboratory filter paper is needed. It should fit a small funnel.

Forceps

These can be purchased very inexpensively. They are used to pick up the stained slides from the Coplin jars. This prevents the stain from getting on the laboratory worker's fingers. Alternately, gloves can be worn.

Funnel

A small funnel is used to filter the Wright's stain to remove unwanted precipitants.

Labels

Adhesive labels are used to label the Coplin jars with the name of the solutions that they contain.

Stain Remover

Dye-Sol cleansing cream (Med-Tek Corporation, Troy, MI 48084) is used to remove stains from the hands. Erad-O-Sol (Cam-

Figure 15–22. Coplin jar with a screw-top lid.

bridge Chemical Products, Inc, 6880 NW 17th Ave, Fort Lauderdale, FL 33309) can be used to remove stains from the counter, sink, and glassware. Both of these products are available from Scientific Products (C 6345 and C 6347).

Test Tube Brush

This is an inexpensive brush that is used to scrub out the Coplin jars.

Wash Bottle

A plastic wash bottle filled with distilled water is used to rinse the slides after they have been stained. Running tap water can also be used.

SPECIMEN COLLECTION

See Section 8 for the details of making blood smear slides. The smears should be stained within 2 hours of being made.

PREPARATION OF THE STAIN

1. Thoroughly shake the large bottle of Camco Quick II stain to mix its contents.
2. Take a clean, dry Coplin jar and put a clean funnel lined with filter paper into the jar.
3. Pour a small amount of the stain from the storage bottle into the funnel. Do not let the stain overflow the filter paper, as this defeats the purpose of the filtering. Keep pouring the stain into the funnel until the Coplin jar is filled to the point where the slots inside the jar are fully covered. Put the lid on the Coplin jar. Label the jar "Wright's stain," and write the date of the filtering.
4. Dispose of the filter, using forceps to avoid staining your fingers.
5. Wash the funnel with Erad-O-Sol solution to remove any stain.

STAINING PROCEDURE

1. Place a paper towel on the laboratory bench to protect it from spilled stain.
2. Put the Coplin jar with Wright's stain on the towel. Place a second, clean, empty Coplin jar next to it. Fill the second jar with distilled water.
3. Remove the lid from the Coplin jar containing the Wright's stain and place the microscope slides, with the frosted end up, into the stain. Do not put more than one slide between each set of slots in the jar.
4. Leave the slides in the stain for about 10 seconds. The exact timing is not critical, but the slides should not be left in for more than a minute.
5. Using forceps, remove each slide from the Coplin jar and place it into the Coplin jar with distilled water. Again, be careful not to put more than one slide between each pair of slots in the Coplin jar. Leave the slides in the distilled water for at least 20 seconds. If darker staining is preferred, leave them in up to 1 minute. It is during the distilled water step that the actual cellular staining takes place.
6. Recap the Coplin jar with the Wright's stain. Do not leave it uncapped for a long period of time because the methyl alcohol will evaporate.
7. At the end of the 20 to 60 seconds in the distilled water, use forceps to remove the slides from the jar. Rinse them by squirting them with distilled water from a wash bottle, or rinse them in gently flowing tap water. Make sure that the granular solution is completely washed from the slide. You will notice the granular solution washing away if you watch the slides carefully while rinsing. Do not overwash.
8. Prop the slide against a rack on top of a paper towel. The frosted end of the slide should be down. Let the slide air dry. Never blot a Wright's-stained

smear as you would a Gram's-stained smear because the specimen will come off the slide.

9. Pour the dirty distilled water from the Coplin jar down the sink. Rinse the jar well and place it upside down on a shelf to dry.
10. The slides will be dry and ready to examine in about 3 to 5 minutes.

STAINING RESULTS

When examined under the microscope, the cells in a properly stained blood smear will show the following staining characteristics:

- Red blood cell: pink to orange or tan (Color Plate I, Figs. 1 to 5)
- Lymphocyte: dark purple nucleus and blue cytoplasm (Color Plate I, Figs. 1 and 4)
- Neutrophil: dark purple nucleus and pink cytoplasm (Color Plate I, Figs. 1 and 5)
- Monocyte: lighter purple nucleus, gray-blue cytoplasm, and very fine pink granules (Color Plate I, Fig. 4)
- Eosinophil: bright red-orange granules (Color Plate I, Fig. 2)
- Basophil: dark blue or purple granules (Color Plate I, Fig. 3)
- Platelet: light to dark purple (Color Plate I, Fig. 2)

QUALITY CONTROL

- The Wright's stain should not be used after its expiration date. The stain's buffers will deteriorate, and this will result in poor cellular staining.
- The stain should be filtered about once every 2 weeks using the third Coplin jar. The stain is filtered into this jar, and the date of the filtering is noted on the label of the jar.
- When the level of stain in the Coplin jar gets so slow that the stain no longer fully covers the smeared portion of the slide, discard the stain. Clean the jar with Erad-O-Sol, and prepare new stain as described in the procedure above. Do not get into the habit of adding a small quantity of new stain to older stain in the Coplin jar.
- If when performing a differential you notice precipitated stain on the slide, wash your slides for a longer time. This will usually solve the problem.

COMMENTS

- Wright's stain is occasionally used to stain specimens other than blood. If this is done the specimen must first be fixed with methanol and then dried. It can then be stained as described above. If the specimen is thick, the staining and rinsing times must be increased. We have found that the Hansel stain (see Chapter 21) is superior to the Wright's stain for performing Tzanck smears and for examining for eosinophils. When examining for fecal leukocytes, the methylene blue stain is superior to Wright's stain (see Chapter 21).
- Inadequate rinsing will leave precipitate on the slide.
- Insufficient time in the stain or the distilled water produces very lightly stained white cells.
- Prolonged washing or rinsing of the slide will cause the staining characteristics to fade.
- It may be necessary to restain a slide. This can be done by flooding the slide with methanol to remove the original stain. The slide is then air dried and restained. Obviously it would be better to start with a fresh slide from the same specimen, but this is not always possible.

SECTION 10: Modified, Simple WBC Differential Count and RBC Morphology

BACKGROUND

The examination of a Wright's-stained peripheral smear is an informative test but one that requires a great deal of experience to master. The test should include:

- WBC differential: 100 WBCs are counted, and the percentage of each type of cell is recorded.
- Red blood cell morphology: cell size, shape, and color.
- Notation of RBC and WBC inclusions.
- Platelet estimate.

The most difficult part of the examination is the recognition of the various types of WBCs. In particular, extensive training is required to differentiate lymphocytes, monocytes, atypical lymphocytes, and immature WBC forms. It is also difficult to learn to recognize the various immature RBC forms. Hematology training is a regular part of a medical technologist's (MT or MLT) education; however, it is a skill that can be lost if not practiced. Most clinicians are not given adequate training to be able to perform a proper differential count.

If there is no one on the office staff who has had hematology training, a simplified differential count can be learned. In this procedure all WBCs with a single, nonlobulated nucleus are classified as "mononuclear" cells. The majority of these mononuclear cells are usually lymphocytes, but monocytes, atypical lymphocytes, and other WBCs are included in this group. The simplified differential count also includes the separate counting of segmented neutrophils, bands, eosinophils, and basophils. If a simplified differential count is performed in the office laboratory, a peripheral smear should be sent to the reference laboratory for a full differential count. This will ensure that abnormal forms do not escape notice.

When clinicians order a differential in the office setting, they usually want to know if the patient has more polys than lymphs or if there is a "shift to the left" (i.e., increased bands). The modified differential count provides this basic information.

After some experience with the simplified differential count, the laboratory worker will often recognize abnormal cells but not know what type of cells they are. These may represent anything from an atypical lymphocyte in a simple viral disease to a blast in acute leukemia. Whenever the laboratory worker recognizes such abnormal cells, he or she should bring them to the clinician's attention.

For anyone who has not had previous hematology training, an atlas of blood cells is very useful. Two inexpensive atlases are available: *The Morphology of Human Blood Cells* (Abbott Laboratories, North Chicago, IL 60064), and *Manual of Hematology* (Scope Publications, Educational Services, Unit 9435, The Upjohn Company, 700 Portage Rd, Kalamazoo, MI 49001).

TEST MATERIALS

Alcohol Wipes

These are used to wipe the stain from the underside of Wright's-stained slides.

Coplin Jar

A glass Coplin jar with a screw-top lid should be filled with xylene. Slides are

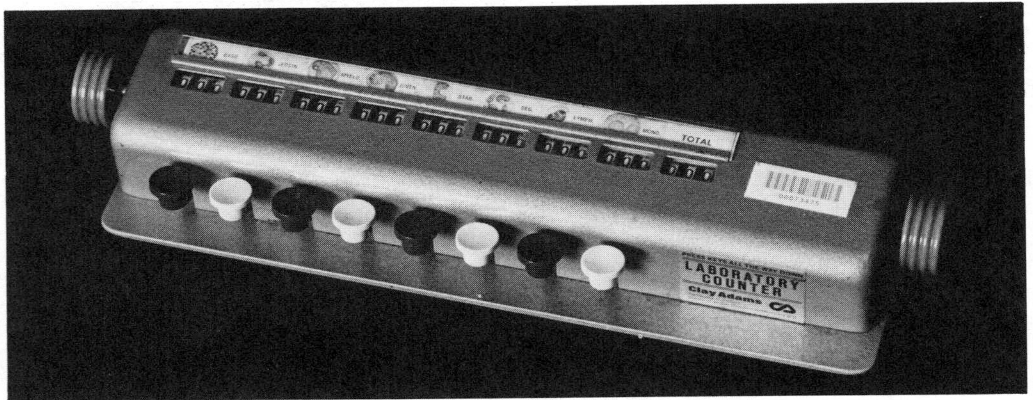

Figure 15–23. Blood cell differential counter.

placed in this jar after the examination to remove the immersion oil (Fig. 15–22).

Differential Cell Counter

These are small counters that record the number of the various WBCs that are seen. When 100 cells are reached, a bell rings. Counters are available with either five or eight keys. The eight-key version is best because it allows the technician to perform a full differential count (Fig. 15–23).

Hand Towel

A small, soft towel should be kept on a hook near the microscope to gently wipe the microscope slides after they have been immersed in xylene.

Immersion Oil

Most microscope manufacturers recommend using standard oil, such as Cargille, type A. Mineral oil is not satisfactory because it has a low viscosity and will therefore seep into the microscope oil objective. Be sure not to buy immersion oil that contains polychlorinated biphenyls (PCB) or other health hazards. The oil is purchased in 4- or 16-ounce bottles. The oil can be transferred to small dropper bottles and should be kept next to the microscope. Small bottles and a plastic transfer pipette can also be used.

Lens Cleaner

A solvent is needed to regularly clean the oil immersion lens on the microscope. Xylene, alcohol, and Kodak lens cleaner can all be used.

Lens Paper

It is important to use lens paper to clean the immersion oil off the 100x objective lens. Gauze or other types of paper can scratch the delicate microscope lenses. The lens paper is purchased in booklets. It can be taken out of these booklets, cut into 2-inch squares, and kept in an empty slide box next to the microscope.

Xylene

Xylene is used to clean microscope slides of immersion oil. Xylene can be purchased in 1-gallon safety cans and should be stored away from other volatile chemicals. It is a respiratory irritant and therefore should be used with good ventilation.

Microscope with Oil Immersion Objective

(See Chapter 10).

SPECIMEN COLLECTION AND PREPARATION

See Section 8 ("Making Blood Smear Slides") and Section 9 ("Wright's Stain").

TYPES OF WBCS COUNTED FOR THE MODIFIED WBC DIFFERENTIAL

Segmented Neutrophil

This cell is referred to as a "seg" or "poly" or "PMN." It has a purple, segmented nucleus and pink cytoplasm with very fine pink-lavender granules. Segmented means that the nucleus has several lobes that are connected by very thin threads. Sometimes the lobes are folded back on themselves. It may then be impossible to clearly see the connecting thread (Color Plate I, Figs. 1 and 5).

Hypersegmented Neutrophils

Most neutrophils have from two to four segments to their nuclei. The number of segments averages three per cell. When you find a single six-lobed neutrophil or three five-lobed neutrophils (Fig. 15–24), you should report "hypersegmentation" on the differential report. Hypersegmentation is the earliest peripheral smear sign of a megaloblastic anemia. This finding appears almost 3 months before ovalomacrocytes are seen in this anemia. It is also possible to have hereditary hypersegmentation.

Band Neutrophil

A band is an early form of neutrophil. It is important to recognize bands because a large number of them are referred to as a "shift to the left." The nucleus is purple and the cytoplasm is pink, but the segmentation of the nuclear lobes is not complete. Instead, the lobes are connected by broad bands instead of narrow threads (hence the name). The band must be wide enough to see both of its edges and nuclear material in between (Color Plate I, Fig. 1). Often this gives the nucleus a horseshoe appearance. Sometimes a monocyte will have a band-shaped nucleus. The difference is that the monocyte has a blue cytoplasm while the band has pink cytoplasm. If you are unsure whether to call a cell a band or a seg, convention is to count it as a segmented neutrophil.

Eosinophils

This cell has a segmented or banded nucleus, but the granules in the cytoplasm are uniformly large and red-orange in color. The granules usually obscure the nucleus

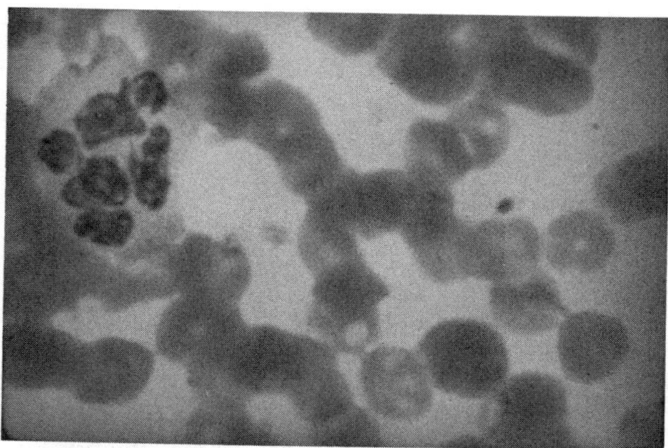

Figure 15–24. Hypersegmented neutrophil. Viewed under oil immersion objective.

and are dispersed throughout the cell (Color Plate I, Fig. 2).

Basophil

These are segmented or band cells that have large, usually prominent dark-purple or purple-red granules. The granules are dispersed throughout the cell and vary in size. No effort is made to differentiate the segmented from the band forms of this cell (Color Plate I, Fig. 3).

Mononuclear Cells

This group of cells is not a distinct cell type. It is a combination of two different types of cells, *monocytes* and *lymphocytes* (Color Plate I, Fig. 4). For the Modified Blood Cell Differential, lymphocytes and monocytes are not differentiated. In most smears lymphocytes will be the predominant cell type. Mononuclear cells do not have segmented nuclei and have few if any granules in the cytoplasm. The rare azurophilic purple granules in lymphs (Fig. 15–25) are quite distinct from the many granules that are seen in neutrophils, basophils, or eosinophils. The cytoplasm of these cells range from very pale blue to dove gray.

Normal lymphocytes vary in size from that of a RBC to many times larger. The nucleus is purple and has a "smudged" appearance. (It is helpful to imagine a charcoal drawing that has been smeared.) The nucleus may be round, unevenly shaped, or kidney bean-shaped. The nucleus has no lobulation. The cytoplasm can be scant or abundant and ranges in color from very pale (almost the color of the background field) to bright blue (Color Plate I, Figs. 1 and 4).

Monocytes are large cells. The nucleus is round, indented, or deeply lobulated. The cytoplasm has a ground glass appearance due to many fine, dustlike granules. The cytoplasm is bluish gray and may be vacuolated (Color Plate I, Fig. 4).

TOXIC ALTERATIONS OF NEUTROPHILS

Neutrophils may show various alterations. These can be difficult to identify and are not a necessary part of a simple differential.

Toxic Granulation

In patients with severe infections, inflammatory diseases, or cancers, the neutrophils may have dark granules in the cytoplasm (Color Plate I, Fig. 5). It is important to differentiate these toxic granulations from precipitated stain. Check the area around the cell for precipitated stain before identifying toxic granulation.

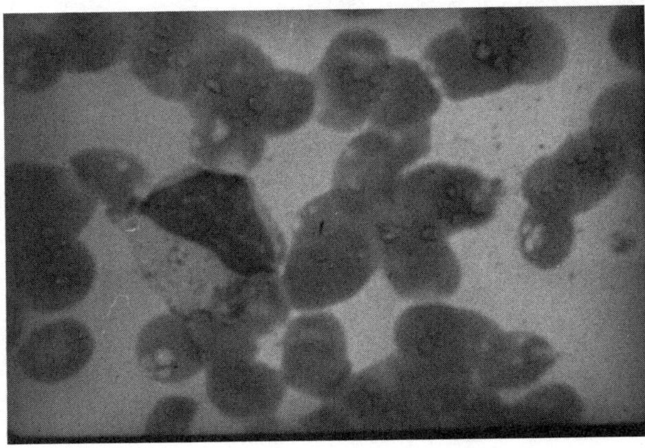

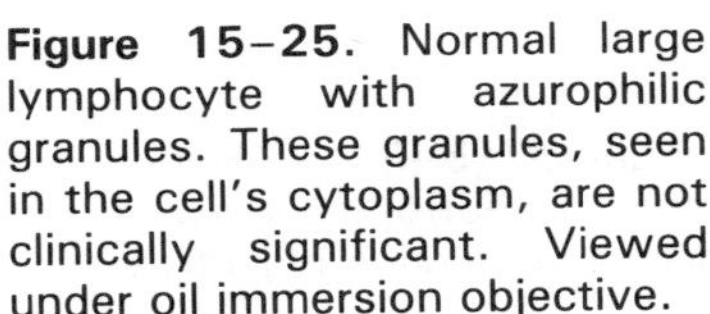

Figure 15–25. Normal large lymphocyte with azurophilic granules. These granules, seen in the cell's cytoplasm, are not clinically significant. Viewed under oil immersion objective.

Döhle Bodies

These are very small, round or oval, light-blue inclusions found in the neutrophil's cytoplasm. They represent residual RNA. They are seen in severe infections, burns, scarlet fever, and thrombocytopenia purpura (Color Plate I, Fig. 5).

Vacuoles

Vacuoles (small holes) represent areas of phagocytosis. They are seen in severe infections and inflammatory diseases. They are usually seen in combination with toxic granulation and Döhle's bodies.

RBC MORPHOLOGY

Cell Size

It may be difficult to get a sense of the RBC size. In the hospital laboratory technicians are able to correlate the RBC size with the mean corpuscular volume from the cell counter. This is not possible in the office laboratory that uses manual hematology methods. There are a wide range of RBC sizes. Normocytes are normal-sized cells and range from 7 to 8 μ in diameter. Their mean corpuscular volume (MCV) ranges between 80 and 97 μ (Figs. 15–26, 15–27A, 15–33A, 15–35A, 15–36B). These cells should dominate the field in a normal smear. Microcytes are smaller cells, with a diameter of less than 6 μ. Their MCV is less than 80 μ (Fig. 15–27B and D). These cells make up from 0% to 3% of normal smears. Macrocytes are large cells with a diameter of greater than 9 μ and an MCV greater than 97 μ (Fig. 15–27C). Round macrocytes that are between 98 and 106 μ MCV are usually either reticulocytes or can be seen in liver disease. Oval macrocytes can be as large as 150 μ MCV and are seen in megaloblastic anemias.

Variations in Cell Size

Anisocytosis (Fig. 15–27) is seen in many types of anemias. It is always important to examine the peripheral smear because a normal MCV may be due to one population of large cells and a second population of small cells. This pattern of anisocytosis is seen in the early treatment of iron deficiency anemia, when the microcytic cells are mixed with newly released large reticulocytes.

Color/Hemoglobin Concentration

The RBCs color is due to the cell's hemoglobin concentration. This concentration is measured in the RBC indices as the mean corpuscular hemoglobin concentration (MCHC). The MCHC ranges from 32% to 36% hemoglobin in normal cells. Normo-

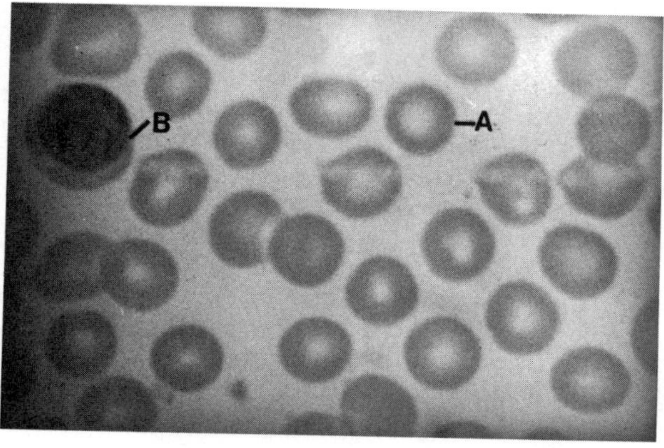

Figure 15–26. **A.** Normal red blood cells and **B.** a small lymphocyte. Viewed under oil immersion objective.

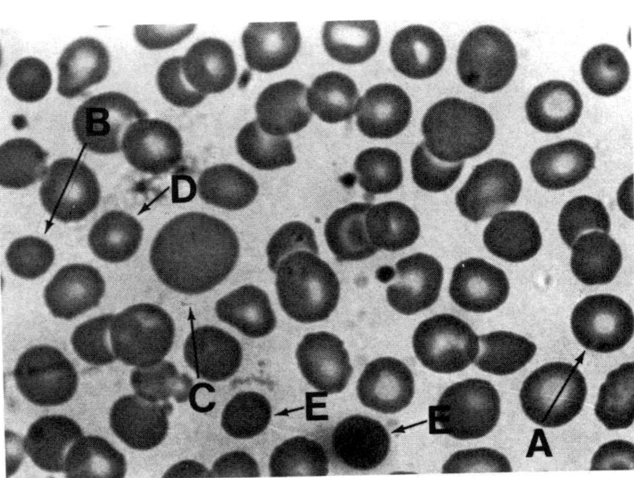

Figure 15–27. Anisocytosis. **A.** Normocyte (normal size), normochromic (normal hemoglobin concentration). **B.** Microcyte (small size), spherocyte (looks like increased hemoglobin concentration). **C.** Macrocyte (large size). **D.** Microcyte (small size), normochromic (normal hemoglobin concentration). **E.** Spherocyte; notice the lack of central area of pallor. Viewed under oil-immersion objective.

chromic cells have a pale central area that takes up about one third of the cell (Figs. 15–27A and D, 15–36B). Hypochromic cells are pale and have a central area of pallor that takes up greater than one third of the cell (Fig. 15–28A). The central pallor can take up to four fifths of the cell in severe hypochromic anemias. The MCHC index for hypochromic cells is less than 32%. Hypochromic cells make up from 0% to 3% of the cells in a normal peripheral smear.

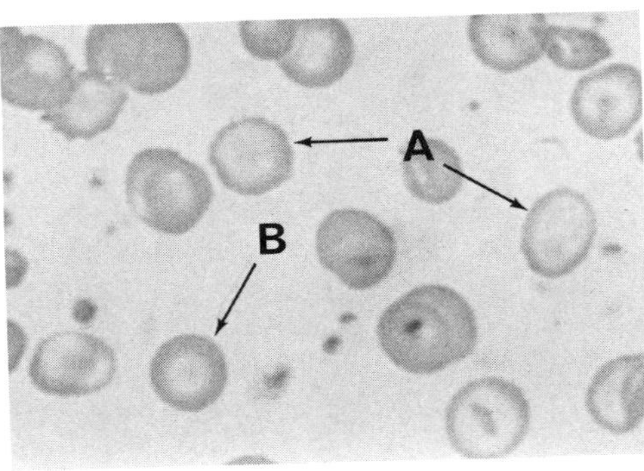

Figure 15–28. A. Hypochromic cells. Note the large area of central pallor in many of the cells. **B.** A cell with almost normal central area of pallor. Viewed under oil-immersion objective.

Spherocytes lack a central area of pallor and therefore appear "hyperchromic." This appearance is due to the fact that the cells are spherical rather than the usual biconcave-disc shape. There is, in fact, no such thing as hyperchromasia (Figs. 15–27B and E, 15–33B).

Reticulocytes in Wright's stained specimens have a bluish hue that is called polychromasia (Color Plate I, Fig. 5). Reticulocytes can account for up to 1.5% of the cells in a normal smear.

RBC Shapes

Normally shaped cells are round with a small round central area of pallor (Figs. 15–27A, 15–33A, 15–36B). There are a variety of abnormally shaped cells:

Acanthocytes

These cells have long, thin, irregularly placed spines sticking out from the cell (Fig. 15–29A). Regular smears may have up to 1% acanthocytes. A high number of these cells are seen with beta-lipoproteinemia. Variable numbers are seen with alcoholic cirrhosis, and low numbers are seen in postsplenectomy patients. They can be seen in hemolytic anemia due to pyruvate kinase deficiency, in neonatal hepatitis, and after the administration of heparin.

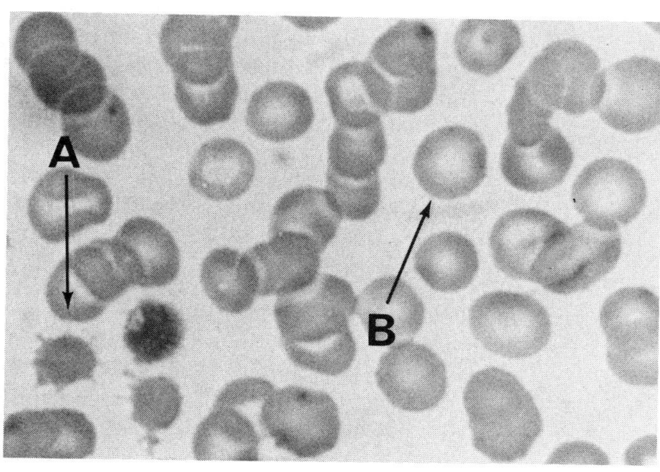

Figure 15–29. A. Acanthocyte. **B.** Normocyte, slightly hypochromic. Viewed under oil-immersion objective.

Crenated Cells (Burr Cells or Echinocytes)

These RBCs have evenly distributed spicules (crenations) on the surface (Fig. 15–30A). These are usually artifacts caused by slow drying of the smear in humid weather. Normal smears may have up to 10% crenated cells. They are also seen in uremia and pyruvate kinase deficiency.

Elliptocytes

These cells are sausage or rod shaped (Fig. 15–31A). They may make up to 3% of the cells in a normal blood smear. In thalassemia, sickle cell trait, or hemoglobin C trait, they may constitute up to 10% of the RBCs. If there are greater than 25% elliptocytes, the patient probably has hereditary elliptocytosis.

Ovalocytes

These are slightly elongated, oval cells (Figs. 15–31B, 15–36C). They can make up to 3% of cells in a normal smear. They are also seen in microcytic anemias or along with teardrop cells in myelofibrosis. If

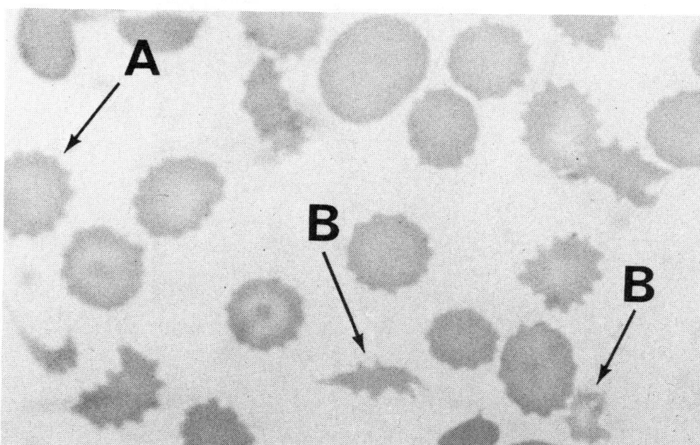

Figure 15–30. A. Crenated cell. **B.** Schistocytes. Viewed under the oil-immersion objective.

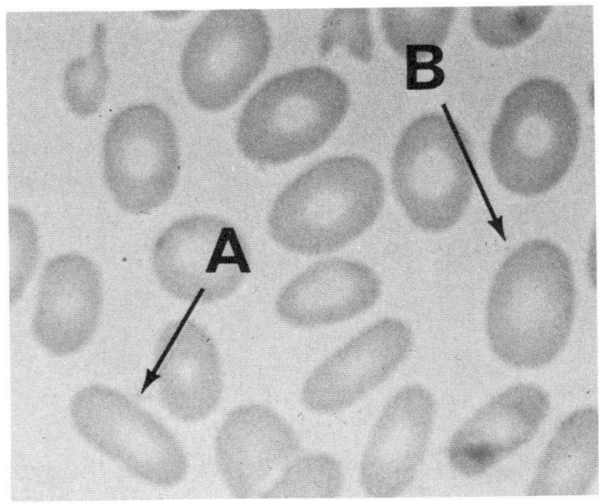

Figure 15–31. A. Elliptocyte. **B.** Ovalocyte. Viewed under the oil-immersion objective.

there is greater than 25% ovalocytes, the patient probably has a hereditary ovalocytosis. Very large ovalocytes (i.e., macro-ovalocytes) are seen in the megaloblastic anemias.

Schistocytes

These fragments of RBCs appear in a variety of shapes, including triangles, helmets, or irregular pieces of cells (Fig. 15–30B). Up to 1% of cells in a normal differential count may be such fragments. Increased numbers are seen with disseminated intravascular coagulation (DIC), microangiopathic hemolytic anemia, prosthetic heart valves, valvular stenosis, hypertension, uremia, burns, bleeding peptic ulcers, aplastic anemia, pyruvate kinase deficiency, and in a normal newborn.

Sickle Cell (Drepanocyte)

These are crescent-shaped cells with irregular spines (Fig. 15–32). They must have at least one pointed end and often have a holly leaf appearance. These are seen in sickle cell anemia. Rarely, they are also seen in sickle cell trait, sickle cell-B thalassemia, and hemoglobin S–C disease.

Spherocytes

These cells are round and small and have no central area of pallor. They are not usually found in normal smears (Figs. 15–27E, 15–33B). They are seen in hereditary spherocytosis, isoimmune hemolytic anemia, ABO erythroblastosis fetalis, hemolytic transfusion reactions, severe burns, and microangiopathic hemolytic anemia.

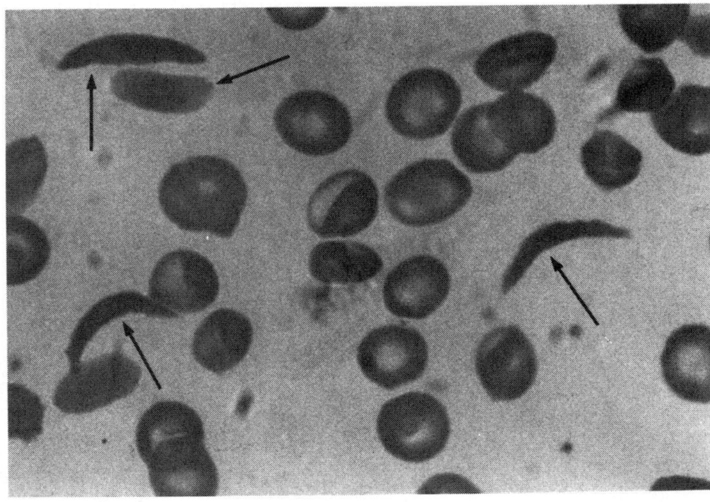

Figure 15–32. Sickle cells. Viewed under the oil-immersion objective.

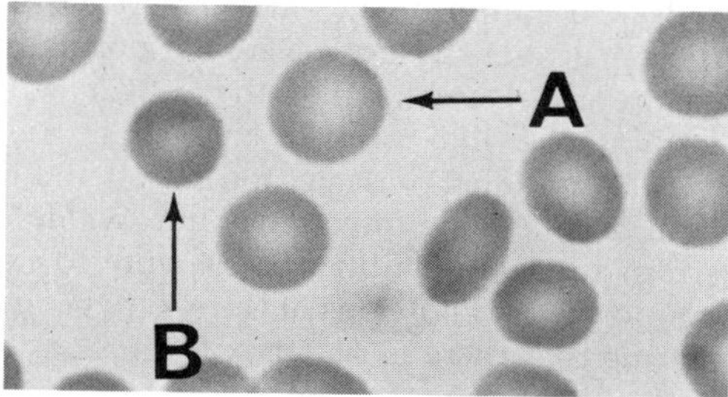

Figure 15–33. A. Normal cell. **B.** Spherocyte. Viewed under the oil-immersion objective.

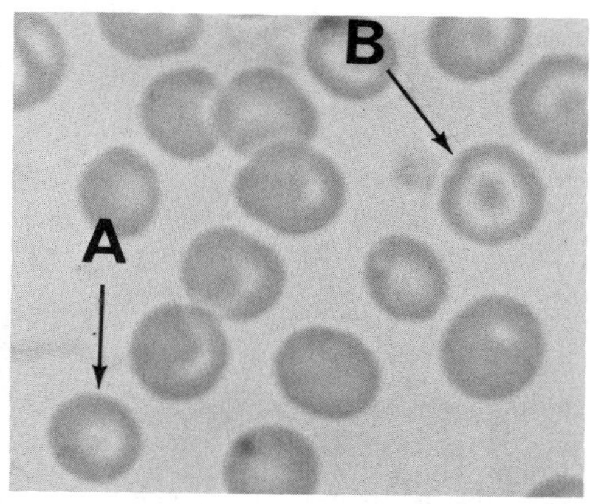

Figure 15–35. A. Normal cell. **B.** Target cell. Viewed under the oil-immersion objective.

Stomatocytes

These cells have a slitlike area of central pallor and appear to have a "mouth" (Figs. 15–34A, 15–40B). They can be seen as artifacts in normal smears or are seen in neoplastic diseases, cardiac illnesses, liver disease, or phenothiazine ingestion.

Target Cells (Codocytes)

These cells have a dark center surrounded by a light ring in the area that is usually the central area of pallor (Fig. 15–35B). They may be artifacts but are also seen in thalassemia, hemoglobin C disease, sickle cell anemia, postsplenectomy, iron deficiency, liver disease, and familial lecithin-cholesterol acyl transferase deficiency.

Teardrop Cells (Dacrocytes)

These cells have one pointed end and are teardrop shaped (Fig. 15-36A). They are

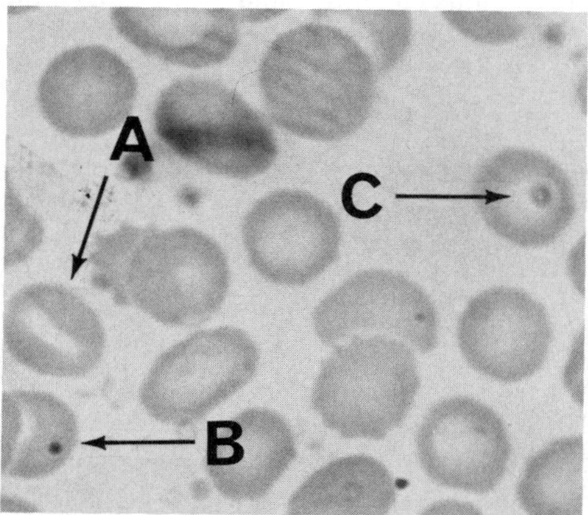

Figure 15–34. A. Stomatocyte. **B.** Howell-Jolly body. **C.** Platelet on top of a red cell. Viewed under oil-immersion objective.

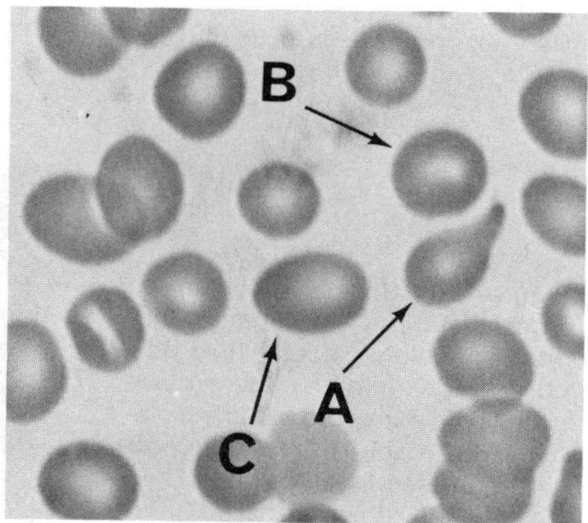

Figure 15–36. A. Teardrop cell. **B.** Normal cell. **C.** Ovalocyte. Viewed under the oil-immersion objective.

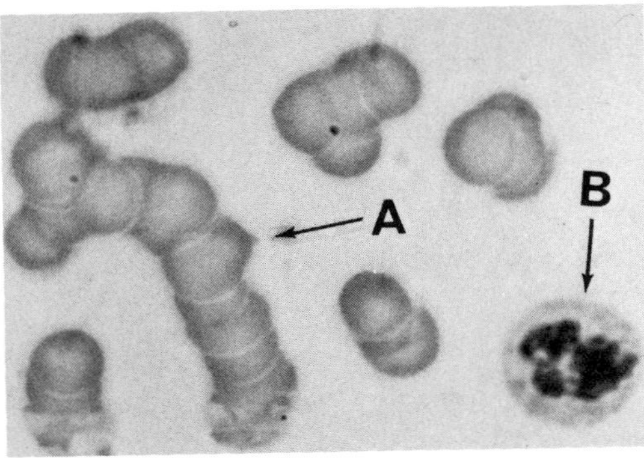

Figure 15–37. A. Rouleaux formation of red cells. **B.** Neutrophil. Viewed under the oil-immersion objective.

seen in thalassemia, myeloproliferative syndromes, and pernicious anemia.

Variations in Red Cell Shape

There is usually a small amount of cell variation in a normal peripheral smear. If there are several distinct types of red cell shapes, the condition is referred to as "poikilocytosis." It is important to note the particular shapes of cells that are seen.

Rouleaux Formation

Cells should appear evenly dispersed when they are properly smeared on a slide. The cells will clump in the thicker portions of the smear, but these areas are not normally examined. Occasionally the RBCs will form stacks (like a roll of coins) (Fig. 15–37A). It is called rouleaux formation and is seen in multiple myeloma and Waldenström's macroglobulinemia. Be sure to examine the thin portions of the smear before stating that there is rouleaux formation.

Agglutination

This is an irregular clumping of the RBCs but without the regular stack-of-coins pattern (Fig. 15–38). It is seen in patients with hemolytic anemia or cold agglutinin disease.

RBC INCLUSIONS

Basophilic Stippling

These are fine or coarse blue granules that are randomly distributed throughout the cell. They represent precipitated RNA. Fine stippling is seen with an increased production of red cells. Coarse granules are seen in lead poisoning, megaloblastic anemia, or other severe anemias (Fig. 15–39A).

Cabot Rings

These are fine filaments that are seen in pernicious anemia and lead poisoning. They are rarely seen.

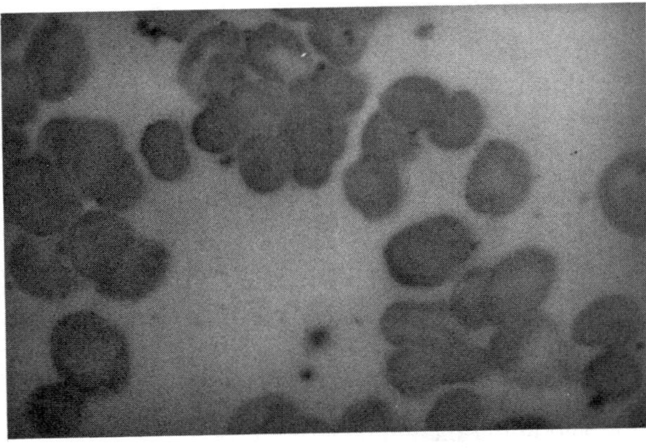

Figure 15–38. This specimen is from a patient with hemolytic anemia. This particular field occurs in the feather edge where you would expect to see the cells close to each other but not clumped (see Fig. 15–41). This "clumping" of cells is called "agglutination." Compare it to the rouleaux formation in Figure 15–37. Viewed under oil immersion objective.

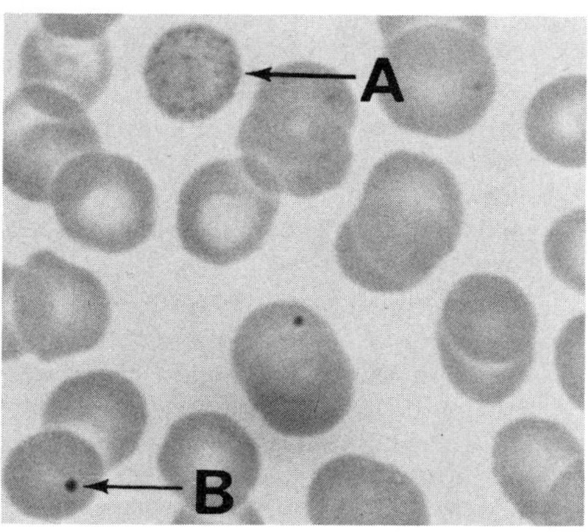

Figure 15–39. **A.** Basophilic stippling. **B.** Howell-Jolly body. Viewed under the oil-immersion objective.

Heinz Bodies

These inclusions cannot be seen with Wright's-stained smears.

Howell-Jolly Body

This is usually a single dark-purple dot in the RBC (Figs. 15-39B and 15-40A). The inclusion is not refractile. It represents remnants of nuclear material. It is seen in pernicious anemia, hemolytic anemias, postsplenectomy, and in megaloblastic anemias.

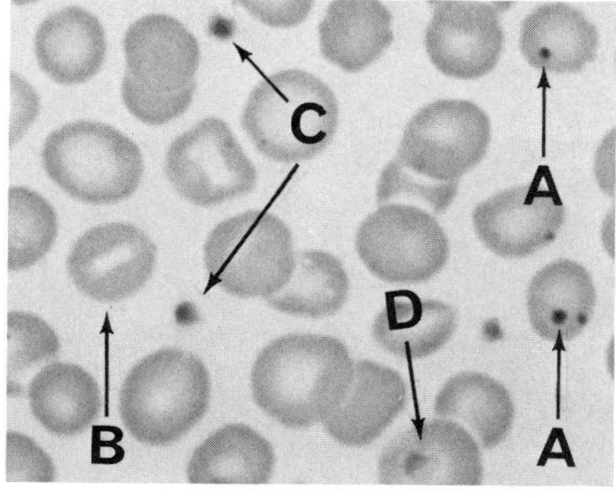

Figure 15–40. **A.** Howell-Jolly bodies. **B.** Stomatocyte. **C.** Platelet. **D.** Platelet on top of RBC. Viewed under the oil-immersion objective.

Malaria

Malaria parasites can be seen in RBCs with Wright's staining. They can take a variety of shapes in addition to the classic signet-ring formation. They may be confused with platelets on top of RBCs. (See *Morphology of Human Blood Cells*, Abbott Laboratories, for excellent drawings.)

Nucleated RBCs

(Color Plate I, Fig. 5) These cells are immature forms of RBCs and are not seen in normal patients. If you see what might be a nucleated RBC, have the slide reviewed. The ''orthochromic'' RBC (or metarubricyte) has a dense, dark-purple nucleus and pink-colored cytoplasm. A ''polychromatophilic'' red cell (or rubricyte) has a dark-purple nucleus and a bluish-gray cytoplasm. The number of nucleated RBCs should be reported as ''number of nucleated red cells/100 WBCs.'' Nucleated RBCs are seen in carcinoma, heart failure, leukemia, liver failure, lymphoma, gastrointestinal hemorrhage, pneumonia, renal failure, sepsis, stroke, and acute pulmonary edema. If there are more than 10 nucleated RBCs per 100 white cells (rare in outpatients), the white cell count is corrected using the following formula:

$$\text{Corrected white count} = \frac{\textit{Initial white count} \times \textit{100}}{100\ (+)\ \text{number of nucleated RBCs/100 WBCs}}$$

Platelets on Top of RBCs

These are not, obviously, true inclusions, but they may be confused with real inclusions. They appear as a central, grainy, purple area surrounded by a clear halo (Fig. 15-40D) (Color Plate I, Fig. 2).

PLATELETS

These are usually small, irregularly shaped cytoplasmic fragments that may clump together. They can look bluish or have fine purple granules. They are varied in size but are usually smaller than RBCs (Color Plate I, Fig. 2, Fig. 15-40C). Giant platelets (about the size of RBCs) should be noted on your report.

PROCEDURE

1. Zero the differential cell counter.
2. Wipe the underside of the slide with an alcohol wipe to remove Wright's stain from the glass. Be sure not to wipe the specimen side of the slide with the alcohol wipe because the specimen will come off.
3. Put the slide, specimen side up, between the slide clips on the microscope stage. Consistently put the frosted end of the slide to either the right or the left side so that you develop a pattern of moving the slide in a single direction.
4. Raise the condenser and open the microscope diaphragm fully. (See Chapter 10, Section 8, "How to Use the Microscope to Examine Stained Slides Under Oil" for a more detailed description of the use of the microscope.)
5. Focus the microscope under low power. Scan the feather edge (i.e., the thinnest part of the specimen), then move the smear to a place where the red cells are close together but not overlapping.
6. Put a drop of oil on the slide where the light from the condenser passes through the slide. Switch the microscope to the oil objective. Be sure not to drag the high-dry objective through the oil.
7. Focus the microscope. If you are unable to focus the slide under the oil objective, you probably have the slide upside down on the stage. Turn the slide over and try again.
8. You will want to find an area of the slide where the RBCs stand alone with only occasional touching or overlapping. The red cell's central area of pallor (Fig. 15-41) should be visible. Figure 15-42 shows an unsatisfactory area. The cells seem flat, with no central area of pallor.
9. When counting the cells, move the slide systematically so that you start on one of the long edges of the slide and move up (or down) toward the other

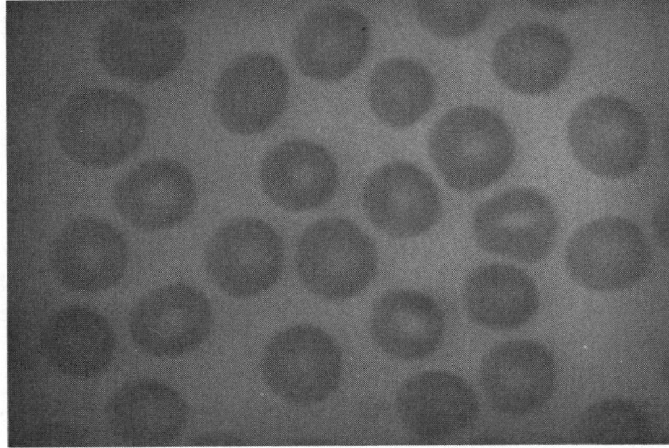

Figure 15-41. Acceptable field for starting differential. Notice cells' central areas of pallor. Viewed under oil immersion objective.

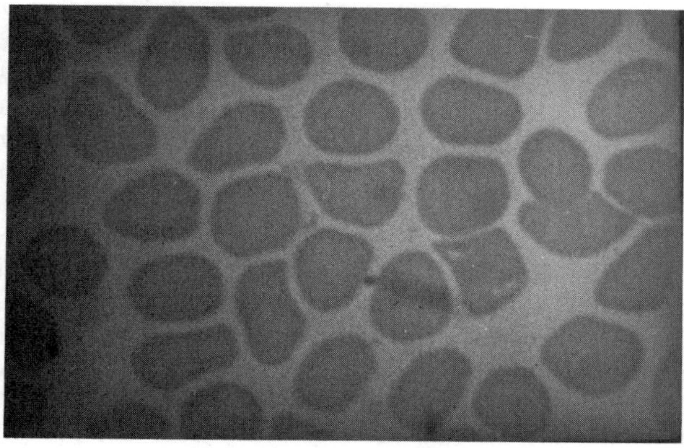

Figure 15–42. Unsatisfactory field for starting differential. Notice how the cells look flat and lack the central area of pallor. Viewed under oil immersion objective.

long edge of the slide. When you come to the other edge, move one microscope field deeper into the smear. This is done by noting the edge of the field at the 3 o'clock or 9 o'clock position and moving the slide sideways so that the cells on the far edge of the original field are moved just out of view. Now reverse the direction of the slide and move toward the opposite long edge (Fig. 15–43). By doing this you are assured that you do not count the same area of the slide twice. You also avoid moving too quickly into the thick part of the smear.

10. Depress the appropriate key on the cell counter whenever you see an intact WBC. If an unusual cell is seen, record this on the differential report as "other cells." Bring this fact to the attention of the clinician. When 100 cells are counted, the bell will ring.

11. Record the number of each types of WBC on the laboratory test form. Be sure that the total number of cells equals 100.

12. Re-examine the smear for the RBC morphology. Look at about ten different fields. Record the RBC size, shape, color (i.e., hemoglobin concentration), and inclusions. If there is a particular morphologic type of red cell, report it as a percent of the total number of cells (e.g., 15% spherocytes).

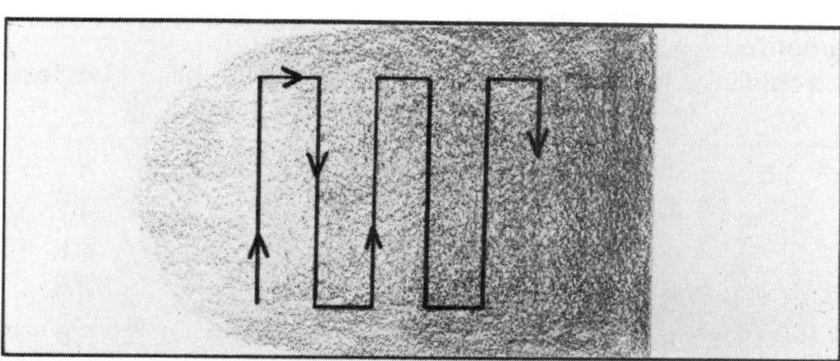

Figure 15–43. Pattern for moving the microscopic slide when performing a WBC differential count to avoid counting the same field twice.

13. Count the number of platelets per oil immersion field. Average this number for ten fields and multiply this number by 20. Report this as a platelet estimate. If the estimate is less than $140 \times 10^9/L$ (i.e., less than seven platelets per field), report "platelets appear decreased." If the estimate is between $140 \times 10^9/L$ and $450 \times 10^9/L$, report "platelets appear adequate." If there are greater than $450 \times 10^9/L$ platelets, report "platelets appear increased." If the platelet estimate is low, examine the feather edge to see if the platelets have accumulated there.
14. Remove the slide from the microscope stage. Wipe the oil objective with special lens paper until no oil appears on the paper.
15. Drop the slide into the Coplin jar with the xylene. Remove it after several seconds and gently wipe the liquid from the slide with a towel.
16. File the slide in the slide storage area or prepare it for the permanent collection (see "Appendix").

TEST INTERPRETATION

The percentage of various types of WBCs varies with the patient's age (Table 15–7). In general, children have a higher percentage of lymphocytes than neutrophils.

Changes in the percentages of the types of cells can be a helpful indicator of disease. When these changes are expressed as percentages of the total number of WBCs, they are referred to as "relative" changes (e.g., "relative lymphocytosis"). The changes can also be expressed as "absolute" by taking the percentages from the differential count and multiplying it by the total WBC count. (For example, if there are 60% lymphocytes and a total WBC count of 11,000, the absolute lymphocyte count is 60% times 11,000 or 6,600.)

Relative Neutrophilia

This is defined as greater than 75% neutrophils in the differential count. The percent bands and percent segmented neutrophils are added to determine the total neutrophil percentage. An increase in neutrophils can be seen with localized bacterial infections (e.g., appendicitis), generalized bacterial infections (e.g., sepsis), stress (which results in epinephrine release), tissue necrosis (e.g., burns, necrotic tumor, or myocardial infarction), acute hemorrhage, acute hemolysis, metabolic diseases (e.g., uremia, acidosis), or myelocytic leukemias.

ABSOLUTE NEUTROPHILIA:

$> 6.5 \times 10^9/L$ (6,500/mm^3)

TABLE 15–7. REFERENCE VALUES FOR WBC DIFFERENTIAL COUNT BY AGE

Age	Segmented Neutrophils %	Band Neutrophils %	Eosinophils %	Basophils %	Lymphocytes %	Monocytes %
Birth	47 ± 15	14 ± 4	2.2	0.6	31 ± 5	5.8
1 week	34	11.8	4.1	0.4	41	9.1
1 year	23	8.1	2.6	0.4	61	4.8
4 years	34 ± 11	8.0 ± 3	2.8	0.6	50 ± 15	5.0
12 years	47	8.0	2.5	0.5	38	4.4
20 years	51 ± 15	8.0	2.7	0.5	33 ± 10	5.0

From Miale J: *Laboratory Medicine, Hematology*. 5th ed. St. Louis, Mo: Mosby; 1977:770. Reprinted with permission.

RELATIVE NEUTROPENIA:

< 40% neutrophils.

ABSOLUTE NEUTROPENIA:

< 1.5 × 10^9/L in children (1,500/mm^3); < 1.8 × 10^9/L in adults (1,800/mm^3)

Relative Eosinophilia

An increase in the percentage of eosinophils above 7% can be seen in allergic disease (e.g., asthma, hay fever, eczema), chronic skin diseases (e.g., psoriasis, ichthyosis), parasitic infections, scarlet fever, ulcerative colitis, tumors (of ovary, bone, uterus), postsplenectomy, collagen vascular diseases, and blood disorders (e.g., sickle cell anemia, polycythemia vera, pernicious anemia).

Absolute Eosinophilia

> 0.4 × 10^9/L (400/mm^3)

Relative Basophilia

An increase in the percentage of basophils greater than 2% is a very unusual finding in the outpatient setting. It has been described in ulcerative colitis, polycythemia vera, chronic myelocytic leukemia, urticaria, and erythroderma.

Relative Lymphocytosis

This is considered to be greater than 47% lymphocytes in an adult. The classic disease for causing a lymphocytosis is infectious mononucleosis. Other viral diseases can also produce a lymphocytosis (e.g., mumps, chickenpox, rubella, viral hepatitis), as can syphilis, tuberculosis, lymphomas, lymphocytic leukemia, and carcinoma.

Absolute Lymphocytosis

> 9.0 × 10^9/L in infants (9,000/mm^3) < 7.0 × 10^9/L (7,000/mm^3) in children; > 4.0 × 10^9/L in adults (4,000/mm^3)

Absolute Lymphopenia

< 1.4 × 10^9/L in children (1,400/mm^3); < 1.0 × 10^9/L in adults (1,000/mm^3)

Relative Monocytosis

This is defined as greater than 10% monocytes. It is an unusual laboratory finding but has been described with tuberculosis, leukemia, systemic fungal infections, bacterial endocarditis, and some protozoan infections.

Shift to the Left

This term is used to describe a differential count with greater than 10% band neutrophils. It is called a ''shift to the left'' because the band and other immature neutrophil keys are to the left of the segmented neutrophil key on the differential cell counter. A shift to the left indicates that the bone marrow is rapidly producing neutrophils. It is seen with infection, hemolysis, and hemorrhage.

COMMENTS

- It is difficult to make good blood smears in patients with severe anemia, polycythemia, or agglutinated RBCs (i.e., rouleaux formation). These can make the differential count inaccurate.
- The technical error in a manual differential count ranges from 10% to 15%. This imprecision is largely due to the fact that only 100 cells are counted.
- WBC differential counts can now be done by office automated instruments. Be sure to check whether your reference laboratory does this test by hand or by an automated technique. The automated instruments differentiate the various cells by their size and staining characteristics. These counts do not always correspond with those done by hand! Some systems do not give band counts. Some automated differentials only report the absolute count of the different types of cells.
- It is important to have a variety of reference sources to help differentiate the WBC types and RBC cell morphology. In

addition to the Abbott and Scope publications mentioned above, the Patrick Ward articles cited below are helpful.

REFERENCES

College of American Pathologists. *EXCEL Manual*. Skokie, Ill: College of American Pathologists; 1981.

Koepke J. Standardization of the manual differential leukocyte count. *Lab Med* 1980;11:371.

Napoli V, Nicholis C, Fleck S. Semi-quantitative estimate method for reporting abnormal red blood cell morphology. *Lab Med* 1980;11:111.

Ward P. Investigation of microcytic anemia. *Postgrad Med* 1979;65(1):235.

Ward P. Investigation of macrocytic anemia. *Postgrad Med* 1979;65(2):203.

Ward P. Investigation of non-poikilocytic, normochromic, normocytic anemia. *Postgrad Med* 1979;65)3):233.

Ward P. Investigation of poikilocytic normochromic, normocytic anemia, round and elongated forms. *Postgrad Med* 1979;65(4):215.

Ward P. Investigation of poikilocytic normochromic, normocytic anemia, spiculated forms. *Postgrad Med* 1979;65(4):229.

SECTION 11: Modified Westergren Erythrocyte Sedimentation Rate

BACKGROUND

Anticoagulated whole blood is made up of blood cells suspended in plasma. When such blood stands for a period of time, the cells settle out. Several factors affect this rate of settling, including the cell shape and the plasma fibrinogen and alpha$_2$-globulin levels. The erythrocyte sedimentation rate (ESR) is a measure, under controlled conditions, of the rate of RBC settling. A variety of ESR techniques have been described, but not all of these procedures are reproducible. The modified Westergren sedimentation rate described here can be peformed easily in the office laboratory. It is more reliable than other available types of sedimentation techniques.

TEST INDICATIONS

The ESR is a nonspecific test. It is useful in following the course of a wide range of infectious, inflammatory, and hematologic diseases. The test, however, is not conclusive evidence for any single disease and may be elevated by such nonspecific factors as advanced patient age. Furthermore, the test is not highly sensitive. Many of the diseases that can produce an increase in the ESR do so in some but not all patients with that disease.

The ESR is dramatically affected by the RBC shape (e.g., sickle cells, spiculated cells), marked variation in red cell size (anisocytosis), severe anemia, or hemolysis. These effects cannot be quantified. The ESR is therefore of no value in evaluating patients with any of these conditions.

TEST MATERIALS

- ESR Tubes: Controlled pipetting was required to fill older ESR tubes. We recommend that office laboratories use the newer ESR tube systems that are more easily filled. One such system (Scientific Products Dispette II Modified Westergren ESR tubes; product No. B-4515) includes disposable plastic tubes and saline reagent reservoirs. Other modified ESR systems use sodium citrate, instead of saline, as a diluent (Fig. 15–44).

- ESR Rack: This rack holds the ESR tubes during the procedure. The rack must be level for proper testing. Either purchase a rack with a built-in level (i.e., Scientific Products No. B-4515-9A), or buy a separate, spirit-filled level (i.e., Curtin Matheson No. 152-876) (Fig. 15–44).
- Plastic transfer pipettes.
- 2 × 2-inch nonsterile gauze pads: These are used to cover the tops of the blood tubes and the reservoirs when you uncap them. This prevents splashing.
- Vinyl gloves
- *Test-Tube Rack:* A small, inexpensive test-tube rack can be used for holding many things in the laboratory.
- *Timer:* A timer that runs for at least 60 minutes and has a loud alarm is needed. It may also be useful to purchase an inexpensive pocket timer. This can be used

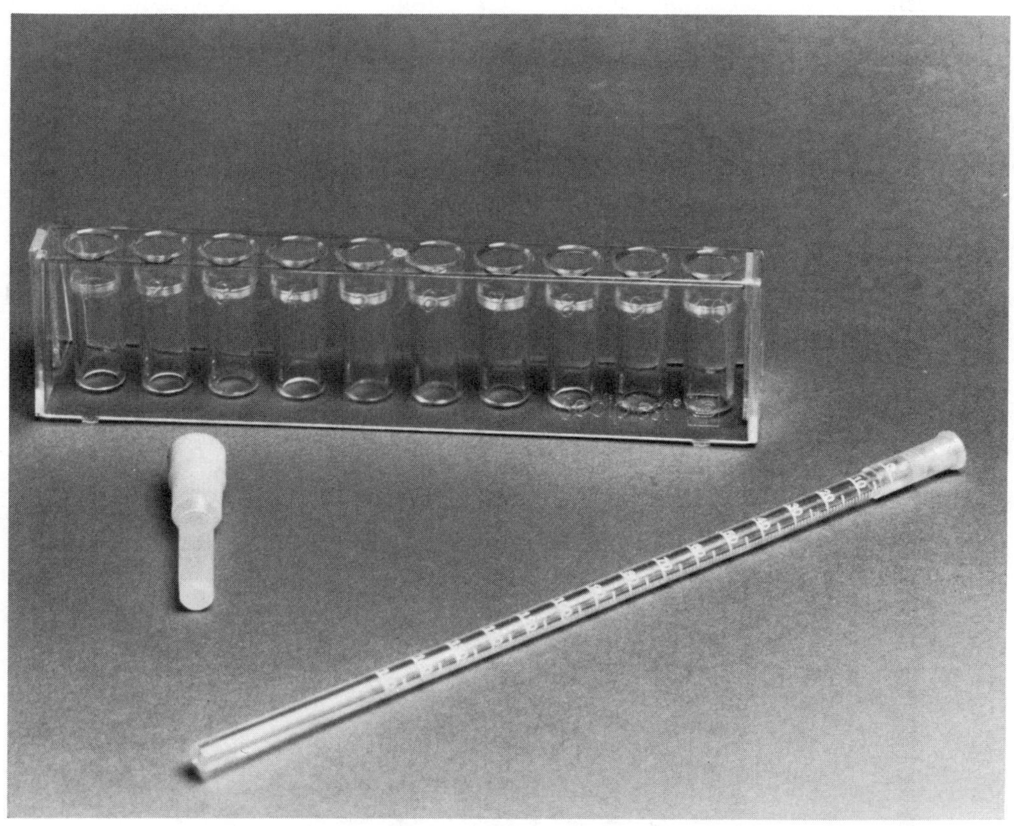

Figure 15–44. An Erythrocyte sedimentation tube, a tube rack, and reagent reservoir are used for the ESR test.

when the laboratory worker must leave the laboratory after setting up the test. The small timer can be set for 50 minutes so that when the alarm goes off 10 minutes remain in which to return to the laboratory to read the ESR.

- *Aliquot Mixer:* This mixer is used to thoroughly mix the blood sample before performing the test (Fig. 15–1).

SPECIMEN COLLECTION

1. Draw a venous blood sample into a regular size purple-top (EDTA) tube.
2. Write the patient's name and the time of the specimen collection on the tube.
3. If the blood remains at room temperature, the test must be performed within 2 hours of specimen collection. If the tube is refrigerated, the specimen can be tested within 12 hours of its collection. However, refrigerated blood must be brought to room temperature before performing the sedimentation test.

TEST PROCEDURE

1. Check that the rack is level. If not, adjust it until the bubble is in the center of the circle in the level. (This can be done by using the adjustable feet, if available, or by placing small pieces of

Figure 15–45. A plastic pipette is used to fill the reagent reservoir with blood.

Color Plates

COLOR PLATE I

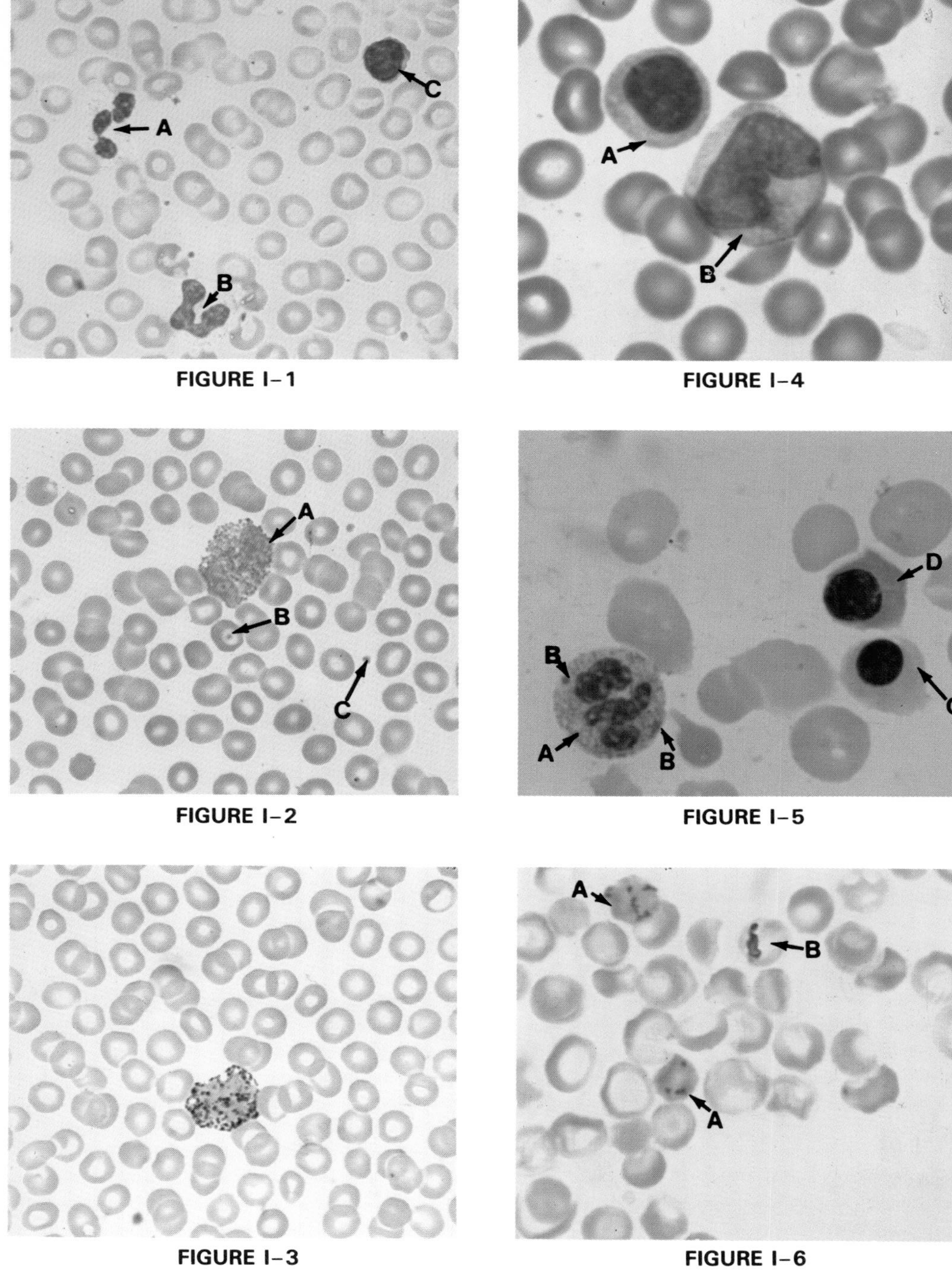

FIGURE I–1

FIGURE I–4

FIGURE I–2

FIGURE I–5

FIGURE I–3

FIGURE I–6

COLOR PLATE I

FIGURE I–1. Peripheral blood smear; Wright's stain. **A.** Segmented neutrophil. **B.** Band neutrophil. **C.** Small lymphocyte. (Viewed under oil immersion)

FIGURE I–2. Peripheral blood smear; Wright's stain. **A.** Eosinophil. **B.** Platelet on top of a red blood cell. **C.** Platelet.(Viewed under oil immersion)

FIGURE I–3. Peripheral blood smear; Wright's stain. Basophil. (Viewed under oil immersion)

FIGURE I–4. Peripheral blood smear; Wright's stain. Mononuclear cells. **A.** Medium size lymphocyte. **B.** Monocyte. (Viewed under oil immersion)

FIGURE I–5. Peripheral blood smear; Wright's stain. **A.** Neutrophil with toxic granulation. **B.** Dohle body. **C.** Orthochromic nucleated red cell (cytoplasm color is same as surrounding red cells). **D.** Polychromatic nucleated red blood cell. The color of the cytoplasm is much bluer. When the term "polychromasia" is used to describe red blood cells, it means that the cytoplasm of the cells is bluer than regular red cells. (Viewed under oil immersion)

FIGURE I–6. Reticulocyte smear; new methylene blue N stain. **A.** Reticulocytes. **B.** Red blood cells with confusing refractile artifacts. (Viewed under oil immersion)

COLOR PLATE II

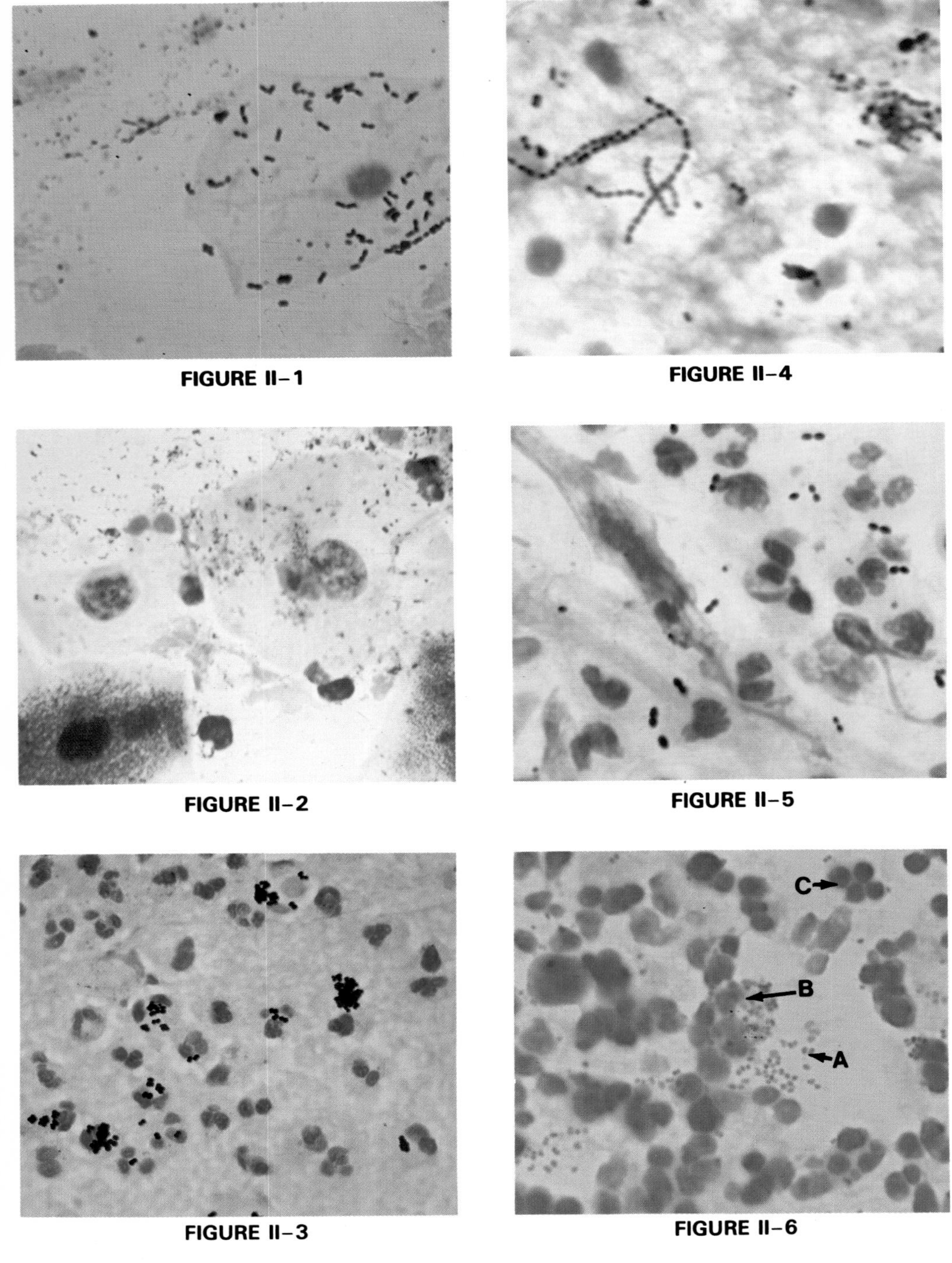

FIGURE II–1

FIGURE II–4

FIGURE II–2

FIGURE II–5

FIGURE II–3

FIGURE II–6

COLOR PLATE II

FIGURE II–1. Gram stained squamous epithelial cell. The nucleus is small and dark pink or red and the cytoplasm is a light pink. While this is a squamous cell, the staining characteristics are similar for white cells. Some gram positive mouth flora is stuck to the surface of the cell. (Viewed under oil immersion)

FIGURE II–2. These are improperly stained cells. The nuclei are purple rather than dark pink or red. (Viewed under oil immersion)

FIGURE II–3. Polys and gram positive cocci in clusters, characteristic of staph infections. (Viewed under oil immersion)

FIGURE II–4. Gram positive cocci in chains or streptococci. (Viewed under oil immersion)

FIGURE II–5. Polys with gram positive lancet shaped diplococci. This is characteristic of an infection with *S pneumonie.* (Viewed under oil immersion lens)

FIGURE II–6. Gram's stain of urethral smear. **A.** Extracellular gram-negative diplococci. **B.** Intracellular-negative diplococci. **C.** Neutrophil. (Viewed under oil immersion)

COLOR PLATE III

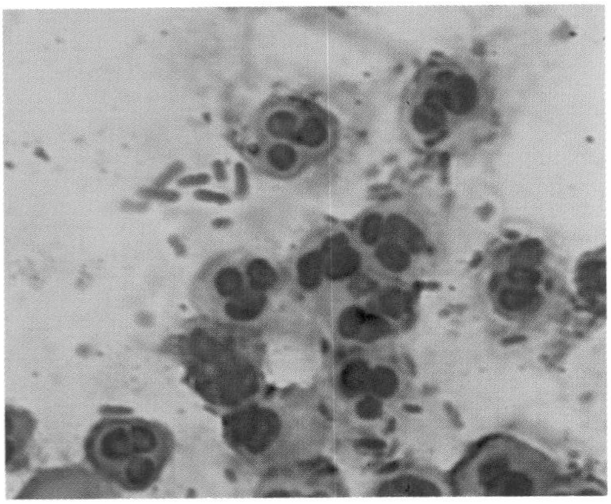

FIGURE III–1

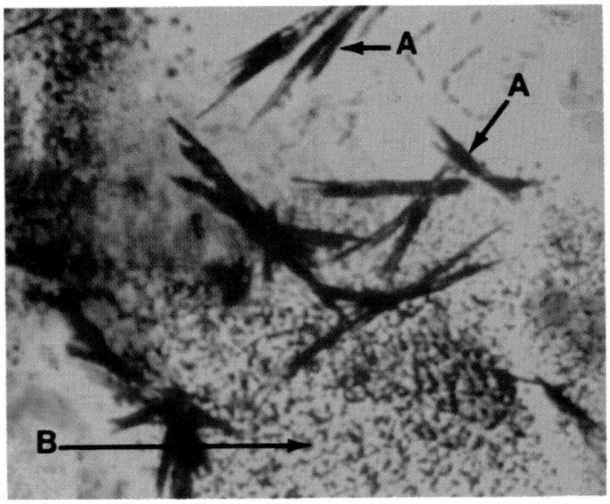

FIGURE III–4

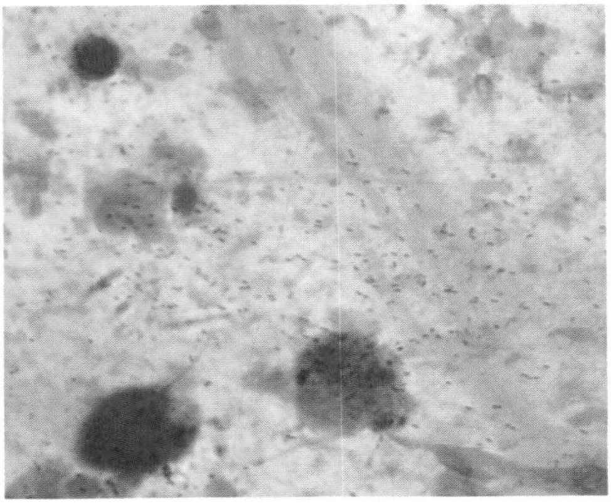

FIGURE III–2

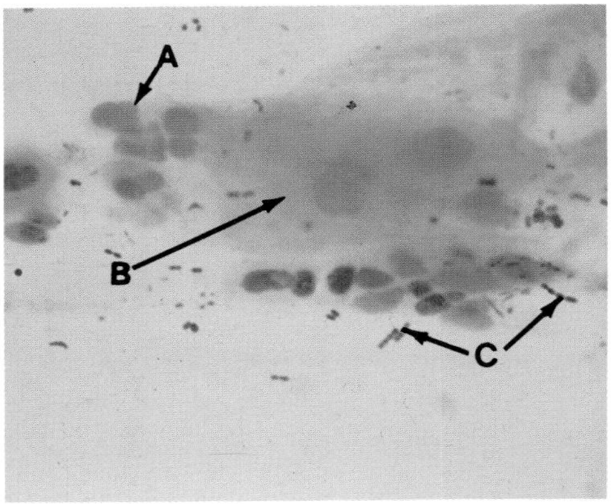

FIGURE III–5

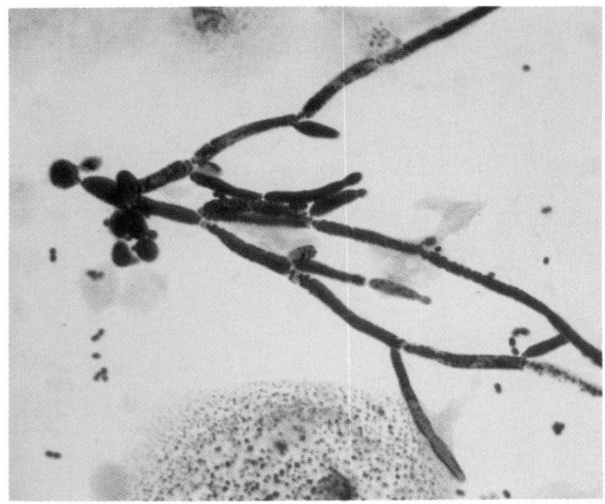

FIGURE III–3

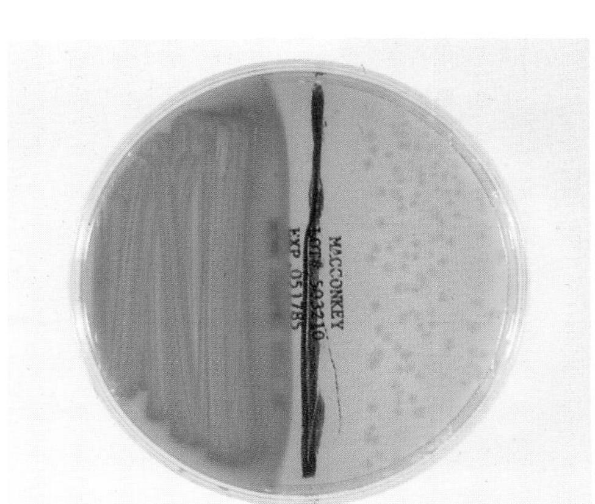

FIGURE III–6

COLOR PLATE III

FIGURE III–1. Polys and gram negative rods. (Viewed under oil immersion)

FIGURE III–2. Polys and gram negative pleomorphic rods which are characteristic of *H influenzae.* (Viewed under oil immersion)

FIGURE III–3. Gram's stain of the pseudohyphae and buds of *C albicans.* Notice how much smaller the gram positive bacteria is. (Viewed under oil immersion)

FIGURE III–4. Artifacts. Here you see both the crystal sheaves (A), and the cocci-like artifacts (B) which are frequently seen on slides that were stained when the slide was still hot. (Viewed under oil immersion)

FIGURE III–5. Polys (A), epithelial cells (B), and mixed flora (C). No conclusion about infecting agent can be drawn from this specimen. (Viewed under oil immersion lens)

FIGURE III–6. MacConkey agar plate. **Left.** Lactose positive (red or purple) organisms. **Right.** Lactose negative (colorless) organisms.

COLOR PLATE IV

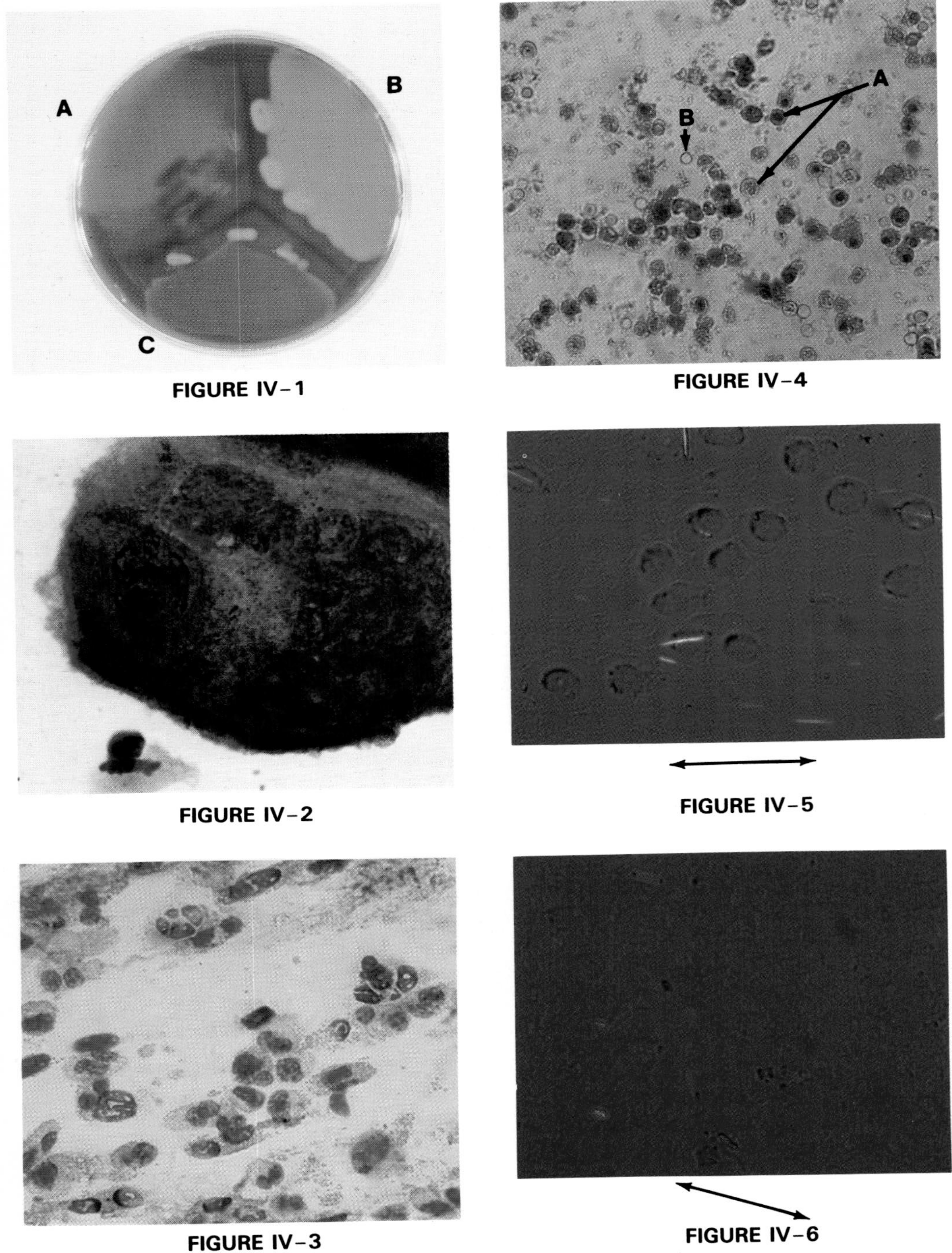

FIGURE IV–1

FIGURE IV–2

FIGURE IV–3

FIGURE IV–4

FIGURE IV–5

FIGURE IV–6

COLOR PLATE IV

FIGURE IV–1. Hemolytic patterns on sheep blood agar plate. **A.** Alpha hemolysis. **B.** Beta hemolysis on the entire surface. **C.** Beta hemolysis in the stabs only.

FIGURE IV–2. Tzanck prep. Hansel stain of a multinucleated giant cell. (Viewed under oil immersion)

FIGURE IV–3. Hansel stain of sputum showing; sheet of eosinophils. (Viewed under oil immersion)

FIGURE IV–4. Methylene blue stain of stool. **A.** White blood cells. **B.** Red blood cells. (Viewed under high-dry)

FIGURE IV–5. Compensated polarization. The arrow marks the direction of the long axis of the full wave compensator. Note the color of the needles which are parallel and perpendicular to this axis. The needles are monosodium urate, or gout crystals. (Viewed under high-dry)

FIGURE IV–6. Compensated polarization. The arrow marks the direction of the long axis of the full wave compensator. Note the color of the needles which are parallel and perpendicular to this axis. These are calcium pyrophosphate or pseudogout crystals. (Viewed under high-dry)

paper under the appropriate corners of the rack).

2. Put on gloves.
3. If the blood has been refrigerated, allow it to come to room temperature.
4. Set the specimen on the aliquot mixer for 2 minutes. Alternately, the blood tube can be inverted *gently by hand* 60 times.
5. Cover the top of the purple-top tube with a 2 × 2-inch gauze, and gently remove the stopper to avoid splashing the blood. Place the tube in the test-tube rack. Check for clots.
6. Remove the top of the reservoir. Do not discard. Set the reservoir in the rack until you refill it.
7. Using a plastic transfer pipette, draw up some blood from the well-mixed sample.
8. Hold the reservoir in the other hand and fill it with blood up to the marked line (Fig. 15–45).
9. Put the pipette back into the purple-top tube.
10. Put the top back on the reservoir and rock it on the aliquot mixer for 2 minutes, or 60 times by hand.
11. Cover the top of the reservoir with a 2 × 2-inch gauze pad and gently remove the top.
12. Take the Westergren sedimentation tube and push the bottom end into the reservoir. You will need to twist and turn the tube as you push. Do this until the blood flows into the overflow chamber at the top of the tube (Fig. 15–46).
13. Place the reservoir with tube into the level rack (Fig. 15–47).
14. Set the timer for 60 minutes. Write the time on the test form. If you have a pocket timer, set it for 50 minutes so that you can return to the laboratory on time.
15. Remove the transfer pipette from the

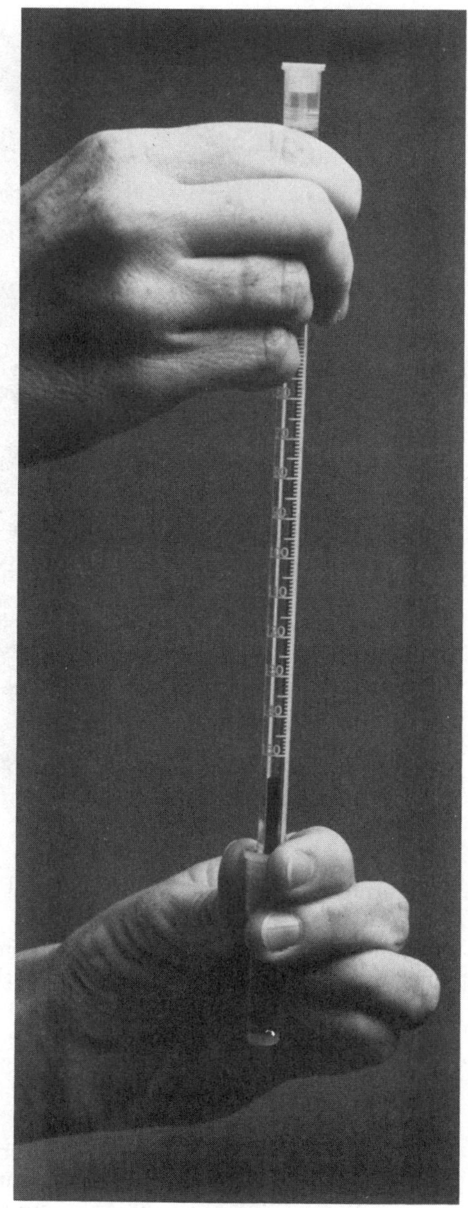

Figure 15–46. The sedimentation tube is inserted into the reagent reservoir. This pushes the blood into the tube.

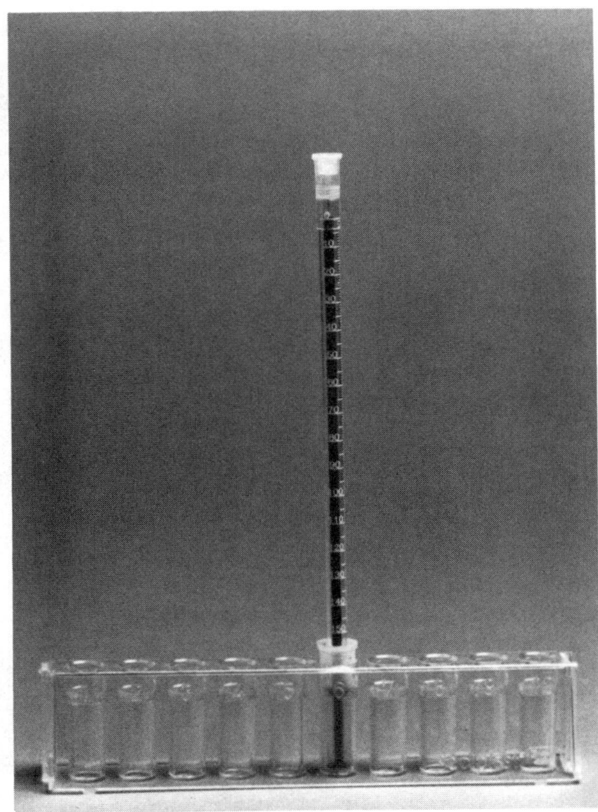

Figure 15–47. Sedimentation tubes must be stable while the red blood cells are settling.

purple-top tube and throw it into the contaminated softs container.

16. Put the top back on the blood tube, and put it into the refrigerator. Save the specimen in case something happens and you need to repeat the test.
17. After exactly 60 minutes, read the ESR. This is the point on the sedimentation tube where the column of RBCs is cleary demarcated from the plasma layer. If this point is not clearly marked, read it as the point where the full density of RBCs first appears. Report the value as "number of millimeters per hour, modified Westergren ESR."
18. Remove the Westergren tube and reservoir from the rack and discard them into the contaminated softs disposal jug.

TEST INTERPRETATION

The reference values for a Westergren sedimentation rate vary by age and sex:

- Children: 0 to 10 mm/hour
- Males under 50 years: 0 to 15 mm/hour
- Males between 50 and 65 years: 0 to 20 mm/hour
- Males over 65 years: 0 to 38 mm/hour
- Females under 50 years: 0 to 20 mm/hour

- Females between 50 and 65 years: 0 to 30 mm/hour
- Females over 65 years: 1 to 53 mm/hour

Table 15–8 lists diseases associated with an increase in the ESR.

QUALITY CONTROL

- Be sure that the rack is level, out of direct sunlight, and is not affected by vibrations at any time during the test.
- Duplicate samples can be run on blood from a purple-top tube until the laboratory worker has mastered the technique. You should be able to reproduce the test to a ± 3-mm value. It is not necessary to routinely run duplicate samples.
- The ESR is very dependent on red-cell shape and size and the patient's hematocrit. Abnormalities in any of these three values will make the ESR results uninterpretable. We therefore suggest that every patient who has ESR testing also have a hematocrit and a peripheral smear. These can both be done from the purple-top tube.

SOURCES OF ERROR

Falsely Elevated ESR

- Tilting the sedimentation tube by only 3 degrees will cause a 30% increase in the ESR. Be sure that the sedimentation rack is level and that the tubes are properly positioned
- Vibrations from laboratory equipment.
- Anemia (There is no adequate correction

TABLE 15–8. CAUSES FOR AN ELEVATED SEDIMENTATION RATE

Markedly Elevated ESR (100 mm by Westergren Technique)	Moderately Elevated ESR (Elevated But < 100 mm)
Tumors	Tumors
Multiple myeloma	Malignant tumors with tissue necrosis
Lymphoma	Infections
Leukemia	Viral hepatitis
Sarcoma	Tuberculosis
Carcinoma	Other acute or chronic infections
Infections	Collagen diseases
Severe bacterial infection	Rheumatoid arthritis
Sepsis	Other
Viral pneumonia	Acute glomerulonephritis
Collagen diseases	Anaphylactoid purpura
Temporal arteritis	Normal pregnancy after third month
Other	Menstruation
Severe renal disease	Ectopic pregnancy
Ulcerative colitis	Lead poisoning
Biliary cirrhosis	Arsenic poisoning
	Hyperthyroidism
	Hypothyroidism
	Myocardial infarction
	Rheumatic fever

factor to adjust the ESR for a low hematocrit)

- An elevation in the room temperature above 25 °C
- Macrocytic RBCs settle faster than normocytic cells, and this will elevate the ESR

Falsely Lowered ESR

- Microcytic RBCs settle slower than normocytic cells and therefore lower the ESR
- Polycythemia (There is no correction factor to adjust for an elevated hematocrit)
- If the testing is done too long after the blood is collected, the sample may show a lower ESR

COMMENTS

- Other office methods of performing the ESR have been shown to be less reproducible than the modified Westergren technique. This includes the undiluted Westergren (i.e., whole blood not diluted with sodium citrate) and the Wintrobe technique. Hospitals may perform a zeta sedimentation rate, which is done by centrifuging blood in a special centrifuge. There are no mathematical formulas to correlate the sedimentation rates determined by different techniques. Most clinical data on sedimentation rates have been determined by the Westergren technique, and we therefore strongly recommend its use.
- There are three stages in red cell sedimentation. For the first 10 minutes the cells settle very slowly. For the next 40 minutes the rate increases, but it again slows during the final 10 minutes of the test. The rate of settling is dependent on the RBCs as well as the levels of plasma fibrinogen and globulins. It is the changes in these two proteins that result in most of the clinically significant ESR elevations.
- This test is very sensitive to subtle variables, such as tube position, room temperature, vibration, and sample dilution. These variables must be carefully controlled to ensure reproducible testing.

REFERENCES

Berlin DC, Morse E, Weinstein A. Whither Westergren—the sedimentation rate reevaluated. *J Rheumatol.* 1981;8:331.

SECTION 12: Sickle Cell Screening Test

BACKGROUND

Sickle cell anemia is an inherited, autosomal-recessive trait that is a major health problem in black patient populations. The disease is the result of a genetic change in the hemoglobin molecule. In normal RBCs the hemoglobin is type A. In patients with sickle cell *disease*, most of the hemoglobin is type S. This abnormal hemoglobin causes the RBCs to "sickle" when the blood is deoxygenated. These sickled cells are unable to pass through the microcirculation and therefore result in microinfarcts in bone, lung, spleen, or kidney. In the United States 0.15% of black children have sickle cell disease.

Patients with sickle cell *trait* have only one gene for hemoglobin S. The other gene is for normal hemoglobin A. These patients have about 35% hemoglobin S in their RBCs. The remaining hemoglobin is type A. Their RBCs do not sickle except under extreme deoxygenation. These patients have little increased illness risk and are usually asymptomatic but may have sickle cell crises when under anesthesia or with vigorous exercise. They may also have painless hematuria and urinary-concentrating deficits because of kidney microinfarcts. Eight percent of the black population in the United States has the sickle cell trait.

Screening for sickle cell trait and sickle cell disease can be easily performed in the office laboratory. A positive screening test should be further evaluated with a hemoglobin electrophoresis to differentiate among sickle cell trait, sickle cell disease, and other hemoglobinopathies.

Any use of a screening test for sickle cell should be combined with an effective genetic counseling program. Telling patients that they have sickle cell trait must be coupled with patient education about the usually benign course of the heterozygous condition, the serious implications of the homozygous disease, and the implications for having children.

For routine office screening, the Sickledex test (Ortho Diagnostics, Raritan, NJ) has been shown to be effective. The test is based on the relative insolubility of hemoglobin S in a buffered solution with a reducing reagent. This results in a turbid solution, while normal hemoglobin produces a transparent solution

TEST INDICATIONS

- Routine screening of all black patients
- Routine screening of black pregnant patients

TEST MATERIALS

A Sickledex kit includes all needed materials for 12 sickle cell tests. This includes the test solution, test reagent powder, disposable test tubes, microcapillary pipettes, pipette bulbs, and a test tube holder. The kit requires refrigeration at 2 °C to 8 °C.

Parafilm

SPECIMEN COLLECTION

Fresh whole blood from a finger stick or venous blood that has been anticoagulated with EDTA, cirtate, oxalate, or heparin can be used. If the blood is not immediately tested, it should be refrigerated. *Do not use clotted blood.* Testing can be done on refrigerated blood for up to 14 days. It is easiest to draw a purple-top (EDTA) tube for sickle cell testing.

REAGENT PREPARATION

1. Remove the foil-sealed top from a vial of Sickledex reagent powder.
2. Add the contents of the vial into one of the plastic bottles of Sickledex test solution.
3. Place the cap tightly onto the solution bottle, and shake the bottle for several seconds to completely dissolve the reagent powder.
4. Write the date that the reagent powder and test solution were mixed together on the side of the bottle. The reconstituted solution is good for 60 days. Discard it after the 60-day period.
5. The reconstituted solution requires refrigeration. It should be allowed to come to room temperature before the testing is done.

TEST PROCEDURE

1. Allow the reagent solution to come to room temperature.
2. Determine the patient's hematocrit (Section 15–4).

3. Place 2 mL of Sickledex test solution into a glass test tube. The solution will come up to the red line on the test tube provided in the kit.
4. Put on gloves.
5. Insert the marked end of the capillary pipette into the pipette bulb.
6. Insert the glass pipette into well-mixed whole blood. By capillary action, collect blood up to the mark on the capillary tube. This will be a blood volume of 0.02 mL.
7. Wipe the outside of the capillary tube with a 2 × 2-inch gauze. Do not touch the gauze to the open tip of the tube.
8. Add the measured whole blood into the Sickledex solution by placing a finger over the pipette bulb hole and gently squeezing the pipette bulb. Put your finger back on the hole and suck up the solution several times to wash the blood out of the capillary tube.
9. If the patient's hematocrit is less than 20, add a second aliquot of whole blood so that 0.04 mL is added.
10. Cover the top of the test tube with Parafilm and invert the tube several times to mix well.
11. Place the test tube in the test-tube rack.
12. Examine the test tube after 6 to 15 minutes. A positive test is indicated by a turbid (i.e., cloudy) solution. For a positive test, the ruled lines on the back of the test-tube rack should not be visible through the turbid solution. For a negative test the solution is either transparent or only slightly turbid. The lines can be easily seen through the solution of a negative test (Fig. 15–48).

QUALITY CONTROL

Positive and negative controls should be run each day that a sickle cell screening test is performed. A positive control (i.e., anticoagulated blood from a patient with sickle cell disease or sickle cell trait) can be stored in the refrigerator for up to 1 month. Negative controls are readily available from whole blood that is collected from other patients.

SOURCES OF ERROR

False-Negative Results

- An infant's blood primarily contains fetal hemoglobin. An affected infant will therefore not produce a positive test until 3 months of age. At that time there is a sufficient percentage of hemoglobin S to give a positive reaction
- Inactive reagent. Do not use the test kit after the expiration date
- Insufficient blood-sample size used for a patient with a hematocrit less than 20
- Testing of old, unrefrigerated blood
- Recent transfusion. Ten percent hemo-

Figure 15–48. Sickledex test. The solution on the left is opaque, indicating a positive test. Lines are visible through the solution on the right, indicating a negative test.

globin S is required to give a positive test. A patient with sickle cell trait who has recently received a blood transfusion may have less than this 10% level of hemoglobin S

False-Positive Results

- Elevated protein (as in multiple myeloma) will cause flocculation of the solution, which can appear to give a positive test.
- Other hemoglobinopathies, such as hemoglobin C Harlem or C Georgetown, will give a positive test.

COMMENTS

- Because this test may be done infrequently, it is best to batch all samples collected during the week and to run the samples on one week day. Collect samples in purple-top tubes and refrigerate them until testing.
- A positive test must be followed with hemoglobin electrophoresis to differentiate sickle cell disease from sickle cell trait or from other hemoglobinopathies.
- Testing should not be done on children less than 3 months of age because of the false-negative results.
- The peripheral smear should not be used as a screening test for sickle cell trait or disease. Patients with sickle cell disease may show a wide variety of distorted cells on a stained peripheral smear, including crescent-shaped, ovalocytes, or target cells. Many RBCs may also show inclusions, such as Howell-Jolly bodies. However, the percentage of sickled cells will vary from patient to patient. In patients with sickle cell trait, it is uncommon to find any sickled cells on a peripheral smear.
- A previously employed sickle cell screening test involved mixing 2% sodium bisulfite with a whole-blood sample and observing for RBC sickling. This test is not recommended because it is much less sensitive than the Sickledex test in identifying patients with sickle cell trait. Furthermore, the reagents for the test must be prepared fresh each day, the test requires 24 hours to perform, and microscopic interpretation is often difficult because of the presence of normal but crenated RBCs. (See the sensitivity and specificity problems in the Appendix for a comparison of the bisulfite test and the Sickledex test.)

REFERENCES

Ballard MS, Radel E, Sakhadeo S, Schorr JB. A new diagnostic test for hemoglobin.S. *J Pediatr.* 1970;76:117.

Loh WP. Evaluation of a rapid test tube turbidity test for the detection of sickle cell hemoglobin. *Am J Clin Pathol.* 1971; 55:55.

Schneider RG, Alperin JB, Lehmann H. Sickling tests: Pitfalls in performance and interpretation. *JAMA* 1967;202:419.

Product Insert: Sickledex, Ortho Diagnostics, Raritan, NJ 08869, 1976.

SECTION 13: Reticulocyte Count

BACKGROUND

Reticulocytes are immature RBCs that have extruded their nucleus but contain residual RNA and protoporphyrin. They are released from the marrow into the peripheral blood and then take 1 to 2 days to synthesize the residual basophilic staining material. When the synthesis is complete, the cells are considered mature RBCs.

On a Wright's-stained peripheral smear, these young blood cells may appear larger than normal red cells and are bluish in color. This is called "polychromasia" (Color Plate I Fig. 5). With special stains a pattern of filaments and granules (i.e., a "reticular pattern") can be seen (Color Plate I, Fig. 6).

The normal RBC has a life span of 120 days; therefore about 1% of the RBCs are destroyed and replaced each day. Since the newly released cells are reticulocytes, about 1% of the RBCs in a normal peripheral smear are reticulocytes. A reticulocyte count is the ratio of reticulocytes compared to the total number of RBCs that are present.

One of the first questions that a clinician must answer when faced with an anemic patient is whether the anemia is due to a decrease in the marrow's production of RBCs or to an increase in the rate of RBC destruction. The reticulocyte count helps with this question because it gives a measure of the marrow's "response" to an anemia. In an anemia due to the marrow's underproduction of red cells, the reticulocyte count will be low. In an anemia due to a shortened red cell life span (i.e., as with hemolysis or blood loss), a healthy marrow will respond with an increase in red cell production. An elevated reticulocyte count will then be seen.

Reticulocyte counts can be done by counting 1,000 cells or by using a Miller disc.

TEST INDICATIONS

- To help diagnose the cause of a newly discovered anemia
- To follow the response to therapy for a previously diagnosed anemia
- To follow the course of a chronic hemolytic anemia

TEST MATERIALS

Reticulocyte Stain

Several stains can be used to identify reticulocytes. These stains are referred to as "supravital" because they stain the cell while it is still living. We recommend the use of new methylene blue N stain. The stain should be purchased in the liquid form. It is available as a Unopette test kit (catalogue No. 5821; Becton Dickinson, Rutherford, NJ 07070). The stain can also be purchased in small amounts from your local hospital laboratory. The composition of commercially available stains can vary greatly; therefore you should only buy stain that is certified by the US Biological Stain Commission. The stain can be stored in a large brown dropper bottle. It should be filtered before use to remove particulate matter.

Small funnel

Small laboratory filter paper

Single Cell Counter

Small hand-held counters are needed to count the reticulocytes and the total number of RBCs. Scientific supply companies sell expensive counters, but it is considerably less expensive to buy lap counters from an athletic shop. They count from 0 to 9,999 and can be quickly reset to zero. Two counters are needed for this procedure.

Microscope

See Chapter 10.

Microscope Slides

Regular-sized frosted-end slides can be used.

Miller Disc (Optional)

This is a small glass disc that fits into the microscope eyepiece. Etched on its surface is a small calibrated square within a larger square. The advantage of using a Miller disc is that it saves time and improves accuracy. It is available from microscope manufacturers for about $40 to $50 (American Scientific Products Microscopy Technical Center, 800-323-3102; or Cambridge Instruments, PO Box 123, Buffalo, NY 14240.) When ordering, it is important to indicate whether you have regular or wide-field oculars.

SPECIMEN COLLECTION

A specimen from a well-mixed tube of EDTA-anticoagulated blood (i.e., a purple-top tube) should be used. This sample is stable for reticulocyte staining for 24 hours at room temperature. The tube should be rocked for 2 mintues on a rocker before being used. This ensures thorough mixing. Alternately, the tube can be gently inverted 60 times by hand.

TEST PROCEDURE

1. Peform a microhematocrit on the collected specimen.
2. Record the hematocrit value on the patient's laboratory slip.
3. Filter a small amount of the reticulocyte stain into a clean centrifuge or a pediatric red-top tube. Use the small funnel and filter paper.
4. Place seven drops of the filtered reticulocyte stain into a second centrifuge tube or a pediatric red-top tube.
5. With a small pipette, transfer seven drops of the well-mixed blood sample into the tube that contains the seven drops of filtered stain.
6. Mix the stain and the blood together by squeezing the pipette bulb and drawing the mixture up into the pipette. Keep the pipette in the tube, as it is used later in the procedure.
7. Let the tube of blood and stain stand at room temperature for a minimum of 15 minutes. The cells require this time to pick up the stain.
8. After 15 minutes but before 30 minutes, remix the blood and stain by squeezing the pipette bulb. Remixing is crucial because the reticulocytes have a different specific gravity than mature RBCs and will separate out while the tube is standing.
9. Place three glass slides on the bench top with the frosted side facing up.
10. With the pipette and bulb, place a drop of the blood/stain mixture near the frosted end of each of the glass slides.
11. Prepare the slides as described in Section 15–8. These smears usually cover more of the slide than regular blood smears. This larger pattern is due to the lower viscosity of the blood/stain mixture.
12. Allow the slides to air dry.
13. There are two methods of cell counting: **Counting 1,000 cells:**
    a. Place a dried, stained, slide on the microscope stage. Move the condenser all the way up and focus under low power.
    b. Find an area of the slide where the RBCs are evenly distributed and are not touching each other.
    c. Put a drop of oil on the slide in the area where the light from the condenser is passing through the slide.

d. Move the oil objective into place and refocus. Be sure not to pass the high-dry objective through the oil drop.
e. Examine the slide for reticulocytes. They will appear as bluish-green cells with dark strands of reticular material (Color Plate I, Fig. 6A). A reticulocyte is any RBC that contains two or more particles of blue-stained material. More mature cells will be bluish green but will have no darkly staining inclusions. Some mature cells will look like reticulocytes because of artifacts that form on the cell during the drying process. These artifacts can be differentiated from reticulum because the blueness disappears and the patch of blue becomes refractile when you focus up and down (Color Plate I, Fig. 6B). Cells with these artifacts are not counted as reticulocytes.
f. Make sure that both counters are set to zero.
g. Looking at the microscope field, imagine that it is divided into four quadrants, such as a pie divided into quarters. Count the quarters, one at a time. Count all four before moving to a new field. While you are counting, you will be constantly focusing up and down to distinguish reticulum from refractile artifacts.
h. Count the total number of RBCs (both mature and reticulocytes) in one quarter of the field using the counter held in your right hand.
i. Count the number of reticulocytes in the same quarter on the counter in your left hand.
j. When you have finished counting one microscopic field, move the slide to a new field. Move in a systematic manner so that you do not count the same field twice.
k. Continue counting until 1,000 total RBCs have been recorded on the counter in your right hand. Record the number of reticulocytes counted per 1,000 RBCs on a piece of paper (i.e., the number on the counter in your left hand.)
l. Reset both counters to zero.
m. Repeat the counting process with a second slide. If the two results agree within 10%, continue with step n. If the two do not agree within 10%, repeat the count with the third slide and use the two values that agree within 10%. If no two slides give a value within 10%, repeat the entire process above from step 4.
n. Average the two values that agree within 10%.
o. Calculate the "uncorrected reticulocyte count" by dividing the average number of reticulocytes per 1,000 cells by 10. This will give the percentage of reticulocytes.
p. Calculate the "corrected reticulocyte count (see below)."

Miller Disc-Counting Procedure:

a. Following the manufacturer's instructions, place the Miller disc into the microscope eyepiece. Once placed, it can remain in the eyepiece.
b. Place a dried, stained slide on the microscope stage. Move the condenser up and focus under low power.
c. Find an area on the slide where the RBCs are evenly distributed but not touching. Note the appearance of the grid through the eye piece.
d. Put a drop of oil on the slide

where the light from the condenser is passing through.

e. Move the oil objective into place and refocus. Be sure not to pass the high-dry objective through the oil drop.

f. Examine the slide for reticulocytes. They will appear as bluish-green cells with dark strands of reticular material (Color Plate I, Fig. 6A). A reticulocyte is any RBC that contains two or more particles of blue-stained material. More mature cells will be bluish green but will have no darkly staining inclusions. Some mature cells will look like reticulocytes because of artifacts that form on the cell during the drying process. These artifacts can be differentiated from reticulum because the blueness disappears and the patch of blue becomes refractile when you focus up and down (Color Plate I, Fig. 6B). Cells with these artifacts are not counted as reticulocytes. While you are counting, you will be constantly focusing up and down to distinguish reticulum from refractile artifacts.

g. Set both lap counters to zero.

h. Using the lap counter in your right hand, count the total number of RBCs (including reticulocytes) in the small square.

i. Using the lap counter in your left hand, count the number of reticulocytes in both the large *and* small squares.

j. When you have finished counting the squares, move the slide to a new field. Move in a systematic manner so that you do not count the same field twice.

k. Continue counting until you have counted a total of 111 RBCs on your right-hand lap counter.

l. Look at the counter in your left hand. Stop if the total number of reticulocytes is 5 or more. If you have counted less than 5 reticulocytes, continue the counting process (steps h through j) until you have counted a total of 222 red cells on the counter in your right hand. Record the number of reticulocytes (left hand counter) on a piece of paper.

m. Reset the counters to zero and repeat the counting process (steps h through l). Make sure that you count the same total number of RBCs for each count (i.e., do not count 222 cells one time and 111 the other).

n. The two numbers should agree within 10% of each other. If they do not, repeat the count on a third slide and use the two values that do agree within 10%. If no two slides agree within 10%, repeat the entire process above from step 4.

o. Average the two numbers.

p. If you counted 111 cells, move the decimal point one space to the left. For example, if there were 25 reticulocytes counted, the "uncorrected reticulocyte count" is 2.5%. If you counted 222 cells, divide the average by 2 and then move the decimal point one space to the left. For example, 25 reticulocytes counted; divided by 2 = 12.5. Move the decimal point one space to the left, 1.25%.

14. The uncorrected reticulocyte percentage must be corrected for the individual patient's hematocrit using the formula:

$$\text{Corrected reticulocyte percent} = \text{uncorrected reticulocyte count} \times \frac{\textit{Patients Hct}}{45}$$

This correction factor is necessary because the reticulocyte percentage is a function of both the number of reticulocytes and the total number of RBCs. Therefore a decrease in the total number of RBCs (i.e., by hemolysis) could result in an increase in the uncorrected reticulocyte percentage even if the number of reticulocytes remained the same. This increase in the reticulocyte percentage would give the false impression of an active marrow response even though the marrow had not really responded—only the number of mature RBCs had decreased. The correction factor takes this decrease in the number of RBCs into account by using the patient's hematocrit in the calculation.

15. A second correction factor should be made if there is evidence that the marrow is very rapidly turning out young cells. In this case the reticulocytes will take more than 1 day to mature because they leave the marrow at an earlier stage in their development. The reticulocyte count is dependent on both the actual number of reticulocytes and on how long each cell remains a reticulocyte. Therefore a second correction should be performed if nucleated RBCs or "shift reticulocytes" (Color Plate I, Fig. 5C and D) are seen on the peripheral smear. The latter are large, bluish, non-nucleated, RBCs that can be seen on the Wright's stain. The correction involves dividing the corrected reticulocyte percent (as calculated above) by the number of days that are required for maturation of the reticulocyte as shown below:

Hematocrit Level	Maturity in Days
40-45	1.0
35	1.5
25	2.0
15	2.5

TEST INTERPRETATION

- Standard reference values for the corrected reticulocyte count are:
  Newborn to 2 weeks of age: 2.5% to 6.5%
  Male older than 2 weeks: 0.5% to 1.5%
  Female older than 2 weeks: 0.5% to 1.5%

- In general, an increased reticulocyte count indicates an active marrow response to an anemia that is due to shortened RBC survival time. This is seen following an acute hemorrhage or with hemolysis (i.e., sickle cell anemia, hereditary spherocytosis, autoimmune hemolytic anemia). In massive hemorrhage the reticulocyte count is usually 10% or greater. In a chronic hemolytic anemia, it can be persistently elevated to between 5% and 10%. An elevated reticulocyte count can be used to follow the response to treatment of a nutritional anemia. For example, after iron therapy is started for an iron deficiency anemia, the reticulocyte count begins to rise in 2 days and peaks at 2 weeks. Therefore when it is used to document a response to therapy, it should be checked within the first 2 weeks.

A low reticulocyte count reflects a failure of the marrow to respond properly to an anemia. This may be secondary to a marrow nutritional deficiency (i.e., iron, folate, or vitamin B_{12}), to a chronic disease (i.e., uremia, liver disease, chronic infection, or a collagen vascular disease), aplastic anemia, or marror infiltration (i.e., a myelophthisic process).

QUALITY CONTROL

There is no good way to routinely perform quality control on this procedure. When first setting up this procedure in your laboratory, it is a good idea to send several blood samples to your reference laboratory and to compare their results with the results from your own office laboratory.

SOURCES OF ERROR

Falsely Elevated Counts

1. Failure to correct the count for an anemia
2. Counting cells with refractile artifacts as reticulocytes
3. Insufficient mixing of the sample
4. Counting red cells with inclusions as reticulum. These inclusions include:
   a. Howell-Jolly bodies: These are nuclear remnants that appear as well-defined deep-purple dots. They can also be seen on the Wright's-stained peripheral smear (Figs. 15–39 and 15–40)
   b. Heinz bodies: These inclusions are denatured hemoglobin. They appear lighter in color than reticulum and are seen in the periphery of the cell. They are not seen with Wright's staining
   c. Pappenheimer bodies: These are purple-staining iron deposits. They occur in small clusters and are visible with Wright's staining

Falsely Decreased Counts

- Blood that contains high concentrations of glucose will not take up the reticulocyte stain.
- Insufficient mixing of the sample.

COMMENTS

- This test may seem long and complicated. There is a temptation to take shortcuts, such as eyeballing a field and saying that it contains about 100 cells. The test has a high statistical error, so any such shortcuts should be discouraged.
- The lower the reticulocyte count the greater the amount of error. There is a ± 25% level of imprecision when the corrected count is 1% to 4%. If the count is 5% or greater, this error decreases to ± 10%. The higher the count the better the precision.

REFERENCES

Fannon M, Thomas R, Sawyer L. Effect of staining and storage times on reticulocyte counts. *Lab Med.* 1982; 13:7.

Hillman RS, Finch CA. Erythropoiesis: Normal and abnormal. *Semin Hematol.* 1967;4:327.

Hillman RS, Finch CA. The misused reticulocyte. *Br J Hematol.* 1969;17:313.

Prouty HW. Correcting the reticulocyte count. *Lab Med.* 1979;10:161.

National Committee for Clinical Laboratory Standards, Proposed Standard H16P. Villanova, PA. 1985.

CHAPTER 16

Prothrombin Time

INTRODUCTION

The most common coagulation tests in clinical practice are the prothrombin time (PT), activated partial thromboplastin time (APTt), platelet count, and bleeding time. The procedure for a manual platelet count is described in Chapter 15, Section 7. Of the other coagulation tests, the PT is the only one commonly used in the outpatient setting. The APTT is most commonly used to monitor heparin therapy in hospitalized patients. Bleeding time is occasionally used to study platelet function; however, it is rarely the preferred test for common clinical problems.

The PT is time required to initiate clotting when tissue thromboplastin is mixed with blood. It is a measure of the function of the external clotting pathway.

Until very recently we did not recommend that office laboratories do PT times. This was because it can be a tricky test to perform, and test errors have very serious implications for the patient. Most PTs are done with an instrument called a fibrometer, which measures clot formation either optically or electromechanically. Even with such an instrument, PT testing can be difficult because the reagents have short shelf lives, there is great variation in the thromboplastin reagents, the test is sensitive to differences in specimen handling, and the test temperatures must be controlled to within a tenth of a degree centigrade. All of these factors led us to recommend that office laboratories not do PT testing.

On the other hand, a reliable PT would be a very useful test for an office laboratory. It is the standard test to measure the therapeutic effect of Coumadin (sodium warfarin). This anticoagulant is used in patients with cardiac disease, strokes, and those with either recurrent deep-vein thrombosis or pulmonary emboli. An office PT would make monitoring the anticoagulant therapy in these patients much simpler. In addition, it would permit the rapid assessment of the level of anticoagulation in patients who presented with bleeding problems (i.e., new bruising). A final reason for doing an office PT is that special specimen handling is required. Most laboratories recommend that the test be performed within 6 hours of being drawn and that the specimen be placed on ice until the time of testing.

A new method to do PTs has recently been introduced (Coumatrak, DuPont Medical Products, Barley Mill Plaza, Wilmington, DE 19898). This test uses whole blood from a finger stick and individually wrapped reagent cartridges (Fig. 16–1). The 1988 instrument cost is $1,175.

Test Indications

- Any patient with a history of abnormal bleeding, especially those with liver disease, malabsorption, or malnutrition
- Monitoring the degree of anticoagulation in patients who are treated with warfarin (i.e., Coumadin)
- To assess the level of hepatic function in patients with acute hepatitis
- PT is *not* considered to be a useful preoperative screening test for healthy patients undergoing surgery; nor is it considered a useful test in the screening of otherwise healthy adults

Specimen Collection

1. This test is specifically designed to use a whole-blood specimen from a finger stick. (See Chapter 12 for the details of this procedure.) Be sure to wipe the first drop of blood away with gauze. It may contain tissue thromboplastin, which could interfere with the test. Also be sure to quickly apply the drop of blood to the reagent cartridge. If this is not done within about 5 seconds, the PT may be artificially reduced.
2. Venous whole blood drawn into a plastic syringe can also be used for this test. No anticoagulant should be used. If a syringe is used, it is essential that the blood be immediately applied to the reagent cartridge.

Test Procedure

The exact procedure recommended by the manufacturer should be followed.

1. Remove the reagent cartridge from the refrigerator and permit it to come to room temperature (about 5 minutes).

Figure 16–1. Coumatrak: Whole blood Protime Instrument (Courtesy of DuPont Medical Products).

2. Remove the reagent cartridge from the pouch and insert the cartridge into the test instrument.
3. Perform a finger stick. Wipe away the first drop.
4. Apply the next drop of blood to the cartridge. This step will be a bit awkward in the beginning. Ask the patient to relax his or her arm and to let you move his or her finger to the specimen area on the cartridge. As soon as the blood touches the cartridge, it is drawn by capillary action into the reagent chamber, where PT is measured by a laser.
5. Within 1 minute PT and the PT ratio (i.e., patient's value compared to a normal control value of 12.0 seconds) are displayed on the instrument monitor.

Test Interpretation

There has been a great deal of confusion about PT testing. A broad range of values has been called "normal" by different laboratories. These differences were largely due to differences between thromboplastin reagents. The thromboplastins in use today are less responsive than those that were used in the early studies on the treatment of deep-vein thrombosis. Clinicians have therefore unknowingly overanticoagulated patients.

In an effort to correct this problem, the International Committee on Thrombosis and Hemostasis and the International Committee for Standardization in Hematology have recommended the adoption of a uniform system for reporting PTs and the calibration of this system to the World Health Organization's (WHO) standard thromboplastin.

Table 16–1 gives the current recommendations by the National Heart, Lung, and Blood Institute (NIH) for desired PT ratios (using two reagent systems) in the treatment of a variety of conditions. Rabbit-brain thromboplastin is the reagent that is now used by most laboratories in the United States. Looking at Table 16–1, it is obvious that if rabbit-brain thromboplastin is used, a PT ratio of 2 represents serious overanticoagulation. The notion that 2.0 was a desirable therapeutic goal in the United States came from an earlier period when a thromboplastin other than rabbit brain was commonly in use.

The Coumatrak reports the PT in any of three ways: in seconds (based on rabbit-brain thromboplastin), in a standard ratio (also based on rabbit thromboplastin), and in the International Normalized Ratio (INR). The INR system is based on the WHO's more sensitive standardized throm-

TABLE 16–1. RECOMMENDED THERAPEUTIC PT RATIOS FOR ORAL ANTICOAGULANT THERAPY

Condition	I Rabbit-Brain Thromboplastin	II International Normalized Ratio
Treatment of deep-vein thrombosis	1.3-1.6	2.0-2.0
Treatment of pulmonary embolism	1.3-1.6	2.0-3.0
Prevention of embolism in patients with atrial fibrillation, tissue heart valves	1.3-1.6	2.0-3.0
Prevention of embolism in patients with prosthetic heart valves	1.6-1.8	3.0-4.5
Prevention in patients with recurrent emboli	1.6-1.8	3.0-4.5

boplastin reagent. Table 16–1 gives the desired Rabbit thromboplastin ratios in the first column and the equivalent INR value in the second column.

It is very likely that your patients will have PT testing done in several laboratories (i.e., office, reference, and hospital). Be sure that you know what type of PT has been done and how it is reported. Overcoagulation can kill!

Comments

- The instrument detects when an inadequately sized drop of blood has been placed on the cartridge. Since the cartridges are somewhat expensive, it is important to get a good drop of blood the first time.
- A normal and high control are available, but they are stable for only 1 day after reconstitution.
- Some states require that PTs be done with a normal control as well as in duplicate. In these states the reagent costs for the Coumatrak system will probably be prohibited
- The Coumatrak result is unaffected by variation in blood hematrocrit.

REFERENCES

Lucas FV, Duncan A, et al. A normal whole blood capillary technique for measuring prothrombin time. *AM J Clin Pathol.* 1987;88:442-446.

Hirsh J, Levine MN. The optimal intensity of oral anticoagulant therapy. *JAMA.* 1987;258:2723-2726.

CHAPTER 17

Urinalysis

SECTION 1: Introduction

The urinalysis is likely to be the most commonly performed test in the primary care office laboratory. Abnormal urine findings are seen in many renal diseases as well as in a wide range of diseases of other organ systems. But urine is far from glamorous, so its examination is often neglected.

Clinicians may come to doubt the findings of a urinalysis in the hospital setting. The least trained nursing aide is often responsible for the urine collection, and the least experienced technician is frequently responsible for its examination. In contrast, an excellent urinalysis can be done in the office setting. Urine is an ideal specimen because it is readily obtainable, easily tested, and provides a great deal of useful information.

TEST USES

A urinalysis has many uses in the diagnosis and management of disease. A few of the more common uses in office practice include:

- Diagnosing Urinary Tract Infections: Patients with dysuria or urinary frequency should have a complete urinalysis to look for pyuria and bacteriuria. Many clinicians treat suspected urinary tract infections based on the microscopic urinalysis findings alone and do not require a urine culture.
- Health Maintenance: The urinalysis is an inexpensive test that is useful as a routine screening procedure in periodic health examinations. Almost every insurance or pre-employment physical requires at least a dipstick urinalysis.
- Pregnant Patients: Toxemia and diabetes are two of the commonest high-risk problems in pregnancy. A dipstick urinalysis for protein and glucose at each office visit is generally recommended.
- Diabetes: With the advent of inexpensive whole-blood glucose analyzers and more reliable and accurate whole-blood dipsticks, testing for glucosuria is no longer the gold standard for diabetic care. The urinalysis remains important because even mild proteinuria may be an important indicator of diabetic nephropathy.
- Hypertension: A urinalysis should be a regular part of the laboratory testing of hypertensive patients. It can indicate both renal causes for hypertension as well as renal injury from the elevated blood pressure. Proteinuria is often the first indicator of hypertensive nephropathy.
- Sports Physicals: Every child participating in sports should have a screening urinalysis. There is a benign condition called "athletic pseudonephritis" that is important to recognize. Patients with this condition may have red blood cells (RBCs), protein, white blood cells (WBCs), red cell casts, white cell casts, renal tubular cell casts, and granular casts in their urine, all without identifiable renal pathology. These findings are

seen for up to 3 days after even mild exercise in some patients. It is important to differentiate this group of patients from those in whom abnormal urinalysis findings indicate renal pathology.

COMPLETE URINALYSIS

This chapter describes the procedure for a complete urinalysis. This includes:

- A macroscopic examination of the urine for color, turbidity, and specific gravity.
- Dipstick testing for bilirubin, blood, glucose, ketones, leukocyte esterase, nitrite, pH, protein, specific gravity, and urobilinogen.
- A microscopic examination of a spun-urine sediment for cells, casts, crystals, and bacteria.

We recommend that all urine that is collected for testing be obtained with an appropriate "clean-catch" technique. This helps to eliminate specimen contamination from skin or vaginal secretions.

CONTROVERSIES

There are several controversies about the proper technique for performing a complete urinalysis.

- **Spun *v* Unspun:** Some people have argued that it is best to examine unspun urine for the microscopic urinalysis. We advise instead that all microscopic urinalyses be done on spun urine sediment. An unspun sample is unlikely to show casts, which can be one of the most important urinalysis findings.
- **Gram's Staining:** Some laboratories have found that it is useful to do a Gram's stain sediment in patients with pyuria. We have found that this test is rarely necessary to make appropriate clinical decisions. We do, however, perform a Gram's strain on urine from male patients with py-uria to rule out a *Neisseria gonorrhoeae* infection.
- **Sediment Staining:** Some laboratories stain all sediments before examination. We have found that staining is helpful in classifying some cellular elements if they cannot be identified unstained, but this is not needed for most routine urinalyses.
- **Quantification:** Research articles on urinary tract infections frequently use hemocytometers to count cellular elements and other microscopic findings. However, in clinical practice such counting is time consuming and of little practical significance. Modified chamber-counting slides are available (See Fisher Scientific Products catalog). It has not been shown that these more expensive alternatives to plain microscopic slides provide more clinically useful information. To provide some standardization to the microscopic findings, we routinely spin the same volume of urine (12 mL), use a plastic transfer pipette to put 2 drops of sediment on a slide, and use a 22 × 30-mm coverslip. We feel that this provides adequate standardization for clinically relevant information. A cup of coffee will dilute the urine enough to negate even the most compulsive attempts at microscopic quantification.

PERSONNEL TRAINING

A trained medical technologist or medical laboratory technician (MLT) will be able to perform a complete urinalysis. It is nevertheless very helpful for the clinician to examine the urine sediment to get a clearer picture of the findings than is possible by reading the test report. The technician should understand that a clinician who wants to look at the sediment is not "checking up" on the staffs' skills.

Office nursing staff can be quickly trained to perform the macroscopic and dipstick urinalysis as well as to prepare the sediment for examination by the clinician. This is a timesaver in practices that have no full-time laboratory worker.

It requires much more time to learn how to do an accurate microscopic examination. If you want to teach staff to do microscopics, we suggest that they study this chapter (especially the photographs) and then study additional atlases (see references). They will need to spend some one-on-one time at a teaching microscope. This can probably be arranged through the laboratory at your local hospital.

REFERENCE SLIDES

Interesting urine sediments can be preserved for future reference as teaching slides. This is an excellent teaching aid for any office in which medical students, residents, or nurses are trained. The technique for preparing these slides is described in the Appendix.

REFERENCES

Each laboratory should have a small reference library of microscopic urinary sediment findings. Charts, booklets, and books are available from a variety of sources:

A wall chart and a *Primer of Microscopic Urinalysis* by Meryl Haber are available free of charge from: Advertising Department, ICL Scientific, 18249 Euclid St, Fountain Valley, CA 92708.

A wall chart and *Modern Urine Chemistry* are available from your local Ames representative or by writing to Ames Division, Miles Laboratories, Inc, PO Box 70, Elkhart, IN 46515.

For under $5 a copy of *An Atlas of Urinary Sediment* is available from Abbott Laboratories, North Chicago, IL 60064.

Haber M. *Urinary Sediment: A Textbook Atlas.* Chicago, IL: Education Products Division, ASCP; 1981.

Graff Sister L. *Handbook of Routine Urinalysis.* Philadelphia, PA: Lippincott; 1983.

SECTION 2: Safety and Infection Control

Urine is normally sterile and therefore poses little hazard to the laboratory worker. However, urine can be infectious in patients with certain illnesses, and because of this the laboratory worker should use care in its handling. Viral hepatitis, pinworms, and *N gonorrhoeae* infections are all transmittable by the handling of infected specimens. Since you do not know who is infected and because there is a common "aesthetic distaste" to accidently spilling urine on your hands, wearing gloves is recommended.

All urine spills should be wiped up with a disinfectant, such as 5% Staphene. At the end of the day all bench tops should be wiped down with the disinfectant solution. The laboratory should have two special waste containers with lids, foot pedals, and removable baskets, which should be lined with two heavy plastic liner bags. One container is to be used for glass only (i.e., pipettes, microscope slides, and broken glassware). The second is used for urine specimen collection cups. Each container should be labeled with a permanent marker to alert the cleaning staff so that they can avoid injury from broken glass.

Some laboratory personnel prefer to use white laboratory coats to protect their clothing from spills. These can be laundered with the clinician's white coat.

Handwashing is the most important procedure in preventing the spread of infections within the laboratory. An antiseptic soap should be used after handling any infectious specimen. Care should be taken to clean under the fingernails. The last procedure of the day for the laboratory worker should be to wash the hands well. If laboratory personnel have open cuts on their hands, we suggest that they wear gloves when handling specimens.

SECTION 3: Equipment and Supplies

Several pieces of equipment are needed to perform a complete urinalysis. The refrigerator is used to store urine specimens until they are examined. The cold temperature prevents the overgrowth of bacteria in the specimen. A centrifuge is used to spin the urine sample for the microscopic examination. Either a standard centrifuge (Fig. 11–1) or a Statspin (Figs. 11–3 and 11–4) can be used. A microscope is needed to examine the urine sediment for cells, casts, and crystals. These pieces of equipment are described more fully in Chapters 10 and 11.

Other reagents and supplies are also needed for the urinalysis. These can be found in catalogs such as Scientific Products or Fisher Scientific. See Appendix, Section 1, for the addresses of these distributors.

Acetic Acid

A 5% solution of acetic acid is used to acidify urine samples in which amorphous phosphate crystals have formed. The acid can be purchased in half-liter containers as 99% glacial acetic acid. The 5% solution is prepared by mixing one part of the acetic acid with 19 parts distilled water. It may be more convenient to obtain small quantities of 5% solution from the local hospital laboratory.

Acid Bottles

These glass bottles are used to store acid reagents. Bottles are available in 120 and 250 mL sizes. They are fitted with polyethylene stoppers.

Antiseptic Soap

An antiseptic soap container with a pump handle should be placed next to the sink so that the laboratory worker can wash his or her hands quickly.

Biohazard Container

All office laboratories should have a system for the disposal of biologic materials. The Terminal Biohazard System (W.G. Whitney Co, Skokie, IL) includes a container, a lid marked with a biohazard symbol, autoclavable plastic bags, and bag closures. This system should be used to store contaminated wastes, such as infectious samples. The filled bags can be autoclaved by the hospital or reference laboratory.

Centrifuge Tubes

Disposable 14- or 15-mL graduated polystyrene centrifuge tubes are used to spin most urine samples. These can be washed and reused, but they are inexpensive and their reuse is not recommended. Capped centrifuge tubes can be purchased, but these are not necessary.

Coverslips

Glass coverslips are used when examining the urine sediment. These protect the high-dry objective from urine, which can otherwise etch the glass surface of the objective. The 22 × 30-mm size works well.

Dipsticks

Up to ten tests can be performed on urine samples by using these reagent strips. They are manufactured by several companies.

Disinfectant
A disinfectant should be used to clean the counter tops in the laboratory each day. 5% Staphene is adequate for cleaning off bench tops and wiping up spills.

Distilled Water
This can be purchased in gallon jugs in most drug stores.

Forceps
Small, inexpensive forceps are used to pick up the reagent tablets for several of the laboratory tests. These can be purchased from scientific supply companies.

Grease Pencils
Both red and black grease pencils should be available. They are used to write the patient's name on the urine collection cup and the centrifuge tube.

Icotest
This is a semiquantitative test for bilirubin in the urine (Ames Division, Miles Laboratories, Inc, PO Box 70 Elkhart, IN 46515). It is more sensitive than the dipstick test for bilirubin.

Lens Cleaner
A solvent is needed to periodically clean the microscope objectives. Xylene, alcohol, or Kodak lens cleaner can be used. If any urine gets on the objective, it should be wiped clean immediately. Urine can etch the glass surface of the objective.

Microscope Slides
Standard plain (unfrosted) 1 × 3-inch slides are preferable. With practice, two urine specimens can be placed on one slide by placing one specimen at each end of the slide.

Minicath
Davol Single Use female cath kit is a graduated plastic specimen tube with a sealing cap and a No. 8 French urethral catheter. (CR Bard Inc, 100 Sockanossett Crossroad, Cranston, RI 02920; 401-463-7000; Fig. 17–1).

Pipettes
Plastic transfer pipettes are available, and they have several advantages over glass pipettes. They do not require a rubber bulb, they are not easily broken, and urinary protein will not adhere to their sides, as can happen with glass pipettes.

Refractometer
This instrument is used to measure urine-specific gravity. Hand-held units cost about $500. Units with a stand cost about $700. It has been our experience that they last about 3 years but then get a huge bubble that is too expensive to repair. We therefore recommend dipstick specific gravity testing instead.

Sediment Stain
Several urine sediment stains are available. Sedi-stain is a concentrated stain produced by Clay-Adams. Kova stain produced by ICL Scientific is a stabilized Sternheimer-Malbin stain. Both help in the differentiation of various cellular elements found in the urine sediment.

Sulfasalicylic Acid, 20%
This solution is used to perform the protein precipitation test. It can be purchased as a 20% solution, but only a small quantity is needed. It is therefore better to obtain the solution from the hospital or reference laboratory. It should be kept in a small acid bottle near the urine centrifuge.

Test-Tube Rack
Two small rubberized test-tube racks are needed. One should be placed on the bench near the centrifuge and the other near the microscope.

Urine Dipstick-Control Solutions
Dipsticks should be checked against controls to be sure that the reagent pads are effective. Solutions with various concentra-

tions of urine substances, such as glucose and protein, should be used.

Urine Specimen Cups

Clear, leak-proof, polypropylene containers with tightly fitting lids should be used for the urine collection. This allows you to store the sample in the refrigerator for later testing without worrying about spillage. Both sterile and nonsterile containers are available.

Urinometer

This device can be used to measure the urine specific gravity; however, it is impossible to test small amounts of urine. It has a scale from 1.000 to 1.060 and costs $10 to $20. With the availability of refractometers and dipstick testing, we do not recommend using a urinometer.

Waste Containers

Two metal containers with attached lids and foot pedals are needed. One is used for urine specimen cups. The second is used only for the disposal of glass.

Wash Bottles

Plastic wash bottles (125 and 250 mL) with spouts are inexpensive. They should be filled with distilled water and labeled with an adhesive sticker.

SECTION 4: Specimen Collections/Handling

Urine specimens can be collected by a variety of techniques. Some techniques are less likely to result in contamination of the specimen. The most common of these techniques is the clean-catch midstream specimen. Clinicians often want a microscopic urinalysis after seeing the results of the dipstick test. If a clean-catch technique is used for the first specimen collection, then that sample can also be used for the microscopic examination or urine culturing. We therefore suggest that all urine testing be done on clean-catch specimens.

COLLECTION TECHNIQUES

Adult Female Clean Catch

The patient is told the following: "Wash the hands well. Sit on the toilet, and separate the labia (genital lips) with one hand. Using a povidone-iodine pad or any other nonalcohol 'wipe,' clean the urethral area from front to back. Repeat this a second time. With the labia still separated, release a small amount of urine. Momentarily stop the urine stream. Hold the collection container up to the urethral opening and restart the urine flow." Nursing supervision may be required for young or disabled patients. If the patient has a heavy vaginal discharge, a tampon should be inserted before the urine sample is collected. This will prevent the contamination of the specimen with the vaginal discharge. If that does not reduce the vaginal contamination, a minicatheterization should be performed.

Minicatheterization

A convenient system for obtaining catheterized urine specimens is available (see above). This system (Fig. 17–1) includes a plastic tube with a screw-on top and a 5-inch long, thin bladder catheter. This system comes in sterile packaging and is quick, fairly inexpensive, and nontraumatic to the patient. To use the minicatheterization system, the urethral meatus is first cleaned with a povidone-iodine gauze. Using sterile gloves, the nurse should then pull the catheter out of the tube so that it is maximally extended. The

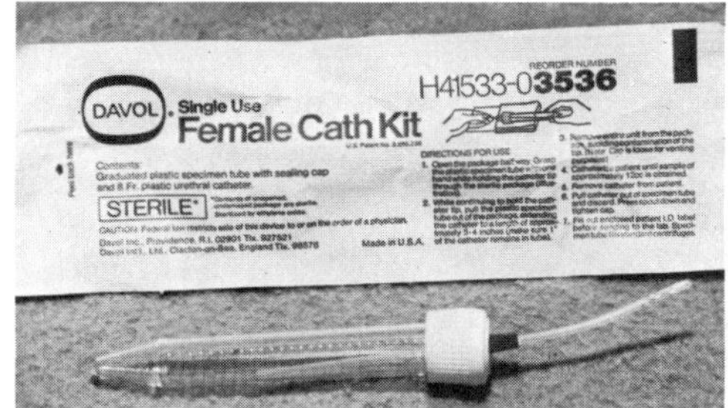

Figure 17–1. Minicatheterization kit.

hand holding the catheter should remain sterile. The other hand should then be used to separate the labia. Gently insert the catheter until urine flows into the collection tube. If there is resistance to passing the catheter, ask the clinician for assistance. If no urine is returned, even with maximal catheter insertion, then the bladder is probably empty. Remove the catheter. Close the top of the minicatheter system. Take the specimen to the laboratory.

Adult Male Specimen

Urinary tract infections, which are so common in females, are relatively uncommon in males. The complaints of dysuria in a male patient usually indicate a gonococcal or nongonococcal urethritis, or prostatitis. Many of these patients have minimal penile discharge, which makes it difficult to obtain a satisfactory urethral smear for Gram's staining. We therefore routinely request that two urine specimens be collected from all male patients with genitourinary complaints. It has been found that males will frequently have many white cells in the first specimen. There are usually far fewer, if any, white cells in the second (i.e., midstream) specimen. If there are more than 5 white cells/high power field (HPF) in the spun sediment of the first specimen, a Gram's stain should be performed on the sediment to look for the presence or absence of gram-negative diplococci. A GC culture may be taken from the sediment of the first specimen if it has not been refrigerated. A regular urine culture, if required, should be done from the second (midstream) specimen. (See the Desai paper on p. 326 for a more detailed discussion.)

Two urine collection cups are required. The nurse should mark one cup "No. 1" and the other "No. 2" with a grease pencil on the side (not the top.) In addition, the nurse should draw a line around cup No. 1 at about the 15-mL mark. Both cups should then be given to the patient with the following instructions: "Pull back the foreskin with one hand. Clean the urethral opening with the povidone-iodine pad. Wipe the area dry using the sterile gauze. The first portion of the urine stream (the first 15 mL) should go into cup No. 1. Stop the stream. Put down cup No. 1. Then pick up cup No. 2. Restart the stream and collect the rest of the specimen in cup No. 2." Both specimens are processed routinely, except be sure not to refrigerate cup No. 1.

Preprostatic/Postprostatic Massage

In male patients complaining of prostatitis, it is helpful to have a preprostatic and a postprostatic massage sample. Since the symptoms may be vague enough to suggest urethritis, such a patient may end up having three specimens examined—the first 15 mL, the midstream, and a postmassage specimen. The postmassage specimen is collected in the same way as the midstream, that is clean catch, except this time the first few drops go into the toilet.

Infant Bagged Specimen

The skin around the genital area is cleaned with alcohol and then dried with gauze. A self-adhesive urine collection bag is securely attached (Fig. 17–2). The infant can then be given a bottle to promote urination. The bag should be removed as soon as a sample is obtained. This method suffers from frequent contamination with skin and bowel bacteria.

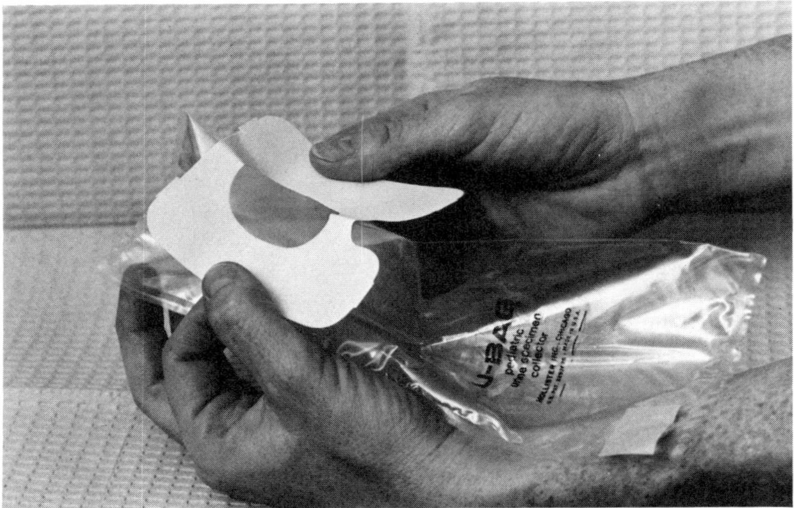

Figure 17–2. A urine collection bag is used to collect infant specimens.

Suprapubic Aspiration

The lower abdomen is cleaned with povidone-iodine and then with alcohol. A 5-mL syringe with a 1.5-inch 21-gauge needle is used for the aspiration. The needle is directed perpendicular to the skin at a midline point 1 cm above the pubic rami. Up to three passages of the needle should be made as back pressure is put on the syringe. If no urine is obtained, the procedure should be stopped. Failure to collect urine by this method indicates that the bladder is empty. This is the only method of urine collection in an infant that ensures a noncontaminated specimen.

SPECIMEN HANDLING

After the urine is collected, it should be taken immediately to the laboratory for processing. If the urine is to be examined later, it should be stored in a tightly covered container and placed in the refrigerator. It must be examined within 2 hours. Urine left standing at room temperature for even one-half hour may show bacterial overgrowth and cellular deterioration.

TWENTY-FOUR HOUR URINE COLLECTION

The kidney excretes a variety of chemicals into the urine. Quantification of some of these substances over a 24-hour period is used in the diagnosis of gout, renal function, hormone-secreting tumors, and metabolic disorders. The chemical analysis of these specimens is performed by a reference laboratory. The reference laboratory will provide the necessary containers and preservatives for the 24-hour collection. Check whether the urine should be refrigerated throughout the specimen collection period. Instruct the patient to:

1. Void the first specimen of the collection day into the toilet as usual.
2. For the rest of the day, place all urine immediately into the special collection container.
3. Also put the first specimen of the second day into the container.
4. Bring the urine to the office laboratory in the early morning of the second day.

RENAL STONE COLLECTION

When the diagnosis of a renal stone is made, the patient should be instructed to strain all urine until the stone passes. A handkerchief makes a convenient strainer. If a stone is obtained, it should be sent to a reference laboratory for analysis. Special techniques are used in the analysis, including optical crystallography, x-ray diffraction, and infrared spectroscopy.

SECTION 5: Urinalysis Procedure

This section describes the procedure for the complete urinalysis (i.e., the macroscopic, dipstick, and microscopic examinations). These tests are usually performed as a single procedure (i.e., in a continuous sequence). The details of each test and its interpretation are given in the sections that follow (Sections 6 through 22). The quality control for all of the urinalysis procedures is described in Section 23.

1. Take a freshly collected urine specimen or a specimen that has been refrigerated for no longer than 2 hours. (Never allow a urine specimen to remain at room temperature for more than a few minutes before testing or refrigeration).
2. Record on the laboratory report the patient's name, collection time, and the type of specimen (clean catch, minicatheter, pediatric bag, or suprapubic aspiration).
3. Write the patient's name on a clean, plastic, centrifuge tube with a grease pencil.
4. Mix the urine sample well, then pour about 12 mL of urine into the labeled centrifuge tube. If there is less than 12 mL in the cup, pour out all but about 1 mL of urine, and return it to the refrigerator for later culturing. If there is less than 12 mL of urine available for centrifugation, note this fact on the laboratory slip.
5. Check the specimen for a cloudy precipitation with a pinkish orange color. This is due to amorphous urate crystals that precipitate when the urine is refrigerated. These crystals must be cleared before proceeding. To clear this precipitate:
   a. Turn on the hot water tap.
   b. Hold the tube in the flowing hot water. Rotate it. After a few minutes the specimen will start to clear. Hold it in the hot water until the whole specimen is clear.
6. Hold the urine up to the light and evaluate the specimen for turbidity. It may be necessary to hold the specimen against a printed background. Record the turbidity as clear (transparent), hazy (light turbidity when held to the light), cloudy (obviously turbid but the print is visible), or opaque (print cannot be seen).
7. Record the color of the specimen: colorless (looks like water), straw (very pale yellow), yellow, yellow-orange, red, red-pink, red-brown, brown, blackish brown, black, blue, blue-green, green, or milky.
8. If a refractometer is used to determine the urine specific gravity, perform this test with a drop of urine from the test tube (Section 8).
9. Remove one urine dipstick from the dipstick bottle. Immediately replace the cap on the dipstick bottle.
10. Dip the dipstick into the urine in the centrifuge tube, being sure that all of

the reagent pads are covered with urine. Note the position of the sweep second hand of the clock in the laboratory so that the reaction time can be accurately measured. Do not put the dipstick into the original specimen cup, as this may contaminate the specimen and make the urine culture less accurate.

11. Immediately remove the dipstick from the urine to avoid dissolution of the test reagents on the dipstick. While removing the dipstick, run its edge along the opening of the tube to remove excess urine from the dipstick.
12. Hold the dipstick horizontally to prevent any mixing of the chemicals on the adjacent reagent pads.
13. Hold the dipstick alongside the color chart on the dipstick bottle and carefully match the colors at the appropriate test time, as stated on the bottle.
14. Record the dipstick test results on the laboratory slip.
15. Place the specimen tube in the centrifuge and balance it with a second centrifuge tube filled with an equal amount of water. Spin the specimen at about 2,500 rpm for 6 minutes or with a Statspin for 45 seconds.
16. Allow the centrifuge to come to a complete stop. Do not brake the centrifuge abruptly, as this will remix the specimen.
17. Remove the tube from the centrifuge. If there is a large amount of pinkish-orange sediment in the bottom of the tube, remix the specimen and clear the urate crystals as described in step 4 above.
18. If the dipstick test for protein was positive, perform a sulfasalicylic acid test to confirm this finding:
    a. Using a Pasteur or plastic pipette, remove 4 to 5 mL of urine from the sample tube and place this into a second clean, clear, plastic centrifuge tube.
    b. Using a new pipette, add 1 mL of 20% sulfasalicylic acid to the tube containing the 4 to 5 mL of urine.
    c. Mix the tube and examine for the amount of protein that was precipitated by the acid (Table 17–1).
    d. Record the degree of proteinuria on the laboratory slip.
19. Decant the urine from the original centrifuge tube by quickly inverting it over the sink. Do not slowly pour off the urine, as the sediment is then more easily washed away.
20. Turn the tube right side up. About 0.5 mL of urine will remain in the tube along with any sediment.
21. Mix the sediment and the small amount of residual urine by tapping the tube against the edge of the counter or snapping the tube with the nail of the third finger. If you use

TABLE 17–1 SSA PROTEIN TEST VALUES FOR PROTINURIA

Finding	Report	Protein Concentration
Clear (no turbidity)	Negative	0
Slight turbidity	Trace	1-10 mg%
Turbidity, print can be read	1+	15-30 mg%
Turbidity, print is visible	2+	40-100 mg%
Turbidity and fine precipitate	3+	150-350 mg%
Flocculation, solidification	4+	Over 400 mg%

a transfer pipette you can mix the sediment by squeezing the bulb several times.

22. With a plastic or Pasteur pipette, place two drops of the well-mixed sediment onto an unfrosted microscopic slide. (Some laboratory workers may find it more convenient to carefully pour a drop of the sediment from the tube onto the slide. It is more difficult to control the amount of urine if it is poured.) If the drop is placed toward one end of the microscope slide, a second specimen can be placed at the other end of the slide. This efficient use of slides cuts down on laboratory expense.
23. Carefully place a coverslip over the drop. Try to avoid getting air bubbles between the specimen and the coverslip.
24. Lower the microscope condenser and set the light source for minimum light.
25. Place the slide on the microscope stage and focus under the low-power objective (x10). (See Chapter 10, Section 7 for a detailed description of how to use the microscope.)
26. Move the stage until the edge of the coverslip is in view.
27. Examine the area along the entire four edges of the coverslip. Look carefully for casts. Move the focus control back and forth. Not all the sediment is in a single focal plane. Casts are frequently found along the border of the coverslip. If they are seen, switch to the high-dry objective (x40) to more easily identify the type of cast. Report the number of casts per low power field (LPF).
28. After examining the four edges under low power, quickly scan the rest of the coverslipped area with the low-power objective to get a general impression of the specimen. Particularly note the presence of squamous epithelial cells. Large numbers of these indicate a contaminated specimen.
29. Switch to the high-dry objective (x40) and examine about ten fields; count the various elements and get an average count per field. Report the microscopic findings:
    a. Per low-power field: casts.
    b. Per high-power field: RBC, WBC, squamous epithelial cells, transitional epithelial cells, renal tubule cells, crystals, and oval fat bodies.
    c. Presence: *Trichomonas vaginalis*, clue cells.
    d. Use 1 + to 4 + (1 +: occasional form noted; 2 +: noted in every field; 3 +: large amount in every field; 4 +: full field): mucous threads, yeast buds, amorphous material, bacteria, and spermatozoa.
30. If the sediment is to be stained, add a drop of the stain to the remaining sediment in the centrifuge tube. Allow this to set for 1 minute, then re-examine as in steps 21 to 29 above. We do not advise examining the stained specimen first. The stain often produces precipitants that resemble bacteria and therefore make part of the microscopic examination difficult. Always examine the unstained sediment first.
31. If a urine culture has been requested, perform this test on the remaining refrigerated urine (Chapter 18).

SECTION 6: Urine Turbidity

Urine turbidity is the gross appearance of the urine. It is important to examine the urine in a clean and transparent tube to accurately assess turbidity. The results are described as:

- Clear: transparent
- Hazy: a slight turbidity seen when the specimen is held to the light
- Cloudy: obvious turbidity but print can be read through the specimen
- Opaque: print cannot be read through the specimen

Turbidity is the result of particulate matter in the urine. This can be caused by a variety of substances:

- Amorphous phosphate and carbonate crystals: These white crystals precipitate out in alkaline urine
- Amorphous urate crystals: These pinkish-orange crystals precipitate out when the urine is refrigerated
- Bacteria
- Fecal contamination
- RBCs
- WBCs
- Clumps of squamous epithelial cells
- Mucous threads
- Prostatic fluid
- Spermatozoa
- X-ray contrast media
- Yeast

SECTION 7: Urine Color

Freshly voided urine is normally yellow and clear. Dilute urine may be very light in color, while concentrated specimens can be orange. Color changes can be due to many factors, such as the urine's concentration, medications, and various diseases. Some color changes may be clinically significant, but most often abnormal colors are caused by drugs or food that the patient has ingested. It is important to take a good food or drug history in any patient whose urine is abnormal in color.

- **Colorless/straw-colored (very pale yellow)**
  very dilute urine
- **Yellow**
  Normal, due to the pigment urochrome
- **Bright yellow**
  Acriflavine
  Vitamin C
  Nitrofurantoin
  Phenacentin
  Vitamin B_{12}
- **Orange**
  Concentrated urine
  Bilirubin (there will be yellow foam)
  Urobilinogen (no foam)
  Chrysophanic acid
  Ethoxazene (Serenium)
  Santonin
  Sulfasalazine (Azulfidine)
  Inandione
  Phenazopyridine (pyridium, Azo Gantrisin, Azo Gantanol)
- **Red/red-pink/red-brown**
  Acetphenetidin (various metabolites)
  Amidopyrine
  Anisindione (Miradon)

Anthracene derivatives (cascara, rhubarb, senna, all when pH is greater than 7)
Beets
Blackberries
Bromsulphalein (BSP, when pH is greater than 7)
Chlorzoxazone (Paraflex)
Crayons
Deferoxamine mesylate (Desferal)
Diphenylhydantoin
Emodin (pH greater than 7)
Eosine
Hemoglobin
Myoglobin
Phenindione (Hedulin)
Phenolphthalein (in laxatives)
Phenothiazines
Phensuximide (Milontin)
Porphyrins
RBCs
Rifampin

- **Brown/blackish-brown/black**
  Argyrol
  Bilirubin (yellow foam)
  Chelidonium
  Chloroquine
  Fava Beans
  Feces (collection contamination or fistula)
  Florazolidine metabolites
  Homogentisic acid (alkaptonuria)
  Acidification of hemoglobin pigments
  Iron sorbitol
  Melanin
  Metronidazole (Flagyl)
  Methocarbamol (Robaxin)
  Methyldopa
  Chronic phenacetin ingestion
  Phenol poisoning
  Phenylalanine metabolites
  Phenylhydrazine
  Primaquine
  Pyrogallol
  Tyrosine metabolites
- **Blue/green/blue-green**
  Amitriptyline
  Anthraquinone
  Arbutin
  Biliverdin
  Water-soluble chlorophyll
  Flavine (acridine antiseptics)
  Indicans
  Indigo blue
  Indigo carmine
  Methylene blue (in over-the-counter analgesics)
  Pseudomonas infections
  Resorcinol
  Salol
  Tetralin (degreasing agent)
  Thymol
  Toluidine blue (Blutene)
  Urinary derivatives of tryptophan indols
  Triamterene (Dyrenium)
- **Milky**
  Pyuria
  Amorphous phosphate crystals (Alkaline urine)

SECTION 8: Urine Specific Gravity

The specific gravity is a measure of the amount of dissolved solutes in a given volume of urine. It is therefore an indication of how concentrated or dilute the specimen is. A normal patient's urine will vary between 1.001 (dilute) and 1.030 (concentrated).

Traditional discussions of the urine specific gravity have focused on renal diseases, such as diabetes insipidus, in which the kidney is unable to properly concentrate the urine. This is an uncommon problem, and the test is not often used in the office laboratory for this purpose. Instead the specific gravity is usually used to evaluate patient hydration, especially in small children with diarrhea or vomiting. It is also used to help in the interpretation of other urine tests. For example, a finding of only 5 WBC/HPF in a patient with dysuria might be clinically significant if the urine has a low specific gravity. Likewise, a urine culture with a low colony count might be indicative of an infection if the urine sample was dilute. Finally, a pregnancy test can be falsely negative if the urine is too dilute (i.e., < 1.010). There are three methods for determining the specific gravity in the office laboratory.

DIPSTICK

A dipstick test for urine specific gravity is available (Multistix SG, Ames Division, Miles Laboratory, Inc, Box 70 Elkhart, IN 46575). The test is easily performed as a part of the routine dipstick urinalysis and costs only a few cents per test. It is the recommended test method in most office laboratories. Results are read in steps of 0.005. The results correlate to within 0.005 of the specific gravity determined by refractometer. For greatest accuracy the urine-specific gravity should be increased by 0.005 if the urine pH is greater than 6.5. Moderate amounts of protein (greater than 100 mg/dL) may produce elevated readings. Highly buffered urines or those containing greater than 1% glucose may show slightly lower readings than obtained by the refractometer. This can be corrected for by using the formula SG = dipstick reading + glucose mg/dL /1,000 × 0.005. The test is not temperature dependent.

URINE REFRACTOMETER

This instrument measures the urine's refractive index, which is related to the specific gravity. The test is easily performed, and only one drop of urine is required. The main disadvantage is that the instrument and stand with a light source cost over $700. This is beyond the budgets of many office laboratories.

Procedure

1. Clean the surface of the refractometer prism with distilled water from a squeeze bottle.
2. Wipe the prism dry with a 2 × 2-inch gauze pad.
3. Place the cover over the prism surface. Be sure that the cover is pressed flat against the prism.
4. Pour a drop of urine from the unspun centrifuge tube into the notched part of the prism cover. The urine will seep down and cover the prism surface.
5. Turn the light in the stand on. If a handheld refractometer is used, it should be pointed toward a light source, such as a window.
6. Look through the refractometer and read the value from the specific gravity scale at the point where there is a sharp division between the light and dark areas.
7. If there is no sharp dividing line, repeat from step 1 above.
8. Clean the prism surface with a squirt of distilled water and then wipe it with a 2 × 2-inch gauze pad. Leave the gauze between the prism and cover until the next test.

URINOMETER

Some laboratories may be equipped with a urinometer. This is not a preferred method for testing the urine specific gravity. This method requires a large volume of urine (at least 15 mL), which may be difficult to obtain from some patients. This is especially true for those dehydrated infants in whom the specific gravity is important. Furthermore, the test is time consuming, the scale is difficult to read, and the test is temperature dependent. If you have a urinometer, we recommend that you switch to dipstick specific gravities.

CONCENTRATED URINE

An occasional patient will have an extremely high specific gravity (greater than 1.035). In this case, either the dipstick or refractometer can be used to determine the urine's specific gravity if the urine sample is diluted 1:1 with distilled water. The actual specific gravity is then calculated by doubling the last three digits of the observed value. For example, if the diluted specimen gives a specific gravity of 1.020, the specific gravity of the original specimen is 1.040.

REFERENCE:

Guthrie R, Lott J, Kriesel S, and Miller I. Does the Dipstick Meet Medical Needs for Urine Specific Gravity? *J Fam Pract.* 1987;25:512-514.

SECTION 9: Dipstick Examination

A variety of chemical tests on urine samples can be rapidly performed using a dipstick (Table 17–2). These are plastic strips on which reagent pads have been placed. Up to ten tests can be performed in 2 minutes with a dipstick. This has greatly improved and speeded up the process of performing the urinalysis. However, the dipsticks are so convenient that it is easy to become cavalier with their use. For example, they are often stored or handled improperly, resulting in reagent deterioration. We have found that 80% of the dipsticks stored at the nursing station in our clinics failed to pass the quality control checks on at least one of the chemical tests. People are also cavalier in the timing of the dipstick reactions. The timing must be done precisely to ensure proper test results.

There are a variety of dipsticks available from several manufacturers. In some patients you may need only one or two tests on a dipstick. For example, in a pregnant patient you could test for just glucose and protein. However, we have found it less confusing to keep a single type of dipstick in the office. Choose one that includes all of the tests you are likely to need. This is simpler than having several types of bottles of different dipsticks for different clinical indications. Having just one type of dipstick is also less expensive when the cost of quality controls is considered.

The test descriptions that follow are based on the Ames Multistix (Ames Division, Miles Laboratories, Inc, Box 70 Elkhart, IN 46575). The information will be relevant to the use of other dipsticks, but specific information regarding these other dipsticks should be found in their product inserts.

All dipsticks should be kept in a cool, dry place. Do not keep them in a cabinet over a light source. The heat from the light may cause a rapid deterioration of the dipstick reagents. Whenever you open a bottle, recap it immediately. Dipsticks will deteriorate rapidly if the lid is left off. The dipsticks should be kept in the original (brown glass, or light-proof) bottle. If there is any discoloration of the test pads, especially the ketone reagent area, throw them away. Do not use the dipsticks after their expiration date. The date is printed on the side of each bottle.

TABLE 17-2. URINE DIPSTICKS

Test Name	Company	Available Tests									
		glu	*bili*	*ket*	*SG*	*Blood*	*pH*	*Prot*	*Uro*	*nit*	*leuk*
Ketostix	Ames[a]			X							
Chemstrip[b]	Bio-Dynamics			X							
Keto-Dia Stix	Ames	X		X							
Chemstrip GK	Bio-Dynamics	X		X							
Uri Stix	Ames	X						X			
Chemstrip GP	Bio-Dynamics	X						X			
Lab Stix	Ames	X		X		X	X	X			
Chemstrip 5L	Bio-Dynamics	X		X		X	X	X			X
Bili Labstix	Ames	X	X	X		X	X	X			
Bili Labstix-SG	Ames	X	X	X	X	X	X	X			
Chemstrip 6L	Bio-Dynamics	X	X	X		X	X	X			X
Multistix	Ames	X	X	X		X	X	X	X		
N-Multistix	Ames	X	X	X		X	X	X	X	X	
Multistix SG	Ames	X	X	X	X	X	X	X	X		
Chemstrip 7L	Bio-Dynamics	X	X	X		X	X	X	X		X
Multistix 8 SG	Ames	X	X	X	X	X	X	X		X	
Multistix 9	Ames	X	X	X		X	X	X	X	X	X
Chemstrip 9	Bio-Dynamics	X	X	X		X	X	X	X	X	X
Multistix 9 SG	Ames	X	X	X	X	X	X	X		X	X
Multistix 10 SG	Ames	X	X	X	X	X	X	X	X	X	X

Abbreviations: bili, bilirubin; glu, glucose; ket, ketone; leuk, leukocyte esterase; nit, nitrite; prot, protein; uro, urobilinogen

[a]Ames Stix, Ames Division, Miles Laboratories, Inc, Box 70, Elkhart, IN 46575.

[b]Chemstrips, Bio-Dynamics/bmc, 9115 Hague Rd, Indianapolis, IN 46250; (800-428-5076).

SECTION 10: Bilirubin

RBCs have about a 3-month life span. When the cells are finally broken down, their hemoglobin is transported to the liver, where it is changed to bilirubin and excreted in the bile. In severe liver disease or in an obstruction of the biliary tract, the bilirubin cannot be normally excreted. It then increases in the blood and is excreted in the urine. The patient often reports seeing dark yellow or brown urine before discovering jaundice.

Before the availability of urine dipsticks, the presence of urinary bilirubin was made by the "foam test." Dark urine was shaken. If the foam that formed was yellow, the urine was presumed to contain bilirubin. Testing is now greatly improved. The dipstick test can detect bilirubin in the urine at concentrations as low as (0.2 to 0.4 mg/dL). The Icotest is a second test for urinary bilirubin that can be easily performed in the laboratory and is more sensitive than routine dipstick testing. It detects concentrations of bilirubin as low as 0.05 mg/dL.

Bilirubin rapidly decomposes when exposed to light or heat. The urine sample should therefore be tested soon after collection.

TEST USES

- Early detection of jaundice
- Early detection of infectious hepatitis
- Screening test in patient with acute right-upper quadrant pain
- Screening test in patients with an occupational exposure to hepatotoxins

SOURCES OF ERROR

- False-positive dipstick tests: presence of ethoxazene, phenazopyridine (Pyridium), phenothiazines, chlorpromazine
- False-negative dipstick tests: presence of ascorbic acid (vitamin C)
- False-negative Icotest tests: Pyridium and Serenium can mask a positive test reaction because they produce a yellow color in the urine
- False-positive Icotest tests: presence of chlorpromazine in large amounts

ICOTEST PROCEDURE

The test kit (Ames 2591) includes 100 reagent tablets and reaction mats. The tablets should be stored at a temperature below 86°F and protected from light, heat, and moisture. The reagent bottle should be recapped immediately after use. Tablet deterioration is indicated by a tan color.

1. Place a reaction mat on a clean surface.
2. With a pipette, place five drops of fresh urine on the mat.
3. Take one reagent tablet from the bottle and place it on the center of the mat. Immediately recap the bottle.
4. Drop two drops of distilled water onto the tablet so that the water runs off the sides of the tablet and onto the mat.
5. After 30 seconds examine the mat for a blue or purple color, which indicates the presence of bilirubin. A pink or red color may form and indicates a negative test.

SECTION 11: Blood

The dipstick test for blood will detect free hemoglobin, myoglobin, or intact RBCs. Its sensitivity is greatest in dilute urine, when the test is able to detect 0.015 to 0.062 mg/dL free hemoglobin or 5 to 20 intact RBCs/μL. The urine should be tested within 4 hours of its collection.

CAUSES FOR POSITIVE TESTING

Blood

Viral infections, fever, dehydration, exercise, menstrual blood contamination, urinary tract infection, renal stone, glomerulonephritis, polycystic kidney, urinary tract tumors, benign familial hematuria, sickle cell trait, malignant hypertension, subacute bacterial endocarditis, urethral irritation, coagulation disorders.

Hemoglobinuria

This is much less common than hematuria. It is often due to the intraurinary lysis of RBCs in dilute urine rather than to true hemoglobinemia. If hemoglobinemia is suspected, the patient's serum or plasma should appear pink. Hemoglobinemia is seen when there is intravascular red cell destruction, as with the hemolytic anemias.

Myoglobinuria

The urine of these patients tests positive by dipstick, but no RBCs are seen on the microscopic examination. The serum is not pink, as is seen with hemoglobinemia. Myoglobinemia is seen with severe physical exertion, trauma to muscles, and some rare familial diseases.

SOURCES OF ERROR

False-Positive Results

- Microbial peroxidase associated with urinary tract infections
- Bleach

False-Negative Results

- High urine specific gravity or proteinuria reduce the test reactivity
- Ascorbic acid concentrations greater than 0.28 mmol/L (5 mg/dL, conversion factor 0.057 for mg/dL to mmol/L) may produce a negative test if there is only a trace of blood.

COMMENTS

Red urine with a negative dipstick indicates the presence of a red substance other than blood. This is commonly seen after a person eats beets.

SECTION 12: Glucose

Testing of glucosuria is useful in many clinical situations. It is helpful in screening patients for diabetes and is usually done on each prenatal visit. It is also frequently the first indication that someone is diabetic. However, whole-blood glucose sticks and glucose analyzers have largely replaced urine glucose testing for the home monitoring of the diabetic patient.

There are two basic chemical tests used to evaluate for glucosuria. The first is the glucose oxidase reaction that is used on some dipsticks and tapes (Tes-Tape, Clini-Stix, Diastix, Multistix, and Chemstrip). This test reaction is specific for glucose. The second test is the Clinitest, which is based on the change of cupric sulfate to cuprous oxide by a reducing substance. Glucose is only one of several reducing substances that will give a positive test result with the Clinitest. (See Section 24 for a description of the Clinitest procedure and Table 17–3 for a comparison of method results.)

In the past these various tests were reported as negative, trace, or 1 + to 4 +. This practice resulted in a great deal of confusion because there was up to a fivefold difference in the glucose concentration between the 1 + result of one test and the 1 + of another test. The American Diabetes Association has now recommended that all test results be reported in grams of glucose per deciliter (or % glu). If a metabolic problem is suspected in a very young child (< 1 year), check the urine with both a dipstick

TABLE 17–3. GLUCOSE TEST VALUES

	Negative	Trace (%)	1+ (%)	2+ (%)	3+ (%)	4+ (%)
Clinitest (Ames) two drop	0	1/2	1	2	3	5 or more
Clinitest (Ames) five drop	0	1/4	1/2	3/4	1	2 or more
Tes-Tape (Lilly)	0	—	1/10	1/4	1/2	2 or more
Multistix (Ames)	0	1/10	1/4	1/2	1	2 or more
Diastix (Ames)	0	1/10	1/4	1/2	1	2 or more
Chemstrip (Bio-Dynamics)	0	—	1/10	1/4	1	—

glucose and with Clinitest. This is to see if there are other reducing sugars besides glucose in the urine. If reducing substances other than glucose are present, the dipstick would be negative, but the Clinitest would be positive. This could be due to the presence of a disease such as galactosemia.

SOURCES OF ERROR

There are several substances that produce false results with the glucose oxidase test:

Ascorbic Acid

Vitamin C is used by many patients as an over-the-counter treatment for respiratory infections. It is also found in foods, fruit drinks, multivitamins, tetracycline, and iron preparations. This substance can produce false-negative test results in patients with low concentrations of glucose in the urine.

Ketones

A moderate concentration of ketones in the urine can produce false-negative glucose testing if only small quantities of glucose are present.

Salicylates

In large doses (over 2 g/day), salicylates can cause false-negative glucose oxidase testing.

Pyridium (phenazopyridine)

This drug can result in either false-negative or false-positive testing. It turns the urine orange, which makes the dipstick almost impossible to read.

SECTION 13: Ketones

Three ketones are produced in the body and may be excreted in the urine: acetoacetic acid (20%), acetone (2%), and beta-hydroxybutyric acid (78%). Testing for ketones in the urine is most frequently requested to see whether a diabetic patient is in ketoacidosis. In diabetic patients the urine ketone test is positive before ketones can be detected by serum testing.

It is also common to find positive urine ketones in nondiabetic patients. This can be seen in any ill patient who has not eaten and who therefore has had an insufficient dietary intake of carbohydrates. In particular, positive ketone testing is seen in children with fever, in patients with vomiting or diarrhea, or in persons on a ketogenic diet. Pregnant women often have a positive ketone reaction several hours after eating.

In the past a positive dipstick was followed by a quantitative test called an Acetest. Dipsticks are now more sensitive and provide a reliable quantitative result. No further testing is therefore required.

SOURCES OF ERROR

False-negative tests can be seen with aspirin use. False-positive tests of a trace or less can be seen with a high specific gravity, low pH, highly pigmented urines, chlorpromazine, large concentrations of phenylketones, levadopa metabolites, and phthalein compounds.

SECTION 14: Leukocyte Esterase

Neutrophils contain esterases that are not present in normal urine. This dipstick test measures the presence of such leukocyte esterases. The strength of the reaction is proportional to the number of WBCs in the urine. While this test will detect the presence of white cells in the urine, a positive test *does not* ensure that the white cells are urinary in origin! They may also come from the vagina in female patients.

It has been established that this test has clinical utility in hospital settings. Traditionally one of the routine tests for all hospitalized patients was a complete urinalysis. This meant that a great deal of time was spent processing and examining "negative" urines. A variety of studies have shown that when screening asymptomatic patients, the blood, protein, nitrite and leukocyte esterase dipstick tests are adequate markers to identify which urines need a microscopic examination. If any of the previously mentioned four are positive, the urine should be examined. Similarly, when screening asymptomatic patients in the outpatient setting (such as routine physicals, diabetics, pregnant patients), leukocyte esterase, as well as blood, protein, and nitrite testing can serve as a marker for those specimens that require microscopic examination.

However, in any patient with genitao urinary symptoms, it is important to routinely perform a microscopic examination. Many symptomatic women drink large quantities of water in an effort to "flush out" the infection. Their urine then becomes very dilute (< 1.010). Only a few WBCs in very dilute urine can be significant.

A positive leukocyte esterase in a symptomatic patient does not automatically indicate a urinary tract infection. Such a patient may have a vaginal/cervical infection instead. A microscopic urinalysis and a vaginal wet preparation may help determine the cause of the symptoms.

SOURCES OF ERROR

False-Negative or Decreased Reaction

- Elevated glucose concentrations (> or equal to 3 g/dL)
- High specific gravity
- Keflex
- Keflin
- High concentration of oxalic acid
- Tetracycline

False Positives

- Oxidizing agents

SECTION 15: Nitrites Test

This dipstick test is based on the conversion of dietary nitrates to nitrites by certain bacteria. Not all bacteria perform this conversion, but the common outpatient pathogen *Escherichia coli* does. *Staphylococcus saprophyticus,* the second most frequent cause of urinary tract infections in otherwise healthy women, does not, so the dipstick would be negative.

The test, when positive, can be useful. However, to have a reliable result the urine must have incubated in the bladder for at least 4 hours. This may sound easy, but it is often difficult to achieve. Remember, one of the symptoms of a urinary tract infection is urinary frequency! Also, symptomatic patients tend to drink large amounts of fluid, which results in short bladder incubation times.

Any degree of pink is considered positive. The intensity of the color is not proportional to the amount of bacteria.

SOURCES OF ERROR

The test has only a 40% accuracy rate when urine is not retained in the bladder. It has an 80% accuracy rate when incubated in the bladder for a minimum of 4 hours.

False Negatives

- Urine not in the bladder for 4 hours
- The pathogen does not convert nitrates to nitrites (e.g., *S saprophyticus*)
- Dietary nitrates are absent
- Ascorbic acid in concentrations of greater than 1.43 mmol/L (25 mg/dL, conversion factor 0.057 for mg/dl to mmol/L)
- Urine with a high specific gravity

SECTION 16: pH

The pH is a measure of the free hydrogen ions in solution. The urinary pH is an indicator of both the patient's overall acid-base status and the kidney's ability to handle acid and bicarbonate. The normal range for the urinary pH is from 4.8 to 7.8. Dipsticks measure pH in a range of 5.0 to 8.5 (5.0, 6.0, 6.5, 7.0, 7.5, 8.0, 8.5). The test is not affected by the various buffers in the urine, such as phosphates. For proper pH measurement, the urine should be dipsticked as soon as possible after collection. If there is a long delay, the sample may lose CO_2 and become more alkaline.

USES FOR pH TESTING

Crystals

The urinary pH is helpful in identifying urine crystals. Some crystals form only in acidic urine, while others form in alkaline urine. Amorphous urates are seen in urines with a pH of 6.0 or less, while amorphous phosphates are seen in urine with a pH of 6.5 or greater. Pathogenic crystals are only found in acidic urine.

Renal Stones

One form of treatment for patients with recurrent kidney stones is to change the urinary pH so that the stones no longer can form. Phosphate, carbonate, and ammonium biurate stones form in alkaline urine. Uric acid and calcium oxalate stones form in acidic urine.

Acid-Base Status

The kidney and the lung are the two organs that maintain the body's pH. The urinary pH is therefore an indicator of the body's pH and the kidney's response to that pH. For example, a diabetic patient with severe ketosis will be acidotic. The kidney responds by excreting acid, and the urinary pH will therefore be low. Remember, a low pH is acidic. A high pH is alkaline. Dipsticks do not measure below a pH of 5.0.

FOODS AND DRUGS AFFECTING URINARY pH

Acidic Urine

Meat, protein, cranberries, ammonium chloride, methionine, methenamine mandelate, and acid phosphate.

Alkaline Urine

Citrus fruit, sodium carbonate, potassium citrate, and acetazolamide.

SECTION 17: Protein

A small amount of protein is normally excreted in the urine (up to 0.5 g/day). This normal physiologic proteinuria is not detected by dipstick testing. The normal urinary proteins are albumin, globulins, and the Tamm-Horsfall protein, which is excreted by the renal tubules.

Proteinuria is seen in a wide variety of renal diseases as well as in such common conditions as fever. The primary urinary protein is albumin, which is small and therefore easily passes into the glomerular filtrate. In some illnesses other proteins predominate, such as the Bence-Jones protein found in multiple myeloma.

The dipstick test for protein is based on the reaction between albumin and bromphenol. The test is therefore most sensitive for albuminuria. If other forms of proteinuria are suspected, a sulfosalicylic acid (SSA) test should be performed. Almost all forms of urine protein precipitate with this acid test. The SSA test should be done on any urine with a dipstick test that is positive for protein. This is because the bromphenol test can be falsely positive in alkaline urine.

The average adult daily urinary output is between 1 and 2 L. A rough idea of the 24-hour urine protein excretion can be determined by multiplying the dipstick mg/dL concentration by 10. For example, a 100 mg/dL dipstick reading would reflect about 1 g of urinary protein per day.

CAUSES OF PROTEINURIA

Exercise

This can result in up to 300 mg/dL proteinuria.

Fever

Up to 3 g/day of protein can be seen. This should clear within 3 days of the acute illness.

Postural (Orthostatic) Proteinuria

Three percent of normal people have this condition, which is characterized by proteinuria only when the person is standing. To test for this cause of proteinuria, the patient is instructed to urinate before sleeping, then to collect the first sample just after rising in the morning and a second sample several hours after being upright. Only the second urine sample should show proteinuria, and there should be no significant sediment findings (i.e., casts or cells). When quantified, there is usually less than

1.5 grams of protein excreted per day. These patients have been followed for many years and do not develop renal disease.

Constant Proteinuria

Patients with constant proteinuria show protein in both the first morning and the ambulatory urine specimens. There are usually no significant findings on the microscopic examination. If less than 2 g/day are spilled, the condition is usually benign. In patients with over 3 g/day of proteinuria, a renal biopsy often shows membranous glomerulopathy or focal glomerulosclerosis. This condition may go unchanged, may resolve, or may develop into nephrotic syndrome.

Diabetic Nephropathy

Proteinuria may be the first evidence of diabetic changes in the kidney.

Toxemia in Pregnancy

Proteinuria is one of the three common markers of this disease.

Congestive Heart Failure

Up to 0.5 g/day of protein can be seen. This proteinuria is believed to be secondary to an increase in the renal venous pressure.

Hypertension

Proteinuria is often the earliest evidence for vascular changes in the kidney. It is also commonly seen with malignant hypertension.

Renal Disease

Almost all intrinsic renal diseases can produce proteinuria. However, proteinuria may not be seen in obstructive nephropathy, renal stones, tumors, or renal congenital malformations.

Tumors

Tumors of the renal pelvis or bladder may produce up to 1 g/day of proteinuria.

Drugs

Proteinuria can be seen as a side effect of many drugs. Gold therapy for rheumatoid arthritis is a classic example.

SOURCES OF ERROR

False-Positive Dipstick Test

Urine that is heavily contaminated with vaginal secretions, highly buffered alkaline urines, and contamination of urine with quaternary ammonium compounds.

False-Negative Dipstick Test

Dilute urine.

False-Positive SSA Test

Tolbutamide, x-ray contrast, turbid urine, penicillin, nafcillin, oxacillin, sulfisoxazole.

COMMENTS

- It is possible to have casts as well as cellular urine elements with proteinuria.
- Any patient with unexpected proteinuria who is not found to have postural proteinuria should have a 24-hour urine collection for total protein, albumin, and creatinine studies. A serum albumin and creatinine test should also be done.
- Proteinuria of 0.15 to 0.5 g/day is considered minimal, and 0.5 to 4.0 g/day is moderate. Greater than 4 g/day represents significant proteinuria, and the patient is at risk for developing the nephrotic syndrome.
- Table 17–1 describes the mg % values for the various levels of proteinuria found on SSA testing.

SECTION 18: Urobilinogen

The bilirubin that is found in bile is altered by colonic bacteria to form urobilinogen and urobilin. Most of this is ultimately excreted in the stool, but a small amount is reabsorbed in the colon and can then appear in the urine. Levels of 0.1 to 1.0 Ehrlich units/dL are considered normal in the urine. An increase in the urinary concentration indicates either an increase in the breakdown of RBCs or an inability of the liver to handle the urobilinogen that is produced.

The dipstick test for urobilinogen has a range of 0.1 to 12.0 Ehrlich units/dL. The testing should be done immediately after the specimen is collected because the substance is quickly degraded, especially if exposed to light or acidic urine.

COMMENTS

- Causes for an increased urobilinogen: hemolytic anemias, pernicious anemia, malaria, hepatitis, portal cirrhosis, congestive heart failure, infectious mononucleosis, alcoholic liver disease.
- False-positive tests can be caused by para-amino salicylic acid, azo dyes, porphobilinogen, sulfonomides, procaine, 5-hydroxindoleacetric acid, indol, methyldopa.
- The slightest hepatic insult results in an increase in the urobilinogen. It is therefore a sensitive test for liver disease.
- See Table 17–4 for comparison of urine bilirubin and urobilinogen.

SECTION 19: Microscopic Examination of Cellular Elements

Characteristics of cells found in the urinary sediment are found in Table 17–5.

SQUAMOUS EPITHELIAL CELLS: (REPORT AS NUMBER/HPF)

These cells originate from the distal one third of the urethra and, additionally in women, from the vagina. They are the largest cells seen in the urinalysis, with a diameter of 30 to 60 μ. The cell's cytoplasm can be either smooth or wrinkled. The nucleus is small, round, and dense. The nucleus is about the diameter of a RBC and is usually centrally placed in the cell (Fig. 17–3). Squamous cells may be folded on themselves or rolled up like a cigar, in which case they may look like a urinary cast. When stained, the cells have a purple nucleus and a pink-to-violet cytoplasm.

It is not uncommon to see a large number of squamous epithelial cells (greater than 5/HPF) in a specimen from a female patient. This is usually due to vaginal contamination and can be avoided by inserting a tampon before collecting the urine or by bladder catheterization. If squames are present in a catheterized specimen, it is due to the scraping off of those cells as the catheter slides through the urethra. If the specimen contains large numbers of both squamous cells and white blood cells, request a vaginal wet preparation to check for a vaginal infection. In patients with *Gardnerella vaginalis* vaginitis, clue cells may be seen in the urine specimen (Fig. 17–4).

TABLE 17–4. CLINICAL CORRELATION OF URINE BILIRUBIN AND URINE UROBILINOGEN

	Health	Hemolytic Anemia	Hepatic Disease	Biliary Obstruction
Urine bilirubin	None	None	±	Increased
Urine urobilinogen	Normal	Increased	Increased	Normal/Low

TABLE 17–5. CHARACTERISTICS OF CELLS SEEN IN URINARY SEDIMENT

Cell	Size	*Shape*	*Other Characteristics*
Squamous cell	30-60μ	Flat	Largest cell seen. Nucleus about the size of a RBC
Transitional epithelial	20-30μ	Spherical, polyhedral caudate	Has a large central nucleus. It can swell to look like a ball. These cells look like vaginal parabasal cells. If large numbers are seen, consider vaginal contamination. Check a vaginal specimen
Parabasal cell	15-20μ	Oval or round	Can look like a swollen transitional cell. Signifies vaginal contamination
Renal tubular epithelial	15-25μ	Polyhedral	Has a large nucleus-to-cytoplasm ratio. These can look like vaginal basal or parabasal cells. If numerous, consider contamination with vaginal cells. Check a vaginal specimen
Basal cell	8-10μ	Round	Has a large nucleus to cytoplasm ratio. Signifies vaginal contamination
Oval fat body	15-25μ	Round	Has refractile vacuoles. They can be polarized if ingested lipids are cholesterol
RBC	7.5μ	Biconcave	Has no nucleus. They can look granular if they are crenated
WBC	12-15μ	Round	The nucleus can be obviously segmented or may show no segmentation but rather moving cytoplasmic granules (i.e., a ''glitter cell'')
Macrophage	20-30μ	Round	Has many vacuoles

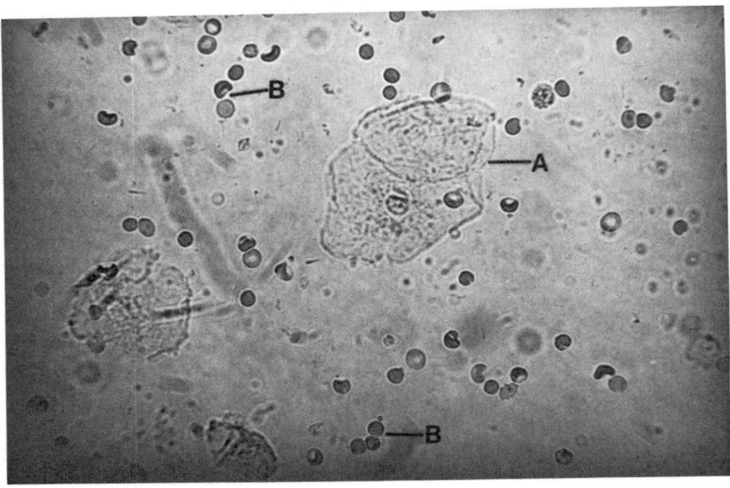

Figure 17–3. A. Squamous epithelial cells. **B.** Red blood cells. Viewed under high-dry objective.

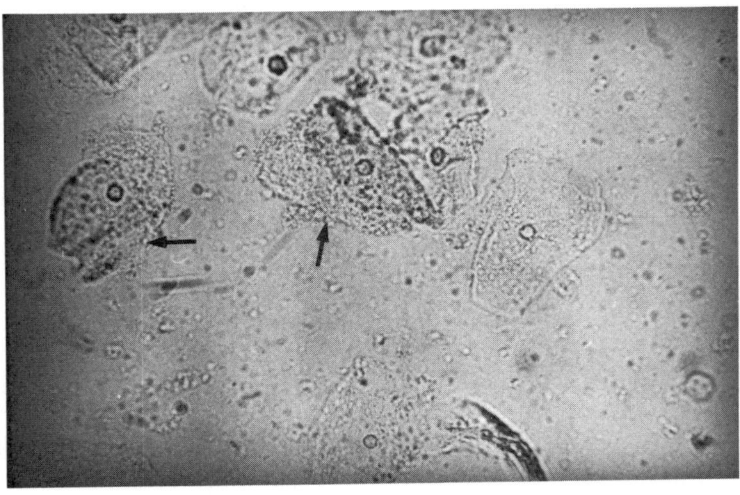

Figure 17–4. Clue Cells. These can be seen in the urine of both males and females. Viewed under high-dry objective.

It is uncommon to find large numbers of squames in specimens from male patients. If they are found, the specimen should be sent for cytologic study to rule out a urinary tract tumor.

Urethral irritation, such as that caused by a chronic indwelling catheter, can cause squamous epithelial cells to be found in the urine. Some such cells may have more than one nucleus. Any urine with such nuclear atypia should be sent for cytology to rule out a malignancy (Fig. 17–5).

TRANSITIONAL EPITHELIAL CELLS: (REPORT AS NUMBER/HPF)

These cells line the entire urinary tract from the renal pelvis to the midportion of the urethra and are normally sloughed in small numbers into the urine. They are 20 to 30 μ in diameter and vary in shape from round to pear shaped to spheroidal. They occasionally have long tails and a tadpole shape (so-called caudate transitional epithelial cells). Transitional cells are able to absorb water, and when they do they take on a "swollen" appearance—looking like a basketball. The nuclei are large, round, and centrally placed. The nucleus is about half the diameter of the cell (Figs. 17–6 and 17–7). When stained these cells have a dark-blue nucleus and pale-blue cytoplasm.

Parabasal cells from the vagina (Fig. 17–8) look just like transitional epithelial cells. If you see a large number of what might appear to be swollen transitional cells in a specimen from a female, it is extremely helpful to examine a vaginal specimen from that patient. This will help you differentiate the cells.

Large numbers of transitional epithelial cells or cells with atypical nuclei may indicate the presence of a transitional cell cancer. Urinary cytology is then indicated.

RENAL TUBULAR EPITHELIAL CELLS: (REPORT AS NUMBER/HPF)

Renal tubular epithelial (RTE) cells line the tubules of the kidneys and are seen normally in small numbers because of sloughing of the cells into the urine. They are 15 to 20 μ in diameter and have a large nucleus that may be either centrally or eccentrically placed. The cells are polyhedral or cuboidal in shape. They have a 1:1 nucleus-to-cytoplasm ratio. The cells do not swell and therefore do not take on the spherical appearance of transitional epithelial cells (Figs. 17–9 and 17–10). When stained they have a dark nucleus with an orange-to-pur-

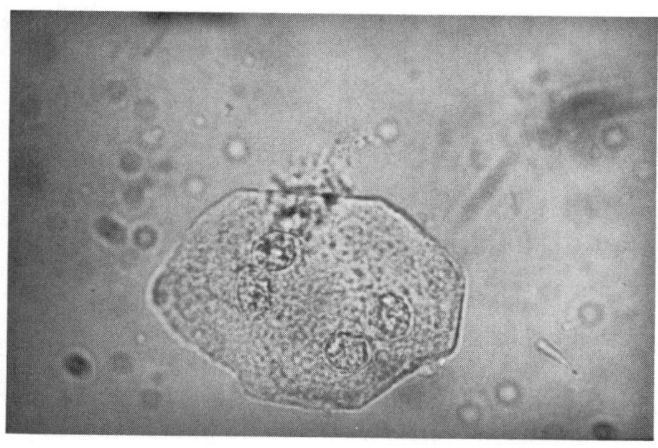

Figure 17–5. Multinucleated squamous cell. Viewed under high-dry objective.

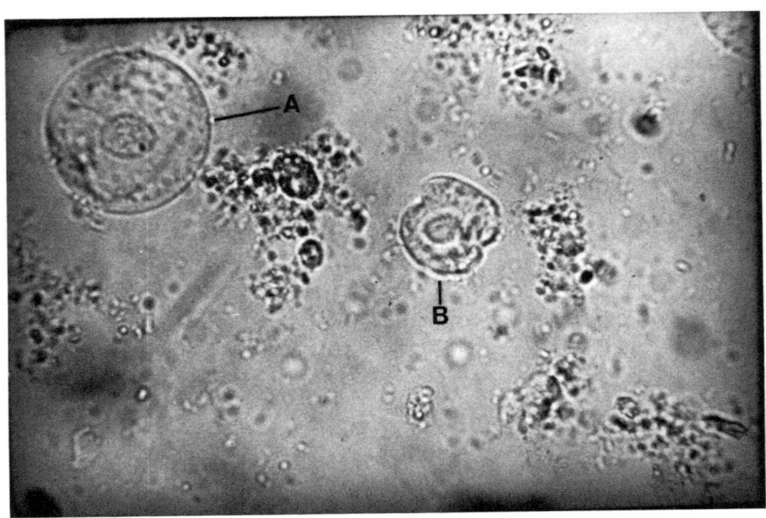

Figure 17–6. Transitional cells. **A.** Cell swollen with fluid. **B.** Transitional cell without fluid absorption. Viewed under high-dry objective.

ple cytoplasm. It may be difficult to distinguish these cells from transitional epithelial cells or degenerated WBCs. In addition, small parabasal cells (Fig. 17–8) or basal cells (Fig. 17–11) can look like RTE cells. Examine a vaginal specimen for any woman who appears to have more than a few RTE cells. Urinary sediment staining is also helpful.

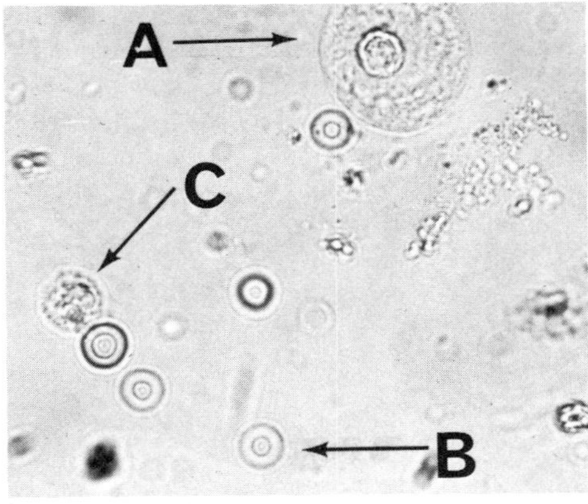

Figure 17–7. A. Transitional epithelial cell. Note the swollen or basketball shape. **B.** RBC. **C.** WBC. Viewed under high-dry objective.

Large numbers of RTE cells are seen in patients with generalized viral diseases (measles, viral hepatitis, cytomegalic inclusion disease) or patients with acute tubular damage. Renal diseases in which these cells are seen include pyelonephritis, the diuretic phase of acute tubular necrosis, nephrosclerosis, necrotizing papillitis, and acute renal-graft rejection. Some toxins (heavy metals) and drugs (salicylates) will also result in sloughing of these cells into the urinary sediment.

OVAL FAT BODIES: (REPORT AS NUMBER/HPF)

Oval fat bodies (OFB) are renal tubular epithelial cells that are filled with highly refractile fat droplets (Fig. 17–12). The cells appear round and swollen, and the nucleus is only rarely visible. The lipid in the drop-

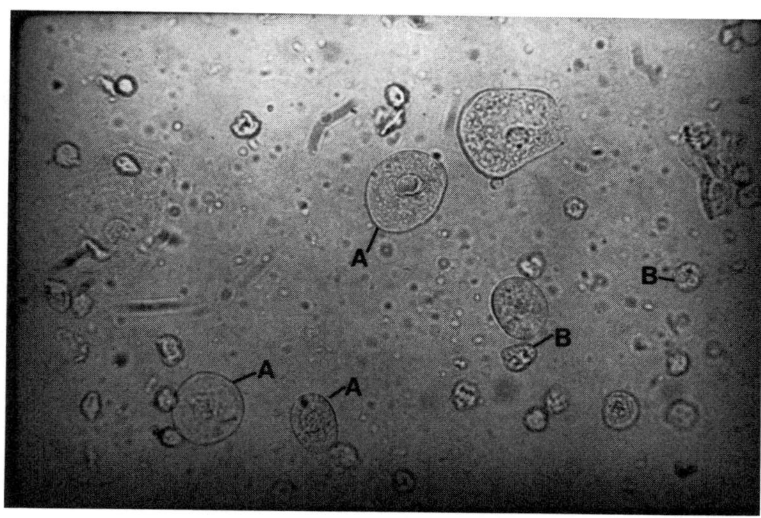

Figure 17–8. A. Vaginal parabasal cells. **B.** White cells. **C.** Normal squamous cell. Viewed under high-dry objective.

lets may be absorbed by the cells from the glomerular filtrate or may coalesce from the cell's own cytoplasm as the cells degenerate and are sloughed into the urine. If the fat droplets contain cholesterol esters, they will give a Maltese cross appearance when viewed with a polarizing microscope (Fig. 17–26). The Maltese cross is referred to as "anisotropic" and is seen only in urine containing either cholesterol or those specimens contaminated with starch granules. If the fat droplets contain primarily triglycerides, they will not appear birefringent with a polarizing microscope, but they can be identified by a fat stain such as Sudan III or IV.

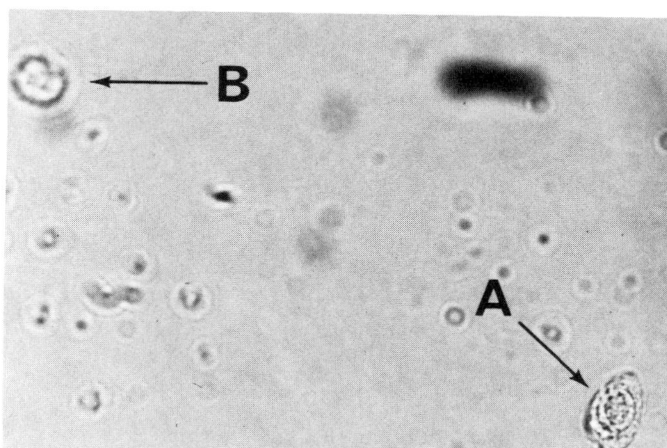

Figure 17–9. A. Renal tubule cell. Note the clear central nucleus. **B.** WBC (slightly out of focus) for size comparison. Viewed under high-dry objective.

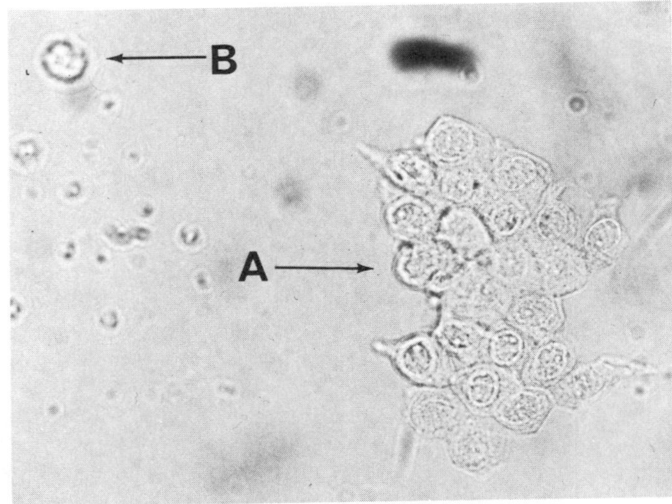

Figure 17–10. A. Clump of renal tubal epithelial cells. Note the central nucleus. **B.** WBC. Viewed under high-dry objective.

The presence of oval fat bodies indicates pathology. They can be seen in nephrotic syndrome, diabetic nephropathy, lipoid nephrosis, and toxic renal tubular nephrosis.

There is one cell that looks a great deal like an OFB. Patients with paroxysmal nocturnal hemoglobinuria (PNH) spill hemosiderin granules into the urine, and these can be ingested by cells (Fig. 17–13). These

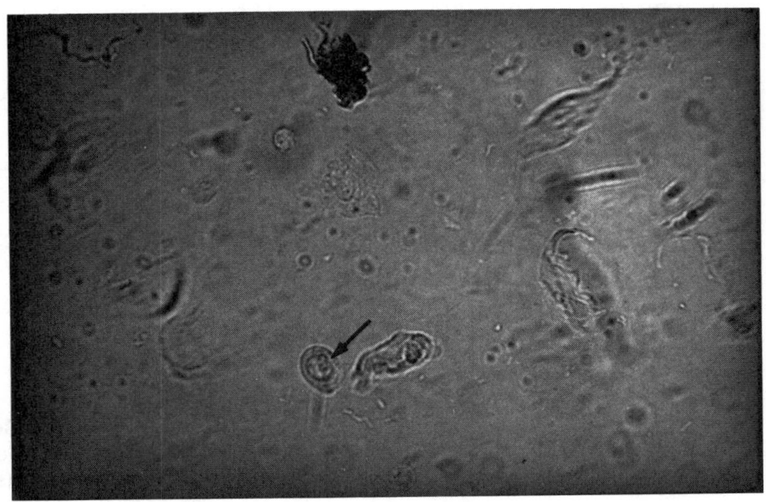

Figure 17–11. Vaginal basal cell. Viewed under high-dry objective.

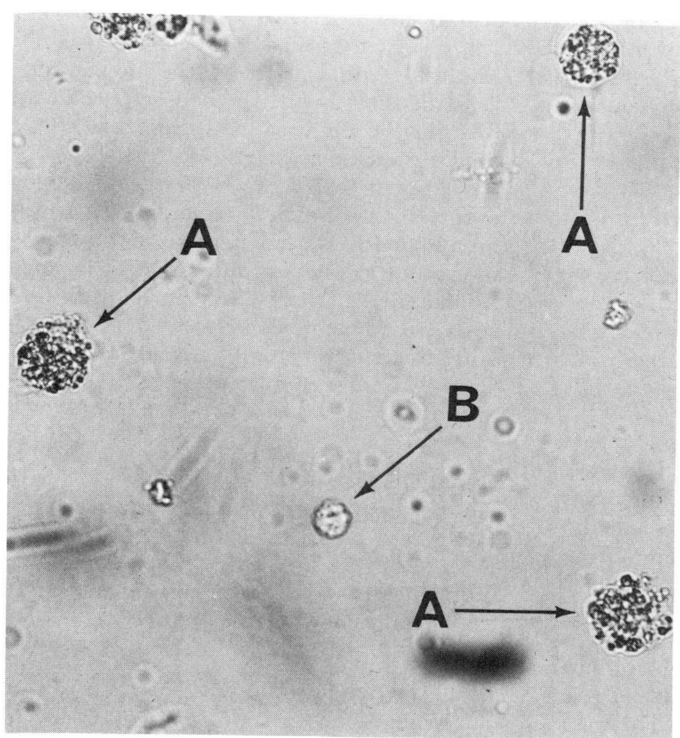

Figure 17–12. A. Oval fat bodies. **B.** WBC. Viewed under high-dry objective.

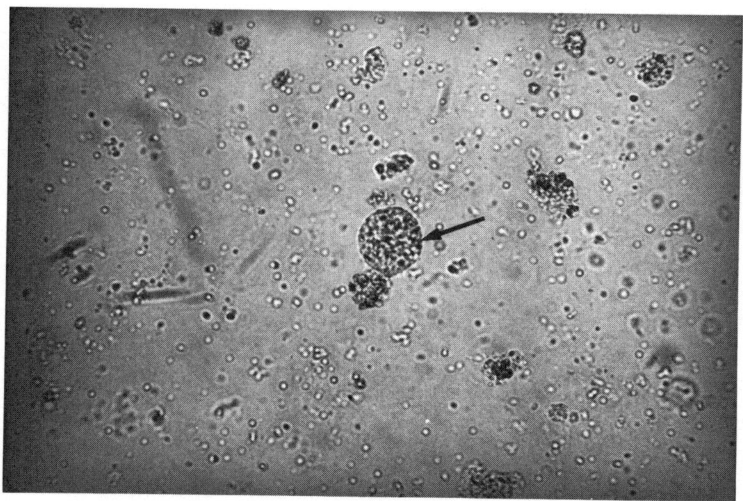

Figure 17–13. Cell filled with hemosiderin. Viewed under high-dry objective.

cells are about the same size as OFBs and have a refractile appearance; however, they do not polarize and do not take up the Sudan stains. Clinical history will help you to properly identify these cells.

RBCs: (REPORT AS NUMBER/HPF)

These cells are easily recognized in urine as pale, round, biconcave discs. They are usually agranular and have no nucleus (Fig. 17-14). The color of stained RBC varies with the urine pH (acid, light purple; neutral, pink; alkaline, dark purple).

In hypertonic urine the RBCs will shrink and become crenated. Crenated RBCs have a granular appearance that can be mistaken for WBCs except for the cells' small size (Fig. 17-15). In hypotonic urine the cells swell and break open, leaving only faint circular outlines of the empty cell wall. RBCs (Fig. 17-16), may be confused with small air bubbles or oil droplets, but they are generally not as refractile as these artifacts and will not vary in size. They can also be confused with yeast buds (Fig. 17-17), in which case a KOH smear is useful. The KOH will lyse the RBCs but not the yeast buds. If any of the cells are sickled, this may be the first indication that a patient has sickle cell trait (Fig. 15-32).

From 0 to 2 RBC/HPF is considered normal for both males and females. A large number of RBCs in a female patient is commonly due to vaginal contamination during a menstrual period. This can be seen for several days before and after the patient reports that she is actually menstruating.

RBCs are seen in a variety of other conditions.

Renal Disease

Glomerulonephritis, tumor, infection, lupus nephritis, kidney stone, renal venous thrombosis, trauma, polycystic kidney, hydronephrosis, acute tubular necrosis, and malignant nephrosclerosis.

Lower Tract Pathology

Infection, stones, tumors, and strictures.

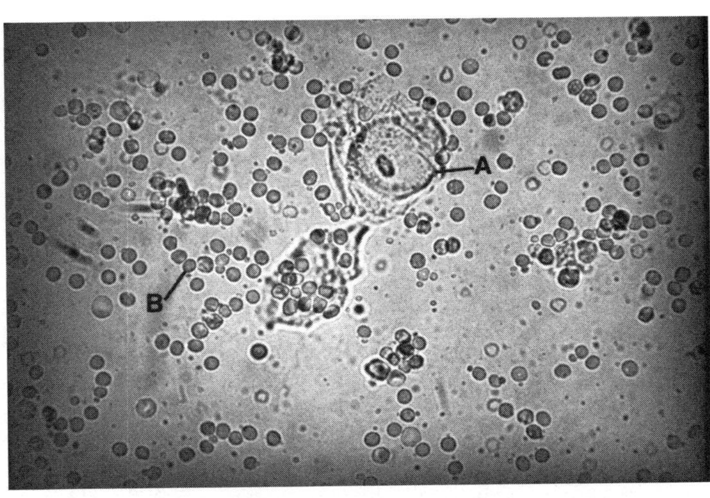

Figure 17-14. A. Squamous cell. **B.** RBCs. Viewed under high-dry objective.

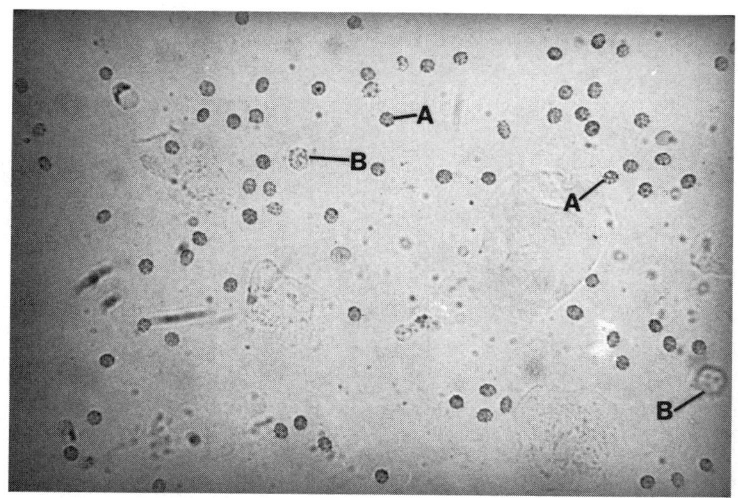

Figure 17–15. A. Crenated red cells. Notice the granular appearance. **B.** WBC. Viewed under high-dry objective.

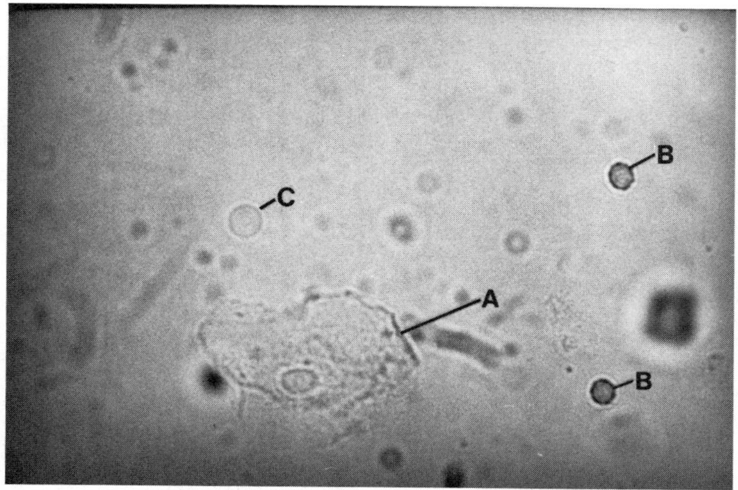

Figure 17–16. A. Squamous cell. **B.** RBC. **C.** Ghost cell. Viewed under high-dry objective.

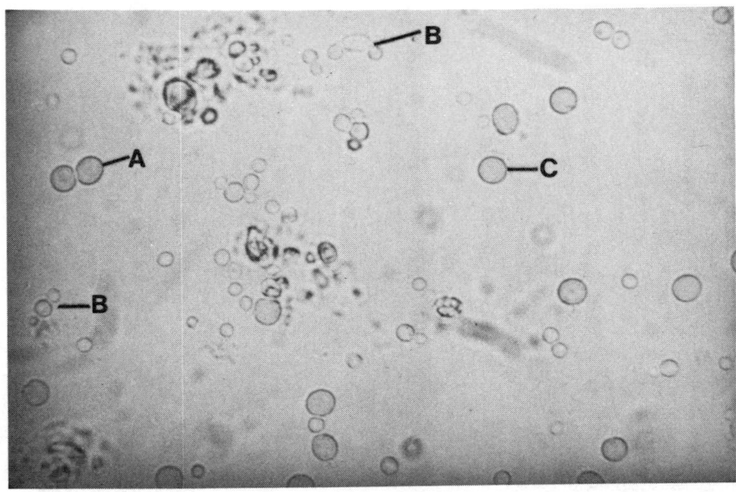

Figure 17–17. A. Two RBCs. B. Budding yeast. C. RBC. Viewed under high-dry objective.

Extrarenal Conditions
Appendicitis, colorectal cancer, salpingitis, diverticulitis, fever, malaria, subacute bacterial endocarditis, sickle cell trait, malignant hypertension, polyarteritis nodosa, scurvy, and exercise.

Drugs
Salicylates, sulfonamides, anticoagulants, and methenamine.

POLYMORPHONUCLEAR NEUTROPHIL: (REPORT AS NUMBER/HPF)

The polymorphonuclear neutrophil (PMN, poly, seg) is the most common leukocyte found in the urine and is about 12 μ in diameter (i.e., 1.5 times the size of an RBC) (Fig. 17–18). You can often see the segmented nucleus of the neutrophil amid the cytoplasmic granules. Sometimes, however, the nucleus is not visible, and what you see are cytoplasmic granules that seem to move (i.e., the so-called "glitter" cell).

When stained the PMN's nucleus is red and purple and the cytoplasm is violet with dark granules. Glitter cells do not take up the stain well and appear light blue to almost colorless.

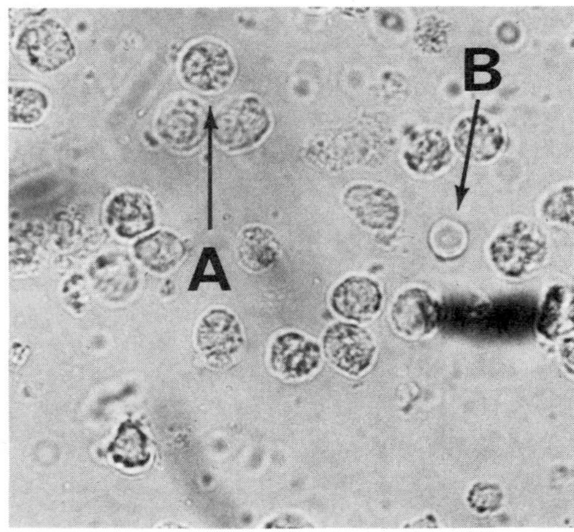

Figure 17–18. A. WBC. B. RBC. Viewed under high-dry objective.

From zero to five WBCs per high-powered field is considered normal. However, in very dilute urine, even this low number may represent significant pyuria in a symptomatic patient. Increased numbers of PMNs are seen in almost all renal and urinary tract diseases as well as with high fever, strenuous exercise, and appendicitis.

The most common cause of pyuria in women is cystitis or vaginitis/cervicitis. Suspect vaginitis/cervicitis if large numbers of squamous epithelial cells are also present. In males pyuria is most commonly due to urethritis or prostatitis. Suspect *N gonorrhoeae* urethritis in any male with greater than five WBCs per high-power field on a routine urinalysis. The first 15-mL specimen from a male with urethritis but who has very little urethral discharge will often show an increased number of WBCs. The midstream specimen will show fewer cells. If prostatitis is suspected, a postprostatic massage specimen should show an increase in the number of PMNs as well as the presence of lipid-laden macrophages.

In cases of simple urinary tract infections, WBCs can be seen in the urine for up to 5 days after the start of treatment. Despite the presence of these cells, the urine culture is usually sterile by the second day of appropriate antibiotics.

"Sterile" pyuria is seen in a variety of renal diseases, including acute glomerulonephritis, lupus nephritis, and the urethral syndrome caused by chlamydia trachomatis. Although often mentioned, sterile pyuria is rarely caused by renal tuberculosis.

MACROPHAGES

These are large mononuclear WBCs that are about twice the size of a PMN. They contain numerous cytoplasmic vacuoles of various sizes. It is important to distinguish these cells from RTE cells (large central nucleus) or OFBs (refractile vacuoles). Urinary macrophages are often seen in small numbers and have no established clinical significance (Figs. 17–19). Figure 17–20 shows the size relationship between different cells.

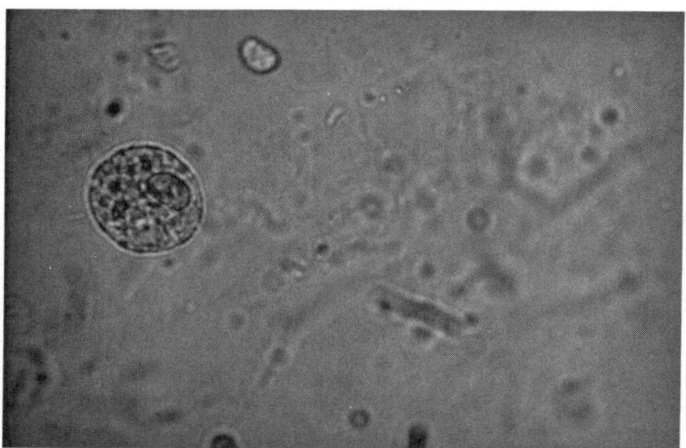

Figure 17–19. Macrophage in the urine. Notice all the vacuoles. Compare to the renal tubule cell in Figure 17–9. They can be about the same size. The RTE cell has one central nucleus and no vacuoles.

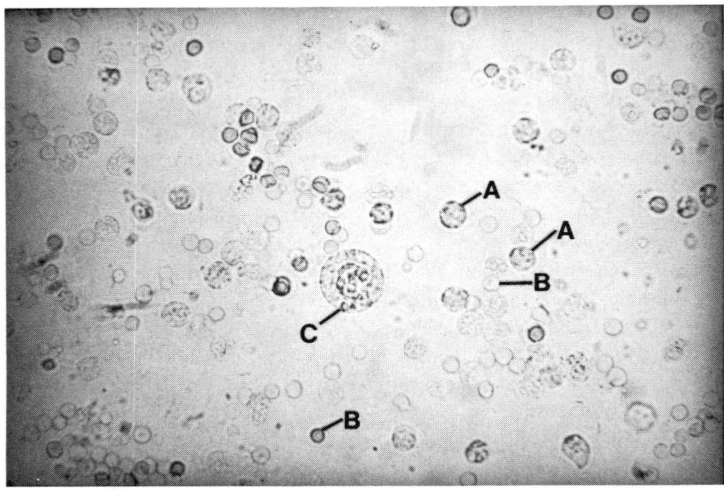

Figure 17–20. Size relationship. **A.** White cell. **B.** Red cell. **C.** Macrophage. Viewed under high-dry objective.

SECTION 20: Urinary Casts

Casts may be the most helpful finding in a microscopic urinalysis. They are cylindrically shaped elements that are molded by the kidney's tubules. They are therefore evidence of what is actually happening in the kidney and can be called a "poor man's renal biopsy." Casts are made up primarily of a special protein called Tamm-Horsfall protein, which is secreted into the urine by the epithelial cells that line the kidney tubules. Mixed into this protein matrix can be a variety of cells or materials that are present within the tubules when the cast is formed. Casts are named by the elements that are included in the cylindrical protein matrix.

Casts can also be further categorized by their shape and size. These characteristics are determined by the portion of the tubule in which the cast is formed. Convoluted casts have a spiral shape and are formed in the distal convoluted tubule. Broad casts are very wide and are formed in the dilated collecting ducts of patients with end-stage renal disease and are therefore referred to as "end stage" or "renal failure" casts. Narrow casts are formed in tubules whose lumina have been reduced. Cylindroids are casts that are tapered at both ends. This finding is not clinically significant, as cylindroids can be found in both health and disease.

The formation of casts is favored by a decrease in urine flow, low urinary pH, and a high solute or protein concentration in the glomerular filtrate. Casts tend to float to the top of a spun urine specimen. It is therefore important to mix the sediment well after decanting the supernatant. Some casts break up easily. For this reason urine samples should be examined as soon as possible after the specimen collection. Casts often dissolve if the urine specific gravity is less than 1.003.

The presence of casts in the urine is called cylindruria.

ACELLULAR CASTS

Hyaline Casts

These casts are made up of the basic protein matrix without any other inclusions. They are pale, transparent, and smooth. They may be difficult to see because their refractive index is close to that of urine (Figs. 17–21 and 17–22). They are best visualized with a low light intensity and with urine-sediment staining. When stained they appear pink to light purple. It is helpful to focus up and down while scanning the microscope field to find these casts. Hyaline casts dissolve easily in dilute and alkaline urine. The specimen should therefore be examined promptly.

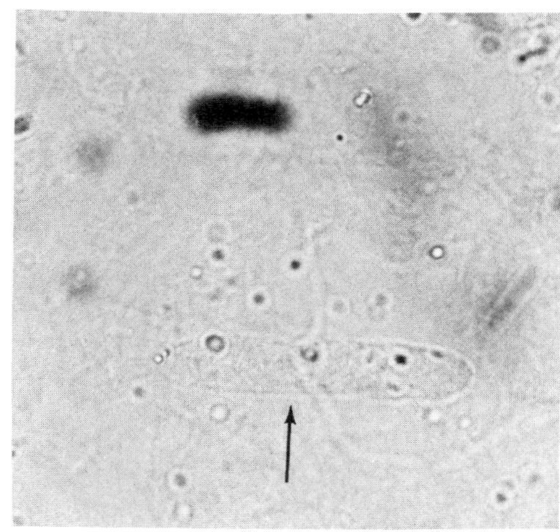

Figure 17–22. Hyaline cast in center. Notice how difficult it is to distinguish from the background. Viewed under high-dry objective.

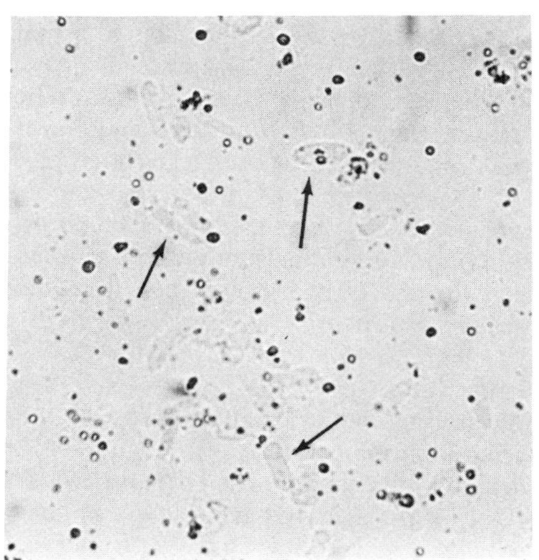

Figure 17–21. Hylaline casts. Notice how difficult it is to see them. Viewed under low-power objective.

Hyaline casts are frequently present in small numbers in healthy patients' urine (one or two per low-power field). Severe emotional or physical stress can lead to an increase in the number of hyaline casts. This does not indicate renal pathology. Exercise, fever, postural changes, and diuretic therapy may all result in an increase in the number of hyaline casts. It is not uncommon to find hyaline casts without proteinuria. Persistent proteinuria with cylindruria is often pathologic. This finding may precede obvious renal disease by many years. The diseases that are associated with hyaline casts are acute glomerulonephritis, pyelonephritis, malignant hypertension, any chronic renal disease, congestive heart failure, and diabetic nephropathy.

Granular Casts

These are acellular casts in which granules of various sizes can be seen. The granules may be found throughout the cast or local-

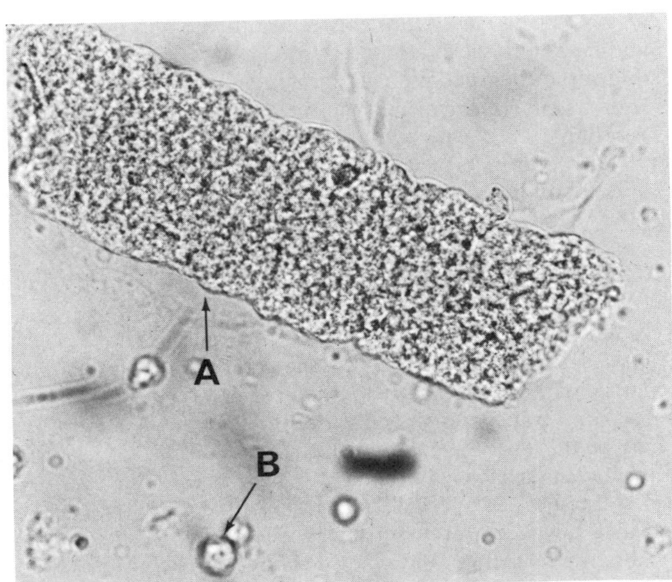

Figure 17–23. A. Granular cast. **B.** WBC (out of focus) for size comparison. Viewed under high-dry objective.

ized to one part (Fig. 17–23). The granules may be small (fine granular casts) or large (coarse granular casts), but this differentiation is not considered to be of any clinical significance. The presence of the granules makes these casts easier to see than hyaline casts. When stained the protein matrix appears pink to light purple and the granules are a dark purple.

Like hyaline casts, granular casts can be seen in a variety of nonspecific conditions, such as viral diseases, fever, physical activity, or stress. In these conditions the granular material is derived from proteins that are present in the glomerular filtrate. Granular casts are also seen in many types of intrinsic renal disease. In these cases the casts often begin in the kidney as cellular casts but change to granular casts as the cellular inclusions degenerate.

Waxy Casts

These casts are easily seen because of their high refractive index and glassy or waxy appearance. They are often broad and stubby, with sharp, broken-off ends. Their sides are parallel but with occasional fragmentation because their matrix is brittle (Fig. 17–24). These casts are believed to form by the very slow breakdown of cellular casts in urinary systems with a long transit time. They may be colorless or a pale tan in color. When stained they are light purple and have a much sharper outline than do hyaline casts.

Waxy casts are an indicator of chronic renal diseases, such as diabetic nephropathy, renal amyloidosis, or chronic glomerulonephritis. They are not seen in normal urinary sediment.

Fatty Casts

These casts contain fat droplets or OFBs. The lipid droplets have a high refractive index and are therefore easily identified (Fig. 17–25). A cast may have a few droplets or may be packed with lipid. With polarized microscopy, a cholesterol-containing droplet shows a Maltese cross pattern (Fig. 17–26). When these casts are stained, the matrix appears pink, but the fat remains

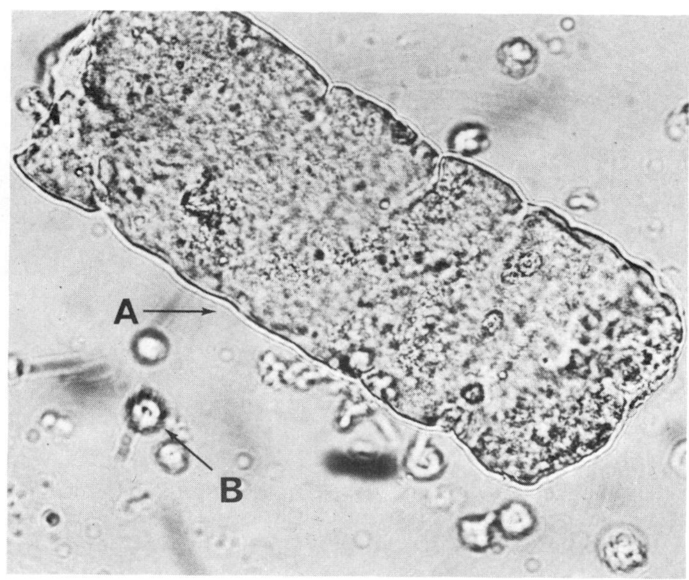

Figure 17–24. A. Waxy cast. Notice the cracks in the side. **B.** WBC (out of focus) for size comparison. Viewed under high-dry objective.

unstained. This can be helpful in differentiating fatty casts from granular casts.

Fatty casts are seen along with free-fat bodies and OFBs in renal diseases such as the nephrotic syndrome, diabetic nephropathy, lipoid nephrosis, and toxic renal-tubular nephrosis.

Bile Casts

These casts are hyaline casts in which bile has been mixed into the protein matrix. They appear as yellow, transparent casts. They are seen in liver diseases in which there are also large quantities of bilirubin or urobilinogen in the urine.

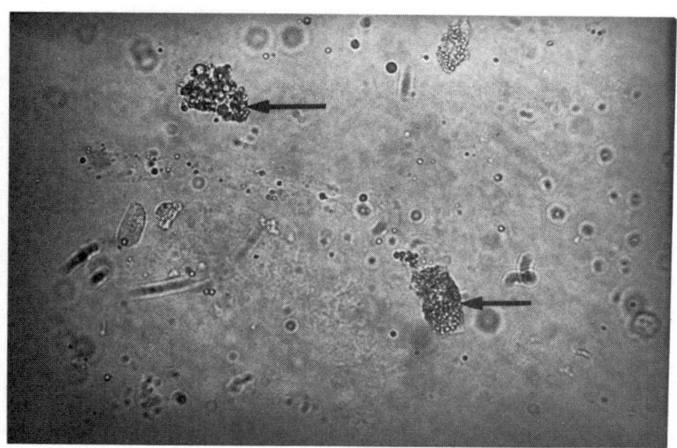

Figure 17–25. Fatty cast. Viewed under high-dry objective.

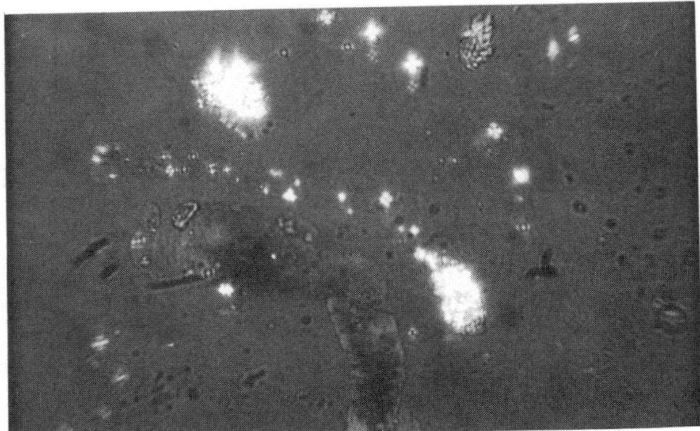

Figure 17–26. The same fatty cast viewed in Figure **17–25** but with polarized light. Note the Maltese crosses. Viewed under high-dry objective.

Hemoglobin Casts

These are transparent, brownish-red casts without inclusions. They are found in diseases in which a high concentration of free hemoglobin is present in the glomerular filtrate, such as the hemolytic anemias.

Cellular Casts

RBC Casts

These are casts in which RBCs are trapped in the protein matrix as the cast forms in the tubules. RBCs can be identified within the cast (Figs. 17–27 and 17–28). These casts often have a reddish brown color, but they may also be colorless. The RBCs often degenerate within the cast and produce a granular background, but distinct RBCs must be identified to classify the cast as a red cell cast. If there are no intact cells, the cast is called a "blood cast." The distinction between a RBC cast and a blood cast is not considered to be clinically significant. When RBC casts are stained, the cast matrix appears pink and the intact cells are lavender.

RBC casts are formed when there is a disruption in the glomerular basement membrane, so that RBCs pass into the urinary filtrate. These casts are seen in acute glomerulonephritis, renal vein thrombosis, renal thromboemobolism, flank trauma, toxic nephritis, lupus nephritis, polyarteritis nodosa, Goodpasture's syndrome, renal infarction, acute pyelonephritis, and focal glomerulitis. Proteinuria is commonly seen when RBC casts are present.

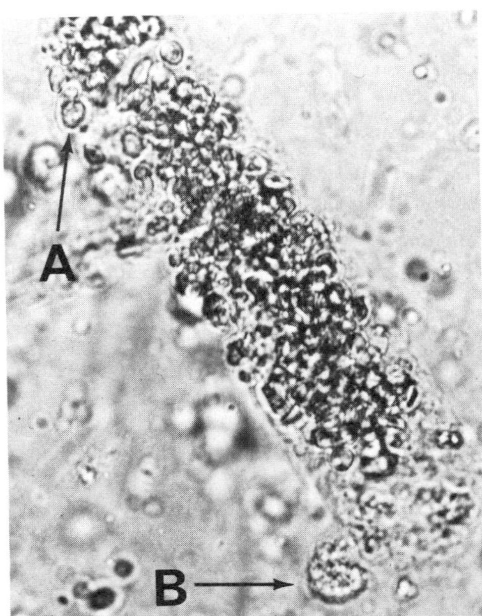

Figure 17–27. A. RBC cast. Note the intact red cells, especially at the top of the cast. **B.** WBC for size comparison. Viewed under high-dry objective.

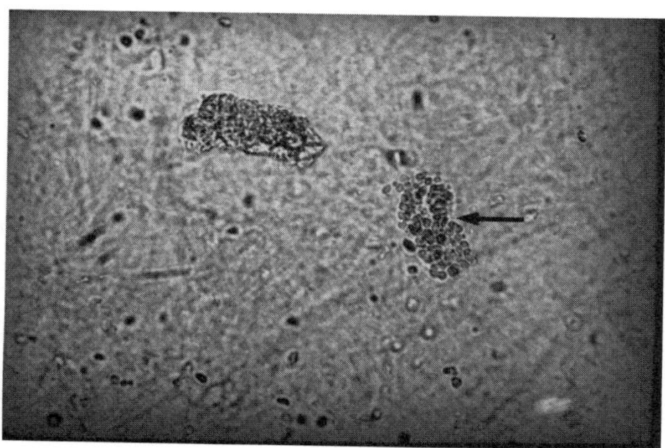

Figure 17–28. A piece of a red cell cast. Viewed under high-dry objective.

WBC Casts

Polymorphonuclear neutrophils are present within the protein matrix of these casts (Fig. 17–29). The neutrophils are often degenerated, with indistinct cell borders and fragmented nuclei. This may make their identification difficult. Staining is helpful because the multilobed nuclei of the neutrophils stain a dark blue. The granular cytoplasm of the WBCs stains a light red and makes the protein matrix of the cast appear pink to light purple.

WBCs may enter the tubules by two mechanisms. In diseases associated with a disruption of the basement membrane, the WBCs can migrate directly from the glomerulus into the filtrate, as do the RBCs in RBC casts. In diseases that produce an interstitial kidney inflammation (e.g., pyelo-

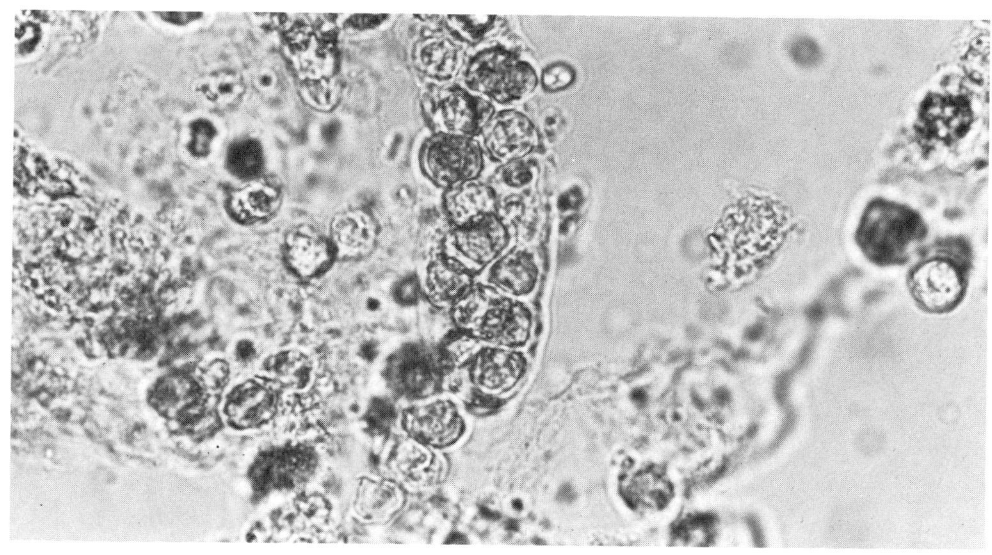

Figure 17–29. WBC cast. Viewed under high-dry objective.

nephritis), the WBCs migrate through the lining cells of the tubules and into the tubule lumen. WBC casts are seen in many renal disorders, including pyelonephritis, lupus nephritis, glomerulonephritis, interstitial nephritis, and the nephrotic syndrome.

RTE Casts:

These casts have RTE cells embedded in the protein matrix. The cells have a high refractive index, large nuclei, and only a little cytoplasm. The cells are often polyhedral in shape. They may be lined up in rows covering the entire cast surface, or only a few cells may be seen within the cast (Fig. 17–30). When stained the cells have a dark-purple nucleus and red cytoplasm, while the cast matrix is pink to light purple. If the cells have degernated, it may be difficult to distinguish an RTE cast from a WBC cast.

RTE cell casts are present in diseases that destroy the tubular epithelium. This is seen with various toxins, including mercury, ethylene glycol, chemotherapeutic agents, and salicylates. It is also seen with hepatitis and cytomegalovirus infections.

Mixed Cellular Casts:

These casts include several different types of cells—RBC, WBC, and RTE. They are seen with mixed tubular-interstitial renal disease.

Cellular Casts:

It is not uncommon to find casts with definite cellular inclusions but in which the cell type is indeterminant. These are simply called "cellular casts" (Fig. 17–31). It is often helpful to look at the cells surrounding the cast in the sediment to identify the cellular inclusions.

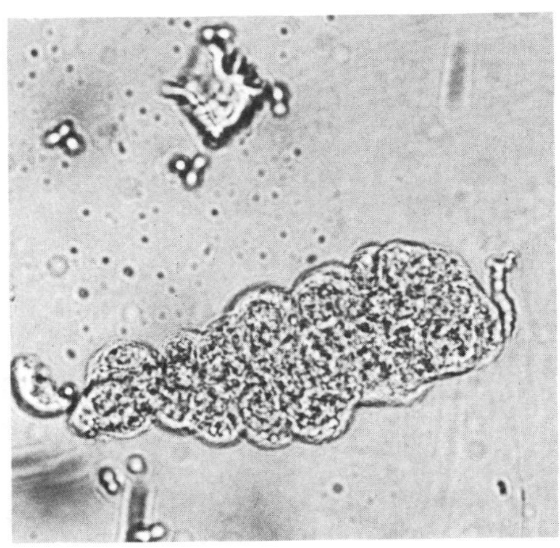

Figure 17–30. Renal tubule epithelial cell cast. If you look very carefully you will see the central nucleus. Viewed under high-dry objective.

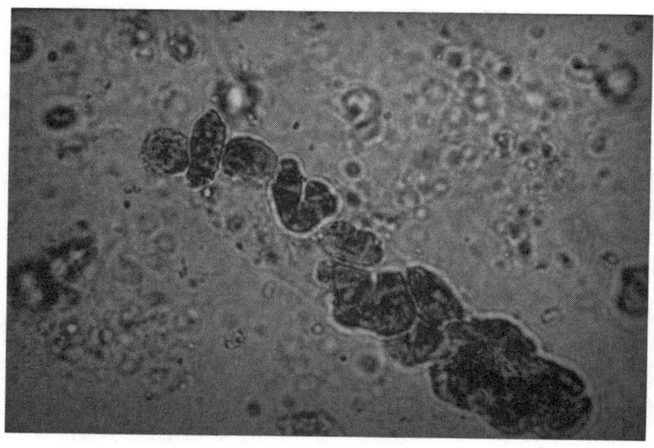

Figure 17–31. Cellular cast stained with bile. The cells are most like renal tubule cells, but it is not possible to see the nucleus. Viewed with high-dry objective.

Other Casts

Crystalline Casts:

Urate and oxalate crystals are occasionally seen on the cast surface and may appear to be a component of the cast itself. These casts have no special clinical significance.

Bacterial Casts:

Bacterial casts are only seen in pyelonephritis and can be identified only if the sediment is stained. In an unstained specimen the casts look like granular casts or WBC casts with a granular matrix. With staining the dead bacteria appear dark purple and can therefore be differentiated from other types of casts. This can be an important finding because granular or WBC casts are found in a variety of illnesses, but bacterial casts are seen only with renal infections.

Artifacts or Pseudocasts

Pseudocasts:

These are any elements in the urine sediment that have a cylindrical shape and therefore resemble casts (Fig. 17–32).

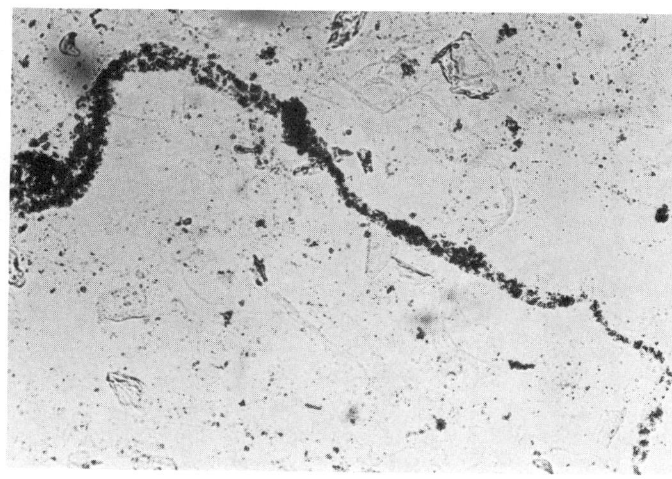

Figure 17–32. Pseudocast made up of amorphous debris caught in a mass of mucous threads. Viewed under low-power objective.

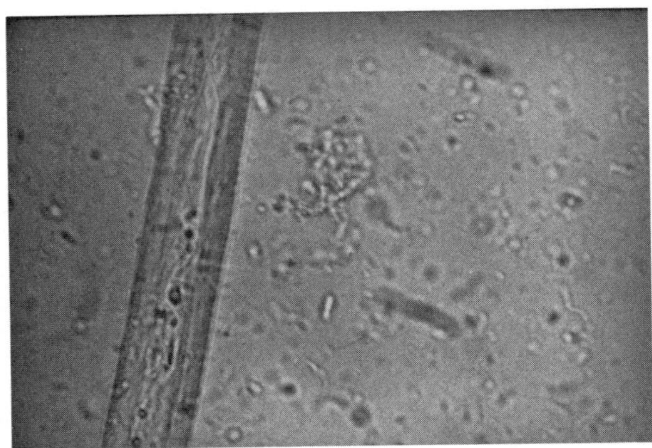

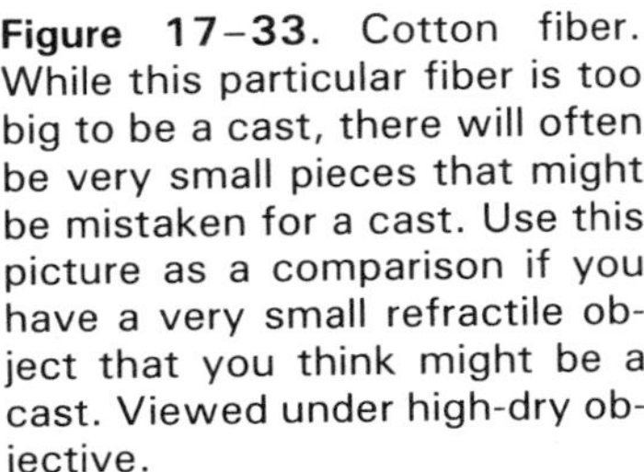

Figure 17–33. Cotton fiber. While this particular fiber is too big to be a cast, there will often be very small pieces that might be mistaken for a cast. Use this picture as a comparison if you have a very small refractile object that you think might be a cast. Viewed under high-dry objective.

Cotton or Wool Fibers:

These are long, flat, and ribbon shaped. They are usually much larger and more refractile than a cast (Fig. 17–33.)

Animal or Vegetable Fibers:

These are usually the result of fecal contamination. Under low power they resemble casts, but they are far more refractile. Under high power they are more striated than cotton or wool (Fig. 17–34).

Hair:

Hair has a sharply defined outline and a dense structure. It too is highly refractile (Fig. 17–35).

Clumps or strings of WBCs, RBCs, or amorphous crystals can all be confused with casts. Clumps of cells or crystals can be differentiated because they have no matrix outline and have irregular edges.

Mucous Threads

Mucous threads are commonly found in urine sediments and are of no clinical significance. They have a low refractive index and take on a variety of shapes. These threads can be easily confused with hyaline casts, but they do not have the smooth, par-

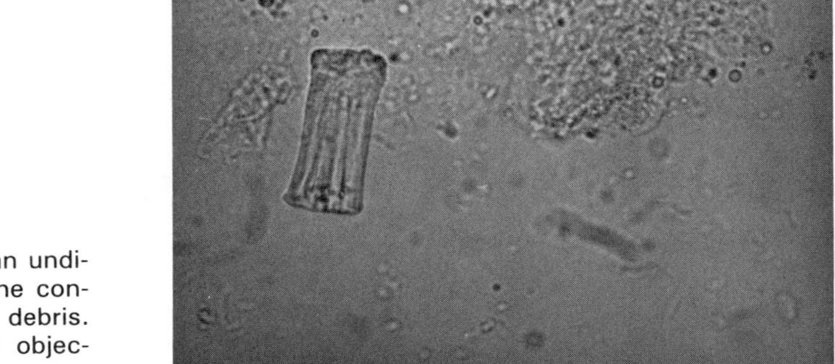

Figure 17–34. This is an undigested food fiber in urine contaminated with fecal debris. Viewed under high-dry objective.

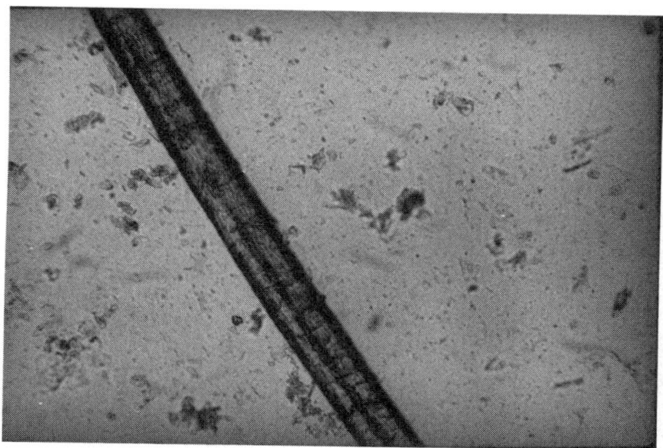

Figure 17–35. This is a piece of hair in a urine sample. Viewed under high power objective.

allel sides and the cylindrical shape of the cast. They are also longer and narrower than a cast and have poorly defined edges (Fig. 17–36).

Parasites

It is uncommon in the office laboratory to find parasites in the urinary sediment; however, pinworm eggs are occasionally seen. The eggs are oval with a granular center and are about the size of an epithelial cell (Figs. 21–6 and 21–7).

Spermatozoa

Spermatozoa can be seen in male patients (after ejaculation) and in female patients (vaginal contamination). Urine is toxic to sperm, so they are usually not motile. Spermatozoa have very small oval bodies, about one fourth the size of an RBC, and long delicate tails (Fig. 17–37).

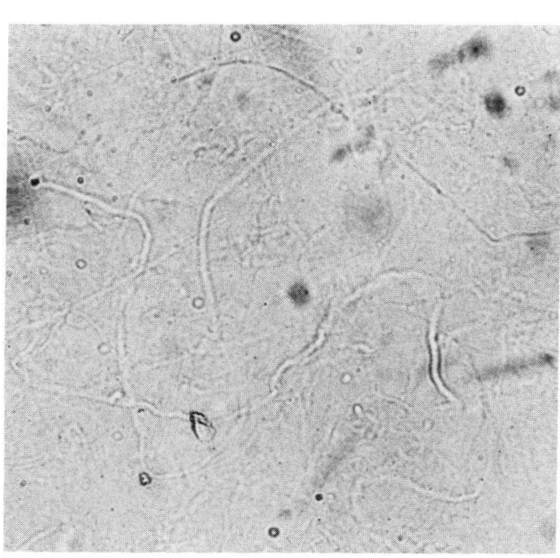

Figure 17–36. Mucous threads. Notice how difficult it is to see them. Viewed under low-power objective.

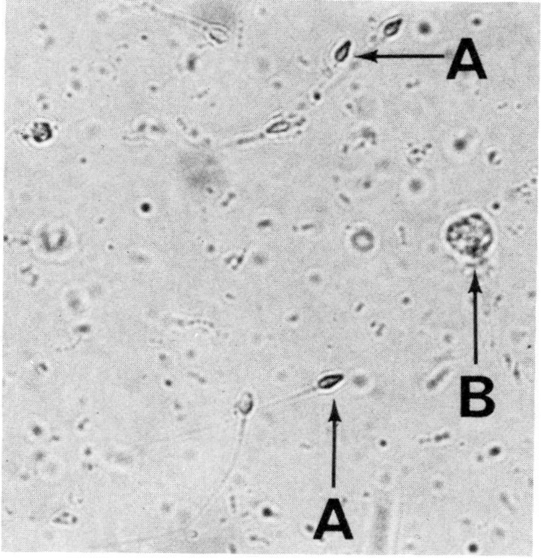

Figure 17–37. A. Spermatozoa. **B.** WBC. Viewed under high-dry objective.

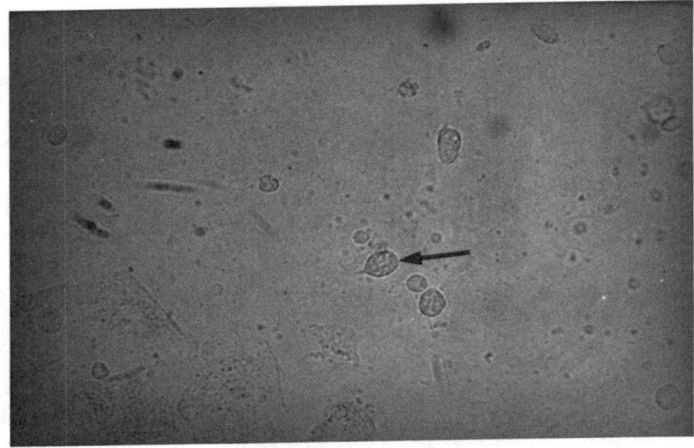

Figure 17–38. Trichomonad. Viewed under high-dry objective.

Trichomonas Vaginalis

This organism can be seen in the urine of male or female patients. It takes a variety of shapes and sizes and may even look exactly like a WBC (Figs. 17–38 and 17–39). The only way to positively identify the organism is to see flagellar movement. Often a rod-shaped bacteria will attach to a white cell, thus mimicking a trichomonad. However, in this case there is no flagellar movement.

Telescoped Sediment

This is the name given to a sediment that shows RBCs, red cell casts, broad casts, waxy casts, OFBs, and fatty casts. This type of sediment can be seen with lupus nephritis or subacute bacterial endocarditis.

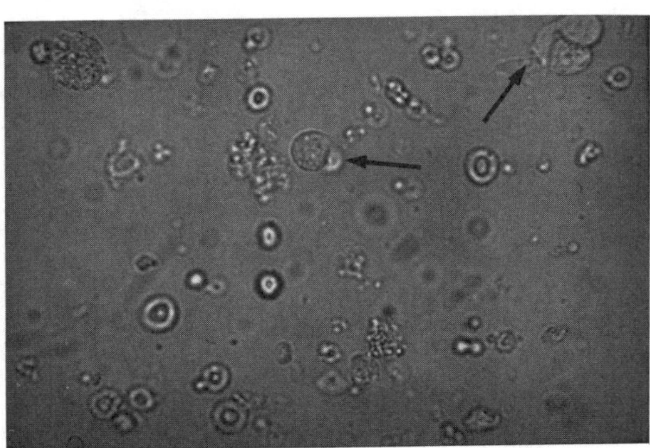

Figure 17–39. Trichomonads. Viewed under high-dry objective.

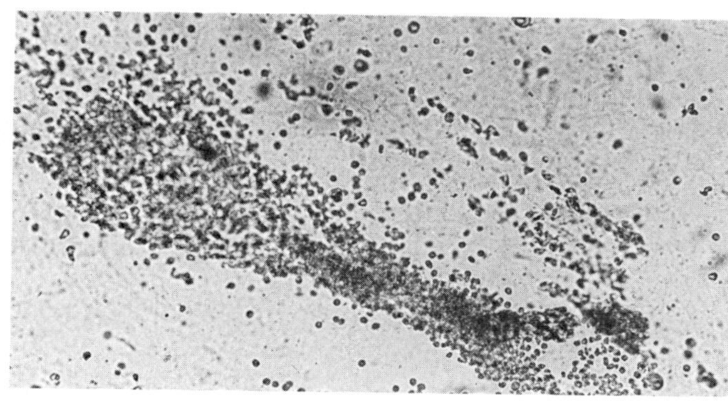

Figure 17–40. Large number of WBCs adherent to a clump of mucous. This is frequently seen in prostatitis. Viewed under low-power objective.

Prostatitis

It is common in prostatitis to find large strands or clumps of mucus completely covered with WBCs (Fig. 17–40). White cells are not even distributed throughout the urine but are caught in these huge strands of mucus.

SECTION 21: Crystals

A wide variety of crystals can be found in the routine examination of urinary sediments (Table 17–6). These crystals are generally of little clinical significance. It is important, however, to be able to recognize the common crystals so that rare pathologic crystals are not overlooked. Identification of the crystals in the urine from patients with kidney stones can be helpful in identifying the type of stone.

All crystals of pathologic significance are found in acidic urine. Therefore if the urinalysis is being performed specifically to check for crystals, it should be with an acidified urine sample. This can be easily done by adding two drops of 5% acetic acid to the centrifuge tube. This acidification is not recommended for the routine urinalysis because the acid may lyse the cellular elements of the urine.

It is difficult to identify crystals without previous experience. A urinary sediment atlas such as Graff's *Handbook of Routine Urinalysis* (Philadelphia, Pa: Lippincott; 1983) is extremely helpful in identifying uncommon crystals.

NORMAL CRYSTALS FOUND IN ALKALINE URINE

Ammonium Biurates

These rarely seen crystals are yellow-brown in color. They are spherical in shape and may have radiating spicules (i.e., the thorn-apple appearance.).

Amorphous Phosphates

These commonly seen, fine, white, granular crystals often precipitate when a urine specimen has an alkaline pH. There may be so many that they obscure the examination

TABLE 17–6. CHARACTERISTICS OF CRYSTALS SEEN IN URINARY SEDIMENT

Crystals	Shape
Normal Crystals Commonly Found in Alkaline Urine	
Amorphous phosphate	Granular—white
Triple phosphate	Coffin lid
Normal Crystals Commonly Found in Acidic Urine	
Amorphous urate	Granular—pink/orange
Calcium oxalate	Octahedral or ovoid
Uric acid	Varied
Pathologic Crystals Found in Acidic Urine (These are rarely seen)	
Bilirubin	Needle or granular—reddish-brown plates, cubes
Cholesterol	Flat plate with notched corner
Cystine	Hexagonal plates
Leucine	Round, oily looking
Tyrosine	Fine needles in sheaves; colorless/yellow; appear black when in focus

of the other elements in the urine sample. These amorphous crystals can be cleared by adding two drops of acetic acid to the specimen (Fig. 17–41).

Calcium Carbonate

These uncommon crystals are small, colorless, dumbbell-shaped granules. They may be difficult to distinguish from bacteria.

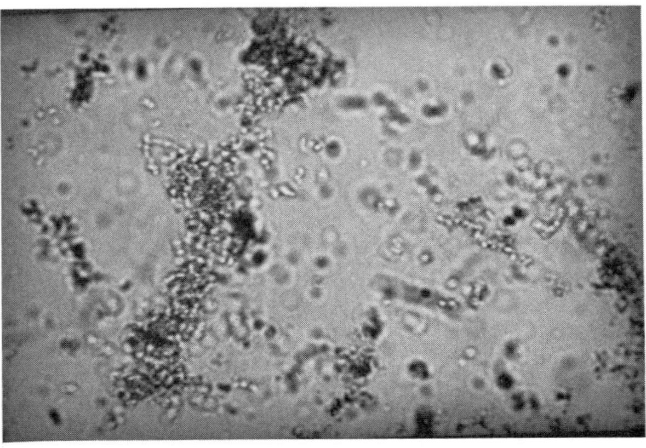

Figure 17–41. Amorphous phosphates. Viewed under high-dry objective.

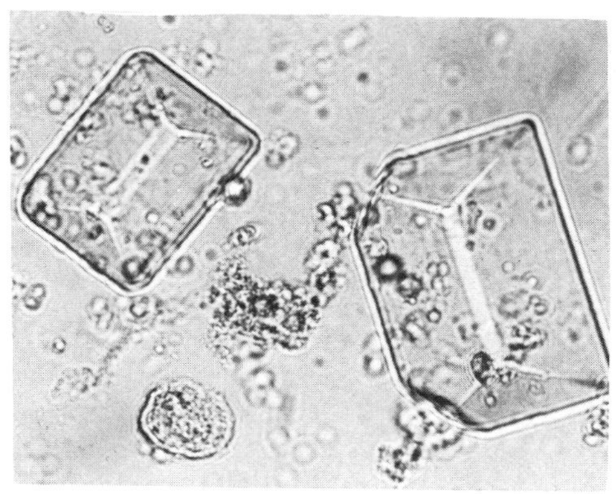

Figure 17–42. Triple phosphate crystal. Note the coffin-lid appearance.

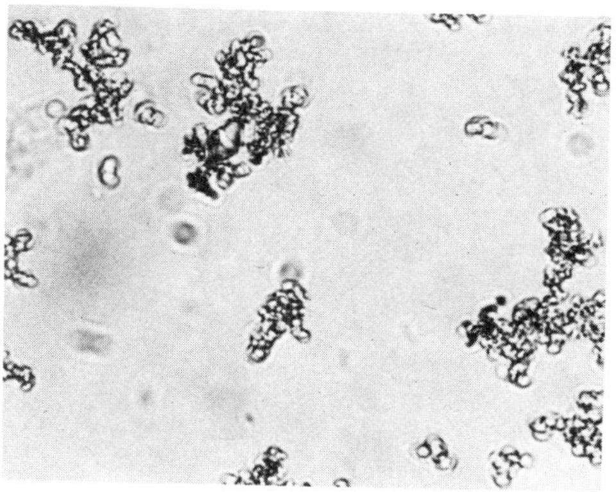

Figure 17–43. Amorphous urate crystals. Viewed under high-dry objective.

Calcium Phosphate

These uncommon crystals are colorless and can appear as either small granules or large, flat, irregularly shaped crystals.

Triple Phosphate

These commonly seen crystals are colorless but very refractile. They have a prismatic shape that is described as looking like a coffin lid (Fig. 17–42).

NORMAL CRYSTALS FOUND IN ACIDIC URINE

Amorphous Urates

These extremely common pinkish-orange, amorphous salts may precipitate when the urine specimen is cooled. They can be so numerous that they obscure the microscopic examination (Fig. 17–43). They can be cleared by gently heating the urine so that the crystals dissolve.

Calcium Oxalate

The dihydrate form of this crystal is the familiar refractile, octahedral crystal (Fig. 17–44). Less commonly seen is the monohydrate form that is ovoid or dumbbell-shaped (Fig. 17–45). It is similar in shape to a pinworm egg except that pinworm eggs are much larger. If there are many calcium oxalate crystals, it may reflect severe renal

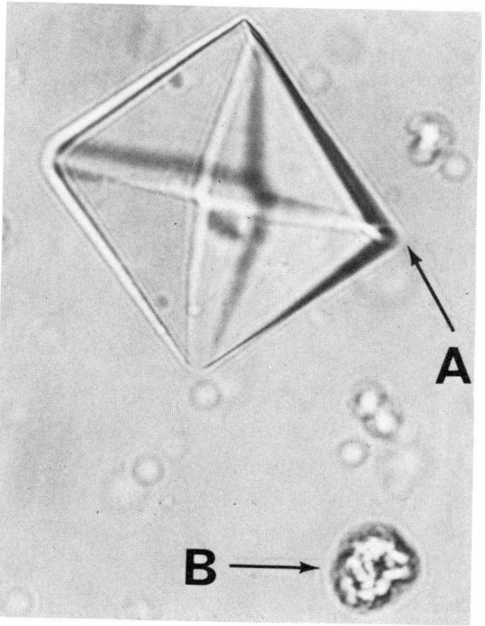

Figure 17–44. A. Calcium oxalate crystal, dihydrate form. **B.** WBC. Viewed under high-dry objective.

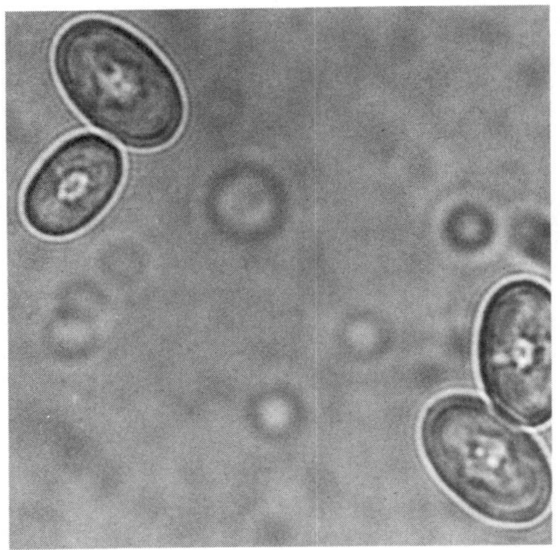

Figure 17–45. Calcium oxalate crystals, monohydrate form. Viewed under high-dry objective.

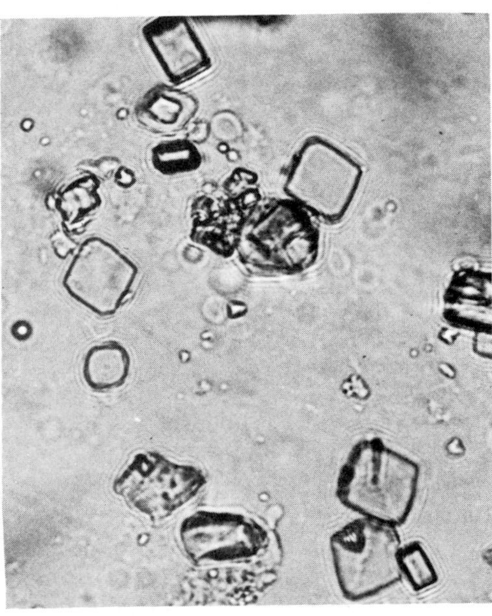

Figure 17–46. Uric acid crystals. Viewed under high-dry objective.

disease. They are also seen in ethylene glycol and methoxyflurane toxicity.

Uric Acid

This is one of the most common crystals seen in urinary sediments. The crystals take on a variety of shapes (rhomboid, six-sided plates, needles, or rosettes) and can therefore mimic other types of crystals (Figs. 17–46 and 17–47). They are colorless (rare) or yellow to red to brown (common).

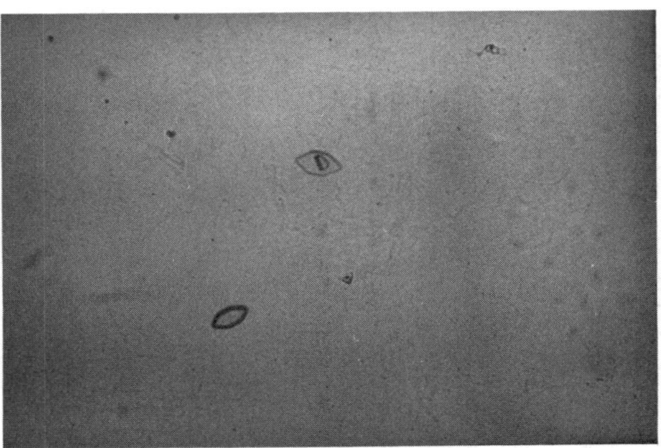

Figure 17–47. Uric acid crystals. Viewed under high-dry objective.

Uric acid crystals have no significance when found in refrigerated urine. When found in freshly voided urine, they may have clinical significance. Gout and acute fever are the commonest causes. Because uric acid can reflect increased protein metabolism, they can also be seen in leukemia. When viewed with a polarizing microscope, they present a stained-glass window effect.

Hippuric Acid

These rare crystals are colorless or very pale yellow. They are shaped like needles, six-sided prisms, or rhomboid plates.

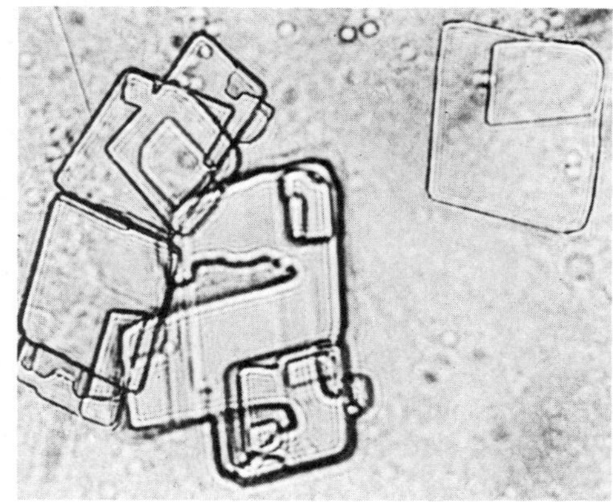

Figure 17–48. Cholesterol crystals. Note the notched corners. Viewed under high-dry objective.

ABNORMAL CRYSTALS FOUND IN ACIDIC URINE

All of these crystals are rarely seen, but when found, they do indicate pathology—renal or other.

Bilirubin

These yellow crystals are either granular or needle shaped. They appear in the urine of patients with obvious jaundice and bilirubinuria.

Cholesterol

These crystals are flat, rhomboidal, transparent plates. They often have one corner broken off, giving them a stair-step shape (Fig. 17–48). They are sometimes seen in patients with nephrotic syndrome.

Cystine

Cystine crystals can be seen in the urine of patients with congenital cystinuria or cystine stones. The crystals are thin, refractile, colorless, hexagonal plates.

Leucine

These are dense, brown, highly refractile, round crystals. They may appear oily and can therefore be confused with free-fat bodies. Starch granules are sometimes mistaken for leucine crystals. Leucine crystals are seen in patients with severe liver disease.

Tyrosine

These crystals are very long, fine, and needle shaped. They often appear in clumps and may be seen along with leucine crystals. They appear in the urine of patients with severe liver disease.

Radiographic Media

These can look like notched cholesterol plates but are identified by the patient's history and the urine's high specific gravity. They can also look like long needles.

Sulfa Crystals

The sulfa drugs that are commonly used today are much more soluble than those used in the past. Sulfa crystals therefore are now rarely seen. Sometimes you will see those crystals in the urine of patients taking either sulfamethoxazole or sulfasoxazole. They are small, colorless, flat, rectangular crystals (Fig. 17–49).

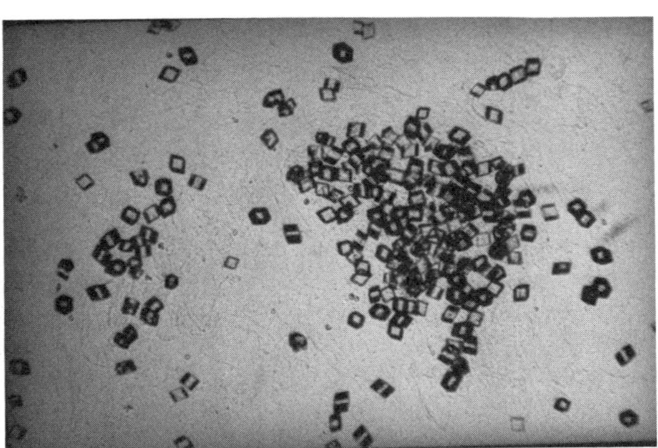

Figure 17–49. Sulfa crystals from a patient on Septra. Viewed under high-dry objective.

CONTAMINANT CRYSTALS

Starch

These small, round, refractile crystals (Fig. 17-50) appear in urine when patients use dusting powder containing cornstarch. Also nurses, physicians, and technicians wearing powdered gloves can contaminate the urine. It is important not to confuse these with leucine crystals.

Talc

These crystals look almost like degenerating epithelial cells, but they are more refractile and lack a nucleus (Fig. 17-51). These come from dusting powder with a talc base.

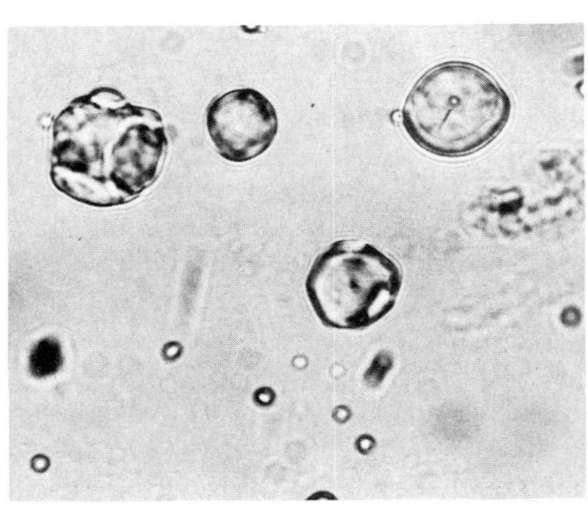

Figure 17–50. Starch crystals. Viewed under high-dry objective.

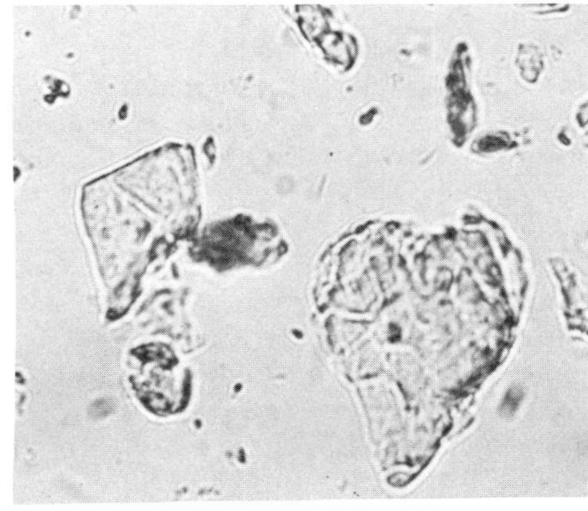

Figure 17–51. Talc crystals (Not squamous cells; note lack of nucleus).

SECTION 22: Other Findings in Urinary Sediment

BACTERIA

The most common type of bacteria found in urine is rod shaped (Fig. 17–52). The other common pathogen is *S saprophytius,* and it looks like cocci in clusters (Fig. 17–53). It is extremely difficult without doing a Gram's stain to identify cocci from small amorphous crystals. Some people feel that amorphous crystals move in a random "Brownian movement," while bacteria move in nonrandom fashion. This is often difficult to distinguish microscopically.

Urine should normally contain no bacteria. When bacteria are found, it may indicate a urinary tract infection. Traditionally, bacteriuria has been used to identify patients with a urinary tract infection. However, pyuria (even without noticeable bacteriuria) is the commonest mark for a urinary tract infection.

In a specimen from a female patient, the presence of bacteria and white cells along with squamous epithelial cells may also indicate vaginal fluid contamination.

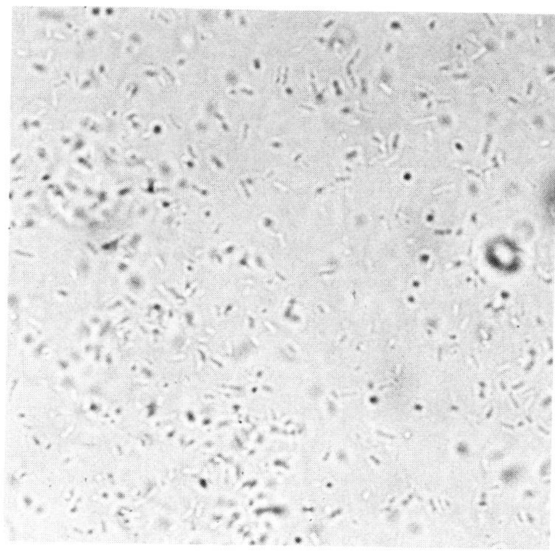

Figure 17–52. Bacteriuria. Viewed under high-dry objective.

There will often be many WBCs but no freely floating bacteria. In this case, look very carefully for a shaggy appearance to the white cells caused by bacteria adhering

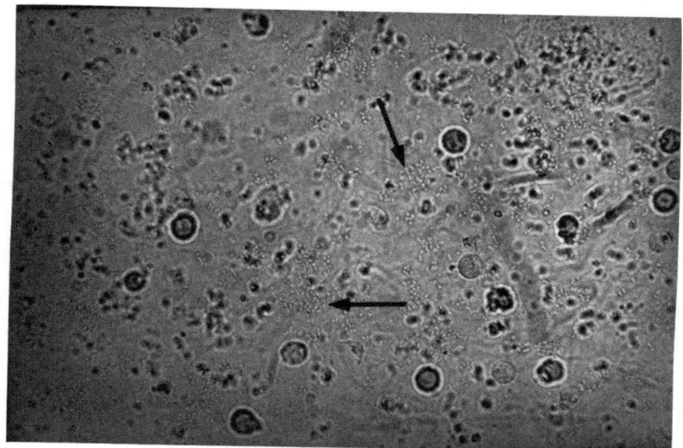

Figure 17–53. Staphylococcal urinary tract infection. Notice the clustering. Viewed under high-dry objective.

to their surface. If there are no white cells but a great number of bacteria, check that the specimen has not been standing at room temperature and become overgrown with bacteria. When reporting the presence of bacteria in the specimen, note in parenthesis whether they are rods or cocci.

YEAST

Budding yeast is the most common form of *Candida albicans* found in the urine (Fig. 17–17). They are round or oval and are 2.5 to 3 μ in size. Pseudohyphal forms are occasionally seen. *Candida* can be due to either vaginal contamination or a urinary tract infection. Such fungal urinary infections are seen in diabetic patients.

Yeast buds may be confused with RBCs. If you tap the coverslip, the RBCs will change shape as they turn. Yeast buds do not. If there is any question, a KOH smear will be helpful because the RBCs, but not the yeast, will lyse.

Torulopsis glabrata is occasionally seen in diabetic patients. The cells are smaller than *Candida* buds and are oval to spherical. No hyphae are seen (Fig. 17–54). These organisms do not generally invade the bladder tissues.

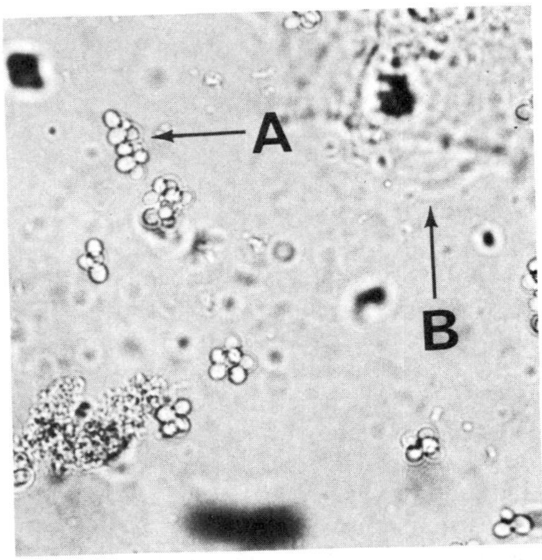

Figure 17–54. (A) Clumps of *Torulopsis glabrata* yeast buds. **(B)** Squamous epithelial cell. Viewed under high-dry objective.

SECTION 23: Urinalysis Quality Assurance

The office laboratory has the opportunity in the urinalysis test to exceed the quality control standards of even good hospital laboratories. This is because the collection techniques can be supervised and the samples can be tested right after their collection. This is in contrast to the hospital setting, where many different nursing personnel are responsible for specimen collection and where the sample may sit for 1 or 2 hours before being taken to the laboratory. We strongly urge office laboratories to routinely perform quality control procedures.

LABORATORY PERSONNEL

The best quality control assurance is to have a properly trained laboratory worker. This is especially important for the microscopic urinalysis, where identification of the various casts, crystals, and cells can be done only with training and previous experience. Sediment atlases are an enormous help and should be readily available at the bench. Laboratory workers and other office staff who perform dipstick testing should be checked for color blindness.

SPECIMEN COLLECTION

Office nurses should be instructed in the proper techniques for collecting urine specimens. Nurses are usually responsible for the collection of catheterized samples and infant bagged specimens. They are also responsible for instructing patients in the collection of midstream clean-catch samples. If repeated specimens show contamination, the clinician may need to reinforce the importance of good collection techniques with the nursing staff. This can be done by showing them the microscopic findings of the urine whose collection they have supervised.

The sooner the urine sample gets to the laboratory, the better. Standing without refrigeration for even one-half hour can result in bacterial overgrowth and the deterioration of fragile casts. Specimen handling in the office can be speeded up by having a pass-through window between the laboratory and the patient bathroom.

URINE SPECIFIC GRAVITY

Multistix-SG (Ames)

The dipstick specific gravity can be checked by squirting the specific gravity pad with distilled water at room temperature. The pad should turn dark blue, corresponding to a specific gravity of 1.000. The quality control solutions for dipsticks have a specified specific gravity and can be used to evaluate the specific-gravity reagent pad. Remember to correct for alkaline urine.

Refractometer

Check to see if there is an air bubble floating in the prism area. This bubble provides temperature compensation and should be shaken into the trap where it belongs. To do this, remove the refractometer from the stand and hold it vertically with the eyepiece down. Then shake gently. The instrument can be checked by placing a drop of distilled water onto the prism surface. The specific gravity scale should read 1.000. If this reading is not seen, clean the prism surface with a weak bleach solution to remove protein build-up, then add a drop of distilled water. If the instrument still does not read 1.000, consult the operator manual for "zero setting" the instrument. The quality control solutions for dipsticks have a specified specific gravity and can be used to check the refractometer.

Dipsticks

The most crucial aspect in a quality assurance program for dipsticks is their proper storage and handling! Dipsticks must be stored in a cool, dry area. The cap should be placed on the bottle as soon as the reagent strip is removed. They must not be used after their expiration date. This must be stressed with any nonlaboratory personnel who do dipstick testing—especially nurses.

There are a variety of different "controls" available. Look in the Scientific Products, Fisher, or Curtin Matheson catalogs under urinalysis and purchase the least expensive controls. Most controls are lyophilized human urine that must be reconstituted with distilled water. Once reconstituted it is usually "good" for 5 days. The product inserts give specific directions and values.

Ames makes a convenient dipstick form of abnormal control (CHEK-STIX). The stick is placed in distilled water; the chemicals leak out into the water, which can then be tested. You use one stick at a time, and the prepared solution is good for 1 day.

An inexpensive negative control is distilled water. The SG would be 1.000. All other pads would be negative. The pH would vary. If urine is tested at nursing stations away from the laboratory, keep the main bottle in the laboratory and give out only ten to 15 dipsticks to the nurses at one time. These can be kept in a regular dipstick bottle with the dessicant pouch. This minimizes the chances of improper storage affecting an entire bottle of sticks. You can run the quality control tests from the laboratory or at the nursing station.

We recommend that you perform daily quality control tests (normal and abnormal) when you first set up your laboratory. We have found that dipsticks are extremely stable if stored and handled correctly. After a year's experience of no "out-of-control" dipsticks, weekly checks are adequate.

When new office personnel are hired, check them out with daily controls until you are satisfied that they are storing and handling the dipsticks properly. The most common error, other than improper dipstick storage, is to too quickly read all of the pads rather than waiting the recommended time.

QUALITY CONTROL PROCEDURE

1. Store control reagents properly, according to the manufacturer's specific instructions. This will vary with the product and will be stated on the product insert. Always stack the reagent boxes so that the box with the earliest expiration date is used first.
2. Note the expiration date of the reagent bottle. Be sure to reconstitute the control solution before this date.
3. Reconstitute the control you choose according to the manufacturer's instructions. An appropriately sized syringe or a graduated centrifuge tube can be used to measure the water for reconstitution. Write the reconstituted date and expiration on the bottle.
4. Allow the solution to sit for the proper time before using.
5. Use the solution as you would a urine specimen to check the dipstick. The specific values for the solution are listed on the product insert.
6. Record the dipstick results in the quality control book (Fig. 17–55).
7. Follow the manufacturer's storage instructions.
8. If the dipsticks show any deviation from the specified solution values, open a new bottle of dipsticks and quality control them. If there are continuing problems, contact the local product representative.
9. Use both normal and abnormal controls.

MICROSCOPIC ANALYSIS

The results of the microscopic urinalysis are reported as "number of elements per field." This is only a semiquantitative test because the volume of sediment and volume of urine under the coverslip will vary. To improve the test's reproducibility, a constant volume of urine should be spun, the time and speed of centrifugation should be constant, the volume of urine in which the sediment is resuspended should be the same, and the same size of sediment drop should be examined each time. We recommend dropping two drops of sediment from a plastic transfer pipette. The volume of urine is dependent on the patient's recent fluid intake. The concentration of microscopic elements can therefore vary greatly.

OTHER QUALITY ASSURANCE MEASURES

A variety of wall charts and other reference materials are available for identifying the elements in the microscopic urinalysis. We recommend that a wall chart be kept near the microscope for easy comparison of crystals and casts. Any office worker who is to be responsible for performing a urinalysis should have adequate training to identify the various microscopic findings. The clinician can often arrange for this training in

URINALYSIS CONTROLS: Normal or Abnormal (Circle one)

		DIPSTICKS*		CONTROLS*		CONTROL TEST RESULTS (Write in from product insert)									
Date	Technician	Cont No.	Expiration Date	Cont No.	Expiration Date	glu	bili	ket	SG	blood	pH	protein	uro	nit	leuk

*List Brand

Figure 17–55. Dipstick quality control form.

the urinalysis section of the local hospital or reference laboratory. Finally, enroll in a proficiency testing program that provides color slides of microscopic urinalysis findings and a urine sample to be tested as part of their regular office laboratory program.

SECTION 24: Clinitest Two-Drop Technique

The Clinitest is a urine test that should be available in any office with pediatric patients. Clinitest can be used to check for glucose in the urine, but its usefulness extends to detecting other reducing substances, such as galactose, fructose, lactose, and pentose. It does not detect sucrose, which is not a reducing sugar. These other reducing sugars can be seen in the urine of infants with such rare metabolic diseases as galactosemia.

The traditional Clinitest procedure uses five drops of urine and is able to detect levels of glucose up to 2%. The two-drop technique described here can detect levels of up to 5% glucose and can therefore be used in patients with high glucose concentrations on whom dipstick testing is inadequate. The test is not recommended for routine use by patients in their homes because it is neither as convenient nor as safe as the dipstick and Tes-Tape methods.

The two-drop Clinitest can be purchased as a complete kit with reagent tablets, reaction tubes, and a dropper (Ames 2114). Extra reagent tablets can be purchased separately in boxes of 36 (Ames 2112). The tablets will degrade if exposed to light, heat, or moisture. The bottle should therefore be recapped as soon as a tablet is removed. The tablets contain sodium hydroxide, which is a poison. They should be handled with care.

TEST PROCEDURE

1. Collect a fresh urine specimen. The urine should be tested immediately, since glucose may be metabolized by urinary bacteria upon standing. A clean-catch specimen is not required.
2. Test the urine with a glucose-specific dipstick and record the results.
3. Place a clear glass test tube or the special Clinitest reaction tube in a test tube rack.
4. Fill the dropper with urine from the collection container.
5. Holding the dropper straight up and down, drop two drops of urine into the reaction tube.
6. Rinse out the dropper.
7. Fill the dropper with distilled water, and holding it straight up and down, put ten drops of distilled water into the tube with the urine.
8. Open the reagent bottle, and with a pair of forceps remove one tablet. Do not use your fingers, since these tablets can burn the skin. Drop the tablet into the tube with the urine and distilled water. Immediately recap the bottle of reagent tablets.
9. Watch the complete reaction in the tube. The solution will bubble and the tube will get hot. Do not shake the tube during the reaction!
10. Fifteen seconds after the bubbling has stopped, gently shake the tube.
11. Hold the tube up to the Clinitest two-drop color chart and record the results as negative, trace, 0.5%, 1%, 2%, 3%, or 5%. If there is greater than 5% glucose, the reaction will show a rapid passing through in which the solution

changes from the bright orange seen at 5% to a dark greenish brown. Record the test result as "greater than 5%" if this is seen.

INTERPRETATION

A positive Clinitest and dipstick indicates glucosuria. If the dipstick is negative but the Clinitest is positive, the urine contains another reducing sugar or a false-positive substance. If you have eliminated false-positive sources, the specimen should be sent to a reference laboratory for special sugar-identification testing.

False-Positive Results

These can be caused by ascorbic acid, cephalosporins, chloramphenicol, levadopa, metaxalone, methyldopa (Aldomet), nalidixic acid, penicillin (in high doses), probenecid, salicylates (in high doses), streptomycin, sulfonamides, tetracycline, or tyrosine.

QUALITY CONTROL

The solutions used to quality control the dipstick testing can be used as positive and negative controls for the Clinitest. Quality control testing should be done whenever a new bottle of reagent tablets is opened. Clinitest controls can also be run whenever the dipstick controls are done. If the Clinitest does not give the expected percent of glucose, discard the opened bottle of reagent tablets and use the controls to check the tablets in a newly opened bottle. Write the date that the bottle is opened on the side of the bottle. Do not use the tablets after the expiration date has passed.

SECTION 25: Urinalysis References

Bradley M, Schummann G. Examination of urine. In *Clinical Diagnosis and Management by Laboratory Methods.* Henry 3 (ed) 17th Ed. W.B. Saunders, Phila, 1984.

Cannon D. Identification and pathogenesis of urine casts. *Lab Med.* 1979;10:8.

College of American Pathologists. EXCEL Manual. Skokie, IL : College of American Pathologists; 1981.

Free A, Free H. *Urinalysis in Clinical Laboratory Practice.* Cleveland, Oh: CRC Press; 1975.

Graff L. *Handbook of Routine Urinalysis.* Philadelphia, Pa: Lippincott; 1983.

Haber M. *A Primer of Microscopic Urinalysis.* Fountain Valley, Calif: ICL Scientific; 1979.

Haber M (ed): *Urinary Sediment: A Textbook Atlas.* Chicago, IL: Education Products Division, American Society of Clinical Pathologists; 1981.

Stamm W, Counts G, Running K, et al. Diagnosis of Coliform infection in acutely dysuric women. *N Engl J Med.* 1982;307:8.

Product Inserts: Multistix 10-SG, Clinitest, and Tek-check. Ames Division, Miles Laboratories, Inc. Box 70, Elkhart, IN 46515.

Two-drop Clinitest, Ames Co., Division of Miles Laboratories, Inc., Box 70, Elkhart, IN 46515.

CHAPTER 18

Microbiology

SECTION 1: Introduction

Microbiology testing methods have undergone dramatic changes since the first edition of this book. We now stand on the brink of the development of a long list of rapid tests which will replace most common bacteriologic cultures. It is not at all unimaginable that in the very near future, rapid tests will be available that will identify most important clinical pathogens.

Nevertheless we have chosen in this chapter to describe the familiar culture methods for streptococcus pharyngitis, urinary tract infections, and gonorrhea. We also provide a detailed Gram's stain section, aware that this is a difficult test. (The Gram's stain is one of the original "rapid" microbiology tests.) Also described is the rapid test for group A beta hemolytic Streptococci. This is the first of the new generation of rapid tests to achieve widespread acceptance.

The advantages of office bacterial testing include:

- Information is often available when the patient is in the office (i.e., rapid streptococcus test).
- Office culture procedures are often less expensive than sending cultures to a reference laboratory.
- The clinician can directly correlate the clinical and microbiologic data.
- Some of the common organisms are very fastidious and do not survive transport to a reference laboratory.
- Some specimens will show contamination with bacterial overgrowth if the sample is transported. This is especially true of urine samples.

The techniques that are described for specimen collection, Gram's staining, and culturing are easily learned by office nursing personnel. The interpretation of Gram's stains and colony morphology on culture plates should be done by the clinician, a laboratory technician, or other office personnel who have been appropriately trained.

PROCEDURES NOT RECOMMENDED FOR THE OFFICE LABORATORY

Some bacteriologic procedures are not recommended for the routine office laboratory. These include:

- Species identification of gram-negative rods: There are kits available for identifying the gram-negative rods, which are based on determining a variety of biochemical characteristics. However, 95% of women with gram-negative rod urinary tract infections seen in primary care are infected with *Escherichia coli.* With the urine culture technique described below, an unusual organism can be either presumptively identified (i.e., *Proteus* species) or at least identified as "unusual." It can then be sent to your reference laboratory for definitive identification and sensitivities.

- Antibiotic Sensitivities: Most outpatient infections can be treated successfully with a knowledge of the common bacterial pathogens and their usual drug sensitivities. The common gram-negative organisms are usually susceptible to ampicillin or sulfonamide therapy. When the clinical situation warrants drug sensitivity testing (i.e., a urine infection that has not responded to initial therapy), the culture-isolated bacterium can be sent to a reference laboratory for antibiotic sensitivities. Most clinicians are unaware of how complicated sensitivity testing can be. The procedure involves a standard amount of inoculum and specified sized zones of inhibition. It also involves time-consuming quality controls, which are not feasible in most office laboratories.

- Stool Cultures: This is an uncommonly ordered test in routine office practice because of the infrequent findings of *Salmonella* or *Shigella* in diarrheal disease. Stool culturing involves using several agar plates, identification of the organism by biochemical testing, and serotyping. These cannot be done in the office setting and should be referred to the state laboratory or your reference laboratory. Furthermore, ''routine'' stool cultures now go far beyond testing for *Salmonella* or *Shigella*—including cultures for *Campylobacter* and *Yersinia*. The diagnosis of *Clostridium difficile* requires special testing for the organism's toxin. This is also inappropriate in the office setting.

- Blood Cultures: These cultures are infrequently needed in the outpatient setting. When needed they can be collected, put into the office incubator, and then taken to the hospital laboratory. The testing that should be done on blood cultures involves full identification and antibiotic sensitivities.

- Sputum Cultures: As described in this chapter, a sputum Gram's stain is often superior to a sputum culture in the diagnosis of a bacterial pneumonia or bronchitis. This is because of the poor growth of common pathogens (*Haemophilus influenzae* and *Streptococcus pneumoniae*) and the overgrowth of the culture by normal respiratory flora. For patients with chronic pulmonary disease (i.e., cystic fibrosis), a sputum culture and sensitivity test can be important, and these samples should be referred to a reference laboratory.

- Tuberculosis Identification by Acid-Fast Staining: This is an uncommon procedure in outpatient medicine. It can be difficult to interpret microscopically, must be confirmed by a culture, and can be time consuming to perform properly. Both the culture and sputum examination should therefore be performed by a laboratory specializing in mycobacteriology. All state laboratories provide this testing.

- Stool Parasitology: With the exception of pinworms, we recommend that all parasitology testing be done by specially trained parasitology laboratories. It is too easy to ''see'' parasites that are not really there.

- Wound Cultures: Anaerobes are often the infecting agent, and office laboratories are not equipped to perform this type of culture. If only staphylococcus or streptococcus are clinically suspected, then these can be cultured in the office on sheep blood agar. (see Section 6).

MAJOR BACTERIOLOGIC EQUIPMENT

Incubator

Incubators are relatively simple devices. The basic incubator is a box with a heating element and a thermostat control. (Fig. 18–1). It is designed to keep the space within the box at a constant temperature. For most

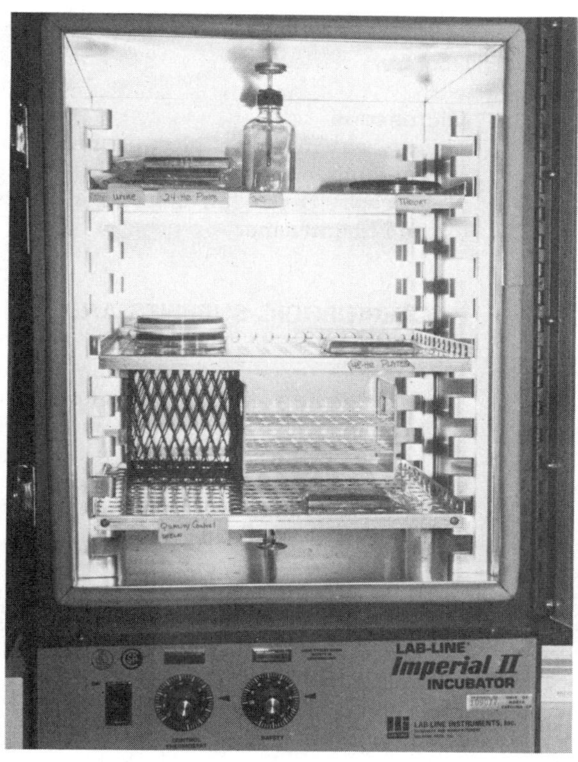

Figure 18–1. Small office laboratory incubator. Note the thermometer, sitting in the culture bottle filled with water, on the top shelf. The culture bottle has a hole in the top where a thermometer can be inserted. This allows for easier temperature reading.

work in microbiology, this temperature is 37 °C, which approximates the temperature of the human body.

Many companies manufacture incubators that are suitable for office laboratory use. The cost for such incubators is usually between $500 and $900. These incubators are described as "dry-heat convection incubators" because a solid element produces the heat and it then spreads throughout the incubator by convection. An office laboratory does not require a water-jacketed incubator in which heated water is used to control the temperature. Neither does it require a CO_2 incubator in which CO_2 is pumped at a constant rate into the atmosphere of the incubator. Good office microbiology testing does, however, require more than the simple cigar-box and electric-light incubators that are sometimes recommended. The smallest incubators have about 0.5 cubic feet of inside space. These are suitable for small offices that do not do much culturing. An incubator with between 1.0 and 1.5 cubic feet of space is suitable for most other office practices.

Incubators usually have two doors, an outer metal door and an inner glass or acrylic door. This inner door allows you to see if there are any plates in the incubator without opening the door and thus altering the inside temperature. This two-door system is, however, of little use because you must open the inner door to read the culture plates. Avoid buying an incubator with only a single glass or acrylic door and no outer door. The sunlight passing through this clear door can cause the temperature in the incubator to rise above an acceptable limit. The incubator door should have a mechanical door latch to ensure tight closure. Check to see whether the incubator door opens to the right or the left, as this may determine where it can be placed in the laboratory.

The incubator should have a thermostat that will control the inside temperature to within ±1 °C of the desired temperature. Most thermostats can be set to any temperature within a range of 30 °C to 60 °C. Such a wide range is not necessary in the office laboratory, where all incubation is done at 37 °C. Many incubators have a hole in the top where a thermometer can be inserted. This enables the temperature to be checked without opening the door. The thermometer that is used should range from −10° to 110 °C or from 20 °C to 110 °C. The thermometer can be placed in a rubber-stopped culture bottle that has been emptied of media and filled with water.

The incubator temperature should be checked and recorded at least once each day (Fig. 8–7). The temperature may vary at seasonal change points (i.e., when you turn the heat on in the fall or the air conditioner in the summer). Slight adjustment of the thermostat is then required.

Laboratory Refrigerator

The laboratory refrigerator is primarily used to store reagents and media plates. A half-sized refrigerator is satisfactory for many laboratories and can be placed under one of the laboratory benches. A full-sized refrigerator may be needed for larger laboratories, especially those that must refrigerate reagents for chemistry testing. The refrigerator must have a freezer unit for colder storage of some reagents. The temperature in the refrigerator should be checked and recorded daily (Fig. 8–7). The acceptable temperature range is between 4 °C and 8 °C. The temperature should be checked and recorded daily with a thermometer-culture bottle unit similar to that used in the incubator. Refrigerators require little maintenance except for defrosting and occasionally cleaning the motor's filter. Food and drinks should not be kept in the laboratory refrigerator.

Microscope

This is a very important piece of office laboratory equipment. See Chapter 10 ("The Microscope") for a full description of its use and maintenance.

BACTERIOLOGIC SUPPLIES AND MATERIALS

There is a variety of small items needed to perform routine bacteriologic procedures. These can be ordered from any of the large scientific supply companies. Fischer Scientific Company and the Scientific Products catalogs are essential books for the office laboratory. (See Appendix; Section 1) Many of the necessary supplies can be found in these two catalogs. Prices often change on these relatively inexpensive small items, so updated price lists are needed.

Alcohol Lamp

A heat source, such as an alcohol lamp or a bunsen burner, is used in the office laboratory for flaming inoculating loops and fixing Gram's-stained smears. An alcohol lamp is cheaper and more convenient for the average office laboratory. It burns either denatured ethanol or methanol. A bunsen burner can be used if gas lines are installed in the laboratory, but it offers little advantages in most procedures (Fig. 18–2).

Figure 18–2. Inexpensive alcohol lamp.

Antiseptic Soap

An antiseptic soap container with a pump handle should be placed next to the sink so that the laboratory worker can wash his or her hands quickly. Wetting the hands before applying the soap seems to decrease the skin-drying effect. We also keep a plunger bottle of handcream near the sink to help prevent chapped hands. We have found packaged surgical scrub brushes, such as E-Z Scrub (Deseret Pharmaceutical Co, Sandy, UT), to also be useful.

Bacteriologic Streaking Loops

The laboratory should have a regular streaking loop and a calibrated loop (Fig. 18–3). Regular streaking loops are made of tungsten or nickel-chromium wire and have plastic, wooden, or metal handles. The loop should be made of platinum wire if it ever comes in contact with the oxidase reagent used in identifying *Neisseria* species. Urine cultures are plated with a special calibrated loop. If a loop becomes bent, it can be straightened with a pair of small pliers.

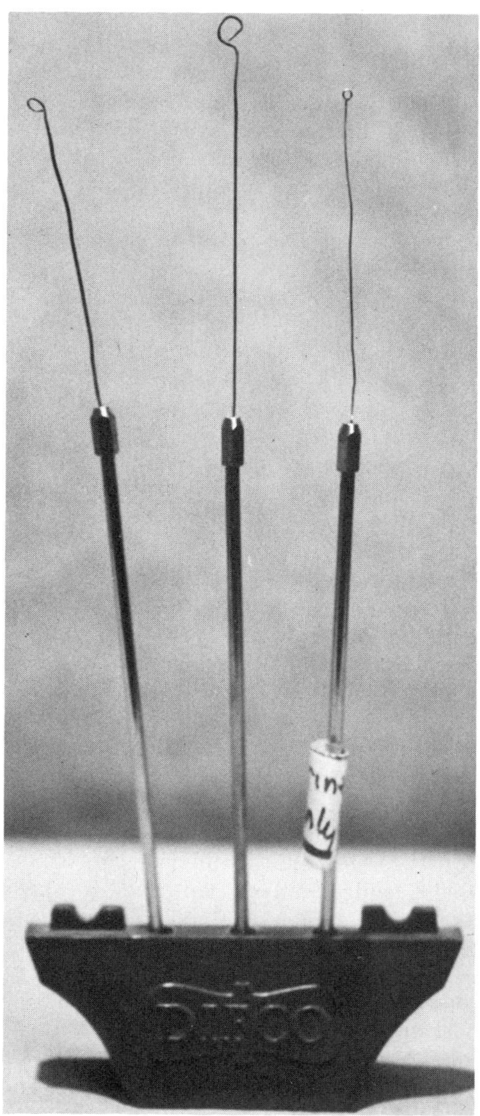

Figure 18–3. Bacteriologic streaking loops and loop holder. The calibrated urine loop has a label around the handle to easily identify it from the others.

Bibulous Paper

This paper comes in small booklets and is used to blot (not wipe) Gram's-stained slides after staining. It is very important to blot the slide or to let it air dry. Wiping the slide will frequently result in removing the specimen. Paper towels can be used for this purpose, but they are not as convenient as bibulous paper.

Biohazard Container

All office laboratories should have a system for the storage of biologically contaminated materials. The Terminal Biohazard System (W.G. Whitney Co, Skokie, IL) includes a container, a lid marked with the biohazard symbol, an autoclavable plastic bag, and bag closures. This system is used to store contaminated waste, such as media plates, blood tubes, or infectious samples. The filled bags should be autoclaved by the hospital or reference laboratory before they are disposed of in the public waste system. (See discussion of laboratory wastes in Chapter 9, "Office Laboratory Safety.")

Coplin Jar

This is a glass or plastic jar that can be used to store reagents. It has grooves along the side to hold microscope slides while they are being stained. Coplin jars with screw-on lids should be purchased, as this will prevent the evaporation of the stored solutions (Fig. 15–22).

Disinfectant

A disinfectant should be used each day to clean the counter tops in the laboratory. A 1:10 dilution of household bleach is the cheapest and most accessible disinfectant, but the mixture may not remain effective over time and it can be a respiratory irritant. 5% Staphene is more stable and is not an irritant. *Always* wear gloves when using a disinfectant.

Grease Pencil

A grease pencil should be used for writing patients' names on the bottoms of the media plates. Using a grease pencil instead of a label permits easier reading of hemolysis on sheep-blood plates. A black grease pencil is preferred because red is difficult to read on the red agar of the sheep-blood plates.

Immersion Oil

Most microscope manufacturers recommend using standard immersion oil, such as Cargille, type A. Mineral oil is not satisfactory because it has a low viscosity and will therefore seep inside the objective. Be sure that you do not buy immersion oil that contains polychlorinated biphenyls (PCB) or other health hazards. The oil is purchased in 4- and 6-ounce bottles. The oil can be put into any small bottle and kept next to the microscope. A small pipette or dropper top can be used to transfer oil onto the microscope.

Lens Cleaner

A solvent is needed to regularly clean the oil immersion lens of the microscope. Xylene, methanol, or Kodak lens cleaner can all be used.

Lens Paper

It is important to use lens paper to clean the oil off the oil immersion objective after each use. Lens paper is also used to clean the ocular lens. Gauze, tissue, or paper towels can scratch the delicate microscope lenses. Lens paper is purchased in booklets. It can be taken out of the booklets, cut into 2-inch square pieces, and kept in an empty box next to the microscope.

Media

Plates should be stored in the refrigerator. They usually are purchased in plastic tube bags and should be stored in these bags to prevent drying out. These bags of plates should be stored on their side or with the medium side up to prevent condensation from collecting on the medium surface. The plates should be placed in the incubator with the medium side up for the same reason. Never use medium that is contaminated, dried out, or older than the expiration date. Expired plates can be used to train office personnel about plating techniques. It takes some practice to learn to streak agar without tearing its thin surface.

95% Methanol

This is available from scientific supply companies. It is used to "fix" Gram's-stain slides as an alternative to "heat" fixing.

Microscope Slides

Slides with a frosted end are preferred because the patient's name can be written on the frosted end in pencil. If nonfrosted slides are used, the slide should be identified with either an adhesive label or by marking with a diamond-tipped marker or grease pencil. Frosted slides are slightly more expensive than plain slides. The standard 3 × 1-inch size is adequate for most laboratory uses.

Pipettes

Five-inch disposable Pasteur pipettes (and a small rubber bulb) or small plastic transfer pipettes are useful.

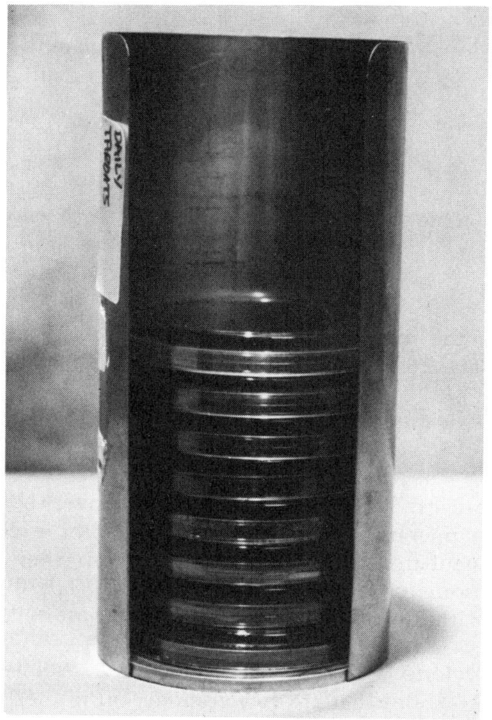

Figure 18–4. Petri dish holder filled with plates.

Petri Dish Holders

A metal holder can be purchased to store petri dishes in the refrigerator. These inexpensive holders secure 15 plates and are superior to just stacking plates one on top of another. The holders have an opening along the side to allow the plates to be easily removed (Fig. 18–4). They are used to store stock culture plates (for quality control procedures) and any positive cultures. By refrigerating positive plates for several days, the clinician can keep the bacteriologic sample until the patient has shown a response to the chosen therapy. This is especially useful for positive urine cultures.

Reference cultures

Each laboratory may want to keep reference cultures of the common bacterial pathogens. These are used in the microbiology quality-control procedures.

Slide Storage

Each laboratory should have a slide-storage area. This is used to store reference slides and routine patient slides. It is important for legal reasons to store the slides in all cases of gonorrhea. Scientific supply companies sell microscope slide drawers that very efficiently hold many slides and allow easy slide retrieval. These drawers are more expensive than the wooden or metal boxes that hold only 20 slides. Another inexpensive method is to store slides in the cardboard boxes that the new slides come in. For routine patient slides it is generally easiest to store slides consecutively by dates. Each slide should be marked with both the patient name and the collection date. The Gram's-stained slides can be filed with the Wright's-stained peripheral smears. (See the Appendix for a description of methods used for making a permanent slide collection.)

Staining Rack

A special rack can be purchased to hold slides while they are being stained, but it is cheaper and easy to make your own. First, measure the length of the sink to be used for staining. Buy two lengths of glass tubing of three eighths-inch diameter that are a little longer than the sink measurement. Cut two pieces of rubber tubing that will fit over the ends of the glass tubing. Each piece of rubber tubing should be 3 inches long. Place the glass rods parallel to each other, and slip the rubber tubing over the ends to connect the tubes together. This rack can then be placed over the sink to hold slides for staining (Fig. 18–5).

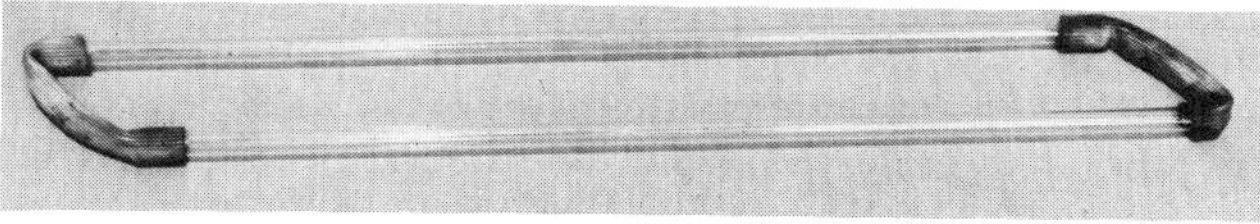

Figure 18–5. Homemade staining rack.

Figure 18–6. Teasing needles.

Stain Remover

Some people use rubber gloves when working with stains. If you choose not to wear gloves you can hold the slides with a forceps, a hemostat, or a clamp-type clothespin. This will help prevent stain from getting on your fingers. Should you forget and stain your fingers or have an accident and spill stain on your clothes, we have found two products to be useful. Dye-Sol Cleaning Cream (Scientific Products, McGaw Park, IL) will remove stains from the skin. Erado-Sol (Cambridge Chemical Products, Ft. Lauderdale, FL) removes stain from clothes, glassware, and counter tops. Both are available from Scientific Products.

Streaking Loop Holder

A plastic loop holder is available (Difco Laboratories, Detroit, MI 48201). It can be kept near the plating area (Fig. 18–3).

Teasing Needle

These are convenient for separating sputum or other thick specimens when they are being streaked onto a microscope slide. They come in both straight and angular points (Fig. 18–6).

Thermometers

An inexpensive mercury thermometer can be purchased in various temperature ranges. The thermometer should cover the temperature range of your refrigerator and incubator. One commonly available thermometer ranges from − 10 °C to + 110 °C with divisions of 1 °C.

Xylene

All slides that are viewed under oil immersion will need to be cleaned before filing. This can be easily done if a Coplin jar (Fig. 15–22) is filled with xylene and left covered near the microscope. The slide can be dropped into the Coplin jar for several seconds and then wiped dry with a cloth. The jar should have a tightly fitting top, as xylene will evaporate and can cause respiratory irritation. Replace the top when the xylene is not being used. Xylene can be purchased in 1-gallon safety cans. It should be stored away from other volatile chemicals (i.e., in the cabinet under the sink).

SECTION 2: Safety and Infection Control

The handling of potentially infectious materials makes safety an important issue in office bacteriology. In general, the organisms found in the office laboratory are not as dangerous as those found in a large microbiology laboratory. Nevertheless the proper handling of infectious materials is necessary to minimize the risk of illness.

(See Chapter 9 for a more in-depth discussion of laboratory safety issues.)

Little concern is often shown in the disposal of contaminated culture plates. It is not acceptable to merely throw them out, untreated, into the office rubbish! It is also impractical to decontaminate them in the office autoclave because of the resulting smell. One reasonable alternative is to place them in a container for contaminated waste and to have them processed by the local hospital microbiology laboratory. A second option is to gently squirt the surface of the agar with disinfectant (i.e., a 1:10 solution of household bleach or a 5% solution of Staphene). This is done over a sink with a squirt bottle. Flood the plate with the disinfectant and let it "soak in." The plates can then be double bagged and discarded with the regular office garbage. The problem with this method is that it creates a strong disinfectant smell in the laboratory. Check with your state laboratory for its guidelines or regulations about culture plate disposal.

It is a good practice to wipe the top of the benches or counters each day with a disinfectant and paper towels. Since these used towels may pick up infectious material, they should be placed in a biohazard container for disposal.

Hand washing is the most important procedure for limiting the spread of infections within the laboratory. Laboratory personnel should wash their hands with an antiseptic soap after handling any infectious materials, contaminated plates, culture swabs, or urine samples. Care should be taken to clean under the nails. The last procedure of the day for the laboratory worker should be to wash the hands well. If the laboratory worker has an open cut on the hand, we suggest that he or she wear rubber or plastic gloves when handling infectious materials.

Some laboratory personnel prefer to wear white laboratory coats to protect their clothing from reagent or specimen spills. These coats may protect clothing; however, they do not protect laboratory workers from infections. If white coats are worn, the clinician should make arrangements to have them washed by the same laundry that washes the clinician's coats.

The laboratory should have a special disposal system for glass products, such as microscope slides, pipettes, and broken glassware. It is best to keep a separate labeled container with two inner plastic bags for this glassware. This will ensure that broken glass is not spilled and that the housekeeping staff will take special precautions with its disposal. This container can be placed next to the microscope so that used slides can be conveniently discarded.

SECTION 3: Gram's Stain

In a time when millions of dollars are being spent on developing rapid microbiology tests, it is important not to forget the humble Gram's stain. Bottone has called the Gram's stain a "century-old quintessential rapid diagnostic test." It is the most important and commonly used stain for identifying clinically important bacteria.

In this section is described the basic Gram's stain procedure, including specimen preparation and interpretation. In Section 4 additional training and reference aids are offered.

MATERIALS

The reagents used in the Gram's stain include Gram crystal violet, Gram iodine, acetone or alcohol decolorizer, and Gram safranin counterstain. These reagents are available from many sources and in many forms, including dry powders, liquid concentrates, and premixed solution. For convenience we recommend the use of premixed Gram's stains. These are available in ready-to-use plastic dispenser bottles (Difco Laboratories, Detroit, MI 48201). Do not use reagents that are out of date.

SPECIMEN COLLECTION

The Gram's stain is used on many types of specimens ranging from organisms grown on a culture plate to those collected directly from patients. Table 18–1 gives the specimen sources routinely encountered in the office laboratory with tips for their collection and slide preparation.

TRANSFER TECHNIQUES

Patient specimens from a variety of sources must be transferred to a slide before it can be Gram's stained. There are several tools and techniques that are routinely used to do this. When a thick material is to be stained (i.e., colonies from a culture plate, pus, etc.), place a drop of sterile saline or sterile water in the middle of the glass slide and mix the specimen into the fluid. This makes the specimen more dilute and therefore easier to stain.

1. *Bacti loops:* Regular streaking loops are made of different metals (tungsten, or nickel chromium or platinum) and seated in a handle. Loops can be used for a variety of transfer activities (e.g., to pick up an individual colony from a culture plate or a drop of urine).
   a. Heat the loop over an alcohol lamp or bunsen burner until it glows red.
   b. Allow the loop to cool.
      (1) If it is an organism on a plate, touch the loop to a single colony and pick it up.
      (2) If it is a fluid, dip the loop into the liquid. Make sure that the loop is full of liquid and not air.
      (3) If it is pus or sputum, touch the loop to the specimen and pick up some of the material. You do not need a large amount.
   c. Touch the loop to the center of the slide and gently smear the specimen onto the slide. If it is liquid (i.e., urine, joint fluid), spread it around so that you get a uniformly thin specimen. If it is thicker material (i.e., colony, pus), mix it with a drop of saline/water previously placed on the slide.
   d. Reheat the loop before you return it to the stand. This will kill any remaining organisms.
2. *Sterile Swabs:* These are 6-inch sterile cotton or artificial fiber-tipped swabs (e.g., dacron, calcium alginate) that can be used to transfer thick material. A loop or a pipette is better than a swab for transferring liquid specimens.
   a. Touch the swab to the material to be transferred.
   b. *Roll* (do not rub) the swab onto the center of the clean labeled slide, touching all surfaces of the swab to the glass.
   c. Make a thick uniform smear. If it is too thick (frequently the case with sputum), place a second slide (label side down) on top of the first slide. Press the slides together and then pull them apart.
3. *Pipette:* This is the easiest way to transfer liquid material. Use a common glass or plastic transfer pipette.
   a. Squeeze the pipette bulb to suck up a small amount of liquid.

b. Deposit the fluid onto the center of the clean labeled slide. Then use the pipette tip to spread out the specimen so that you get a uniformly thin specimen.
c. Dispose of the pipette in the contaminated bag. Save the rubber bulb.

SLIDE PREPARATION/STAINING PROCEDURE

1. Write the patient's name and date on the frosted end of the microscope slide. Prepare a thin specimen. Thick smears are difficult to properly stain. If there is too much material on the slide, thin it out with a teasing needle or place a second slide on top of the first slide and pull the two apart.
2. Allow the slide to dry. It is best to put the slide aside and to allow it to slowly air dry; but if there is a need for quick results, the slide can be carefully dried high above the flame of an alcohol lamp. Some people prefer to place the slide inside the incubator to speed up the drying process.
3. When the specimen has dried, it needs to be "fixed" to the slide so that it does not wash off during the staining process. There are two ways to "fix" the specimen to the slide:
   a. **Heat Fixing.** Hold the dried slide with your fingers. (This prevents overheating.) Pass the slide through a flame several times, specimen side up (away from flame). Touch the back of the slide to your hand. The slide should get warm but not hot! The disadvantages of heat fixing are that overheating can distort the appearance of the bacteria, while underheating can result in the specimen being washed off during the staining procedure. Heat fixing is commonly used and is fast.
   b. **95% Methanol Fixing.** Place a drop of methanol onto the dried specimen on slide. Allow the methanol to completely evaporate. This method is very effective with urine and joint-fluid specimens. It preserves cell morphology better than heat fixing, but it takes longer to dry. Methanol is an irritant, so avoid skin contact.
4. Allow the slide to cool or dry completely. If the slide is stained before the slide is completely cool, the crystal violet will precipitate. This makes interpretation difficult because the precipitated stain looks like small gram-positive cocci.
5. With a grease pencil draw a circle around the specimen on the slide. This line acts as a "well" and minimizes the amount of stain that needs to be used.
6. Place the slide with the specimen side up onto a staining rack over the laboratory sink. Alternately, you can hold the slide with a spring-loaded clothespin or with gloved fingers.
7. Flood the slide withGram crystal violet and allow it to stand for 10 seconds.
8. Wash the slide with running tap water. When doing this be sure to hold the slide parallel to the flow of the water. Do not hold the slide perpendicular to the running water because the force of the water against the slide can result in your specimen being washed off. Tilt the slide to drain off any excess water.
9. Flood the slide withGram iodine and allow it to stand for 10 seconds.
10. Wash the slide with tap water as you did before and tilt the slide to drain off any excess water.
11. Hold the slide at an angle over the sink, and squirt the stained portion

TABLE 18–1. SPECIMEN SOURCES ROUTINELY ENCOUNTERED IN THE PRIMARY CARE SETTING—COLLECTION AND SLIDE PREPARATION

Specimen Type	How to Collect	Gross Appearance and Evaluation	What to Do	Additional Comments
1. Colonies from a culture plate	Use a sterile loop or swab. Pick up a *single* colony. Avoid mixing different colony types	Thickish material REMEMBER: You do not need a big specimen	Mix gently with a drop of sterile water or saline already placed on the slide	Harsh mixing can cause cell distortion Thick smears result in variably colored Gram's stains
2. Sputum	*Adults/older children* 1. Instruct patient to wash out mouth with water; 2. Swallow any saliva before coughing; 3. Breathe deeply and cough up sputum in a few deep coughs. Lightly pounding the back over the lung area prior to coughing may loosen material	Foamy or clear liquid	This is saliva and not sputum! It is a waste of time to examine such specimens. Have the patient collect a more satisfactory specimen	
		Saliva with yellow-green purulent flecks of sputum	Pick up the purulent material by tilting the cup and capturing it onto a swab Avoid the saliva	Thin smears are needed. Two slides pressed gently together and then pulled apart produce a thin smear

	Very young children 1. Use a nasopharyngeal swab 2. Hold the mouth open with a tongue depressor 3. Touch the epiglottis with a sterile swab. A cough should follow with expelled sputum sticking to the swab. Avoid touching the swab to the sides of the mouth. 4. Transfer the specimen to a glass slide	Thick yellow-green material	This is a good specimen. Obtain a sample on the tip of a swab	On even the best slides there are usually both thick and thin areas. Examine the thin areas after staining
3. Pus from an abscess/ wound	If it is a walled abscess it is best to collect pus from the center of the abscess with a syringe If it is a flatter wound use a sterile swab to collect the pus or material	Usually thick pus	Place drop of sterile water or saline in the center of the slide SYRINGE: Very slowly and gently express the material onto the slide. Then use a sterile swab or loop to mix it with the water or saline SWAB:Simply roll the swab onto the slide	If the specimen is too thick, thin it out by pressing two slides together

TABLE 18–1 CONTINUED

Specimen Type	How to Collect	Gross Appearance and Evaluation	What to Do	Additional Comments
4. Joint fluid	This is collected by the physician into a syringe. An anticoagulant (i.e., EDTA) may be added if clotting is a problem Usually the sample is divided with one part being cultured, one for chemistries, one for a cell count, one for Gram's stain, and one for crystal examination If there are only a few drops, use one for Gram's stain and save the rest for culture.	Can be stringy, purulent, or thick and viscous	TUBES: Place in the centrifuge and spin at either urine or blood speed. When finished use a transfer pipette to remove supernatant. Place a drop of the remaining sediment onto the slide. Use the tip of the pipette to make a thin smear of the sediment If there is a small amount of fluid in the syringe, squeeze out one drop for Gram's stain, one drop for crystal examination, and send the rest for culture	Centrifuging is very helpful if there is enough fluid because it increases your chances of seeing bacteria when very few are present
5. Urine	Clean catch catheterized or suprapubic aspirate. Prepare the slide immediately or refrigerate the urine for later examination	Clear or turbid	*Unspun/spun* Mix well and use a sterile loop or pipette to place and spread a drop onto the slide	As with joint fluid, you increase your chances of seeing bacteria by centrifuging the specimen and examining the stained sediment

			A wet specimen prepared for microscopic urinalysis can also be used. Merely remove the coverslip and allow the specimen to air dry. Then stain and examine
6. Urethral discharge	If there is an obvious discharge, pull back the foreskin and collect the drop on a sterile swab If there is no discharge, insert a sterile calcium alginate urethral swab into the urethra (2 cm) and rotate it gently	Roll onto a prepared slide	In patients with *minimal* discharge, the best specimen for evaluation is often obtained by collecting the first 15 mL of urine. Centrifuge this and stain the sediment
7. Other specimens (e.g., cervical secretions, vaginal secretions, nasal secretions)	Follow standard collection procedures and, given the consistency of the specimen, process using the technique described above that seems most appropriate to you		

Addison et al. Gram staining, University of North Carolina, Chapel Hill, 1987. Instructional module supported in part by funds from the Bureau of Health Professions, Grant 1 D32 PE 14025. Used by permission of the author.

of the slide with the decolorizer just until the fluid dripping from the slide has changed from blue to colorless. This is the most crucial step. Do not overdecolorize!

12. Immediately wash the slide with tap water as you did before; tilt the slide to drain off any excess water.
13. Flood the slide with gram safranin and allow it to stand for 20 to 30 seconds. Organisms take up this stain more slowly.
14. Wash the slide with water and then tilt it to drain.
15. Allow the slide to air dry or blot off any excess water with bibulous paper.

SLIDE EXAMINATION PROCEDURE

See Chapter 10 ("The Microscope") for a detailed tutorial on how to use the microscope to examine slides under the oil objective.

1. Snap the low-power objective into place. Put the slide, specimen side up, onto the microscope stage between the clips. Raise the condenser all the way up. Turn the light to a low setting. Be sure that the diaphragm is completely open.

2. Focus and examine the slide under the low power to find a field that has an area of stained specimen. Avoid the areas where the specimen is thick and heavily stained.

3. Place a very small drop of oil on the slide over the area where the light from the condenser passes through the slide.

4. Switch the objective to oil power and refocus. Do not draw the high-dry objective through the oil. Focus the microscope.

5. Spend some time examining several fields. This will give you an idea of the total specimen rather than just a single field.

6. It is important to record what is seen and not what is expected to be seen.

7. When you have finished examining the slide, turn the objective turret so that the oil objective is out of the oil. Then wipe the objective with lens paper to remove any excess oil from the lens.

8. Dip the slide in xylene and very gently wipe the slide clean with a soft cloth. The xylene will dissolve any remaining oil on the slide.

9. File the slide in the slide file.

10. If it is an especially good slide and you would like to keep it for reference, it can be sealed and coverslipped as described in the Appendix.

BASIC TERMINOLOGY

It is often taken for granted that the various terms used to describe the morphology of organisms are familiar to everyone and that everyone uses the terms correctly. For a Gram's stain to be clinically useful, what is actually seen must be described in accurate detail.

1. **Color or gram reaction.**
   a. *Dark purple or blue = Gram positive:* This means that the organism retains the color of the *first* stain.
   b. *Pink/pinkish red = gram negative:* This means that the organism did not retain the original stain when decolorized—rather it took up the color of the Safranin counterstain (i.e., the last staining step).
   c. *Gram variable:* This means that the organism takes up the stains unevenly and therefore appears as a mixture of both dark purple and pink organisms.
2. Shape of Organisms: At this point the focus is on shape. These shapes could be gram positive, gram negative, or gram variable.

a. *Cocci:* These are round, oval, lancet, or kidney bean-shaped organisms (figs. 18–7 to 18–10).
b. *Rods or bacilli:* These organisms are shaped like rods. They can be long or short, fat or thin (Fig. 18–11).
c. *Coccobacilli:* These are organisms that are so small that it is difficult to tell whether they are rods or cocci. They are often called small "pleomorphic rods" (i.e., "many shapes"; Fig. 18–12).
d. *Pseudohyphae/Buds:* These are yeasts. They are much larger than any bacteria (Fig. 18–13).
e. *Filamentous Rods:* These are long, delicate rods that form branches or filaments. They are uncommon (i.e., actinomyces; Fig. 18–14).
f. *Curved Forms:* These are curved rods. These are also rarely seen (i.e., Vibrio species; Fig. 18–15).

3. **Configuration of basic shapes:** In addition to the individual shapes cited above, organisms group or combine in different ways. Some of these configurations are extremely important to note and aid in the presumptive identification of the organism.
   a. *Chains:* Three or more organisms end to end (Fig. 18–7).
   b. *Clusters:* The typical organism here is *Staphylococcus,* which occurs in "grapelike" clusters (Fig. 18–8).
   c. *Diplococci:* These are two cocci paired together. They may be lancet- or almond-shaped (head-to-head) (Fig. 18–9) or kidney bean-shaped (side-to-side; Fig. 18–10).

CONFIGURATION AND GRAM'S STAIN CHARACTERISTICS OF CELLS AND ORGANISMS COMMONLY SEEN IN CLINICAL SPECIMENS

Organisms frequently group together into definite configurations. You should be able

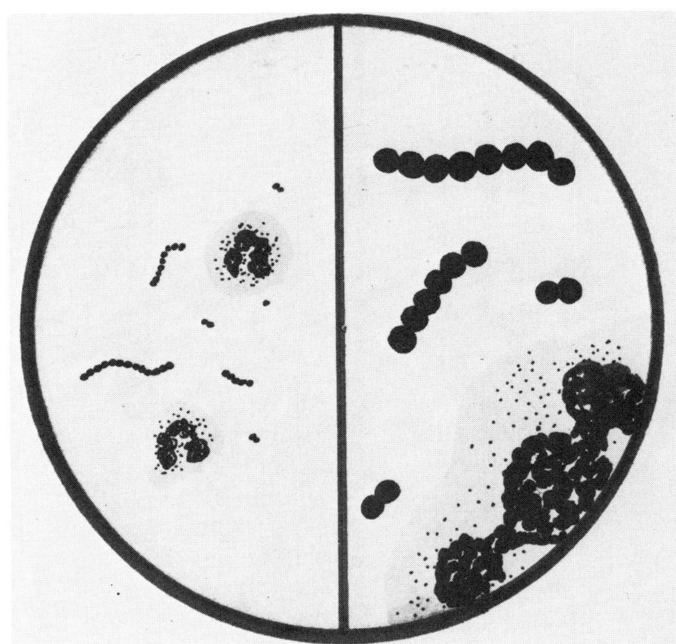

Figure 18–7. Streptococci. **Left.** Approximates oil-immersion view **Right.** Further enlargement. Note the round cells in chains and the segmented neutrophilis.

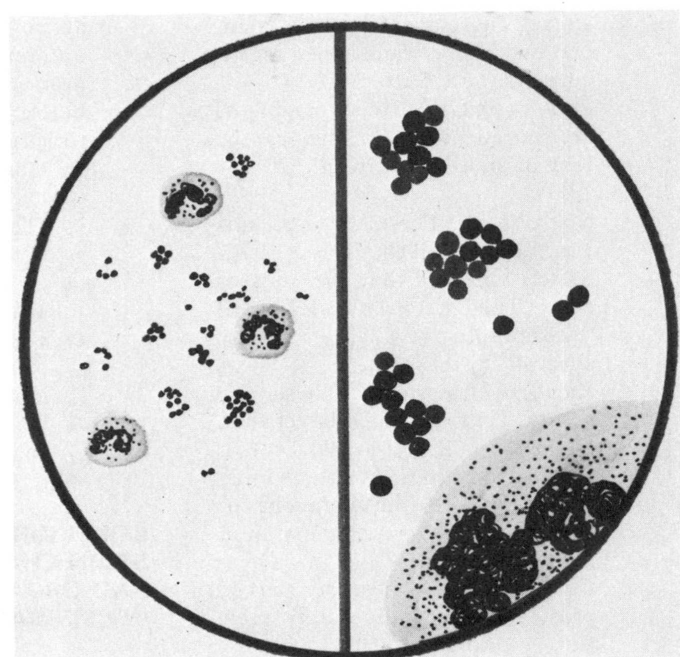

Figure 18–8. Staphylococci, round organisms in grapelike clusters. **Left.** Approximates oil-immersion view. **Right.** Further enlargement. Note the segmented neutrophils.

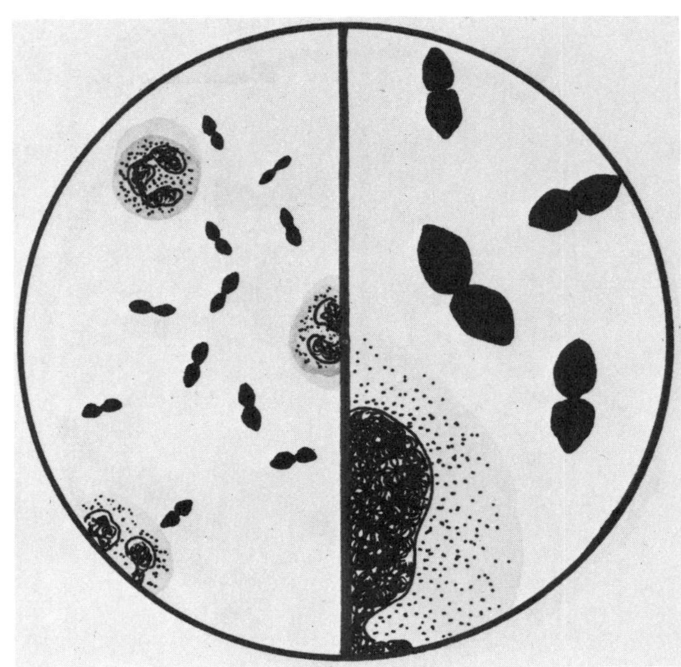

Figure 18–9. Diplococci; paired, lancet-shaped organisms. This configuration is characteristic of *Streptococcus pneumoniae.* **Left.** Approximates oil-immersion view. **Right.** Further enlargement. Note the segmented neutrophils.

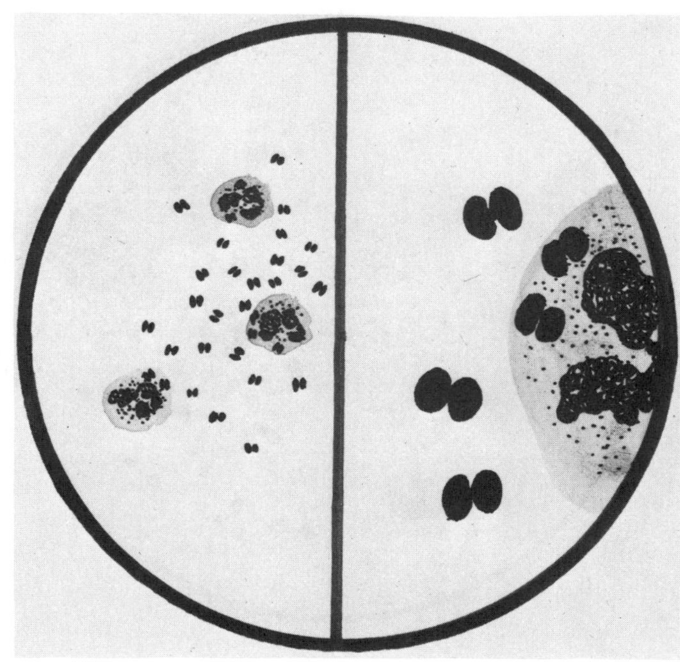

Figure 18–10. Diplococci; paired kidney bean-shaped organisms. This configuration is characteristic of *N. gonorrhoeae.* **Left.** Approximates oil-immersion view. **Right.** Further enlargement. Note the segmented neutrophils.

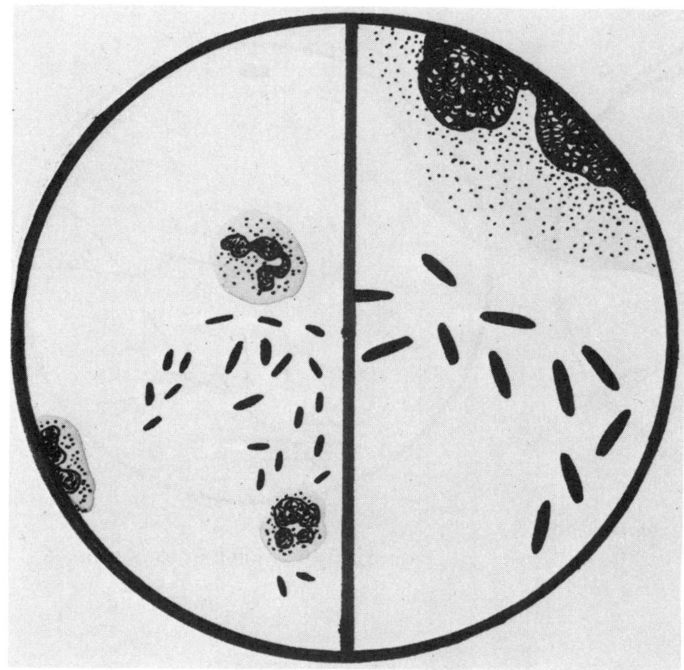

Figure 18–11. Bacilli; rod-shaped organisms. **Left.** Approximates oil-immersion view. **Right.** Further enlargement. Note the segmented neutrophils.

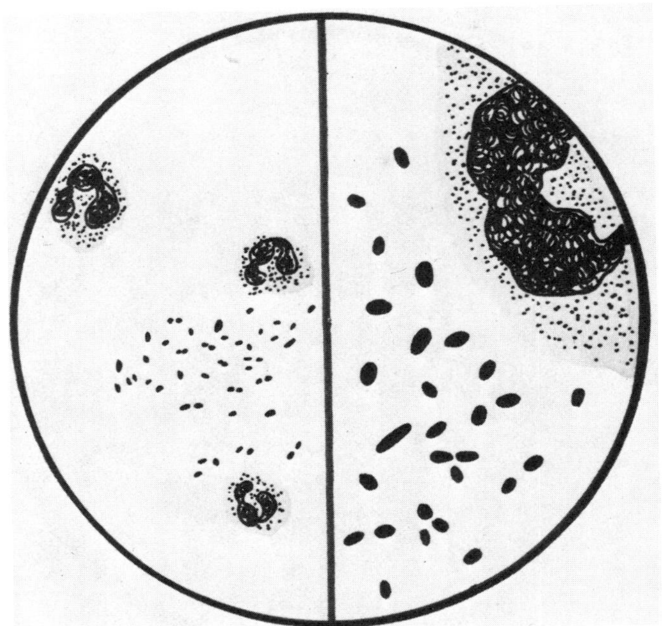

Figure 18–12. Coccobacilli. **Left.** Approximates oil-immersion view. **Right.** Further enlargement. Note the segmented neutrophilis.

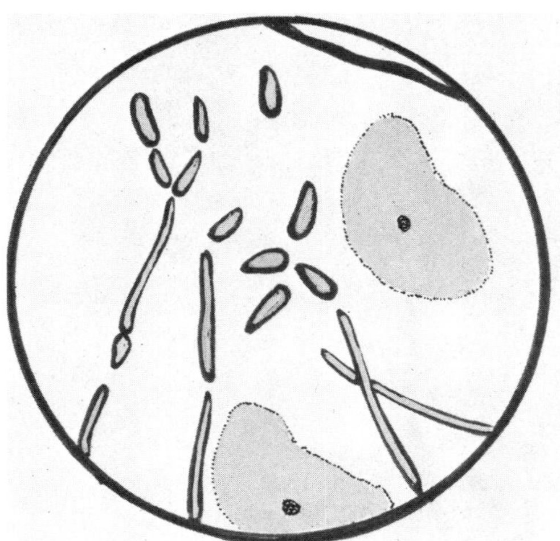

Figure 18–13. Fungal forms. Note the squamous epithelial cells.

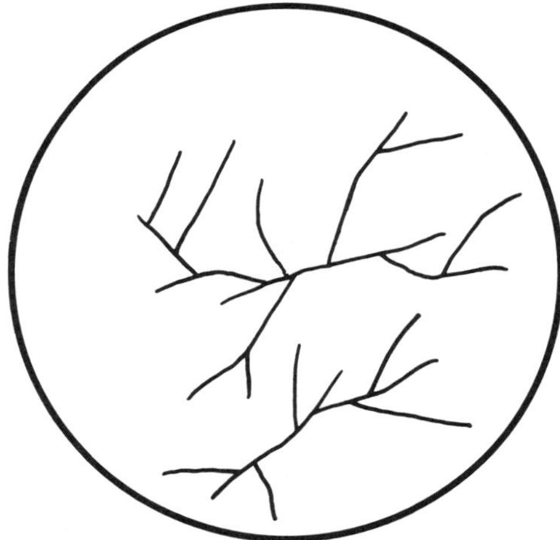

Figure 18–14. Filamentous rods.

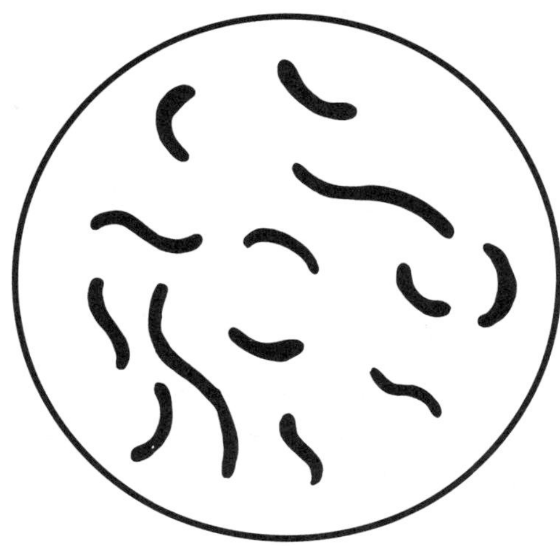

Figure 18–15. Spiral organisms.

to accurately identify these additional shapes and to know what the common organisms are. Table 18–2 lists the most common specimen sources, most likely pathogens, and the slide appearance.

CELLS

When cells are properly stained they have a pink cytoplasm with a darker pink or red nucleus (Color Plate II, Fig. 1). If they are underdecolorized, they will have a purple nucleus and/or cytoplasm (Color Plate II, Fig. 2). This is a clue to a staining problem.

Squamous Epithelial Cell

This is a large cell with a small nucleus (Color Plate II, Fig. 1).

White Cell (poly, segmented neutrophil)

These are significantly smaller than squamous cells and may sometimes be broken apart by the mechanics of making the slide (Color Plate II, Figs. 3 and 5, Color Plate III, Figs. 1, 2, and 5).

Mononuclear Cells

Macrophages or histiocytes are sometimes seen. They are mononuclear as opposed to being segmented and are a little larger than polys. Like the polys and squames, when properly stained the cytoplasm is pink and the nucleus is a darker pink/red.

COMMON ORGANISMS

- *Micrococcus* species
  Gram-positive cocci
  Color: dark purple
  Configuration: usually combinations of two or sets of four cocci
  Special Comments: They are rarely the primary cause of disease. It is important not to confuse them with *Staphylococcus* species. They are most often seen on sputum slides.

- *Staphylococci* species:
  Gram positive cocci
  Color: dark purple
  Configuration: grapelike clusters, pairs, short chains, singly (Fig. 18–8; Color Plate II, Fig. 3).
  Special Comments: The clinically pathologic species that are usually seen are *S aureus* (many sources) and *S saprophyticus* (urine). *S epidermidis,* on the other hand, is usually a contaminant. All of these organisms look alike under the microscope. *Staphylococcus* species have their classic grape cluster configuration only when grown in moist medium (e.g., sputum, joint fluid, culture broth). If you make a smear from organisms taken from a culture plate, you will see a sheet of cocci but not the grapelike clusters.

- *Streptococci* species:
  Gram-positive cocci
  Color: dark purple
  Configuration: Streptococci are obviously round and appear in pairs and/or

TABLE 18–2. COMMON GRAM'S STAIN FINDINGS/INTERPRETATIONS (OIL, 100X OBJECTIVE)

Cellular Findings	Bacteria	Cause
Sputum		
1. Many squames, few WBC	Gram + cocci, Gram − rods	Saliva contamination, inadequate specimen
2. Many WBCs, rare squame	No organisms seen	Viral or mycoplasma pneumonia
3. Many WBCs, rare squame	Lancet-shaped Gram + cocci in pairs (i.e., diplococci)	*Streptoccus pneumonia*
4. Many WBCs, rare squame	Very small gram − coccobacilli	*Haemophilus influenza*
5. Many WBCs, rare squame	Gram + cocci in clusters	*Staphylococcus aureus*
6. Many WBCs, rare squame	Gram + cocci in chains	Group A streptococci
7. Many WBCs, rare squame	Mixed gram + cocci, gram − rods	Mixed infections, seen in patients with chronic obstructive pulmonary disease
Urethral		
1. Few squames, no WBCs	No organisms or few G + cocci	Inadequate specimen
2. Many WBCs	Kidney bean-shaped Gram − diplococci, some intracellular	*Neisseria gonorrhoe* (occasionally *N meningitidis* or *Branhamella catrrehalis*)
3. Many WBCs	No organisms	Nongonoccocal urethritis
Joint		
1. Few to many WBCs	No organisms	Inflammatory arthritis *v.* early septic arthritis
2. Few to many WBCs	Gram + cocci in clusters	*Staphylococcus aureus*
3. Few to many WBCs	Gram + cocci in chains	Group A streptococci
4. Few to many WBCs	Small Gram − coccobacilli	*Haemophilus influenzae* (in children)
Skin		
1. Squames, rare WBCs	Occasional G + cocci	Inadequate specimen
2. Many WBCs, few squames	G + cocci in clusters	*Staphylococcus aureus*
3. Many WBCs, few squames	G + cocci in chains	Group A streptococci
4. Many WBCs, few squames	G + pseudohyphae	*Candida ablicans*
5. Many WBCs, few squames	G − dipplococci	Disseminated *N gonorrhoeae*
6. Many WBCs, few squames	G + cocci, G − rods	Mixed infections as seen in decubitus ulcers

chains (Fig. 18–7, Color Plate II, Fig. 4)
Special Comments: Again, this classic chain configuration is seen only when grown in a moist medium. A smear made directly from organisms grown on a culture plate will show a sheet of cocci.

- ***Streptococcus pneumoniae:***
Gram positive diplococci
Color: dark purple
Configuration: These are head-to-head, lancet-shaped diplococci. They may also appear in short chains. (A lancet is a knife blade; Fig. 18–9 and Color Plate II, Fig. 5)

- ***Neisseria* species:**
Gram-negative diplococci
Color: pink/reddish pink
Configuration: These are side-by-side kidney bean-shaped diplococci. They vary in size (Fig. 18–10, Color Plate II, Fig. 6).
Special Comments: You cannot differentiate *N gonorrhoeae* from *N meningitidis* or other *Neisseria* species by Gram's stain. Special additional tests are needed to speciate these organisms. In clinical specimens the diplococci will frequently appear intracellular (i.e., within the white cell). *Neisseria* is an autolytic organism (i.e., it dissolves itself). It is therefore important to Gram's stain the specimen as soon as its growth is noted on a culture plate. If delayed the organism takes on a more amorphous shape and is much harder to recognize.

- ***Enterobacteriaceae*** *(i.e., Citrobacter, Edwardsiella, Enterobacter, Erwinia, Escherichia, Klebsiella, Providencia, Proteus, Salmonella, Serratia, Shigella, and Yersinia)* ***and Pseudomonas species:***
Gram negative: rods or bacilli
Color: pink
Configuration: These are all medium-sized rod-shaped organisms (Fig. 18–11, Color Plate III, Fig. 1)
Special Comments: It is not possible to differentiate gram-negative enterics from their appearance under the microscope.

- ***Haemophilus influenzae:***
Gram negative: coccobacilli
Color: pink/pinkish red
Configuration: These are very small, thin coccobacilli. They are also called small, pleomorphic (many shaped) rods (Fig. 18–12, Color Plate III, Fig. 2)
Special Comments: H. influenzae can be easily missed in a clinical specimen not only because it is such a small organism but also because it is a pale gram negative and blends into the pink mucous debris in the background material. You must examine clinical specimens very carefully for *H. influenzae.*

- ***Candida albicans:***
Gram positive: pseudohyphae and buds
Color: dark purple
Configuration: The term "pseudohyphae" is used because these hyphal-like forms do not have true septate. *Candida* hyphae have constricted points along the strand (resembling link sausages). These are where the early buds were attached (Fig. 18–13, Color Plate III, Fig. 3)
Special Comments: Candida is usually gram positive. It is much larger than any bacteria. When it is grown on a culture plate (such as Thayer Martin culture for GC), only bud forms are seen. Other fungi are also seen in primary care. These have more difficulty taking up the Gram's stain and may therefore appear as gram variable or barely stained. These are also very large and are not easily confused with bacteria.

- **Artifacts:** Sometimes the crystal violet will precipitate out and form large needlelike crystals, usually in sheaves or small coccilike debris. It is easy to recognize the artifactual character of the crystals. Other debris on the slide may be easier to confuse with gram-positive cocci (Color Plate III, Fig. 4)

PROCEDURE SUMMARY*

1. Clean and label slide
2. Make at least two uniformly thin smears
3. Air dry
4. Fix (heat or methanol)
5. Circle specimen with a grease pencil
6. Gram stain:
   a. Crystal violet for 10 seconds
   b. Wash
   c. Iodine for 10 seconds
   d. Wash
   e. Decolorizer drop by drop until clear
   f. Wash
   g. Safranin for 20 seconds
   h. Wash
6. Dry
7. Examine under the microscope (scan under low, examine under oil immersion)
8. Describe what you see
   a. Bacteria (Gram's staining characteristics and morphology)
   b. Cells, if present
9. Evaluate the slide for the:
   a. quality of Gram's stain technique (overdecolorized, underdecolorized)
   b. Quality of specimen

QUALITY ASSURANCE CHECKS

We have never had Gram's stain reagents fail to perform correctly. We have also asked the supervisor of our reference bacteriology laboratory if the Gram's stain reagents have ever failed to peform correctly. Her answer—an emphatic "no." She did, however, add that she has had personnel fail to perform correctly.

This points out that the usual failure is a human one, not reagents. Since examining Gram's stains may not be done everyday in a primary care office laboratory, we recommend that once a week each laboratory staff who performs Gram's stain (physicians included) Gram's stain a slide with known gram-negative rods and gram-positive cocci. Record this information on a quality control form.

These slides may be purchased (Microscan Gram Stain Control Slide, American Scientific Products, B1089 packages of 50, $39.06), or they may be prepared in the following manner from reference cultures kept in the refrigerator.

1. Obtain a plate growing a pure culture of *E coli* and another plate growing a pure culture of Staphylococcus species or a Streptococcus species from your hospital laboratory. *E coli* is a *gram-negative* rod (therefore pink) and Staphylococcus or Streptococcus is a *gram-positive* cocci (therefore dark purple).

2. Take two pediatric red-top tubes (they are sterile) and pop off the tops. Label one "*E coli*" and the other "Staph" or "Strep."

3. Add about 0.50 mL of sterile normal saline or distilled water to each tube.

4. Using a sterile swab, wipe the swab across the *E coli* plate, collecting a dab of organisms on the swab. You are not trying to pick up a single colony at this point. Put the swab into the tube marked *E coli* and swirl the swab around—mixing organisms into the fluid.

5. Lift the swab out of the fluid (but still keep it inside the tube), and press it against the inside of the tube to wring out the fluid. Dispose of the swab in the "contaminated" trash container.

6. Using a fresh sterile swab, wipe the swab across the Staphylcoccus or Streptococcus plate, collecting a dab of organisms on the swab. Put the swab into the tube marked staph or strep and swirl the swab around—mixing organisms into the fluid.

*Copy this page and tape it to the wall above the laboratory sink.

7. Lift the swab out of the fluid (but still keep it inside of the tube) and press it against the inside of the tube to wring out the fluid. Dispose of the swab in the "contaminated" trash container.
8. Take a sterile pipette and transfer the fluid from one of the tubes into the other. Mix well by squeezing the pipette bulb. You now have a mixture of *E coli* and Streptococcus or Staphylococcus, that is, a mixture of gram-negative rods and gram-positive cocci.
9. Clean and label (quality control slide) a microscope slide.
10. Use the pipette or a cotton swab to transfer a drop of this mixture of gram-positive cocci and gram-negative rods onto the clean and labeled quality control slide. Use the tip of the pipette to spread the drop into a thin smear.
11. Allow the slide to dry and then fix.
12. Check to make sure that all of the Gram's stain reagents are "in date."
13. Stain the slide and examine it microscopically.
14. You should see gram-negative (pink) rods (*E coli)* and gram-positive (purple) cocci (Staphylococcus or Streptococcus).
15. If the quality control slide does not show these results, consult "Quality Control Slide Interpretation" (below).
16. To keep a viable culture available to make the quality control slides, restreak a fresh sheep-blood agar plate with the appropriate organism. Incubate every 2 weeks at 37 °C for at least 12 hours, then refrigerate the plate.
17. A record of the results should be maintained (Fig. 18–16).

QUALITY CONTROL SLIDE INTERPRETATION

While there is usually some variability on all stains (i.e., some of the rods may seem dark or some cocci pink), most of the bacteria should be stained correctly.

1. *Satisfactory Results: pink rods and purple cocci.* This means that the reagents are working properly and that the staining technique was good.
2. *Unsatisfactory Results: blue rods and cocci or pink rods and cocci.* You will have to make additional quality control slides at this point. Make several at one time. Before proceeding, double check the position of the condenser and the amount of light that you are using on the microscope. Insufficient light will give erroneous colors. The condenser should be up and the light should be set on medium to bright.
   a. *Purple rods:* purple rods and purple cocci.
      You have *underdecolorized.* Repeat Gram's staining using a longer time for decolorizing until rods are pink and cocci remain blue.
   b. *Pink cocci:* pink rods and pink cocci.
      You have *overdecolorized.* Repeat Gram's staining using a shorter time for decolorizing until rods are pink and cocci are blue or purple.
      Organisms old. Repeat using 18-to-24 hour culture.

COMMENTS

- The decolorizing step is the most crucial step in the entire staining procedure. Underdecolorizing of gram-negative organisms will make them appear as gram positive. Overdecolorizing of gram-positive organisms will make them appear as gram negative. If the specimen contains cells, their nuclei can serve as a guide to evaluate underdecolorizing, since they

		CRYSTAL VIOLET		GRAM IODINE		DECOLORIZER		SAFRANIN		
Date	Tech	Lot#	ExpDate	Lot#	ExpDate	Lot#	ExpDate	Lot#	ExpDate	Report of QC Slide

Figure 18–16. Gram's stain reagent quality-control form.

will appear red with proper decolorizing. However, you cannot use cells as an aid to evaluate overdecolorization. The decolorizing step can be mastered only with practice. Practice this step by preparing slides from gram-positive and gram-negative stock cultures.

- Each laboratory should have reference slides of common organisms (e.g., *N gonorrhoeae, S pneumoniae, Staphylococcus, E coli, and Haemophilus*). These can be referred to for help in slide interpretation and can also be used for teaching.
- Gram safranin is the most commonly used counterstain, but some organisms are more easily identified by using other counterstains. Basic fuchsin produces a more consistent staining of *H influenzae.* A bottle of this stain can be kept in the laboratory to be used when this organism is suspected. It is used in the Gram's stain procedure instead of the gram safranin, with the same staining time.
- Gram iodine and the decolorizing solution may have shelf lives as short as only 3 months. If the iodine becomes ineffective, every organism will appear gram negative. If the decolorizer becomes ineffective, all organisms will remain gram positive. This can cause unexpected problems in the laboratory. Regular quality control testing will help to avoid these problems.
- There is an inexpensive atlas that has very good Gram's stain and culture plate photographs: *Atlas of Diagnostic Microbiology* by S. S. Schneierson (Abbott Laboratories, North Chicago, IL 60064).
- Another excellent and inexpensive reference (text and pictures) is *Manual of Acute Bacterial Infections,* 2nd ed. by P. Gardner and H. Provine (Little, Brown and Co, Boston, MA).

REFERENCES

Battone E. The gram stain: The century old quintessential rapid diagnostic test. *Lab Med.* 1988;19(5):288-291.

Koneman E, Roberts G. Mycotic disease. In: Henry J, ed. *Todd, Stanford, Davidsohl Clinical Diagnosis and Management by Laboratory Methods.* Philadelphia, Pa: Saunders; p. 1177.

SECTION 4: Sputum Examination

BACKGROUND

Bronchitis and pneumonia are common clinical infections. The usual pathogens causing pneumonia are viruses (*respiratory syncytial virus* in children and influenza in adults), *Mycoplasma pneumoniae,* and *S pneumoniae* (i.e., pneumococcus). Less common pathogens include *Staphylococcus aureus, Klebsiella pneumoniae, H influenzae* and fungi. *S pneumonia* and *H influenzae* are also common pathogens in acute and chronic bronchitis.

The examination of sputum from patients with lower respiratory infections can be the most important data in the patient evaluation. This is because the chest x-ray is rarely diagnostic of the infecting organism, and sputum cultures are not available until 2 days after the specimen has been collected. These cultures are often read as "normal respiratory flora" or contain organisms that are unlikely to have caused the infection. Furthermore, the infecting organism may be fastidious and therefore may not survive the trip to the reference laboratory.

Sputum examination with Gram's staining provides the clinician with an immediate idea of the etiology of the infection at the time when the patient comes to the office. It is therefore the best laboratory aid in managing outpatients with pulmonary infections and in making the initial antibiotic decision.

TEST INDICATIONS

- The evaluation of changes in the sputum of patients with chronic pulmonary disease.
- Determination of the etiologic agent in patients with clinical evidence of bronchitis or pneumonia.

TEST MATERIALS

Most of the equipment needed for a sputum examination has been described in this chapter (Section 3, ''Gram's Stain''). The sputum sample should be collected in a plastic collection cup, such as the ones used to collect urine specimens.

SPECIMEN COLLECTION

Patient instruction is very important in collecting sputum specimens. Several collection methods can be used. (Table 18–1)

Voluntary Expectoration

This is the simplest method to collect a sputum specimen. The quality of the specimen can be increased by having the patient rinse out his or her mouth with water before expectorating. If sinusitis is a concern, the patient should first blow his or her nose to minimize contamination with upper airway secretions; he or she should be advised to swallow any saliva before coughing. The clinician should coach the patient to draw in a few deep breaths and then to give a few deep coughs. Coaching by the clinician will increase the chances of collecting a good specimen and may give you data about the patient's clinical condition.

Induction of Expectoration

Voluntary coughing may not produce sputum in a patient with poor hydration, viscous sputum, or an ineffective cough. In these patients the use of saline with a bronchodilator and ultrasonic nebulization is helpful. This provides moisture, airway dilation, and irritant aerosols, all of which promote sputum expectoration.

Pediatric Patients

Children are difficult subjects from whom to collect a sputum sample. If bronchial disease is suspected, a nasal pharyngeal swab may be adequate. A second method is the ''cough-swab'' technique. The child's mouth is held open with a tongue depressor. The epiglottis is visualized and touched with a clean swab. This will induce a cough, and some of the induced material will be expelled and stick to the swab. The mouth should not be touched with the swab, as this will contaminate the specimen.

Invasive Techniques

Transtracheal aspiration, fiberoptic bronchoscopy, and endotracheal tube suction can all be used to obtain sputum samples on hospitalized patients.

TEST PROCEDURE

1. The patient is instructed in the production of a sputum sample.
2. The sample is collected in a clean cup. Specimens can also be collected in a tissue; however, they quickly become too dry to use.

3. The sample should be grossly examined. If it is just foamy "spit," it is really not worth the time to stain and certainly is not acceptable as a culture specimen. There will frequently be a few strands of purulent sputum floating in a large volume of saliva. Take a serile cotton swab and pick up some of the sputum from the saliva. Spread this thinly onto one or two microscope slides and allow them to air dry. If the specimen is thick, you can spread it on one slide and place a second slide (label side down) over the first. Apply a small amount of pressure and pull apart. This usually provides a satisfactory smear. However, if the specimen is especially thick, you may use a teasing needle to separate the strands of sputum on the microscope slide. Flame the teasing needles before and after use.
4. After the slide has air dried, fix the slides using either heat or methanol and then Gram's stain them.
5. The stained slide should be reviewed for its sputum quality. A satisfactory specimen should have at least 25 white blood cells (WBCs) and less than ten squamous epithelial cells per low-power field. Squamous epithelial cells are shed from the buccal mocosa and reflect saliva contamination. Bronchial epithelial cells are round or rectangular and are the size of a lymphocyte. The presence of macrophages is the best assurance that the specimen is from the lower respiratory tract.
6. The slide should be judged for its staining characteristics. The nuclei of polymorphonuclear cells (PMNs) will be pink if there has been adequate decolorization. Some gram-positive mouth flora may be present to indicate that there has not been overdecolorization.

TEST INTERPRETATION

The presence of many PMNs in the specimen is indicative of an infectious process. A specific bacterial diagnosis can be made only with an abundance of PMNs and a single predominant bacterial species. Fortunately this is often found with outpatient infections (Table 18–2). The common infections are caused by the following organisms:

- *Mycoplasma pneumoniae:* This organism is too small to be seen. It is suspected when the specimen shows lots of polys and no organisms.
- *S pneumoniae:* This organism is classically described as gram-positive, lancet-shaped diplococci (Color Plate II, Fig. 5). Frequently there is not an abundance of the organisms, but they should appear in most fields.
- *H influenzae:* This organism is a very small, gram-negative, pleomorphic rod. It is often difficult to tell whether it is a rod or a coccus, and it is therefore described as a coccobacillus. The stain is sometimes concentrated in the two ends of the rod, giving the appearance of paired gram-negative cocci. *H influenzae* stains lightly with regular gram-safranin counterstain and is often missed or interpreted as background debris. The use of gram basic fuchsin as a counterstain (instead of gram safranin) will make the organism more easily seen (Color Plate III, Fig. 2).
- *N meningitidis and Branhamella catarrhalis* can also cause community-based pneumonia and bronchitis. They appear as intracellular/extracellular gram-negative kidney bean-shaped diplococci and are indistinguishable from each other and from *N gonorrhoeae* on gram stain (Color Plate II, Fig. 6).
- *S aureus:* This organisms is a gram-positive coccus that, when collected from a moist medium such as sputum, will usually show the clasic grape-cluster formation. The organism is often intracellular, which is not a finding seen with the other common bacterial pathogens (Color Plate II, Fig. 3).

- *K. pneumoniae:* This organism is a gram-negative, encapsulated rod. The capsule is not always prominent with Gram's staining. A culture should be done if this organism is suspected (Color Plate III, Fig. 1).
- Mixed flora: Often there will not be a dominant organism, but rather multiple organisms with or without white cells (Color Plate III, Fig. 5).

REFERENCES

Pennington JE. An empiric approach to pneumonia. *Res Staff Phys.* 1976;22(12):36.

Zimet I. Common errors in sputum collection and examination. *Pract Cardiol.* 1980;6:96.

SECTION 5: Urine Culture

BACKGROUND

Urinary tract infections (UTI) are among the commonest bacterial infections. Fifteen percent of all women experience a UTI at least once in their life. Pregnancy, obstructive uropathies, urethral trauma during sexual intercourse, and vesicoureteral reflux all predispose to such infections. One percent of all children have asymptomatic bacteriuria. The significance of this condition is in dispute. Symptoms, pyuria, and a positive urine culture are all indications of a UTI. A urine culture is not the only way to diagnosis a UTI; however, in some clinical situations it is indispensable.

TEST INDICATIONS

- Suspected UTI (i.e., cystitis, urethritis, pyelonephritis, or prostatitis). In the outpatient setting many clinicians choose to empirically treat a symptomatic woman who has pyuria. No culture is done unless there is concern about pyelonephritis, antibiotic resistance, or recurrent infections.
- Screening for asymptomatic bacteriuria (e.g., obstetric patients).
- Follow-up of patients after treatment for a UTI to document sterile urine.
- Evaluation of a child or an adult with unexplained fever.
- Evaluation of enuresis.
- Recurrent UTIs.

TEST MATERIALS

MacConkey Agar Plate

This is a selective medium that supports the growth of gram-negative rods. Those that ferment lactose will produce purple colonies on this medium. Nonfermenters form clear-colored colonies. The plates usually come in a plastic tube wrapper. It is important to retape the tube after removing a plate. This maintains a moist atmosphere and prevents the plates from drying out. Check for the expiration date and store them in the refrigerator. Plates are available from scientific supply companies or from the local hospital laboratory.

Sheep-Blood Agar Plate

This agar supports the growth of a wide range of gram-negative and gram-positive organisms.

Calibrated Streaking Loop

This loop allows the quantification of bacteriuria in a urine sample by delivering a specified volume of urine for streaking. A

0.01-mL loop should be used. This will produce a culture in which one colony represents 100 organisms/mL of urine.

Collection Cups

Studies have shown that the use of common paper cups is adequate for routine office culturing. False-positive tests are rare even though regular paper cups are not sterile. However, we prefer to use inexpensive plastic specimen containers with a screw-on lid. They are better than paper cups because they can be stacked in the refrigerator and are not easily spilled. Specimens can then be stored in the refrigerator during the day, and all of the plating can be quickly done at the end of the day.

Reagents

Hydrogen peroxide and plasma coagulase are needed for organism identification. They can be purchased from scientific supply companies.

SPECIMEN COLLECTION

Adult Female Clean Catch

The patient is told the following: "Wash the hands well. Sit on the toilet and separate the labia (genital lips) with one hand. Using a povidone-iodine pad or any other nonalcohol "wipe," clean the urethral area from front to back. Repeat this a second time. With the labia still separated, release a small amount of urine. Momentarily stop the urine stream. Hold the collection container up to the urethral opening and restart the urine flow." Nursing supervision may be required for young or disabled patients. If the patient has a heavy vaginal discharge, a tampon should be inserted before the urine sample is collected. This will prevent the contamination of the specimen with the vaginal discharge. If that does not reduce the vaginal contamination, a minicatheterization should be performed.

Minicatheterization

A convenient system for obtaining catheterized urine specimens is available (Chap. 17 and Fig. 17–1). This system includes a plastic tube with a screw-on top and a 5-inch long, thin bladder catheter. This system comes in sterile packaging and is quick, fairly inexpensive, and nontraumatic to the patient. To use the minicatheterization system, the urethral meatus is first cleaned with a povidone-iodine gauze. Using sterile gloves the nurse should then pull the catheter out of the tube so that it is maximally extended. The hand holding the catheter should remain sterile. The other hand should then be used to separate the labia. Gently insert the catheter until urine flows into the collection tube. If there is resistance to passing the catheter, ask the clinician for assistance. If no urine is returned, even with maximal catheter insertion, then the bladder is probably empty. Remove the catheter. Close the top of the minicatheterization system. Take the specimen to the laboratory.

Infant Bagged Specimen

The skin around the genital area is cleaned with alcohol and then dried with gauze. A self-adhesive urine collection bag is securely attached. The infant can then be given a bottle to promote urination. The bag should be removed as soon as a sample is obtained. This method suffers from frequent contamination with skin and bowel bacteria.

Suprapublic Aspiration

The lower abdomen is cleaned with povidome-iodine and then with alcohol. A 5-mL syringe with a 1.5-inch 21-gauge needle is used for the aspiration. The needle is directed perpendicular to the skin at a midline point 1 cm above the pubic rami. Up to three passages of the needle should be made as back pressure is put on the syringe. If no urine is obtained, the procedure should be stopped. Failure to collect urine by this method indicates that the bladder is empty. This is the only method of urine collection in an infant than ensures a noncontaminated specimen.

Adult Male Clean Catch

It is important to remember that many men with dysuria have urethritis (gonorrhea or chlamydia) rather than cystitis. (See p. 237) For those male patients in whom a urine infection is suspected, the clean-catch urine technique is the best method to obtain a specimen: "Pull back the foreskin with one hand. Clean the urethral opening with the povidone-iodine pad. Wipe the area dry with a sterile gauze. Without letting the foreskin slide back over the tip of the penis, express the first portion of the urine into the toilet. Stop the stream. Put the remainder of the urine into the collection cup and return it to the laboratory."

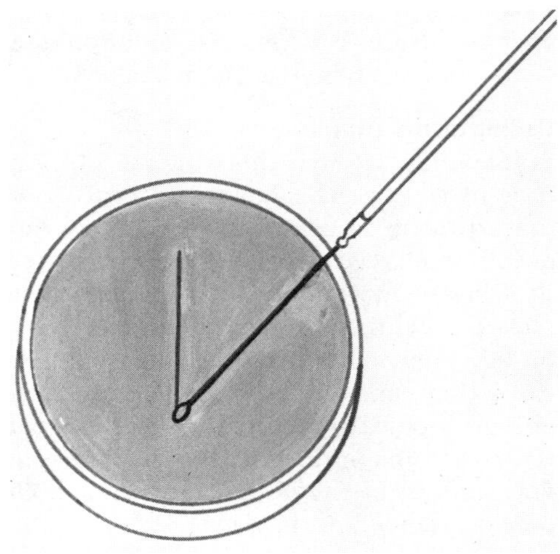

Figure 18–17. Streaking urine specimen down the center of the medium plate.

TEST PROCEDURE

1. The urine sample can be plated immediately after collection, or if the specimen is refrigerated the plating should be done within 24 hours.
2. If it is your laboratory policy, put on gloves before handling the urine. It spills easily.
3. Mark the patient's name and the date on the bottom of the plate using a label or a grease pencil. Do not label the lid, since lids can be mistakenly exchanged between specimens.
4. Make sure that the lid of the urine container is screwed on tightly, and then gently shake the cup to completely mix the specimen.
5. Flame the calibrated urine-streaking loop until it turns red. Then allow it to cool.
6. Reswirl the urine in the cup and dip the loop into the urine to collect a sample in the loop.
7. When you pull the loop out of the urine, check to make sure that there is a complete film of urine in the loop and that there are no air bubbles.
8. The sample is streaked across the center of a sheep-blood agar plate first (Fig. 18–17). The streaked sample is then spread across the plate by streaking back and forth across the original streak (Fig. 18–18). The sheep plate is streaked first be-

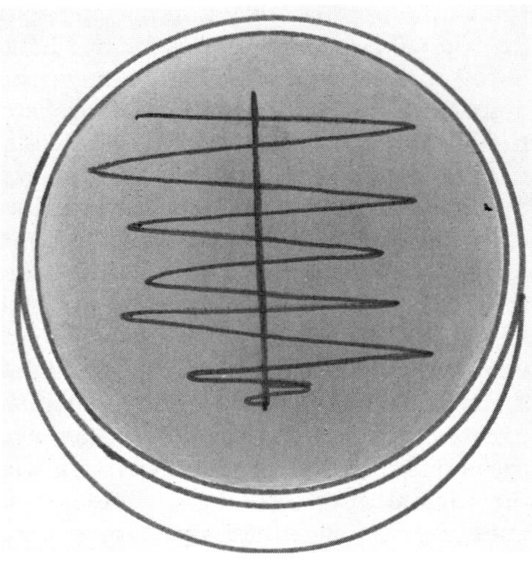

Figure 18–18. Cross-streaking pattern of the urine culture plate.

cause this medium contains fewer inhibitors. If the MacConkey plate is streaked first it would be necessary to reflame and cool the loop. Steps 3 through 5 are repeated with the MacConkey agar plate. The plates do not have to come to room temperature before streaking.

9. Place the plates media side up in the incubator for 18 to 24 hours. If there is no growth within this period, incubate the plates for an additional 24 hours. Gram-positive organisms frequently are not seen at the end of the first 24-hour period.

TEST INTERPRETATION

Colony Count

The colony count is done on the sheep-blood agar plate, not on the selective MacConkey medium, which might inhibit some colony growth. A culture is read as negative if there is no growth on the sheep-blood agar plate at the end of 48 hours. If colonies are found, the number of colonies is multiplied by 100 to give the level of bacteriuria (i.e., the number of bacteria/mL of urine). This assumes that each colony represents 100 bacteria/mL in original sample. There is no hard-and-fast rule about the level of bacteriuria that is significant to indicate a UTI. This is dependent on the way the specimen was collected, how clean the sample appears to be by microscopic examination, the clinical history of the patient (e.g., is the patient asymptomatic or are there symptoms of an active UTI), and how concentrated the specimen is found to be by its specific gravity. All of this information is important to the clinician and must be considered before a level of bacteriuria is considered to be diagnostic of a UTI. A colony count of greater than 100,000 organisms/mL of urine has traditionally been the cut-off point for significant bacteriuria. Recent studies have indicated that colony counts as low as 100 organisms/mL of urine in asymptomatic patientsmay be significant.

PLATE APPEARANCE (See Figure 18–19)

Growth on Both MacConkey Agar and Sheep-Blood Agar Plates

The MacConkey plate is a selective medium for gram-negative rods, so all growth on this plate will be by gram-negative organisms. These gram-negative rods also grow on the sheep-blood agar plate. The sheep-blood plate is used for the colony count because this medium has no inhibitors. If the colonies on the MacConkey plate are purple, the organism is a lactose fermenter (Color Plate III, Fig. 6). In the office laboratory this will usually be *E coli,* which is also the most common organism to cause a UTI. *Klebsiella* is another lactose fermenter that is occasionally seen and cannot be differentiated by colony appearance from *E coli*. *Klebsiella* colonies are said to have a "mucoidy" appearance. However, some stains of *E coli* also look mucoidy. The importance of this characteristic for the clinician is that *Klebsiella* is resistant to ampicillin. If a patient has a lactose positive, mucoidy-looking culture and was treated with ampicillin, the plates should be sent to your reference laboratory for sensitivities. If the colonies are colorless on the MacConkey plate (Color Plate III, Fig. 6) (i.e., nonlactose fermenters), the organism may be either a *Proteus* species, or some other enteric organism.

Some *Proteus* species swarm over the entire sheep-blood agar plate rather than form isolated colonies. *Proteus* also has a characteristic "dirty-sock" smell. If an unusual organism is suspected, it should be sent to a reference laboratory for sensitivities and further identification.

Growth Only on Sheep-Blood Agar Plate

Look carefully at the sheep-blood agar plate because some of the organisms that grow on this plate will form very small and hard-to-see colonies. Growth on this plate but not on MacConkey agar represents gram-

Urine Culture Flow Chart

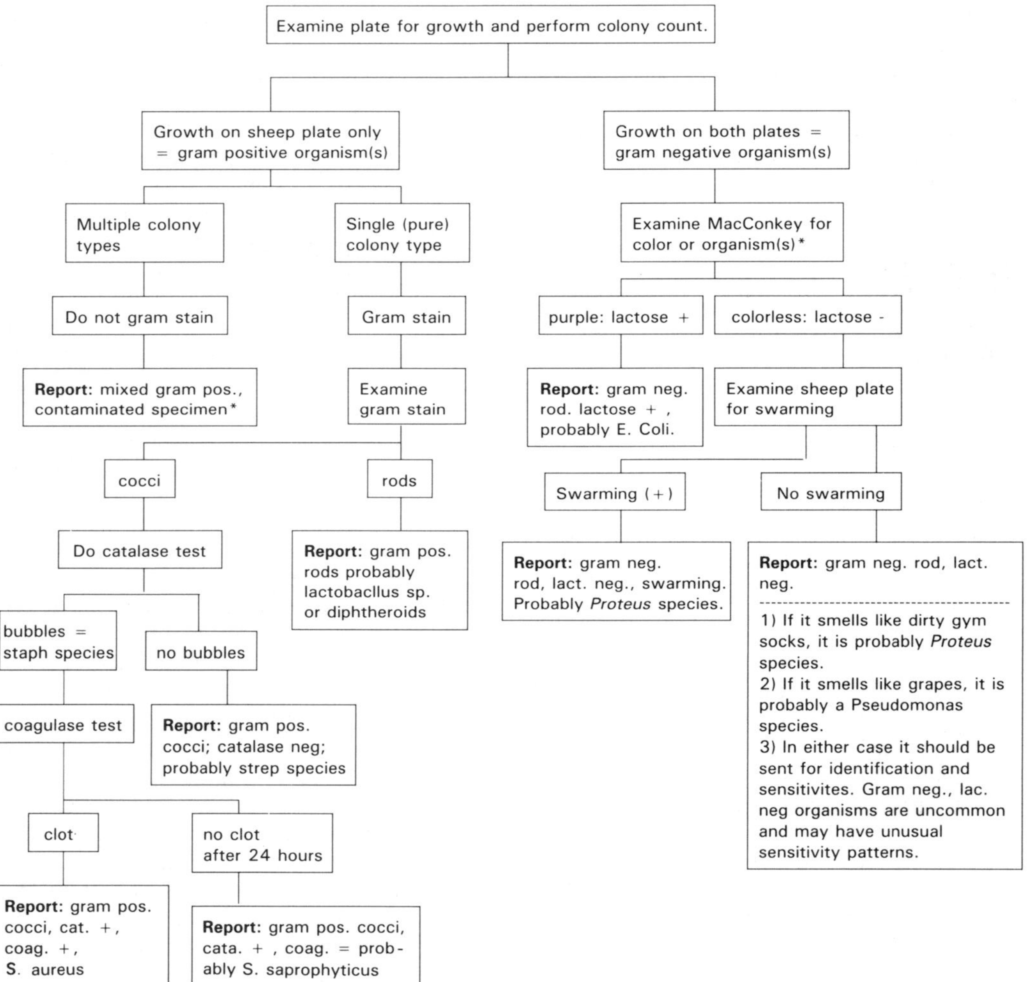

*Multiple gram positive organisms probably indicate contamination while multiple gram negative organisms *may* indicate a mixed infection.

Figure 18–19. Urine Culture flow chart.

positive organisms. These colonies should be Gram's stained. If the organism is a gram-positive rod, the species may be a *Lactobacillus* or a diphtheroid. If the organism is a coccus, it is either a *Staphylococcus* or *Streptococcus* species, and a catalase test should be done to differentiate the two. Often the plate will have several different types of colonies. When this is seen we report "mixed gram positives probable contaminated specimen." There is no need to Gram's stain the colonies from contaminated plates.

CATALASE TEST

1. Heat the loop and then let it completely cool, or use a sterile cotton swab.

2. Place a loopful of inoculum onto a glass slide with a streaking loop, or spread some organisms from the swab.

3. Add a drop of hydrogen peroxide and watch for bubbles. If the inocolum does not bubble, it is probably a *Streptococcus* species, usually *Streptococcus faecalis* (enterococcus). If the inoculum bubbles, the organism is a *Staphylococcus* species. *S. aureus* can be differentiated from coagulase-negative *Staphylococcus* species by the plasma coagulase test.

PLASMA COAGULASE TEST

1. Take a small test tube containing 0.5 mL of rabbit plasma. This reagent can be purchased ready to use from scientific supply companies or reference laboratories.

2. Heat and then air cool a streaking loop, or use a cotton swab for transfer.

3. Scoop up some inoculum from the surface of the sheep-blood agar plate.

4. Add the inoculum to the plasma and stir.

5. Recap the test tube and and place it in the incubator for 4 hours.

6. Examine the test tube after 4 hours and then again at 24 hours. A positive test is indicated by any clotting of the plasma. *S aureus* is coagulase positive. *S epidermidis* and *S saprophyticus* are coagulase negative. In an outpatient setting most staphylococcal UTIs are caused by *S saprophyticus*.

7. Run a positive and negative control (described below) with each test.

QUALITY CONTROL

Media

1. Plates usually come in a plastic sleeve, 10 or 20 plates per sleeve. Make sure that you carefully open the sleeve so that you can close it up with a piece of masking tape. Note the expiration date on both the individual plates and on the label on the outside of the plastic sleeve. *Do not use plates after the expiration date.*
2. Store plates in the refrigerator (2° to 10 °C).
3. While it is optimal to examine all the plates when they first arrive, it is bothersome to remove them from the sleeve and then to put them back. We instead suggest that you check each plate before using it. Look for:
   a. Cracked petri dishes
   b. Uneven filling of the agar
   c. Very pale red color in the sheep-blood plates (too little blood)
   d. Dry or cracked media
   e. Too many bubbles
   f. Organisms growing on the unused plate

   The National Committee for Clinical Laboratory Standards suggests that visual inspection, proper storage, and use before expiration are adequate checks of agar plates.

4. Do not use any plate if there are problems with the media. Notify the manufacturer and keep a record of this correspondence.

Hydrogen Peroxide

This reagent has a very long shelf life if kept in a brown bottle to prevent photodeactivation and if recapped immediately after use. The reagent activity should be checked whenever a catalase test is done. This can be easily done by mixing innoculum from a reference *S. aureus* culture. The presence of bubbling ensures that the reagent is effective.

Coagulase Reagent

If the reagent is purchased from a hospital laboratory, that laboratory will do the necessary quality controls and should notify you that if there is any evidence that the reagent is inactive. Quality controls can also be done whenever the test is done by using an inoculum from a S *epidermidis* culture (no clot) and a S. *aureus* culture (clot forms). If these organisms do not give the specified results, consult your reagent distributor.

COMMENTS

- The culture method described here is an accurate and reliable procedure for culturing urine. However, even with this technique there is a 10% false-negative rate (i.e., 10% of infected urines will show no growth on the culture plates). False-negative tests are due to a low colony count secondary to a rapid diuresis, a patient taking antibiotics at the time that the culture is done, or media that are ineffective in supporting bacterial growth.
- False-positive testing can be due to contamination of the sample with skin or rectal bacteria or the improper storage of urine samples resulting in overgrowth of bacteria in the urine.
- A variety of commercially produced urine culture systems are available. Bacuricult (Wampole Laboratories, Cranbury, NJ) uses a plastic tube whose inner wall is coated with an agar medium. The culture is prepared by pouring urine into the tube to coat the agar surface. Uricult (Medical Technology Corporation, Hackensack, NJ), Clinicult (Smith Kline Diagnostics, Sunnyvale, CA), and Uri-Dip (Scott Laboratories, Fiskeville, RI) employ a similar dip-slide method. An agar-coated paddle is dipped into the urine specimen and then incubated. These various products offer a slight advantage over the quantified loop method in terms of office-staff time. However, they are more costly, less reliable, and do not allow for the precise colony count quantification that routine plating provides.
- A split agar plate is available that contains MacConkey agar on one half and sheep-blood agar on the other half. This allows easier handling of the specimen and is slightly less expensive than the cost of two separate plates.

REFERENCES

Desai K, Robson H. Comparison of gram-stained urethral smear and first-voided urine sediment in the diagnosis of non-gonococcal urethritis. *Sex Trans Dis.* 1982; 9:1.

Schaeffer AJ. The office laboratory. *Urol Clin North Am.* 1980;7:29.

Stamey T. *Pathogenesis and Treatment of Urinary Tract Infections.* Baltimore, Md: Williams & Wilkins; 1980: 9-19.

Stamm W, Counts G, Runniny K, et al. Diagnosis of coliform infection in acutely dysuric women. *N Engl J Med.* 1982; 307:8.

SECTION 6: Group A *Streptococcus* Throat Screens: Rapid Tests and Culture

BACKGROUND

Upper respiratory infections (URI) are the most common clinical problem seen in the primary care practice. They result in 15,300,000 visits to physicians each year and cost $300 million to treat. It has been estimated that every child will experience ten URIs and every adult three to four URIs each year. Many of these patients will have a sore throat as their primary symptom. For patients with a sore throat, the presence of group A streptococcal disease must be considered. Antibiotic treatment of a patient with a ''strep throat'' prevents the development of a peritonsillar abscess, rheumatic fever, and glomerulonephritis. These complications are seen in 2% of patients with proven streptococcal pharyngitis. Recent evidence has also indicated that early treatment may shorten the course of the pharyngitis symptoms.

Streptococcal disease is primarily seen during the months of November to May. Peak ages for patients with this disease are 5 to 15 years. As few as 5% of adults with sore throats will be found to have streptococcal pharyngitis. The carrier rate for streptococci in the pharynx of asymptomatic patients is 20% to 30%. The throat culture, therefore, has an unavoidable false-positive rate of at least 20%. Other causes of a sore throat include viruses, *Chlamydia, M pneumoniae,* and *N gonorrhoeae.* In half of all cases no causative organism can be found.

There has been a great debate in the literature about the role of throat cultures and the role of the newer ''rapid tests'' for strep throat. The specificity of the rapid tests is high, so a positive test means that the patient does in fact have strep throat. The problem with the rapid tests is that many have only a 90% sensitivity. This means that one in ten patients with strep will be missed. Many of these false-negative tests are in patients who have only a few streptococcal colonies when cultured. The significance of a low number of colonies is unclear. Some of these cases probably do represent carriers; however, about one third of the 10% false-negative rapid tests (i.e., 3%) are estimated to truly have streptococcal pharyngitis. Because of the added cost, we consider this rate too low to recommend the routine culturing of all patient's with negative rapid streptococcal tests.

The problem with the rapid test's sensitivity is much less a concern than the well-documented problem that office laboratories have doing throat cultures. There is no doubt that rapid tests are easier for office staff to read than cultures. However, they cost more for reagents and require more hands-on time for the staff. But rapid tests do improve the efficiency of caring for patients with sore throats; they reduce unnecessary antibiotic treatment; and the rapid test results are reassuring to patients. We therefore recommend rapid streptococcal tests over cultures except for those laboratories that must minimize their reagent costs. Both procedures are described in this section.

TEST INDICATIONS

Doing a rapid streptococcal test or a screening culture is the only technique readily available to the office clinician for identifying those patients with pharyngitis who are not infected with group A streptococci and who therefore can be spared the cost, risks, and inconvenience of antibiotic therapy. Specific indications include:

- Patients (especially children) with sore throat; high fever; tonsillar exudate; tender, swollen, neck lymph nodes; or a scarlatiniform rash.
- Patients with a history of rheumatic fever.
- Contacts of patients with proven streptococcal pharyngitis.

TEST MATERIALS

In addition to the routine bacteriologic materials outlined earlier in this chapter, several special items are needed:

1. Tongue depressor.
2. Swabs:
   a. Rapid test: sterile dacron swab.
   b. Culture: sterile cotton swab or Culturette(R) collection system.
3. A rapid strep kit: There are many kits now on the market. Our own studies indicate that colored end points are much easier for office personnel to accurately recognize than agglutination end points. The test should take less than 10 minutes to process. It is advantageous to have a kit in which all reagents can be stored at room temperature.
4. Culture materials:
   a. Five percent sheep-blood agar (TSA) plate. This plate allows the growth of a variety of gram-positive and gram-negative organisms. It is not selective for just group A streptococci.
   b. Bacitracin differentiation discs. These discs ("A" discs) will inhibit the growth of group A streptococci, therefore allowing their differentiation from other bacterial growth on sheep-blood agar plates (Fig. 18–20). There are 0.04 units of bacitracin per disc. Be sure to use a "differentiation disc" rather than a "sensitivity disc." The latter will have a higher concentration of antibiotic and will produce a misleading zone of inhibition. These discs can be purchased from scientific supply companies. They come in cartridges that should be stored in the freezer or the refrigerator.
   c. Neomycin (30 μg) sensitivity discs. This disc (N disc) is used to inhibit the growth of staphylococci that can interfere with the reading of the plates (Fig. 18–20).
   d. Sensitivity disc dispenser (BBL single-disc dispenser 60457, BBL, Cockysville, MD 21030). This is an inexpensive and easy-to-use tool for holding and dispensing

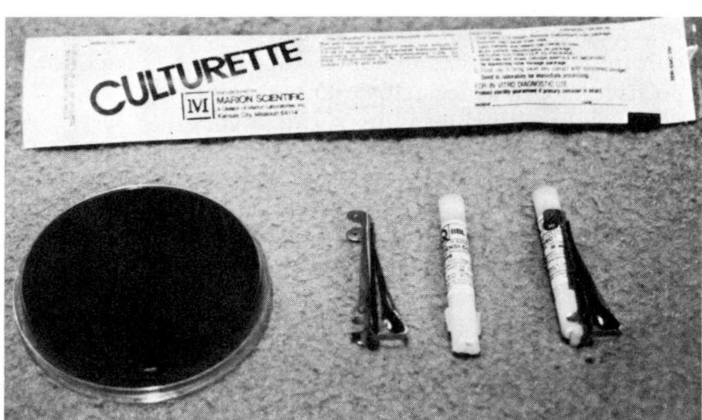

Figure 18–20. Materials for throat culturing: Culturette swab, sheep-blood agar plate, sensitivity disc dispenser, A discs, and N discs.

discs onto the plates (Fig. 18–20). A pair of forceps can also be used for this purpose.

SPECIMEN COLLECTION (for both tests)

1. Depress the tongue completely with a tongue blade. It often helps to ask the patient (especially children) to "pant like a dog." An adult patient can be asked to say "EHH" (as in "less", not "AHH" as in "amen").
2. Using a sterile dacron swab (for the rapid test) or a cotton swab or Culturette for a culture, vigorously swab each tonsillar pillar and the posterior pharyngeal wall. Be sure to swab any visible exudate. The collection should be done in one continuous motion. When done properly it will be uncomfortable for the patient! Culture plates on which there is no growth of any type are the result of swabbing too gently.
3. If a plain cotton swab is used for the culture, it must be immediately streaked onto a sheep-blood agar plate. It may be more convenient to do all of the plating at the end of the day. If this is the case the specimen can be collected with a Culturette system (Marion Labs, Kansas City, MO). This contains a plastic tube with a transport medium that will preserve the specimen for up to 72 hours. Do not refrigerate the specimen prior to plating.

TEST PROCEDURES

Rapid Strep Procedure

This will vary with individual kits. We suggest that you tape the instructions to the wall where you will be performing the test.

Culture Procedure

We recommend a throat-culture method that is different from the one used in most hospital laboratories. Hospital laboratories use a 2-day procedure in which the sample is first incubated for 24 hours. Then isolated colonies of beta-hemolytic streptococci are subcultured with A discs. That two-step procedure requires 48 hours, two culture plates, and extra technician time. We recommend, instead, a single-day, one-step procedure in which the sample is cultured and tested with a differentiation disc on the original culture plate. In addition, we use a neomycin disc to inhibit the growth of staphylococci, which might otherwise produce a confusing hemolytic pattern.

Proper throat-culture technique involves strict attention to specimen collection, plate preparation, and plate interpretation. The interpretation involves reading the colony appearance, the hemolytic pattern, and any disc inhibition.

1. Allow the sheep-blood agar plate to come to room temperature. Leave the plate out of the refrigerator for at least 5 minutes before using.
2. With a black grease pencil draw a line down the bottom of the plate. Half of the plate is used for each patient. Label the bottom of the plate with the patient's name and the date.
3. Streak the specimen over two thirds of the half plate. Twist the cotton swab so that all of its surfaces contact the surface of the medium (Fig. 18–21).
4. Heat the loop until red; then cool it by putting it into an unstreaked portion of the half plate, near the edge of the plate (Fig. 18–22).
5. Streak back and forth across the inoculated area with the cooled loop (Fig. 18–23).
6. Without reheating the loop, make several stabs in the unstreaked portion of the half plate. Do not make large

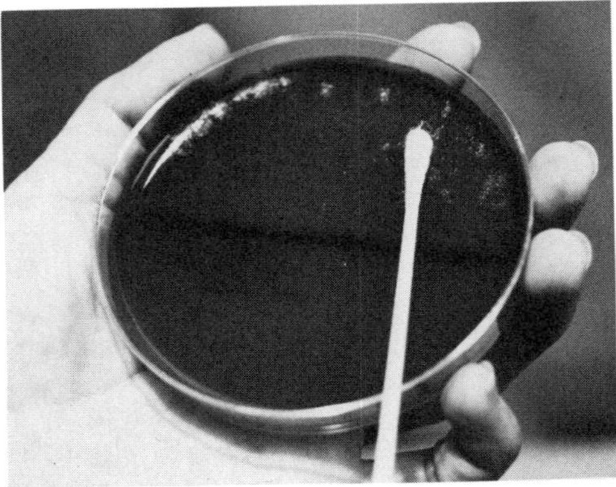

Figure 18–21. Throat culture procedure. Streak the specimen over two thirds of the half plate, twisting the cotton swab so that all of its surface touches the sheep-blood agar plate.

gashes in the medium, just small stabs (Fig. 18–24).

7. Apply the 30-μg neomycin differentiation disc to the center of the streaked area (Fig. 18–25). Gently press the disc with the other end of the disc dispenser so that it sticks to the surface of the medium (Fig. 18–26).

Figure 18–22. Throat-culture procedure. Cool the loop by putting it into the unstreaked portion of the half plate, near the edge of the plate.

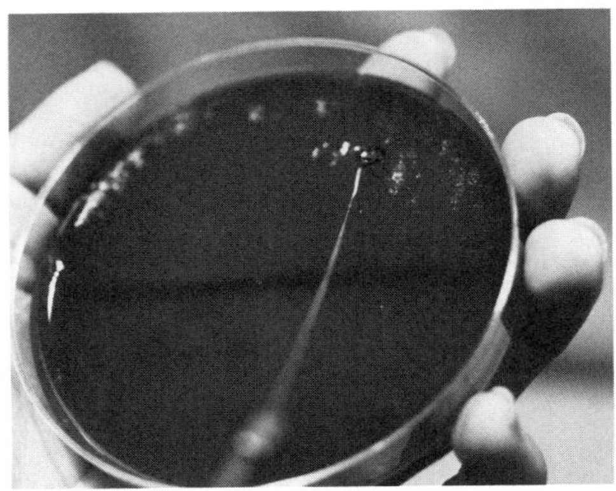

Figure 18–23. Throat-culture procedure. Streak back and forth across the inoculated area with the cooled loop.

8. Place a bacitracin differentiation disc (A disc) within one disc's diameter of the neomycin disc (Fig. 18–25). Gently press the disc so that it sticks to the surface of the medium (Fig. 18–26).
9. Incubate overnight at 35 °C to 37 °C with the plates placed medium side up (i.e., in the "upside down" position).
10. Examine the plates the next morning. It may be impossible to determine the

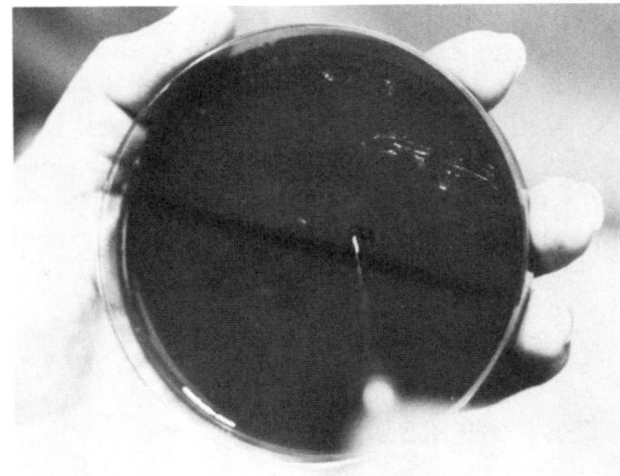

Figure 18–24. Throat culture procedure. Make several stabs in the unstreaked portion of the half plate (without reheating the loop).

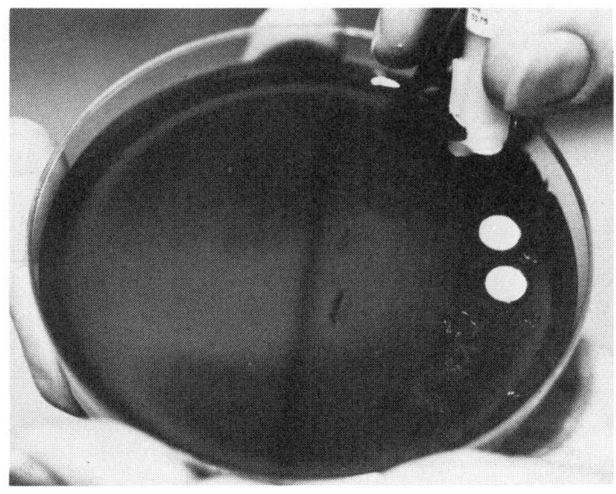

Figure 18–25. Throat culture procedure. Place the discs on the surface of the plate.

hemolytic pattern after incubation of longer than 24 hours because of deterioration of the medium.

11. If the culture is not clearly positive or negative, then subculturing is necessary; or if you also have a rapid test kit, use it to test the organism from the plate. Rapid tests are extremely accurate when used as confirmatory tests.

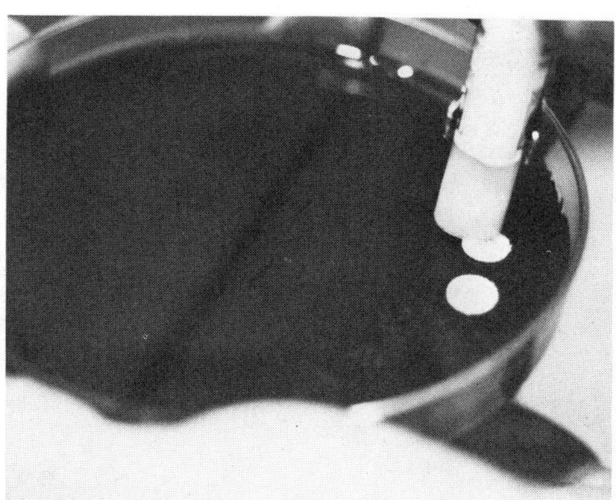

Figure 18–26. Throat-culture procedure. Press the disc with the opposite end of the disc holder so that the disc sticks to the plate when the plate is put in the incubator; media side is up.

Subculture Procedure

1. Heat the loop until it is red, then cool it.
2. If the hemolysis area in question is found within the stab, collect a specimen from this area of the original plate by gently pushing the loop into the stab. If the area in question is on the agar surface, collect a specimen from within the zone of the neomycin disc on the side opposite from the A disc by gently picking up colonies with a loop.
3. Using the loop, streak the specimen along a 2-inch line on the midportion of one half of a sheep-blood agar plate and make several stabs in the uninoculated portion of the plate near the center line (Fig. 18–24).
4. Heat the loop and return it to the stand.
5. Take a sterile cotton swab and go back and forth across the streaked specimen line to smear the inoculum evenly over two thirds of the half plate. Do not streak the area with the stabs (Fig. 18–21).
6. Place discs (Figs. 18–25 and 18–26), incubate, and interpret as for other throat cultures.

INTERPRETATION

Reading a throat culture is difficult and requires proper training and practice. It involves knowledge of hemolysis patterns, disc inhibition, and colony morphology (Fig. 18–27).

Hemolysis

Hemolysis is the clearing of the red blood cells (RBCs) in the medium. It is a function of hemolysins produced by the bacteria. To read the hemolysis pattern, the plates should be studied both by looking at the surface of the medium and by looking through the medium with the plate held up to a light. A fluorescent ceiling light works well for this purpose. The hemolysis pat-

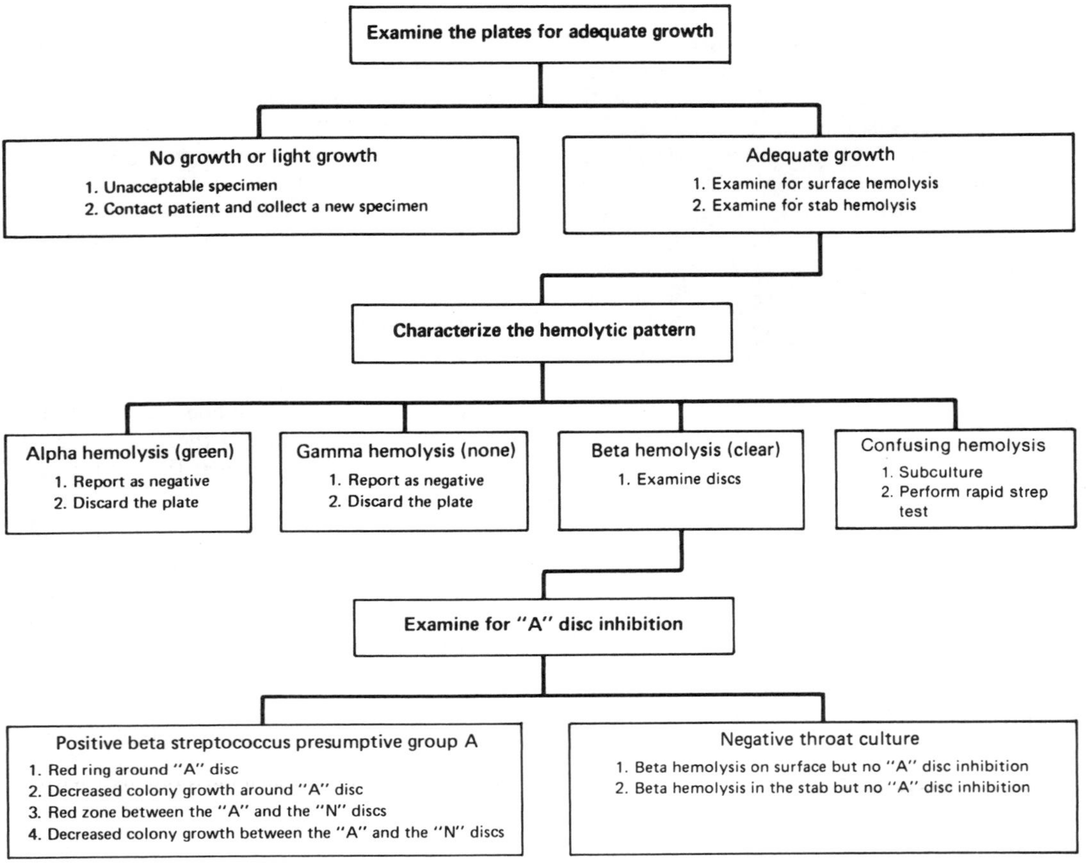

Figure 18–27. Throat-culture interpretation flow chart.

tern should be noted for the colonies both on the surface of the medium and in the stabs. The stabs are areas of reduced oxygen tension. Some hemolysins work best in these relatively anaerobic areas. There are four types of hemolysis:

1. Alpha Hemolysis: This hemolysis appears green both on the surface and in the stabs. It is due to incomplete lysing of RBCs and is seen with a wide variety of normal respiratory bacteria (Color Plate IV, Fig. 1A).
2. Beta Hemolysis: This represents a complete lysis of RBCs and appears as a clearing of the medium from red to translucent. Beta hemolysis may appear on either the surface of the medium (Color Plate IV, Fig. 1B) or only in the stabs (Color Plate IV, Fig. 1C). All group A streptococci produce beta hemolysis, but this type of hemolysis can also be caused by beta-hemolytic staphylococci, nongroup A beta-streptococci, and other organisms.
3. Gamma Hemolysis: This is no hemolysis at all. It will appear as colony growth without any underlying changes in the appearance of the red medium.
4. Confusing Hemolysis: This is hemolysis that is not easy to characterize as either alpha or beta. Hemolysis patterns are of-

ten confusing, especially to a person who is new to reading cultures. This hemolysis may appear golden green rather than just translucent or green. Confusing hemolysis may be due to incubation longer than 24 hours, large gashes in the agar rather than small stabs, incubation for less than 12 hours, or alpha-hemolytic stabs that have golden green areas in the center. For any of these problems, the area in question should be subcultured or examined by a more experienced plate reader.

If the plates show only alpha or gamma hemolysis, they are not growing group A streptococci and should be discarded. There is no need to further examine the plates for disc inhibition or colony morphology. These cultures are reported as "negative for group A beta strep." If the hemolysis is confusing, the specimen should be subcultured and carefully evaluated the next day. If the plate shows beta hemolysis in either the swabs or on the medium surface, then and only then should disc inhibition be studied.

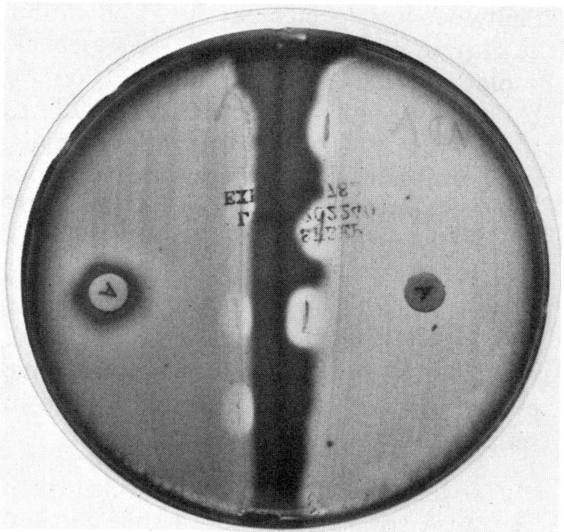

Figure 18–28. An A disc inhibition pattern on cultures showing surface hemolysis. **Left.** Beta hemolysis with inhibition by the A disc. Note the dark ring around the disc. **Right.** Beta hemolysis with *no* inhibition by the A disc. Note how the hemolysis goes right up to the disc.

Colony Morphology

Streptococcal colonies are very small (pinpoint) and are translucent or slightly opaque. With the technique described in this procedure, it is not essential to identify individual colonies. Nevertheless plate readers should be familiar with the appearance of streptococcal colonies. This can be learned by examining reference plates.

Disc Inhibition

The discs used for throat cultures contain antibiotics that inhibit the growth of sensitive organisms (Fig. 18–28). After you have found beta hemolysis, you should look for a zone of inhibition of growth around the disc. If the plate has only one type of organism (i.e., a pure culture), it is quite easy to determine if there is such a zone of inhibition by looking for an area of reduced growth or no growth around the disc. On a pure culture of group A streptococcus that produces surface hemolysis, this zone of inhibition will appear as a dark red area without beta hemolysis around the disc. Any size zone is considered significant! In some pure cultures, you can look for this red zone and need not pay attention to the actual surface colony growth, but some group A streptococci will produce beta hemolysis only in the stabs. For these plates there will be no clear red ring but only an area of reduced colony growth. A zone of inhibition can therefore be characterized by either the decreased growth of colonies on the surface of the medium or the hemolytic changes that those colonies produce in the medium. It may be helpful to examine the plates with a magnifying glass to check for reduced colony growth.

The primary plates used in this procedure are not pure cultures of a single organism. This makes disc reading more difficult. To make this easier we use an additional disc, the 30-μg neomycin (N) disc. This antibiotic inhibits the growth of staphylococcal species but does not inhibit the growth of beta-hemolytic streptococci. By placing the N disc on the plate, you are able to create an area where beta-hemolytic streptococci will grow but where staphylococci will not. This is very valuable because staphylococci can also be beta hemolytic and may grow right up to the A disc, making A-disc inhibition impossible to determine for the streptococcal colonies.

The positive throat culture will show A-disc inhibition. This inhibition will have several appearances on a primary plate in conjunction with the N disc.

- If the culture is an almost pure culture of group A streptococci and if that particular species produces surface hemolysis, the zone of inhibition will appear as a red ring around the A disc. The size of the zone does not matter. No colonies have grown in the area around the A disc, and therefore there is no clearing (i.e., beta hemolysis) in a ring around the disc.
- If the culture is an almost pure culture of group A streptococci but that particular species produces hemolysis only in the stabs, the zone of inhibition will not stand out as a red ring. The entire surface of the plate will be red except in the stabs, which appear translucent. The disc inhibition will appear as reduced colony growth in a ring around the A disc. This ring may be small, and the plates should be examined with a magnifying glass. Hold the plate in your hand and tilt it at various angles to the light source until you can clearly see the fine growth on the agar surface and any areas of inhibition.
- If the plate has a mixed culture of staphylococci and streptococci that produce surface hemolysis, the plate will show a red zone in the area between the N and A discs. You should check for decreased colony growth on the surface of the plate. The area between the two discs will inhibit the growth of staphylococci because of the N disc, and group A streptococci will be inhibited by the A disc.
- If a plate has a mixed culture of staphylococci and a strain of group A streptococci that produces hemolysis only in the stabs, the plate will show only an area of decreased colony growth between the two discs. There will be no red zone between the two discs. There should be heavy growth of pinpoint streptococcal colonies around most of the N disc except in the area close to the A disc.

All of these four patterns would be read as "presumptive group A streptococci."

Reading throat culture plates properly is one of the most difficult aspects of office laboratory work. To understand the technique you should prepare several plates from known cultures of group A streptococci, staphylococci, and nongroup A beta-hemolytic streptococci. Make mixed cultures of these organisms on a sheep-blood plate and handle tham as you would a normal throat culture. These will help you to identify the various patterns described above.

QUALITY CONTROL

The quality control procedure for the throat culture involves both the medium (to ensure that it will support growth) and the discs (to verify reactivity). To do this you will need reference cultures.

1. Record the lot number and expiration date for each lot of plates or discs that are received. Never use the plates or discs after the expiration date.
2. Media:
   a. Store media according to manufacturer's directions (usually in the refrigerator) and use within the expiration date.
   b. Check each new batch of media for signs of hemolysis, cracking, drying, contamination (growth on unused plate), freezing, lots of bubbles on the surface area, leaking, etc.

3. Discs: These need to be checked when they are opened. Store according to the manufacturer's instructions. To check their activity you will need reference cultures of group A streptococci, nongroup A streptococci, and a staphylococci species. You can get these from your hospital microbiology laboratory:
   a. Remove three unused sheep-blood agar plates and the refrigerated reference plate and allow them to come to room temperature.
   b. Turn the fresh plates upside down and, with the black grease pencil, write "beta strep A" on the first plate, "beta strep non-A" on the second, and "staph" on the third.
   c. Using a sterile cotton-tipped swab, collect some inoculum from one of the reference culture plates.
   d. In the appropriate fresh plate, smear the inoculum over most of the surface. Leave a small section uninoculated.
   e. Flame the loop. Cool it, then cross streak the inoculated area.
   f. Without reflaming the loop, make a stab in the unstreaked area of the plate.
   g. Flame the loop until red and then cool. Repeat steps c through f for each of the two remaining organisms.
   h. Place an A disc on each of the streptococcal cultures and press the disc gently into the medium.
   i. Place an N disc on the staphylococcal culture and press it into the medium.
   j. Place the plates in the incubator, medium side up, and incubate for 12 hours.
   k. Place the old reference plates in the refrigerator until the new plates are evaluated the next day.
   l. Remove the new reference plates from the incubator and examine them. The following results indicate that everything is fine.
      (1) Beta strep A: Growth with beta hemolysis and a zone of inhibition around the A disc.
      (2) Beta strep non-A: Growth with beta hemolysis but no zone of inhibition around the A disc.
      (3) *Staphylococcus:* The *Staphylococcus* culture should show growth except for a zone around the N disc.
   m. Record the results of each culture. If they do not match the expected results, open a new batch of discs and repeat the process. If the results do match the expected results, the old reference plates can be discarded and the new plates should be refrigerated. The resulting process should be done every 2 weeks to ensure live reference organisms.

SOURCES OF ERROR

Rapid Strep Tests

False-Positive Results

- Beta-hemolytic streptococcus carrier with a viral infection.
- Over-reading the test's positive end point.

False-Negative Results

- Poor specimen collection technique.
- Under-reading the test's positive end point.

Culture

False-Positive Results

- Beta-hemolytic streptococcus carrier with a viral infection.

- Over-reading of beta hemolysis.
- Over-reading of A disc zone of inhibition.
- Ignoring hemolytic pattern and evaluating only the A-disc zone of inhibition.
- Use of sensitivity-A discs rather than differentiation-A discs.
- Reading a plate after longer than 24 hours of incubation.

False-Negative Results

- Poor specimen collection technique.
- Plates not permitted to come to room temperature before streaking.
- Swab not placed immediately in a preservative or streaked onto a plate.
- Faulty plates.
- Specimen taken after the patient has been taking antibiotics.
- Failure to make stabs.
- Use of outdated A discs and therefore no zones of inhibition.

COMMENTS

Rapid Strep Tests

- Because correctly reading throat culture plates is so difficult, we recommend that rapid tests be performed in the office laboratory instead of cultures.
- It has been suggested that a culture should be done on every patient with a negative rapid test. In practical terms this suggestion means that cultures would be done on approximately 80% of patients. This is an unnecessary expense and defeats the advantages of rapid testing.

Culture

- If there is very little or no growth on the plate, the specimen collection was probably inadequate. The patient should return to the office for a more satisfactory specimen.
- The technique described here does not permit the quantification of beta-hemolytic streptococcal growth into light, moderate, or heavy. Some other methods of throat culturing permit this semiquantification, but the significance for this is unclear. Most clinicians will treat any patient with any amount of group A streptococcal growth.
- If there is any question about the reading of a plate, it should be placed in the refrigerator until a more experienced person can examine it. Do not put it back into the incubator, since further growth and hemolysis will confuse the reading.
- There are various commercial kits available for throat culturing. These include Bact-lab Strep Culture Kits (Mountain View, CA 94042), Biopath Systems (Mountain View, CA 94040), and Smith Kline Diagnostics (Sunnyvale, CA 94086). These systems are more expensive than the culture procedure described here and involve a similar amount of technician time. If you use one of those systems, consider switching to a rapid test.

REFERENCES:

Schwartz RH, Quinnell R. Throat cultures in the office. *Am Fam Physician.* 1980;21:72.

BBL Discs for Differentiation of Group A Streptococci, 88-4030-1, BBL, Cockysville, MD 21030.

SECTION 7: *N Gonorrhoeae* Identification and Culture

BACKGROUND

Despite the use of antibiotics, the incidence of infection with *N gonorrhoeae* (i.e., GC, or the gonococcus) has risen in the United States during the past 20 years. There were 250,000 cases in 1960 and 2 million in 1980.

The peak incidence of the disease in men is between the ages of 16 and 28 years. Only one third of infected men become symptomatic with dysuria. In another one third of men there are no symptoms, but a minimal discharge can be found on examination. The final third of male patients show no signs or symptoms of the disease. Males having intercourse with an infected female partner have a 20% chance of acquiring the disease. Their incubation time from contact to symptoms is 2 to 6 days.

The age-related incidence of gonococcal disease is similar for women. Only 20% of infected women will develop symptoms. The remaining 80% of women are totally asymptomatic.

TEST INDICATIONS

- Screening asymptomatic, sexually active female patients during routine pelvic and prenatal examination.
- Screening patients with a previous history of gonococcal infections.
- Evaluating symptomatic patients (i.e., cervicitis, urethritis, or pelvic inflammatory disease).
- Detection in a patient with known exposure to gonorrhea.
- Follow-up examination after treatment for gonorrhea.
- Conjunctivitis in a newborn infant.
- Women with dysuria or urinary frequency and negative urine cultures.
- Any male with more than five WBCs per high-power field on a microscopic urine examination.

TEST MATERIALS

See Section 2 in this chapter for a description of the routine equipment used in office bacteriology.

Media

There are several types of media available for gonorrhea cultures. The most commonly used is the modified Thayer-Martin medium, which has a vancomycin inhibitor to prevent the growth of many nongonorrheal species. It is a misconception to assume that everything that grows on a Thayer-Martin plate is a pathogenic *Neisseria* species. The Jembec culture system is a very reliable and convenient system for the office laboratory. It employs a modified Thayer-Martin plate, a CO_2-producing tablet, and a gas-impermeable plastic pouch (Fig. 18–29). This system is available from American Scientific Products (J3075-1). Jembec plates can also be obtained from

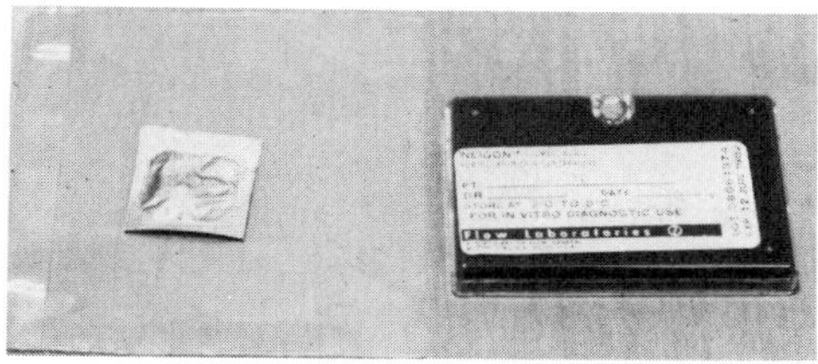

Figure 18–29. JEMBEC culture plate, gas-impermeable bag, and CO_2 tablet in a foil pouch.

some state or reference laboratories. This system eliminates the need to use a candle jar to establish a sufficiently rich CO_2 environment to allow the *N. gonorrhoeae* to grow. The plates (which can be ordered in packages of 20) come in a special bag that permits storage at room temperature for up to 120 days. When a new box of plates is opened, it is best to put each plate with an unopened, foil-covered CO_2 tablet into the plastic pouch and to seal it. This keeps the plate moist and has the convenience of being ready to use as a complete unit. Store the plates in the refrigerator until use, leaving only one or two plates out at room temperature during the day to allow plating at the time of the patient examination. The medium must be brought to room temperature before being used. Do not use plates that have dried out (i.e., that look like chocolate pudding with a skin on top).

Oxidase Reagent

This reagent is used to test any growth on the plates. It can be purchased frozen from the hospital laboratory. Scientific supply companies also sell the reagent in glass ampules and impregnated paper discs. This solution is very irritating to the skin. Avoid contact, and wash your skin immediately if you spill some on yourself.

Synthetic Swabs

Synthetic fiber swabs rather than cotton swabs are used for the specimen collection. The fatty acids in cotton can be toxic to *N gonorrhoeae*. Regular-sized swabs are used for the collection of endocervical and rectal specimens. A smaller swab is used for collecting urethral samples (Fig. 18–30).

Filter Paper

Any regular laboratory filter paper is acceptable.

Loops

A regular streaking loop can be used unless the oxidase testing is done with the discs impregnated with the oxidase reagent. The

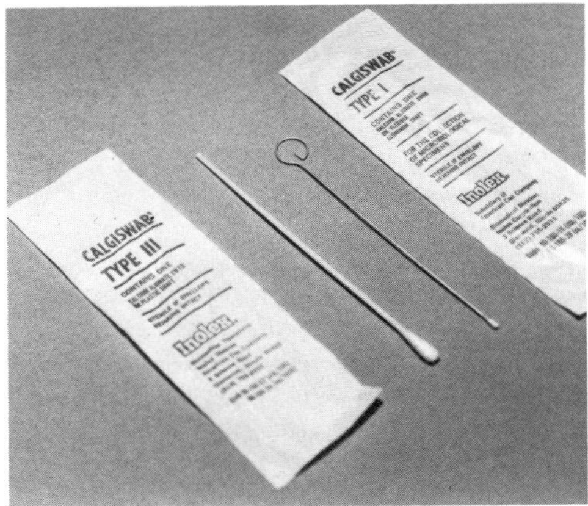

Figure 18–30. Synthetic fiber swabs used for specimen collection. Type III, on the left, is used for collecting endocervical and rectal specimens. Type I, on the right, is used for collecting urethral specimens.

loop will come in contact with the reagent if these discs are used, and the metal in the loop may cause a false-positive test. A special platinum loop is used if oxidase testing is done with reagent discs.

Gram's Stain

See Section 3 of this chapter.

SPECIMEN COLLECTION

The gonococcus is a fastidious organism (i.e., it has very specialized growth requirements). Care should therefore be taken in specimen collection and culturing. Specimens should be plated onto culture plates as soon as possible following collection. The culture plates must be at room temperature when the specimen is plated. A cold plate will kill any gonococci in the specimen. Specimens can be taken from several sources:

Male Urethra

If a male has an obvious penile discharge, the specimen can be collected by milking the shaft of the penis. A drop will form at the urethral meatus. The drop should be collected on a swab. If there is no obvious discharge, insert a sterile calcium alginate swab 2 cm into the urethra. A standard swab is too large to obtain an adequate specimen. An additional swab should be obtained to use in preparing a slide for Gram's staining.

Endocervical Canal

This is the best site for specimen collection in a woman. Moisten a speculum with warm water. Do not use a lubricant, as it may be toxic to the organism. Remove excess mucus from the cervical os. Insert a fiber-tipped swab into the cervical canal and rotate for 20 to 30 seconds. This specimen should be used for culture only. If you wish to Gram's stain a specimen from a woman, carefully take a cervical specimen without contaminating it from the vaginal walls.

Anal Canal

Insert a fiber swab 2 cm into the rectum. Move the swab from side to side for 20 to 30 seconds. Ten percent of female patients will have positive cultures only from the anal canal.

Oropharynx

This is one source for the dissemination of a gonococcal infection. Swab the posterior pharynx and tonsillar crypts with a fiber swab.

Urine

Urine is used for a culture specimen in men with greater than five WBCs per high power field on urinalysis or for women with UTI symptoms but negative routine urine cultures. The urine is centrifuged in the usual fashion. A loopful of the sediment is then plated. *The urine should not be refrigerated before the culturing is done.*

Female Urethra

This collection site should be used in a woman with urinary frequency or dysuria and a negative urine culture. It should also be used for women who have had a hysterectomy, making an endocervical culture impossible. The urethra is striped toward its opening to express any exudate. A sterile loop or small calcium alginate swab (Calgiswab, Inolex Corp, Glenwood, IL) is used to collect the specimen.

Vagina

Prepubescent females develop a vaginal rather than an endocervical infection. A fiber swab is used to obtain a specimen from the posterior vaginal vault.

Conjunctiva

A calcium alginate swab is used to obtain the specimen from the conjunctival surface in a newborn infant with conjunctivitis.

TEST PROCEDURE

1. Allow the medium in the plate to come to room temperature.
2. Write the patient's name, date, and specimen source on a label and attach it to the bottom of the plate (i.e., the media side).
3. Swab the medium with the specimen. *Rotate the swab so that all surfaces of the swab contact the medium.* If cultures are obtained from both the endocervix and the rectum, they can be plated on a single Thayer-Martin plate. The specimens are swabbed on the plate in the pattern of a "C" (for cervical) and "R" (for rectal). One half of the plate is used for each of the two specimens. A throat specimen should not be plated with another specimen from a second site because if the throat specimen culture is positive, it may need to be evaluated by a reference laboratory. Organisms other than *N gonorrhoeae* will grow from a throat specimen on a Thayer-Martin plate.
4. Cross-streak the initial swab to allow for colony isolation. Use a sterile loop for

this cross-streaking. The loop can be sterilized by heating until red in the flame of an alcohol lamp. The loop should be cooled before cross-streaking by sticking it in the medium in an uninoculated portion of the plate.

5. Open the foil pouch and place the CO_2-generating tablet into the plastic well on the plate.
6. Replace the lid on the plate and put the plate in the plastic pouch. Seal the pouch tightly. Place the plate and pouch in the incubator wtih the medium side up. If the bag is not properly sealed, the CO_2 atmosphere is not maintained and the gonococcus will not grow.
7. Check the plates in 24 hours. If there is no growth, add a new CO_2 tablet and return the plate to the incubator. *N gonorrhoeae* is a fastidious organism and may require 48 to 72 hours to produce good colony growth. A plate is read as "no growth" if there is no growth after 72 hours.
8. A Gram's stain should be made as soon as growth is noted on the plate. The organism's morphology deteriorates with prolonged incubation. This can make the microscopic identification very difficult. *N gonorrhoeae* will appear as gram-negative diplococci.

TEST INTERPRETATION

Although much of the literature would lead you to believe that only *N. gonorrhoeae* will grow on Thayer-Martin plates, this is far from the truth (Fig. 18–31). You can ex-

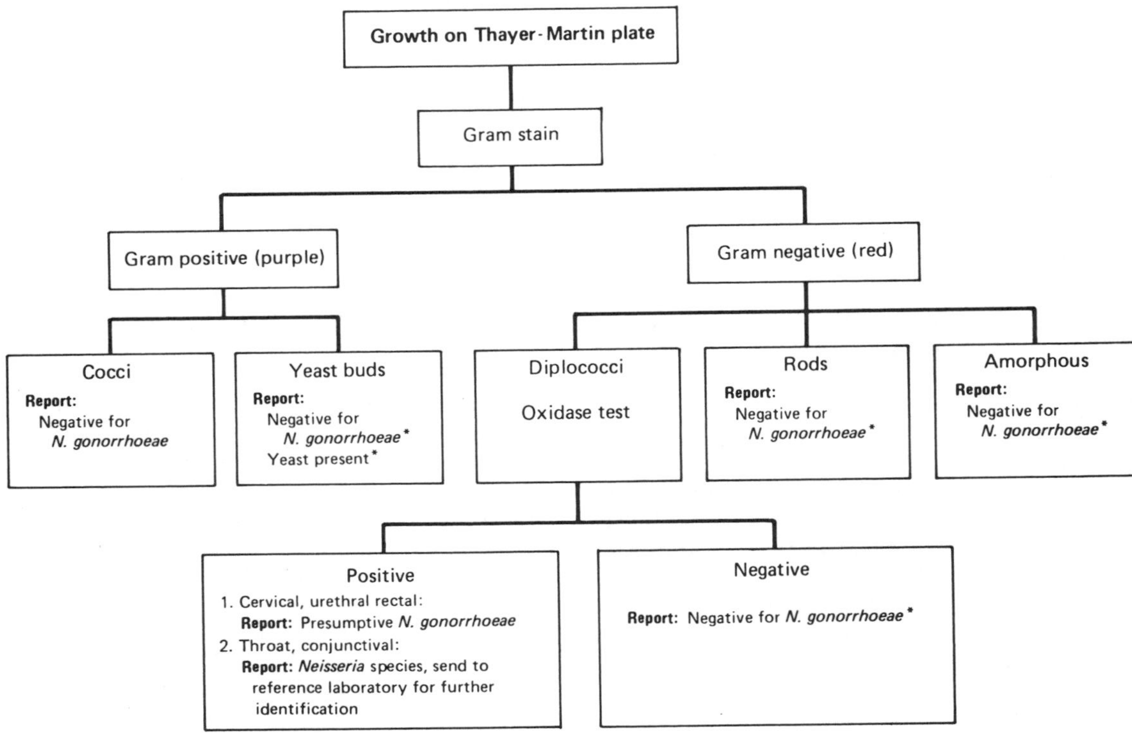

Figure 18–31. Interpretation of the growth on Thayer-Martin medium.

pect staphylococci, *Candida,* and other more confusing organisms to appear. After some experience it becomes obvious that certain types of colonies are not that of *N. gonorrhoeae,* and therefore not all colony growth will need to be Gram's stained. When you are first starting to read gonococcal cultures, however, it is a good idea to Gram's stain all growth that appears on the Thayer-Martin plate. In this way you will gain experience in correlating colony appearance wtih Gram's stain morphology. The gonococcus will produce a translucent to gray colony. The colony size will vary with the age of the colony and crowding of colonies on the plate, but it is usually a small colony (Fig. 18–32). White colonies are not gonorrhea and do not have to be Gram's stained.

Interpretation of Gram's Stain

This is the most difficult step in the procedure. State laboratories usually offer classes in diagnosing gonorrhea. This may prove helpful to you and your staff. Overdecolorizing of common gram-positive organisms will give the impression of a gram-negative organism. In addition to the technical issues of the Gram's stain, the organism itself can present some difficulty. *Neisseria* is an autolytic organism. The longer the organism is on the plate, the less distinct is the cellular morphology on Gram's stain.

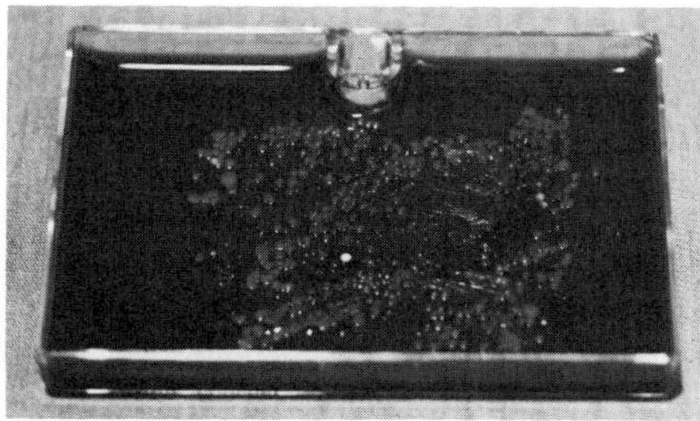

Figure 18–32. JEMBEC plate with *N. gonorrhoeae* growing on the surface.

For the best Gram's stain specimen, make a smear as soon as you see any suspicious growth on the plate. The common species to be seen on the Thayer-Martin plate include:

- *Staphylococcus* species: With proper Gram's staining, gram-positive cocci are noted. With overdecolorizing, they can appear as gram-negative diplococci. One difference is that the cocci are totally round and are not often found in pairs. If overdecolorized, there will frequently be both gram-positive and gram-negative cocci. Staphylococcal species are oxidase negative.
- Yeast: Yeast produces a large white colony on Thayer-Martin medium, but with early growth the colonies may be small and colorless, much like *N gonorrhoeae*. The growth produces a smell similar to baking bread. The Gram's stain will show large purple (gram-positive) buds.
- Rods: Two types of gram-negative rods will occasionally be seen. These are either long and thin or short and fat. They will not be found in pairs.
- Amorphous Growth: This is the most confusing specimen and the one most likely to produce false-positive results. The colony appearance at 24 hours looks much like the gonococcus. At 48 hours the colonies are clear and wet. The Gram's stain shows an amorphous gram-negative field. With an active imagination, a few diplococci can be seen; but the larger part of the field is not made of discrete organisms.
- Diplococci: *N gonorrhoeae* is classically described as two kidney bean-shaped organisms facing each other in a definite pair. They will vary in size depending on the species and strain. If the Gram's stain shows gram-negative (pink) diplococci, the colonies will need to be further identified with an oxidase test. Color Plate II, Fig. 6 shows the organism in a patient's specimen and not from a plate.

OXIDASE TEST

To make the diagnosis "presumptive gonorrhea," it is necessary to perform an oxidase test on all cultures that show gram-negative diplococci. All *Neisseria* species are oxidase positive. There are several *Neisseria* look-alikes that are oxidase negative.

1. If a loop is used, heat the loop until red, and then cool it.
2. With the loop, or a sterile, cotton-tipped swab, remove one or two of the colonies that are suspected of being *Neisseria*.
3. If you used a loop, deposit the colonies on a small piece of filter paper. Note the color of the colonies against the white of the filter paper or cotton swab.
4. If you used a loop, heat it until red and then return it to the loop holder.
5. Drop several drops of oxidase reagent onto the colonies (on the filter paper or on the cotton swab).
6. Observe the specimen for color changes. If there is no change in color, the organism is not *N gonorrhoeae*. If the specimen turns a dark purple, the oxidase test is considered positive. Traditionally a specimen from the penis, endocervix, urethra, or rectum that shows a typical colony appearance, is a gram-negative diplococcus, and has a positive oxidase test is considered to be presumptively positive for gonorrhea. This is usually but not always so! (See "Comments," No. 4, below.) A specimen from the eye with gram-negative diplococci and a positive oxidase test should be sent to a reference laboratory for further testing. Positive cultures from those sites may represent a *Neisseria* species other than *N gonorrhoeae*. If you want to know specifically whether an organism is *N. gonorrhoeae* or whether it is sensitive to penicillin, then send the culture to a reference laboratory for further evaluation.

TEST QUALITY CONTROL

Medium

Each box of culture plates that is opened should be checked by plating a known active culture of *N gonorrhoeae* to see if the medium will support the growth of the organism. Process the plate as you would any specimen, noting the growth, Gram's stain characteristics, and oxidase test results. This information should be recorded in the quality control book in the laboratory. Storage can affect the growth characteristics of the Thayer-Martin plates. Each newly opened box of plates should therefore be checked, even if several boxes are ordered at the same time or have the same lot number. A reference culture for this quality control procedure can be obtained from a positive plate in the office or the hospital or reference laboratory. These laboratories are required to continuously maintain a culture for their own quality control procedures. The clinician can take a plate to the hospital for streaking with a known gonococcal subculture, or the reference laboratory can be asked to send an active subculture to the office in transport medium. If the office Thayer-Martin medium does not support the growth of the subculture, do not perform any patient testing until the problem is corrected.

Oxidase Reagent

It is important to check the oxidase reagent each time an oxidase test is done. This reagent is not very stable, and the clinical significance of a false-negative test is very great. The oxidase reagent can be checked by keeping a positive gonococcal culture in the refrigerator at all times. The cold will kill the organism, but the colonies will continue to give a positive oxidase test. The control test is run in the same manner as described above except that the colony material used is from a known gonococcal reference culture. Staphylococcal species can serve as a negative control.

SOURCES OF ERROR

False negatives can be caused by too low or too high an incubation temperature, poor specimen collection, adverse endocervical pH, loss of CO_2 from the plastic pouch, or antibiotic therapy when the specimen was taken.

False positives are caused by improper Gram's stain interpretation or by assuming that any species from the eye or throat is *N gonorrhoeae*.

COMMENTS

- Because of the availability of methods of incorporating a CO_2 tablet, we have not described the more cumbersome candle-jar technique. There are a variety of other commercially prepared systems for the culturing of *N gonorrhoeae*. These are not much simpler to use and are prone to a higher false-positive rate. The diagnosis of gonorrhea has far-reaching social and legal implications. False-positive or false-negative diagnoses must therefore be minimized. If there is any question about a culture handled in the described fashion, it is best to send the specimen to a reference laboratory for subculture and identification.
- *N gonorrhoeae* and other organisms grown on the selective medium are pathogens. Spread can occur from hand contamination. Use germicidal soap in the laboratory after handling *N gonorrhoeae* plates. Seek immediate medical attention if conjunctivitis develops in any laboratory personnel because this may be gonorrheal ophthalmia.
- A male urethral specimen collected as described above can be used to diagnose gonorrhea by Gram's stain without a culture. This is because there are no other normal gram-negative diplococci in the male urethra, as are found in the vagina. The specimen should be streaked onto a microscope slide and Gram's stained. If intracellular, gram-negative diplococci are found, this is good presumptive evidence for a gonorrhea infection (Color Plate II, Fig. 6). Extracellular gram-negative diplococci suggest an early or chronic gonorrhea infection but should be confirmed with a culture. If many WBCs are seen but there are no organisms, this is probably a nongonococcal urethritis (i.e., a *Chlamydia* or ureaplasma infection). This should, however, be cultured to rule out gonorrhea.
- Recent research has indicated that *N meningitidis* and *B catarrhalis* (formerly *N catarrhalis*) can produce genitourinary disease that mimics gonorrhea in both male and females. This has been best documented in anal cultures from homosexual males; however, heterosexuals may also become infected. *N gonorrhoeae, N meningitidis,* and *B catarrhalis* would not be differentiated by the techniques described here. If there is a suspicion of a *N meningitidis* infection, the positive culture should be sent to the reference laboratory for further identification. Some clinicians may decide to send all presumptively positive cultures for further testing.
- There has been a recent increase in penicillin-resistant *N gonorrhoeae* in the United States. If there is a significant percentage of this in your area, the clinician may want to send all positive culture specimens out for antibiotic sensitivities. Consult your state health department for further information.

REFERENCES

Criteria and Technique for the Diagnosis of Gonorrhea. Atlanta, Ga: Centers for Disease Control, Venereal Disease Control Division; US Department of Health, Education, and Welfare, Public Health Service leaflet 96-552, 1984.

Faur YC, Wilson ME, May PS. Isolation of N. meningitidis from patients in a gonorrhea screening program: A four-year survey in New York City. Am J Pub Health. 1981;71:53

McTighe AH, Patel C, Smith L, et al: Laboratory and clinical aspects of infection with Neisseria gonorrhoeae. *Lab Med.* 1980;11:524.

Miller MA, Millikin P, Griffin PS, Sexton RA, Yousuf M: Neisseria meningitidis urethritis, a case report. *JAMA.* 1979;242:1656.

Riccardi NB, Felman YM: Laboratory diagnosis in the problem of suspected gonococcal infection. *JAMA.* 1979;242:2703.

Smith G. Branhamella catarhallis infection imitates Gonorrhea in a man. *N Engl J Med.* 1988;316:1277.

CHAPTER 19

Obstetrics and Gynecology

SECTION 1: Introduction

There are several easily performed office laboratory tests that provide valuable obstetric and gynecologic information. These tests should be considered for any medical office that regularly sees female patients, such as the specialties of family medicine, obstetrics and gynecology, primary care internal medicine, and adolescent pediatrics.

The tests included in this chapter are the vaginalysis, the urine pregnancy test, and the test for the rupture of amniotic membranes. Papanicolaou (Pap) tests are performed in reference cytopathology laboratories; however, there are aspects of collection and test interpretation that are so important that we have also included them in this chapter. Several other tests that are routinely performed on obstetric patients are described elsewhere in the book. A dipstick urinalysis for glucose and protein should be done at each office visit (Chap. 17). An initial screening urine culture is usually recommended for all obstetric patients because of the high frequency of occult bacteriuria and its associated morbidity and mortality for both mother and fetus (Chap. 18). Finally, the O'Sullivan test, which is a screening procedure for gestational diabetes, is described in Chapter 14.

SECTION 2: The Pap Test

The Pap test is a screening test for cancer of the cervix and its precursor conditions. Cervical cancer is the sixth most common cancer in women and the ninth most common cancer cause of death. It results in 7,700 deaths in this country each year. Women at high risk for cervical cancer include those who had intercourse at an early age, those with multiple sexual partners, those who smoke, those from lower socioeconomic groups, and those from some ethnic backgrounds, including native Americans and blacks. The disease appears to be in some way sexually related. Squamous cell carcinoma rarely occurs in virgins.

The Pap smear is useful as a screening test because it can detect the asymptomatic, precancerous conditions that are known to progress to cervical cancer. The Pap smear has now become a routine part of women's health screening. It has been largely responsible for the decrease in invasive cervical cancer from 44 per 100,000 women in 1947 to 8.8 per 100,000 in 1970. Despite its benefits, many women still have inadequate Pap screening. A survey by the American Cancer Society in 1976 showed that 21% of women older than 18 years had never had a Pap smear and that 50% of women had not had testing done within the previous year.

The majority of cervical cancers arise at the squamocolumnar junction (i.e., the transformation zone). This area of the cervix is where the columnar epithelium of the uterus is transformed into the squamous epithelium that covers the cervix. This junction is usually at the cervical os but moves outward in women who are taking birth control pills, in pregnancy, and with cervical erosions. The junction moves into the endocervical canal with advancing age. The development of invasive cervical cancer begins with dysplasia at an area along the squamocolumnar junction. Dysplasia can be categorized as mild, moderate, or severe, based on the cellular atypia. Dysplasia is not cancer, but it has a 5% yearly rate of progression to cancer. In 30% of patients the dysplasia will regress without treatment, but current medical standards require treatment at this early stage. The next stage in the progression from dysplasia to carcinoma is carcinoma in situ, which involves localized, noninvasive, neoplastic changes in the cells.

The estimates of transit time from precursor lesions to Carcinoma in situ (CIS) and then invasive cancer have had a very wide range, anywhere from four to 40 years. In 1970 Baron and Richart estimated that it took mild dysplasia (cervical intraepithelial neoplasia-1, CIN-1) about 6 years, moderate dysplasia (CIN-2) about 3 years, and severe dysplasia (CIN-3) about 1 year to progress to CIS. In 1978 they estimated a mean duration of CIS of about 10 years, with the lower bound of CIS transit time being 3 years. Using their model, mild dysplasia would take between 9 and 17 years to develop into invasive cancer.

However, in recent years there has been a decrease in the age of patients with dysplasia and with invasive cancer. We are seeing more rapidly developing cancers. A great deal of research is currently aimed at identifying the cause of these faster-progressing lesions. The human papillomavirus (HPV) is thought to be a likely candidate, although there is not sufficient data at the present time. Some reference laboratories are already offering HPV typing in addition to the standard Pap test. HPV types 6 and 11 are the principal ones found in common anogenital warts (condyloma acuminata). HPV 16, 18, 31, and 33 have been associated with dysplasia, CIS, and invasive cervical carcinomas.

Dysplasia and carcinoma in situ are now considered to be a part of a spectrum of disease that is referred to as the CIN system (cervical intraepithelial neoplasia). This new descriptive Pap system is replacing the older system of classifying Pap smears into classes I through V. (See "Test Results" below for clarification.)

Pap smears can be used for detecting a variety of other conditions, including vaginal lesions, endometrial cancer, vaginal infections, and a woman's hormonal status. There are, however, better laboratory techniques than the Pap test for each of these clinical problems.

TEST INDICATIONS

There is currently a controversy about the proper indications for the use of the Pap smear in screening for cervical cancer. The long tradition has been to encourage women to have Pap testing annually because of the low cost and moderate sensitivity of the test. The American Cancer Society has developed newer guidelines based on the fact that there is usually a slow progression from dysplasia to cancer. It has stated that testing should be done every 3 years except for women at high risk for cervical cancer. The Society's current recommendations are:

- To start Pap testing at the age of 20 years or at the onset of sexual activity.
- After two normal Pap smears, to change

from an annual test to one every 3 years until the age of 65 years. However, in the face of changing sexual activity it is important to reassess your patients' risk status, especially with regard to the number of partners.

- After the age of 40 years, pelvic examination should be done annually to detect other gynecologic diseases, but Pap testing should be done only every third year.
- More frequent (unspecified) testing is indicated for women at high risk for developing cancer, including women who have had early sexual intercourse or multiple sexual partners, women with a history of genital herpes, women from a lower socioeconomic group, and women with a previous abnormal Pap test.

It might be argued that the inclusion of these high-risk characteristics keep heterosexually active women in the high-risk category. In 1982 the Canadian Task Force Report clarified and modified its 1976 report by suggesting that all heterosexually active women aged 17 to 35 years should be screened annually.

The American College of Obstetricians and Gynecologists has argued in favor of continued annual Pap testing. Their arguments include the fact that not all invasive cancer shows a slowly progressive course. In addition, they argue that the Pap test has a significant false-negative rate that can be minimized by more frequent testing.

The controversy over Pap frequency has confused both the public and clinicians. There can obviously be no "right" answer because the argument is based on a cost-benefit judgment. Rather than continued debate, energy and dollars might be better spent in providing Pap testing to the many women who receive no regular examination rather than arguing the advantages of 1-year *v* three-year testing.

The clinical uses for Pap testing include:

- Regular testing (1 to 3 years) for every woman over the age of 20 years.
- Regular testing for any heterosexually active woman under the age of 20 years.
- Prenatal testing in all pregnant women.
- Postpartum testing. This should be done at 8 to 12 weeks postpartum rather than at the 6-week checkup because earlier testing may show cellular atypia that is a part of the normal healing process.
- Regular testing for all women taking birth control pills.
- Regular testing in any woman who has had a hysterectomy for the treatment of a gynecologic cancer.
- Testing for any patient with abnormal vaginal bleeding.
- Regular testing in any woman with prior abnormal Pap smear or cryosurgery, cauterization, or laser surgery of the cervix.
- Pap testing of vaginal specimens for women exposed to diethylstilbestrol (DES) to examine for vaginal adenosis and clear-cell carcinoma.

TEST MATERIALS

The supplies needed for performing a Pap smear specimen collection are usually available free of charge from the cytology laboratory that processes the specimens for your office.

Microscope Slides

Two microscope slides with frosted ends are needed. The patient's name should be written on each slide. One slide should be labeled with a "C" for cervical and the other with an "E" for endocervical.

Slide Holders

These are used for returning the slides to the pathology laboratory.

Fixative Spray

This is available in either an aerosol or pump container. A spray fixative is preferred over a solution because of its convenience and because cells may fall off the slide if it is placed in a jar of fixative.

Vaginal Speculum

These come in different sizes (virginal, regular, large) and in disposable plastic or autoclavable metal.

Ayre Spatula

The Ayre Spatula is made of wood and is 7 inches long. It has one end that is used to collect the ectocervical (i.e., cervical) sample and a rounded end to use for collecting a vaginal wall specimen (Fig. 19–1).

Devices for Collecting the Endocervical Samples

- Sterile Swabs: The swabs are used to collect the endocervical specimen. A synthetic fiber swab, such as a Calgiswab III (Inolex Corp, Glenwood, IL) is preferred because this fiber is nonabsorbent. If a cotton swab is used, it should be moistened first with saline and then squeezed to prevent the cells from adhering to the cotton when the swab is streaked on the microscope slide. Swabs are made by winding a fiber in one direction onto a stick. The swabs will retain their shape when used only if they are rotated in the same direction as they were originally wound. This is usually in the clockwise direction.
- Milex spatula: This is a 9-inch long plastic spatula. It has a smooth-curved edge for the ectocervical sample and a longer, pointed end for the collection of the endocervical sample. It is used instead of a fiber swab. The Milex spatula comes in two sizes—medium (white) and small (yellow). These spatulas can be ordered from Milex Products, Inc, 5915 Northwest Highway, Chicago, IL 60639 (800-621-1278; Fig. 19–1).
- Cytobrush: A special brush designed for collecting endocervical cells is now available. It does a much better job of collecting the endocervical specimens than other methods. It has not, however, been adequately studied in pregnant women and therefore is not recommended in this group. It may produce some spotting after use. It can be ordered from International Cytobrush, Inc, PO Box 7733, Hollywood, FL 33081 (800-235-7697) and International Cytobrush, Inc—Canada, 302 Spadina Ave, Suite 406, Toronto, Ontario M5T 2E7 (800-387-9044). Cytobrushes are more expensive than the usual swabs; however, the elimination of

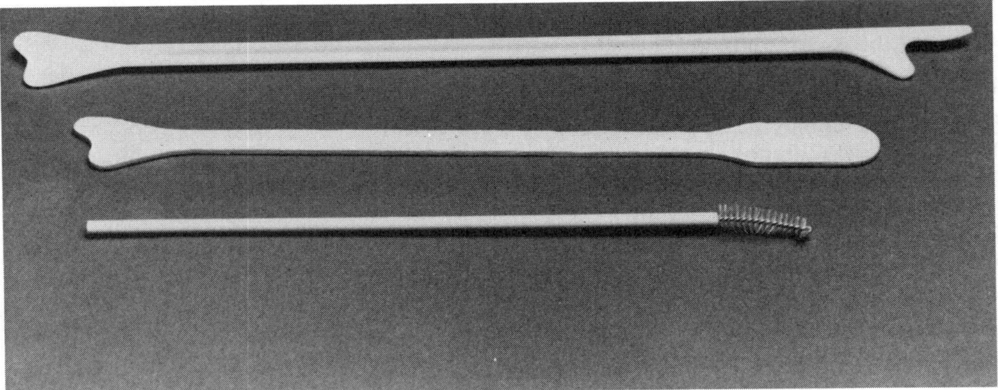

Figure 19–1. A Milex spatula (**top**), an Ayre spatula (**middle**), or cytobrush (**bottom**) may be used for collecting Pap smear specimens.

repeat Pap smears caused by inadequate sampling makes them quite cost-effective. These brushes are not to be used for endometrial sampling (Fig. 19–1).

Result Cards

Every woman who has a Pap test done should be informed of the results even if they are normal. A convenient way to report a normal Pap result is to have printed cards that say: "Your Pap test is normal. Return to the office for your regular examination on (date)." The date can be written into the blank space. Put these in an envelope for mailing to protect patient confidentiality. Abnormal Pap results are best reported by telephone so that the patient's questions can be answered.

SPECIMEN COLLECTION

1. Check that the patient is not menstruating. Menstrual blood will interfere with an accurate cytology interpretation because the red blood cells (RBCs) will overwhelm the slide, making it difficult to see other cells. It is best to advise your receptionist to schedule patients for Pap testing during their midcycle. Patients should also be advised not to douche or use a tampon or any vaginal medication for the 24-hour period before the office visit for a Pap test.
2. Always collect the Pap smear before doing the bimanual examination.
3. Label the microscope slides with the patient's name and a "C" (cervical) and an "E" (endocervical).
4. Moisten a vaginal speculum in warm water. Do not use a lubricant, as this may interfere with satisfactory specimen collection.
5. Place the patient in the usual lithotomy position.
6. Gently insert the speculum and locate the cervix.
7. Examine the vagina and note any abnormalities. Vulvar and vaginal lesions can be scraped and sent for Pap testing, but biopsies are preferred for these suspicious lesions.
8. Before collecting the endocervical specimen, it is important to clean off any mucus from the cervix. Use at least two swabs to do this. If you are collecting a specimen for gonorrhea or beta streptococcal culture, these first two swabs can be used for this. Otherwise, discard the swabs.
9. Next, collect the endocervical specimen. It should be collected before the ectocervical specimen because ectocervical sampling may result in some bleeding, which can interfere with the endocervical sample.
10. Collection of endocervical specimen:
    a. Swabs. If a cotton-tipped swab is used, moisten it with saline and squeeze the tip to remove excess moisture. If an artificial fiber-tipped swab is used, there is no need to premoisten.
        (1) Place the tip of the swab into the cervical os.
        (2) Let it stay there for a few seconds.
        (3) Rotate this swab a full 360 degrees while applying some pressure to abrade some other cells.
        (4) Remove the swab and smear it onto the E slide—making sure that you rotate the entire swab surface across the slide.
    b. Milex Spatula:
        (1) Place the pointed edge of the spatula into the os.
        (2) Rotate it a full 360 degrees while applying some pressure to abrade some of the cells.
        (3) Remove the spatula and smear it onto the E slide,

making sure that you remove specimen from both sides of the Spatula.

c. Cytobrush:
   (1) Gently insert the cytobrush into the endocervix until only the bristles closest to the handle are exposed.
   (2) Slowly rotate one full (360-degree) turn. Do not press down as you have to with a swab because this can cause the endocervix to bleed.
   (3) Remove.
   (4) With moderate pressure, roll and twist the brush across the E glass slide.

11. Immediately spray the slide with fixative. Even 10 seconds of air drying before fixation will distort cells. Monitor pap smear reports for comments like "smear dried before fixation" to evaluate your nurse's technique.
12. Collect the ectocervical specimen using the Ayre spatula or the non-pointed end of the Milex spatula. Tuck the longer lip of the spatula into the cervical os and gently rotate the spatula completely around the cervix.
13. Smear the collected sample evenly onto the microscope slide labeled C and immediately spray the slide with fixative.
14. If a discrete lesion is seen on the cervix, it should be scraped with the spatula and streaked onto a third microscope slide. Label the third slide and make a note to the pathologist about the gross appearance and location of the lesion.
15. Advise the patient that the specimen collection may result in some painless spotting for a day or two.
16. Complete the remainder of the pelvic examination.
17. Tell the patient that the results of her Pap test will be called or mailed to her and the approximate time that the results should be available.

TEST RESULTS

Pap smear cytology should be viewed as a continuum from normal to definite malignancy. The two ends of this spectrum are clear, but the center includes a variety of conditions, including inflammatory responses, precancerous conditions, and early malignancy. A Pap smear should therefore not be considered to be either strictly positive or negative, and patients need to be educated that Pap testing is *not* a yes/no "cancer" test, rather it is a screen for precancerous as well as cancerous lesions.

An older system for cytologic reporting of Pap smears was based on five classes of cells:

1. Class I: Absence of atypical cells.
2. Class II: Slight cytologic atypicalities, no suspicion of malignancy.
3. Class III: Atypical cells indicating marked hyperplasia, dysplasia, or possible malignancy.
4. Class IV: Markedly atypical cells indicating carcinoma in situ or invasive malignancy.
5. Class V: Cytology conclusive for malignancy.

This system is now being replaced by a descriptive report that tells the clinician exactly what cellular changes are seen. This new system places the responsibility on the clinician rather than on the pathologist for deciding what is the clinical significance of particular cytologic findings. This new classification includes such descriptions as:

- Findings inadequate for diagnosis.
- Findings essentially normal.

- Atypical cells present, suggestive of (specify).
- Findings consistent with CIN grade 1 (i.e., mild dysplasia).
- CIN-1 *v* condyloma.
- Findings consistent with CIN grade 2 (i.e., moderate dysplasia).
- Findings consistent with CIN grade 3 (i.e., severe dysplasia or carcinoma in situ).
- Invasive squamous cell carcinoma.
- Endometrial carcinoma.
- Other cancer (specify).
- Condyloma present.
- Atypical Condyloma present.
- Suspicious for CIN.

In addition to the classification, the report will include a variety of other descriptive terms, including:

Benign Reactive Changes

This category covers a range of conditions that the cytology laboratory considers to be totally benign and that do not require follow-up, including squamous metaplasia, epithelial response to inflammation, nuclear enlargement, parabasal cells, and hyperplasia of endocervical cells.

Hyperkeratosis

This condition indicates the presence of keratinized squamous cells without nuclei. If hyperkeratosis is found along the squamocolumnar junction, it is associated with dysplasia in 10% of patients. It is therefore entirely normal 90% of the time. There is some difficulty in making the distinction between benign hyperkeratosis and dysplasia because only surface cells are sampled with the Pap smear, and the deeper cells may be atypical. If the reports of hyperkeratosis persist for 1 year (two or three reports), the patient should be referred for colposcopy. When the referral is made, be sure to indicate that the problem is hyperkeratosis.

Parakeratosis

This condition is similar to hyperkeratosis, except that the nuclei are retained in the cell. If noted, it should be handled in the same manner as hyperkeratosis.

Koilocytosis

This refers to the vacuolization and enlargement of the epithelial cells. It is considered pathognomonic for genital warts. This is evidence for such an infection, even if no warts are clinically seen.

Dyskeratosis

This refers to a mismatch between the nuclear and cytoplasmic maturation of the cells. It is the commonest finding of condyloma but may also represent invasive carcinoma. If no other findings suggestive of carcinoma are found, the reading is usually "suggestive of condyloma."

TEST QUALITY CONTROL

Pap smear staining and cytologic interpretation are specialized techniques that will be done by a reference laboratory. The Clinical Laboratory Amendment of 1988 specifically addressed issues of Pap smear quality control.

It is the responsibility of the referring clinician to check that the reference cytology laboratory is adequately staffed to perform Pap smear interpretation. The laboratory should meet the standards of the American Society of Cytology and the College of American Pathologists. Their standards include staffing by certified cytotechnologists and supervision by a cytopathologist. A cytotechnologist should examine no more than 100 slides per day and should not be paid by the slide. This payment practice encourages serious problems. The laboratory should process at least 25,000 smears per year. The supervisory pathologist must have had cytopathology training. Twenty-five percent of pathologists have had no such training.

SOURCES OF ERROR

False-Negative Test Results

- Inadequate cervical or endocervical sampling technique.
- Sampling from only one site. An ectocervical specimen alone will yield a false-negative rate of 20%. With the addition of the endocervical specimen, this drops to 5%.
- Use of a "vaginal pool" specimen. This specimen technique is associated with a 50% false-negative rate.
- Improper smear fixation.
- Tissue necrosis or infection at the site of an obvious cervical cancer. These conditions can obscure the cytologic diagnosis.
- Errors in specimen handling or laboratory reporting.

False-Positive Test Results

- Presence of cellular atypia due to infection, hormone therapy, or postpartum reparative process if done too soon after delivery.
- Vaginal douching or coitus immediately prior to sampling.
- Errors in specimen handling or laboratory reporting.
- Displacement of abnormal cells from original slide onto another slide during the staining process.

COMMENTS

- If the vaginalysis indicates a large number of white blood cells (WBCs), determine the cause and treat appropriately. Do not send the Pap smear at that time, since severe inflammation can produce some atypia. There is no reason to unnecessarily commit a woman to the work up of an abnormal Pap smear. Have her come back after treatment for her Pap test.
- In addition to noting cellular atypia, the Pap smear report may note the presence of infections with *Candida albicans*, herpes simplex, *Condylomata acuminata*, and *Trichomonas vaginalis*. The Pap test is, however, neither a sensitive nor specific test for these infections. Discuss the presence of these findings with the reporting laboratory.
- It is important to educate patients about the purpose of the Pap test and to discuss the various results with them. To many women a positive Pap means "invasive cancer." They are unaware of the progressive stages of the lesions and of the various treatment alternatives.
- Pap smears can also detect endometrial cancer, especially when an endocervical sample is taken. This cancer is primarily seen in menopausal patients.
- A Pap smear from the vaginal walls near the cervix should be done on women who were exposed in utero to DES (stilbestrol). These patients are at increased risk for vaginal adenosis and clear-cell carcinoma.
- Each laboratory will have a slightly different pattern of Pap smear reporting. Check with the pathologist in your reference laboratory about the classification system. If a report is received with a term that is unclear to you, call the pathologist for clarification.
- Fifty percent of Pap smears that show CIN lesions are found in women with normal appearing cervixes.
- The office laboratory should keep a log of all abnormal Pap results. It should include the patient's name, chart number, collection date, and report description. This will allow the clinician to more easily recall patients who require more frequent follow-up examinations (Fig. 19–2).
- There is a growing concern over the collection of adequate endocervical smears. A number of reference laboratories now assign categories of specimen quality. If

Patient's name	Clinician's name	Date of Pap Test	Descriptive Cytopathologic Report	Follow-up

Figure 19–2. Abnormal Papanicolaou test record.

______________________________ ________ Year % Satisfactory ________

Clinician Year

Month	Satisfactory	Unsatisfactory	% Satisfactory
Jan.	Total ______	Total ______	
Feb.	Total ______	Total ______	
March	Total ______	Total ______	
April	Total ______	Total ______	
May	Total ______	Total ______	
June	Total ______	Total ______	
July	Total ______	Total ______	
Aug.	Total ______	Total ______	
Sept.	Total ______	Total ______	
Oct.	Total ______	Total ______	
Nov.	Total ______	Total ______	
Dec.	Total ______	Total ______	

Figure 19–3. Clinician monthly/yearly adequate specimen-collection record.

the report says "smears dried before fixation," discuss the fixation techniques with your nurse. "Unsatisfactory endocervical specimen" is due to too few endocervical cells. Endocervical cells are one indication that an adequate specimen was taken. Patients with an abnormal history whose endocervical specimens are inadequate should be recalled for a repeat smear.

- One method to monitor the quality of specimen collection is to keep a log of all Pap smears. Each month calculate the percentage of satisfactory specimens by clinician. If the clinician has a low specimen adequacy rate, he or she should review the collection technique (Fig. 19–3).
- There is a high false-negative rate for Pap smear results after the initial smear showing CIN. This can be due to either the regression of the disease or to an inadequate specimen collection. Do not discount abnormal Pap smears as laboratory error or pathologist "over-reading."

REFERENCES

Barrasso R, DuBrux J, Croissant O, Orth G. High prevalance of papillomavirus associated intraepithelial neoplasia in sexual partners in women with cervical intraepithelial neoplasia. *N Engl J Med.* 1987;317: 15

Briggs R. Early neoplasia of uterine cervix. *Obstet Gynecol Surv.* 1979;34:70.

Brinton LA, Fraumeni JF. Epidemiology of uterine cervical cancer. *J Chron Dis.* 1986;39:1051-1065

Canadian Task Force. Cervical cancer screening programs. *Can Med Assoc J.* 1976;114

Canadian Task Force. Cervical cancer screening programs: summary of the 1982 report. *Can Med Assoc J.* 1982;127

Crum, Ikenberg, Richart, Gissman: Human papillomavirus type 16 and early cervical neoplasia. *N Engl J Med.* 1984;310 : 4

Dunn L. Cervical cytologic evaluation. *Postgrad Med.* 1979;65:187

Grubb G. Human papillomavirus and cervical neoplasia: Epidemiological considerations. *Int J Epidemiol.* 1986;15 : 1

Hurt WG. Cervical cytology: Use and follow-up. *J Fam Pract.* 1978;7:579

Jobson V. Cervical intraepithelial neoplasia. *Am Fam Physician* 1981;24:179

Reissman S. Comparison of two papanicolaou smear techniques in a family practice setting. *J Fam Pract* 1988;26: 5

Richart R, Barron B. Screening strategies for cervical cancer and cervical intraepithelial neoplasia. *Cancer.* 1981;47(Suppl 5):

SECTION 3: Vaginalysis*: A Systematic Examination of Vaginal/Cervical Secretions

BACKGROUND

It is misleading to continue to think only of vaginitis when discussing the examination of vaginal/cervical secretions. Cervical secretions are part of the specimen collected from the vagina. It is therefore crucial to realize that a patient's complaint of vaginal discharge may reflect any of a wide spectrum of pathology—from simple uncomplicated vaginitis to PID. Patients with vaginal/cervical infections usually complain of vaginal discomfort, itching, increased discharge, dysuria, and/or foul odor. These infections are also a frequent cause of a female's dysuria and can be mistaken for a urinary tract infection.

The most common etiologic agents for simple vaginitis are *Candida albicans, Gardnerella vaginalis* (formerly *Corynebacterium vaginale* or *Haemophilus vaginalis* or nonspecific vaginitis) and *Trichomonas vaginalis.* These infections can occur singly or in any combination. Cervicitis is a second common problem and can be caused by a long list of pathogens. The most common pathogens are *Neisseria gonorrhea, Chlamydia trachomatis,* and the herpes simplex virus. Allergy, contact irritants, mechanical irritation, and postmenopausal atrophy can also produce vaginal and cervical problems. Again, it is important to remember that the vaginal specimen is actually made up of both vaginal and cervical secretions!

The accurate diagnosis of vaginal/cervical infections or inflammation is usually made by combining information from the medical history (e.g., Did she recently take a course of antibiotics?), the sexual history (e.g., Are there new sexual partners or multiple partners?), the physical examination (e.g., Is there a friable cervix?), and the microscopic examination of secretions. The gross appearance of the vaginal/cervical secretions is often misleading, and multiple infections are very common. The specific diagnosis is therefore best established by in-

*This term was originally coined by Robert Gwyther, MD, Department of Family Medicine, University of North Carolina, Chapel Hill.

cluding a microscopic examination of vaginal/cervical secretions.

The systematic examination of vaginal/cervical secretions (i.e., the "vaginalysis") can provide a great deal of information. The procedure described here involves standardized collection and examination techniques. In addition, this approach begins with a clear image of what is normal (Fig. 19–14). The vaginalysis is more than just a "yes/no" test for *Trichomonas,* yeast, or *Gardnerella*.

TEST INDICATIONS

- A patient with vaginal pain or itch.
- A patient with heavy or foul-smelling discharge.
- A patient with dysuria or urinary frequency, especially if a urinalysis does not show pyuria or bacteriuria.
- A patient with a urine sediment that contains white cells *and* many squamous epithelial cells, thus making it impossible to know whether the patient has a vaginitis or urinary tract infection.
- To determine the course of inflammation in a patient whose Pap smear shows an inflammatory response.
- A follow-up test in a woman after treatment for a vaginal/cervical infection.

TEST MATERIALS

1. Two 6-inch long, sterile, Q-tip brand swabs. (This brand has proven to be the most reliable. Its sturdy wooden sticks are especially important in obtaining a patient-collected specimen.) It is important that the two swabs be used together to ensure a good, cellular specimen.
2. Plastic urine centrifuge tube.
3. Dropper bottle with normal saline.
4. Plain (unfrosted) microscope slides.
5. 22 × 30-mm glass coverslips.
6. Reagents for *Candida* examination: The function of these reagents is to "clear" the specimen of things that might confuse the examiner. For example, the edges of epithelial cells in a sheet of squames can look like pseudohyphae. The reagents clear WBCs, RBCs, trichomonads, and squamous cells from the specimen and permit easier identification of *Candida*.
   a. 10% KOH solution.
   or
   b. Swartz-Lamkin fungal stain. This is available from Dermatologic Lab and Supply Company, Council Bluffs, IA 51501. This is a special mixture of KOH, Parker ink, and other chemicals. It can be used with a 1% Rose bengal solution (available from scientific supply companies) to highlight the Candida against a counterstained background for easier identification.
7. A grease pencil is used to write the patient's name on the specimen tube before taking it to the laboratory.
8. Six-inch wooden applicator sticks.
9. Plastic transfer pipettes.

SPECIMEN COLLECTION BY THE CLINICIAN

Some clinicians prepare the vaginal specimen by placing it directly onto the microscope slide. This results in a sparse, dried-out specimen that makes the diagnosis of *Trichomonas* infection especially difficult. (Trichomonads lose their motility in dry specimens). The following is a standardized collection technique that yields a reliable specimen):

Vaginal Specimen

1. Write the patient's name on a plastic centrifuge tube.

2. Place 1 mL of normal saline, from a dropper bottle, into the plastic tube.
3. Expose the cervix and vaginal mucosa with a speculum. Note the quantity of secretions and the appearance of the vaginal mucosa and the cervix.
4. Hold the two cotton Q-tip swabs together and swab the mucosa along the vaginal walls and the vaginal vault. If secretions have pooled elsewhere in the vagina, swab these areas as well.
5. Quickly place the swabs into the centrifuge tube. Avoid touching the edges of the tube with the cotton swabs so that all of the specimen is placed into the saline in the bottom of the tube.
6. Send the specimen to the laboratory for examination.

SPECIMEN COLLECTION BY THE PATIENT

This self-collection technique yields a sample that is as good as a physician-collected specimen. It does not eliminate the need for a pelvic examination. Rather it may permit more efficient patient care because a symptomatic patient can be instructed to collect the specimen before being seen by the physician. (See patient instructions at the end of this chapter, Fig. 19–26 A–C) This can eliminate the need to perform a second pelvic examination to obtain additional specimens suggested by the vaginalysis' findings (e.g., WBCs in the vaginalysis requiring a test for *Chlamydia* or GC).

TEST PROCEDURE

Following the recommendations of the Centers for Disease Control (CDC), wear gloves while preparing and examining slides.

Slide Preparation

1. Vigorously mix the swabs in and out of the saline. Be sure to mix in any of the specimen that adheres to the sides of the tube (Fig. 19–4).

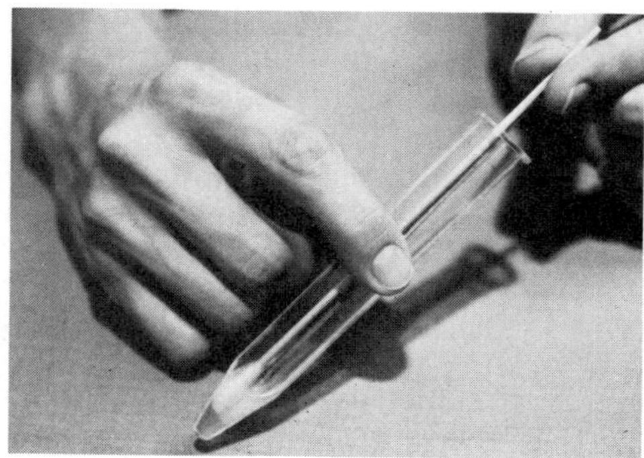

Figure 19–4. Vaginal specimen swabs and saline in a plastic tube.

2. Lay a clean microscope slide on the bench top.
3. Take a plastic transfer pipette or a 5-mL Pasteur pipette with a small rubber bulb and squeeze it several times to remix the specimen.
4. Suck up fluid with the transfer pipette.
5. Deposit *two* drops onto the far right side of the slide. (Some people may choose to transfer the specimen with the collection swabs. If this is done, be sure to get a large, moist drop onto the slide.) Coverslip this area with a 22 × 30-mm coverslip.
6. Deposit *one* drop onto the far left side of the slide.
7. Return the pipette to the plastic tube.
8. Three methods are given for preparing the *Candida* specimen for examination. The use of the two stains provides the easiest specimen for identifying pseudohyphae. None of these methods requires that the specimen be heated, since the cells will be chemically cleared or counterstained while you are examining the saline side of the slide.
   a. 10% KOH: Add one drop of KOH

to the specimen on the left. Take out a clean wooden applicator stick and stir the mixture.

b. Swartz-Lamkin and Rose Bengal stains: Add one drop of Swartz-Lamkin stain to the specimen on the left. Take a clean wooden applicator stick and dip it into the bottle of 1% Rose Bengal solution. Use the reddened end of this stick to stir the mixture.

c. Swartz-Lamkin alone: Add one drop of Swartz-Lamkin stain to the specimen on the left. This will highlight the fungi but not as well as when the Rose Bengal is used as a counterstain.

9. Before you coverslip the second specimen mixture, smell it—the so-called "whiff" test. If there is a fishy odor, the whiff test is positive. This is considered a sign for a *Gardnerella* infection.

10. Coverslip the specimen on the left side of the microscope slide with a 22 × 30-mm coverslip (Fig. 19–5).

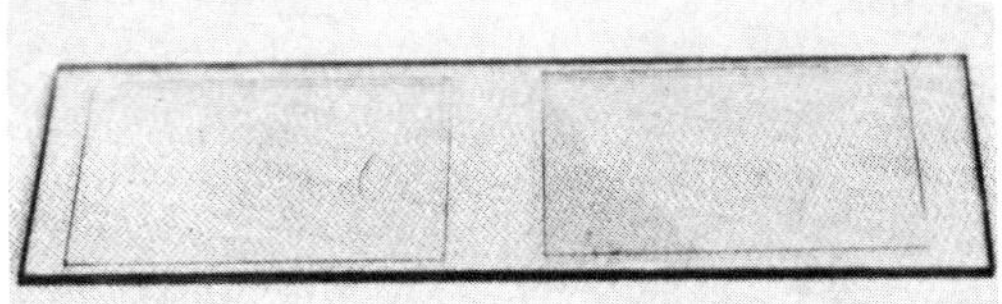

Figure 19–5. Microscope slide with both a specimen-saline and a KOH wet preparation.

MICROSCOPIC EXAMINATION

1. Move the low power (10x) objective into place.
2. Place the prepared slide on the microscope stage between the slide-holding grips.
3. Lower the condenser and turn the light source to its lowest setting.
4. Move the saline-specimen mixture (right-hand side of the slide) so that it is under the objective. Focus the microscope.
5. Scan the saline-specimen mixture under low power to get a general impression of the specimen. Examine at least ten fields. *Do not make any final decisions about the specimen while examining it under low power. Even if you identify one type of vaginal pathogen, continue a systematic examination of the cells, bacteria, etc.*
6. Switch to the high-dry (40x) objective. Refocus and slightly increase the light intensity. Examine at least ten fields. You may immediately see one of the organisms that causes vaginitis. If so, note the finding but *continue* to examine the specimen in the systematic fashion described here. This prevents you from missing a second or third important finding.
7. EVALUATE THE BACTERIA IN THE FLUID AROUND THE CELLS:
   a. *Note the morphology of the bacteria:* Are there rods, cocci, coccobacilli, or a mixture? Easily identifiable rods are the normal dominant flora. These are lactobacilli (Fig. 19–6). Small coccobacilli are indicative of *Gardnerella vaginalis.* This organism is so small that it may be mistaken for background debris (Fig. 19–7). A patient with *Gardnerella* vaginitis usually has only very few, if any, lactobacilli. Sometimes chains of cocci will be seen. These suggest a possible group B streptococcal infection.
   b. *Note the quantity of bacteria.* Are there a few organisms per high-power field ("light bacteria")? Is the field packed ("heavy bacteria")? Is it somewhere in between ("moderate

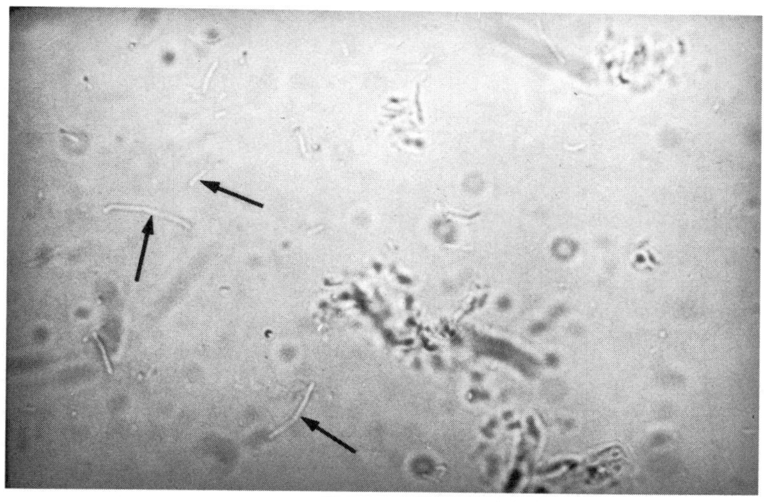

Figure 19–6. Notice the distinctly rod-shaped bacteria in this specimen. These are Lactobacilli. Viewed under high-dry objective.

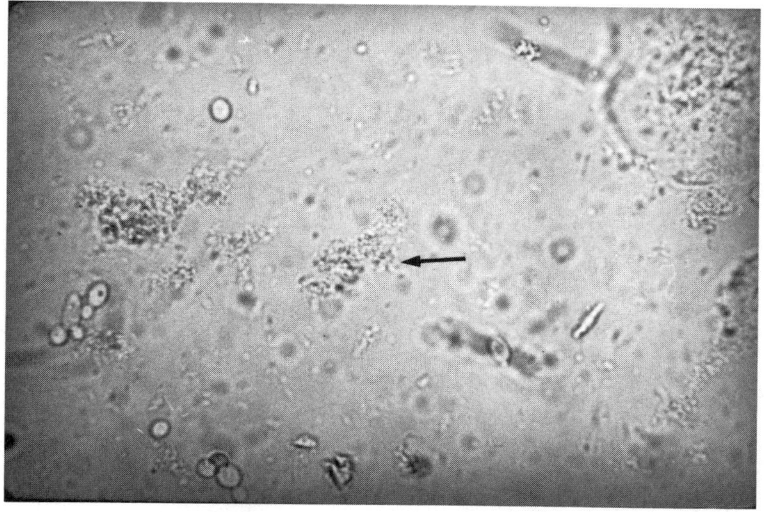

Figure 19–7. In this specimen you do not see the distinct rod-shaped bacteria of Figure 19–6. Rather, you see much smaller bacteria singly or in clumps. This is what Gardnerella vaginalis looks like. Viewed under high-dry objective.

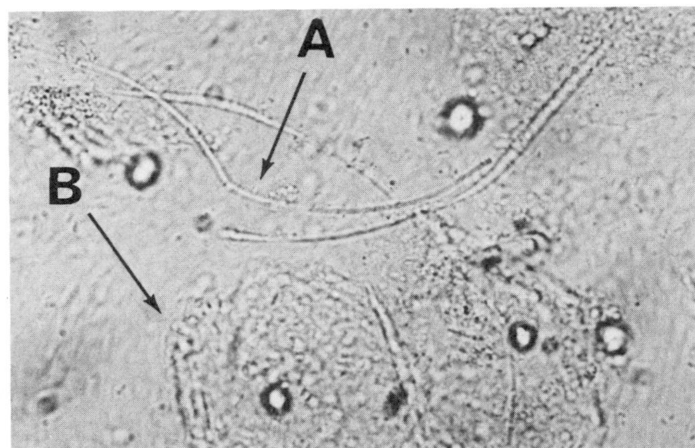

Figure 19–8. A. Leptothrix in a vaginal wet preparation. Note that it is at least twice as long as a squamous epithelial cell. **B.** Squamous epithelial cell. Viewed under high-dry objective.

bacteria'')? It is important to note when no organisms are present. This may indicate that the patient has douched before coming to the office. In this case the cause for the vaginitis may not be found even after a thorough vaginalysis.

You will occasionally see very long, thin rods, at least 2½ times as long as the diameter of a squamous cell. These are most likely *Leptothrix* (Fig. 19–8).

On low power they can be confused with *Candida* pseudohyphae (Fig. 19–9). However, *Leptothrix* is not as wide when viewed under high power, and it is not a branching organism (Fig. 19–10). We have only seen *Lepto-*

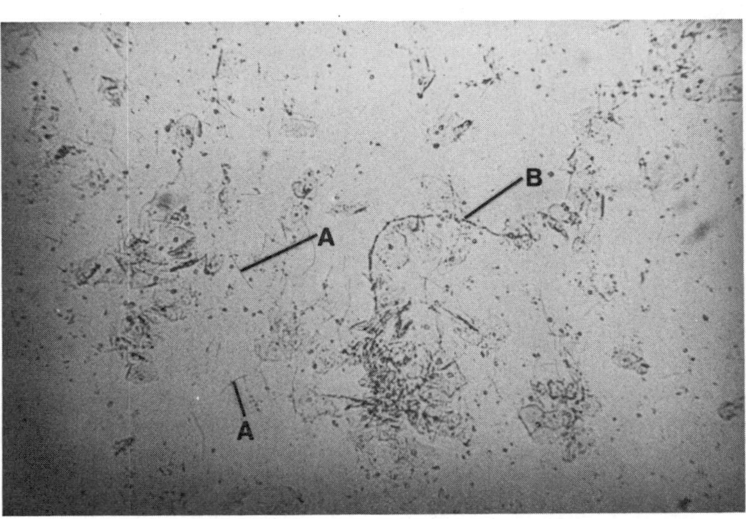

Figure 19–9. A. Leptothrix. **B.** Yeast. Viewed under low-power objective.

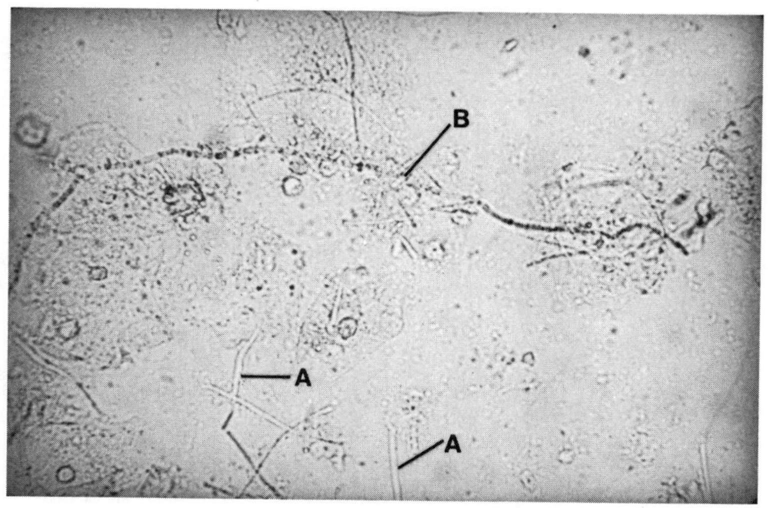

Figure 19–10. A. Leptothrix. **B.** Yeast. Viewed under high-dry objective.

thrix in the presence of *Trichomonas* vaginitis, although, *Trichomonas* is usually seen without seeing *Leptothrix*. Its role in causing vaginitis is unclear; however, it can be used as a marker for a *Trichomonas* infection.

8. EVALUATE THE EPITHELIAL CELLS:
   a. ***Note the type of epithelial cells present:*** The lining of the adult vagina is composed of different maturation layers of squamous epithelial cells (Fig. 19–11). Note the relative thickness of those layers. Also note the size relationship of the various cell types and common organisms in Figs. 19–12 and 19–13.
   b. ***Note the number of epithelial cells per high-powered field.*** Report if there are sheets of squamous cells. Frequently the clinician will suspect a *Candida* infection because of the presence of a white, clumpy discharge. These clumps are often only sheets of sloughed squamous cells with no *Candida*. At other times there will be mats of pseudohyphae mixed in with the squamous cells. It is a good practice to carefully examine the edges of these sheets of squamous cells for protruding *Candida* pseudohyphae.

      Superficial and intermediate squamous epithelial cells are the typical squames that are usually seen in vaginal secretions and in vaginally contaminated urine specimens. They are large cells with small nuclei (Fig. 19–14).

      Parabasal cells come from the deeper layers of the vaginal epithelium (Fig. 19–15). They are considerably smaller than normal squames and have a large central nucleus. They are similar in appearance to the transitional epithelial cells seen in urine sediments (Fig. 19–16). These cells appear in vaginal secretions as a

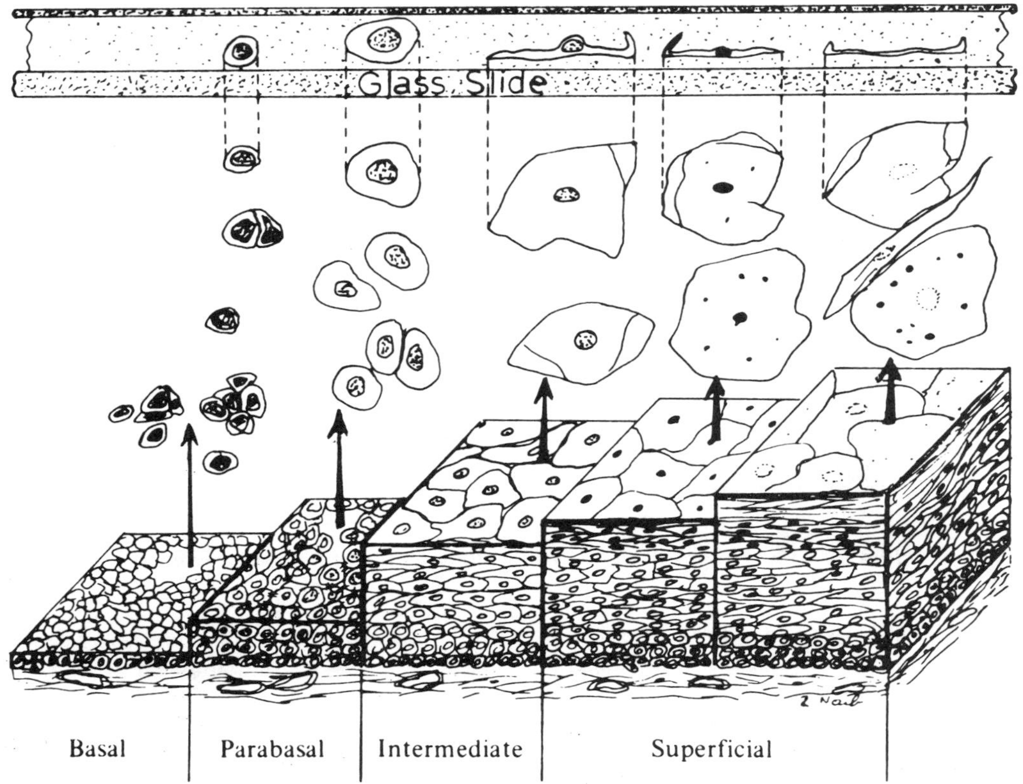

Figure 19–11. Different histologic layers of vaginal stratified squamous eipthelium. (From Naib Z. *Exfoliative Cytopathology,* 3rd ed. Boston, Mass: Little, Brown; 1985.)

result of traumatic exfoliation (i.e., the destructive shedding of the outer cell lining caused by irradiation or irritating douches) or physiologic exfoliation (seen with prepuberty, lactation, or estrogen deficiency).

Basal cells form the deepest cell layer and are smaller than parabasal cells (Fig. 19–17). They are slightly larger than WBCs and have a striking central nucleus. These cells are rarely seen in normal vaginal secretions. They usually indicate a specimen from an extremely atrophic or deeply ulcerated vaginal mucosa.

c. ***Note the character of the squamous cells:*** Normal appearing squamous cells have well-defined edges and look "clean" (i.e., the surface of the cell does not have anything on it and the edges are sharp; Fig. 19–14).

Clue cells are squamous cells with the coccobacilli *Gardnerella vaginalis* stuck to the outer surface of the cell (Fig. 19–18). The bacteria appear to be more heavily concentrated along

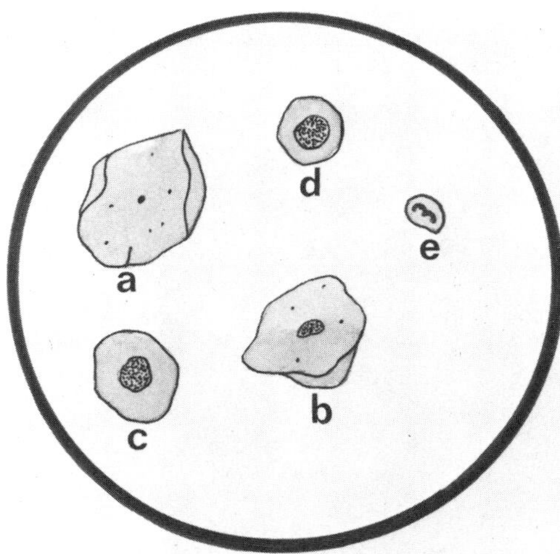

Figure 19–12. Line drawing of cells seen in vaginal wet preparation. **A and B.** Mature squamous epithelial cells. **C.** Parabasal epithelial cell. **D.** Basal epithelial cell. **E.** WBC.

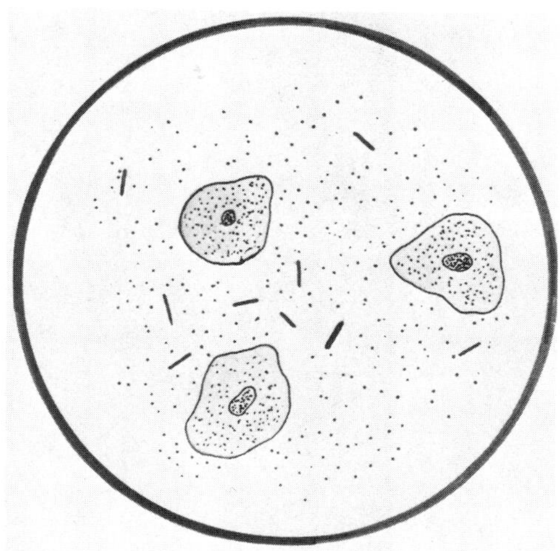

Figure 19–13. Line drawing of clue cells and gardnerella. Note the paucity of normal rods. Approximates the view under high-dry objective.

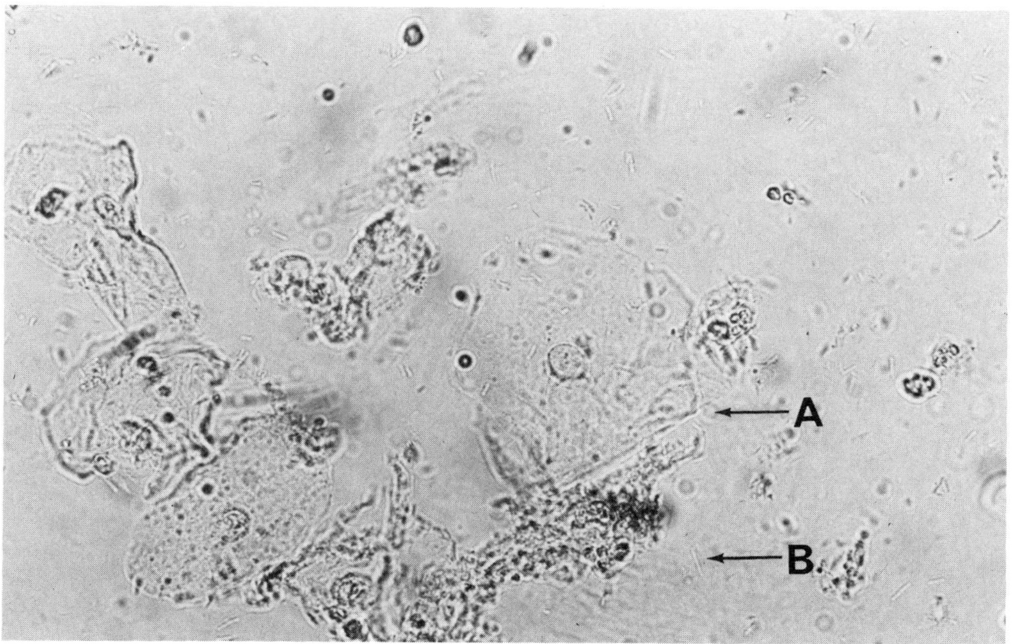

Figure 19–14. Normal vaginal specimen with **A** squamous epithelial cells and **B** lactobacilli. Note how clean these epithelial cells appear compared to those in Figures 19–18 and 19–19 viewed under high-dry objective.

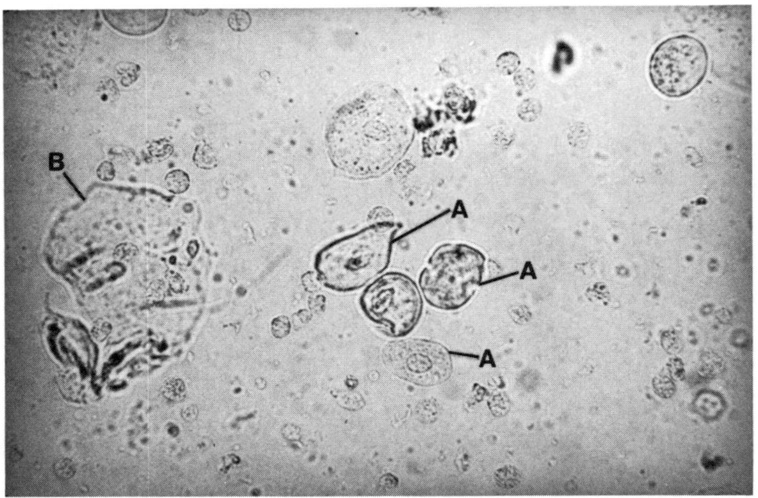

Figure 19–15. A. Parabasal cells. **B.** Mature squamous cells. Viewed under high-dry objective.

the outer cell edge. Frequently only a small part of the entire cell surface will be "clued" (Fig. 19–19). These cells appear refractile. One useful distinction is that normal squamous cells have well-defined (i.e., "clean") edges, while clue cells have ragged edges.

Vacuolated squamous cells are often confused with clue cells. These cells contain small dark spots. When you focus up and down it becomes obvious that the spots are small vacuoles within the cell and not the Gardnerella coccobacilli stuck on the surface of the cell (Fig. 19–20).

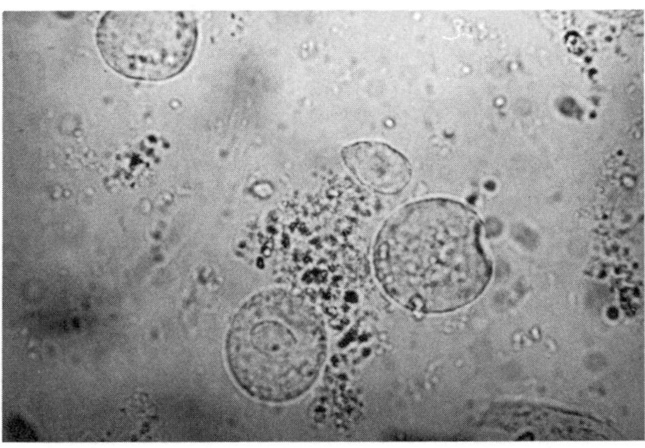

Figure 19–16. Swollen transitional cells from the urinary tract. Viewed under high-dry objective.

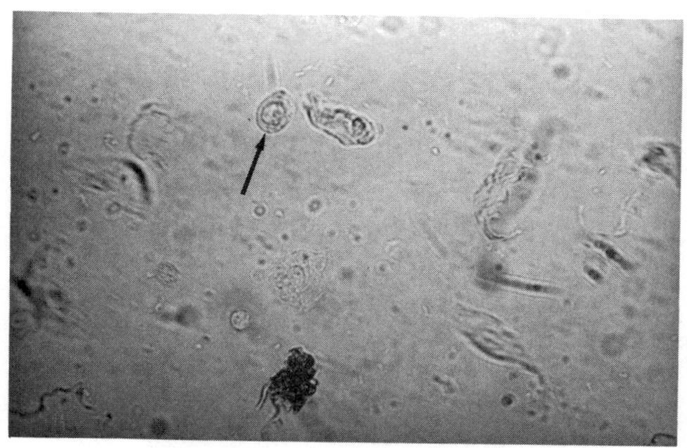

Figure 19–17. Basal cell. Viewed under high-dry objective.

Vacuolated squamous cells are probably degenerating cells. They are often seen in follow-up specimens in women who have been treated with metronidazole or an antifungal medication.

9. EXAMINATION FOR WBCs:

These are small cells with either clearly segmented nuclei or with cytoplasmic granules and an indistinct nucleus (Fig. 19–21). They look the same as white cells in urine. There is little known about the occurrence of WBCs in vaginal secretions and their correlation with either disease or the menstrual cycle. Many specimens from asymptomatic patients have no WBCs. In contrast, the presence of many WBCs is usually associated with either a vaginal or cervical infection. It is unclear whether low numbers of these cells (i.e., one to two per HPF) can be considered normal. We believe that the presence of white cells requires an explanation.

Gardnerella vaginitis is not associated with white cells. Therefore the presence of both clue cells and white cells suggest a second infec-

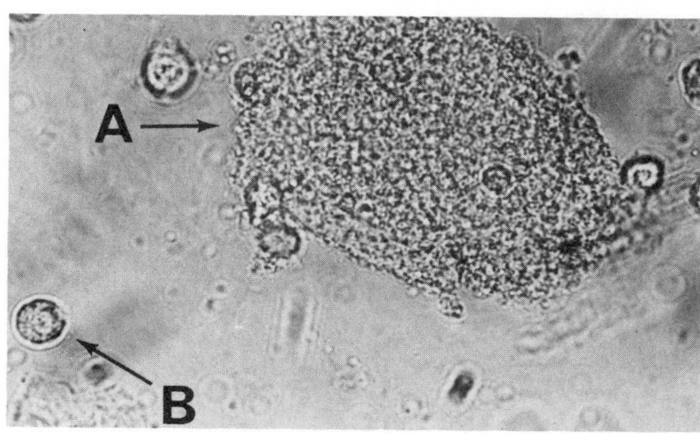

Figure 19–18. A. Clue cell. **B.** WBC in vaginal wet preparation. Viewed under high-dry objective.

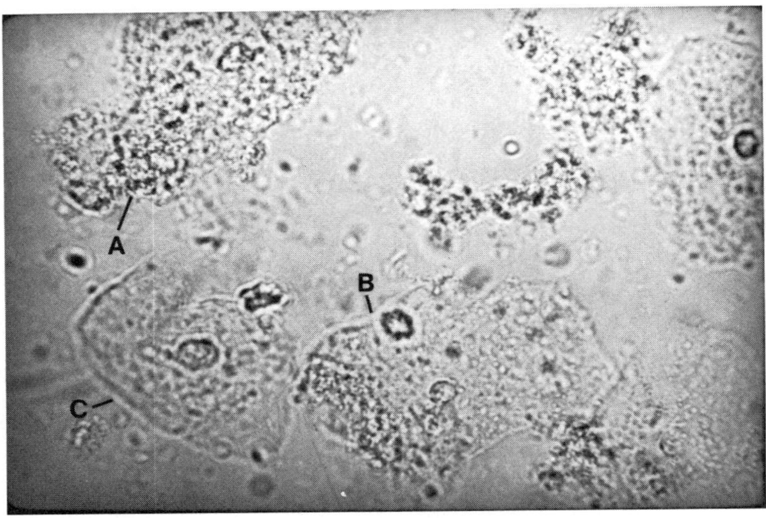

Figure 19–19. A. Clue cell. **B.** Partially clued cell. **C.** Clean squamous cell. Viewed under high-dry objective.

tion in addition to *Gardnerella*. Both *Candida* and *Trichomonas* infections produce a white-cell response. Those infections may be seen with few white cells; however, more commonly you will see many white cells per HPF.

If many white cells are seen but neither *Candida* nor *Trichomonas* are seen, consider evaluating the patient for a cervical infection (i.e., *Chlamydia, N gonnorrhea,* or herpes simplex virus). Other causes of white cells in vaginal secretions include intrauterine devices (IUDs), postpartum reparative process, severe atrophic vaginitis, and an allergic reaction to spermicides or douches. *Any more than five white cells per HPF needs to be explained.*

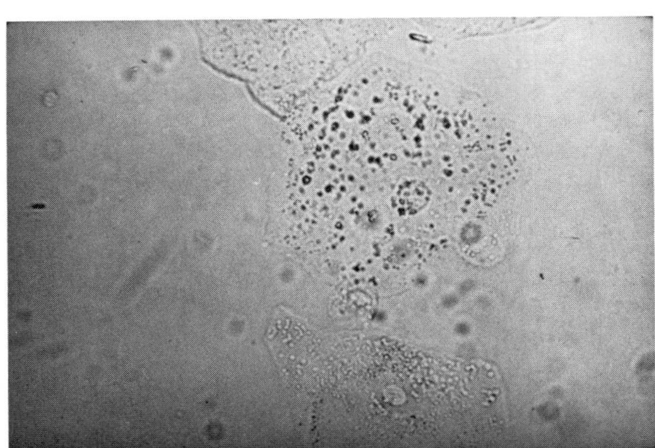

Figure 19–20. Vacuolated normal squamous cells. The vacuoles are small, round holes. Viewed under high-dry objective.

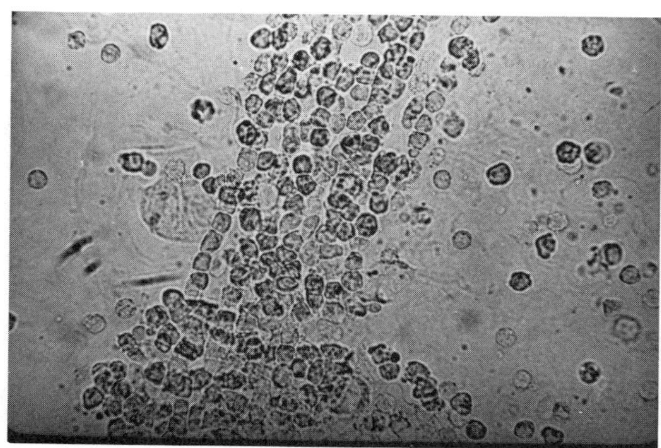

Figure 19–21. WBCs. Viewed under high-dry objective.

10. EXAMINATION FOR TRICHOMONAS:

    These organisms vary in size (from very small to larger than a white cell) and shape (pear shaped, round, or triangular). Photographs are not very satisfactory to show trichomonads because it is this organism's movement that is crucial for making the identification (Fig. 19-22 and 19-23). If they are very motile, it is easy to identify them under the low-power objective (10x). If they are barely motile, it may be difficult to identify them even under the high-power objective (40x). Sometimes they round up and look exactly like white cells. Then it is only the slight flick of the flagella that allows you to make the distinction. *Trichomonas* can provoke a strong white cell response. It is therefore advisable to carefully examine clumps of white cells that may surround and attack a *Trichomonas*. If you see *Leptothrix* on a slide, spend more time looking and you will likely find *Trichomonas*.

11. EXAMINATION FOR *CANDIDA ALBICANS:*

    a. If you have not seen any pseudohyphae while examining the saline-specimen mixture, switch back to the low-power objective, reduce the light, and center the

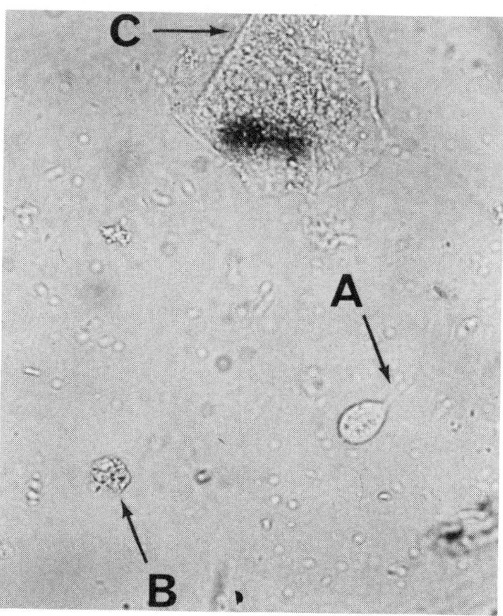

Figure 19–22. A. *T. vaginalis.* The dark line is pointed at the flagellum, which is very difficult to see. When viewed under the microscope, the flagellum would be moving. **B.** WBC. **C.** Squamous epithelial cell. Viewed under high-dry objective.

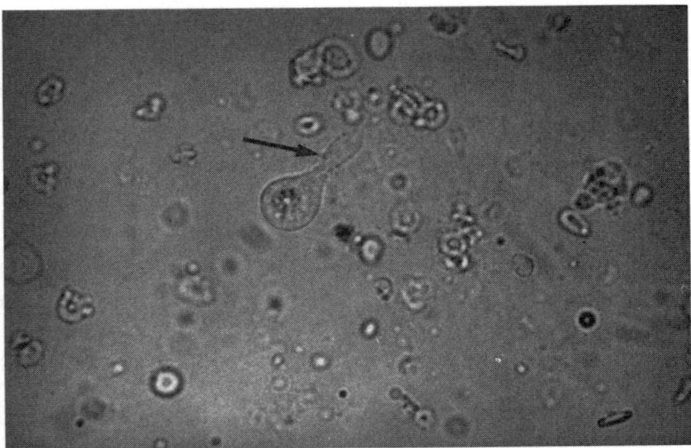

Figure 19–23. Trichomonad. Viewed with high-dry objective.

KOH/stain mixture (left side of slide) under the low-power objective.

b. When you have focused the microscope, move the specimen so that you find the edge of the coverslip. Move into the specimen and quickly and systematically scan the specimen for suspicious elements. This is done by moving up/down a column of the field. When you come to the top or bottom of the coverslip, move the slide over one field's width, then start scanning in the opposite direction. Some elements may look like pseudohyphae under low power but are obvious artifacts under high power.

c. When you find a suspicious element (Fig. 19–24) switch to the high-power objective and refocus. *Candida* pseudohyphae appear as long, somewhat thin, branching organisms (Fig. 19–25). It is important to confirm your low-power finding under high power.

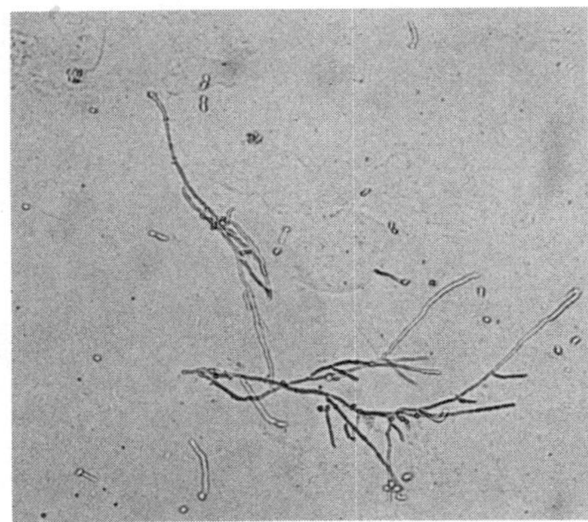

Figure 19–24. *Candida* pseudohyphae in KOH preparation. Viewed under low-power objective.

COMMENTS

- There are a variety of tests that have been proposed for evaluating patients with vaginal complaints. Articles often mention pH testing, bacterial or fungal cultures, and even cytologic staining. The vaginalysis (i.e., a standardized collection and examination procedure) plus a thorough sexual/medical history and physical examination will provide sufficient information to treat most vaginal infections and will provide clues for the specific testing needed to identify cervical pathogens.

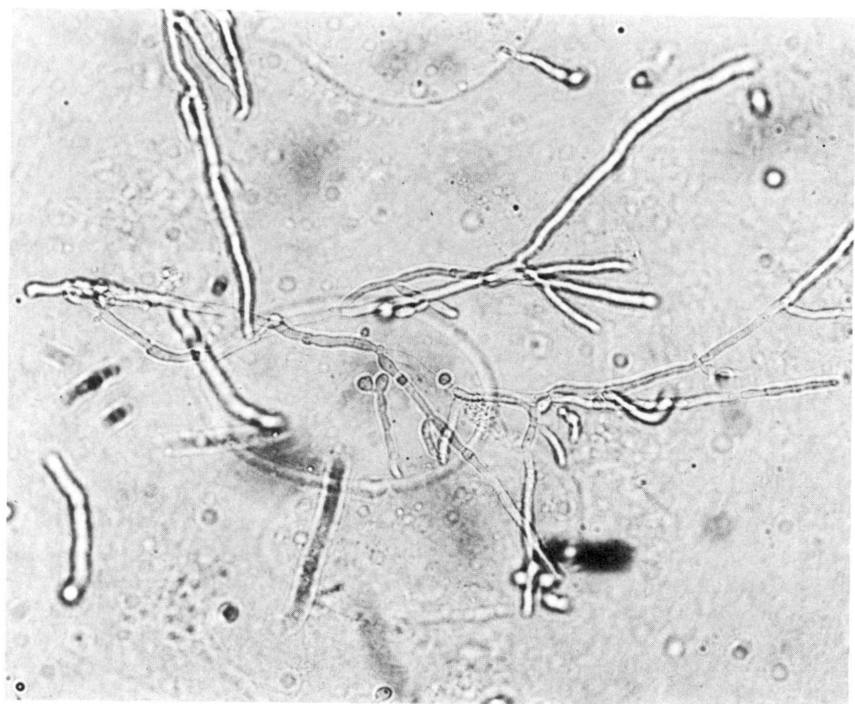

Figure 19–25. *Candida* pseudohyphae in KOH preparation. Same field as Figure 19–24. Viewed under high-dry objective.

- The wet preparation has often been criticized for a high false-negative rate. There has, however, never been a standardized collection and examination procedure. The collection of an adequate sample is crucial. Using two cotton-tipped swabs to collect the secretions yields an abundant specimen. A small amount of normal saline (1 mL) prevents overdilution.
- Remember to consider cervical infections as well as vaginal infections when working up a patient complaining of a vaginal discharge.
- If patient complains of urinary tract symptoms and her urine sediment shows both white cells and more than five squamous cells/HPF, consider performing a vaginalysis. This will help to establish the source of the white cells and to rule out a possible vaginal/cervical infection.
- Because no white cells are evoked by *Gardnerella,* the infection has been more accurately described as a "vaginosis" instead of a vaginitis. This term is being used in the more recent literature.

CASE HISTORIES

To illustrate the essential interplay between the laboratory and other clinical information, we offer several case histories.

Case #1

Background.

A 23-year-old, sexually active, heterosexual female came to her physician for her annual Pap smear and pelvic examination. She expressed some concern to the nurse about a vaginal infection and mentioned a recent radio program on chlamydial infections. The nurse asked her to collect a specimen for a vaginalysis. The laboratory report showed five to ten white cells per HPF, no RBCs, moderate squamous cells, normal appearing flora, no *Candida,* no *Trichomonas,* and no clue cells.

The physician then obtained the history that the patient was on birth control pills and used no barrier methods to prevent sexually transmitted diseases. She had a new sexual partner within the last 2 months.

On physical examination a cervical discharge was noted. The cervix appeared friable and bled when the Pap specimen was collected. The physician collected a specimen for a *Chlamydia* test, which was positive.

Comment.

The presence of the white cells, the friable cervix, and a new sexual partner all led the clinician to suspect *Chlamydia.* The patient's vaginal symptoms were explained by her cervical infection.

Case #2

Background.

A 27-year-old sexually active lesbian saw her physician for problems with vaginal burning and itching. She told the nurse that she probably had a yeast infection. The nurse asked her to collect a specimen for vaginalysis. The vaginalysis showed sparse normal flora, a few white cells, many parabasal cells, and only an occasional normal squamous cell.

This woman was neither perimenopausal nor nursing a baby—the two most common physiologic explanations for the appearance of parabasal cells in vaginal secretions. Because of the sparse flora and the parabasal cells, she was asked if she had recently douched. She said that to cure her suspected yeast infection she had decided to treat herself by douching with a peppermint soap purchased at a local health-food store. It was after the douching that the painful burning occurred.

On physical examination she had red and irritated external genitalia. The vaginal mucosa and cervix were also inflamed. There was no obvious discharge from the cervical os.

Comment.

The presence of parabasal cells clearly indicated that the normal top layer of the epithelium had been sloughed. The clinician knew to seek an explanation for the parabasal cells. Douching with strong chemicals is frequently responsible for these findings.

REFERENCES

Gwyther RE, Addison LA, Spottswood S, et. al. An innovative method for specimen autocollection in the diagnosis of vaginitis. *J Fam Pract.* 1986;23:487-488.

Naib Z. *Exfoliative Cytopathology.* 3rd ed. Boston, Mass: Little, Brown; 1985.

Instruction sheet for patient self-collected vaginal specimen (Fig. 19–26) can be duplicated for your patients.

SECTION 4: Urine Pregnancy Test

BACKGROUND

The diagnosis of pregnancy can often be made by patient history and physical examination. The history may reveal a missed period, recent morning sickness, or tender breasts. The examination may show an enlarged uterus or softening of the cervix (Goodell's sign). However, in early pregnancy the diagnosis may be difficult, and a pregnancy test is required.

Pregnancy tests detect human chori-

Patient Instructions for self-collected vaginal secretions specimen

Please read the following carefully. If you have any questions, the nurse will be happy to assist you.
You will be given:

1. A plastic tube containing a small amount of normal saline solution with your name on it; and
2. A package of 2 sterile 6-inch cotton tipped swabs.

To collect a specimen:

1. Take the plastic tube, the package of swabs and this instruction sheet into the bathroom.
2. Place a plastic tube with the normal saline solution in the tube holder which is near the toilet.
3. Wash and dry your hands.
4. Either sit on the toilet or stand—whichever is more comfortable for you.
5. Open the package of swabs from the stick end (not the cotton tipped end) of the package and remove the 2 swabs.
6. Hold the 2 swabs in one hand, about 2 inches from the wooden end of the swabs. With the other hand, spread apart your genital lips. 19–26 **A**
7. Gently insert the cotton tipped ends deeply into your vagina. *Stop if you feel any discomfort* 19–26 **B.**
8. Move the swabs in a circle 5 times. *You should not feel any pain or discomfort* 19–26 **C.**

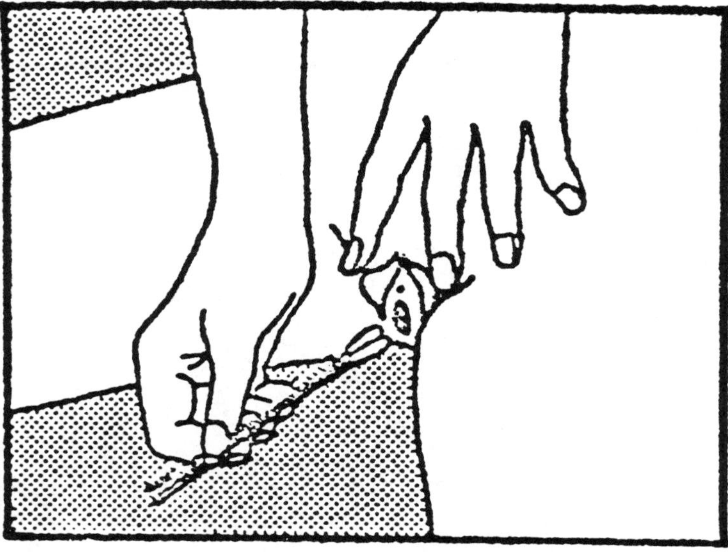

Figure 19–26A. Hold the two swabs in one hand about 2 inches from the wooden end of the swabs. With the other hand, spread apart your genital lips.

(Continued)

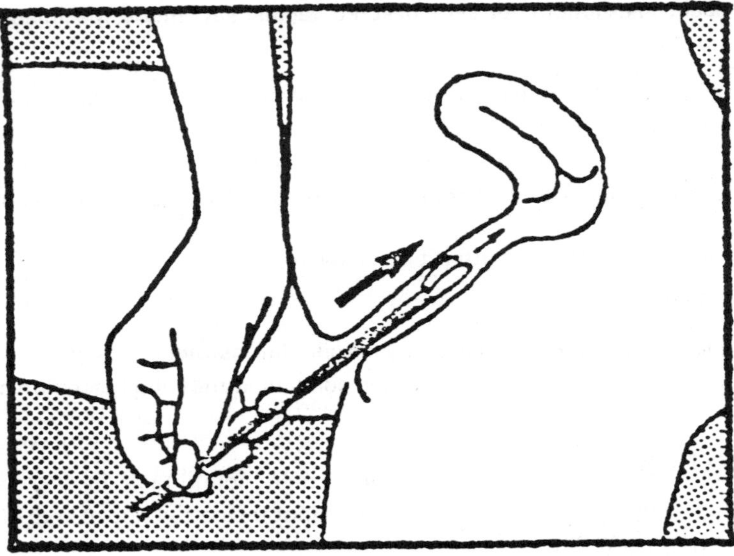

Figure 19–26B. Gently, insert the swabs deeply into your vagina.

9. Remove the swabs and put them both into the plastic tube of saline solution so that the cotton tipped ends are in the solution. Try to avoid hitting the sides of the tube.
10. Wash and dry hands.
11. Give the plastic tube with the swabs to the nurse.

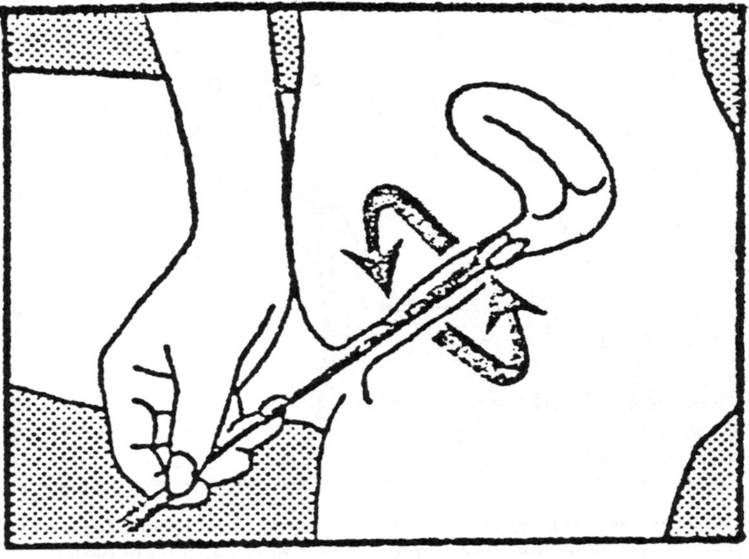

Figure 19–26C. Move the swabs in a circle, five times.

onic gonadotropin (hCG). This hormone is produced by the placenta early in pregnancy and is excreted in the urine. Testing for hCG has undergone a technologic revolution in the past decade. The currently available tests are easier to run, have less problems with interfering substances, have built-in controls, have easy-to-identify endpoints, and can detect pregnancy almost a week prior to the missed menses. While not totally "foolproof," these tests are marked improvements over the insensitive latex agglutination tests that were available 10 years ago.

TEST INDICATIONS

- Confirmation of a diagnosis of pregnancy in the early first trimester.
- Identification of pregnant patients before ordering medications (such as progesterone) or x-rays that could injure the fetus.
- Ruling out pregnancy in women who are to be fitted for an IUD or are being started on birth control pills.
- Evaluating patients with lower abdominal pain to rule out ectopic pregnancy.

SPECIMEN COLLECTION

- The patient who calls the office and asks for a pregnancy test should be requested to collect a first-morning urine specimen in a clean container. Neither a clean-catch technique nor a sterile container is required. Early morning specimens are preferred, since the urine will be concentrated and will therefore have a higher concentration of hCG.
- Any urine sample collected later in the day can be used for pregnancy testing. Commonly this specimen is collected during the patient's visit to the office. It should be noted that the test may be falsely negative if the urine-specific gravity is too low.

TEST MATERIALS

Many of the current generation of pregnancy tests rely on enzyme-linked immunosorbent assays (ELISA). In a typical test monoclonal antibodies to hCG are fixed to filter paper. When the patient's urine is passed through the filter, any hCG that is present binds to the antibody. Next, a solution containing a second antibody to hCG and that has been bound to a color development system is added. This binds to a different portion of the hCG, which in turn is bound to the antibody on the filter paper. Finally, a color reaction is chemically initiated. The presence of color indicates that hCG is present and therefore that the patient is pregnant.

TEST PROCEDURE

The details for performing a pregnancy test will vary with the particular test kit that is purchased. Specific instructions are included with each package insert and should be closely followed. Some procedural comments apply to most brands of pregnancy tests:

1. A specific gravity test should be performed on all urines used for pregnancy testing. This will aid the clinician in evaluating negative results. A urine with a specific gravity less than 1.010 may lead to a false-negative test, especially early in pregnancy.
2. Hold all reagent droppers or pipettes perfectly upright when delivering a drop of reagent or sample. This ensures reproducibility of the volume delivered.
3. Do not mix the reagents from different kits or lots.
4. Some kits have shortened expiration dates if they are not stored in the refrigerator. If so, be sure to write the calculated expiration date on the kit box after it is opened.

5. All reagents must be brought to room temperature before the testing is done. In many practices the test-kit box is kept in the refrigerator during the night and then is put onto the bench top first thing each working morning.
6. Mix each reagent vial before use. This keeps the reagents from settling out.
7. Do not mix up the caps for the different reagent bottles. This can lead to contamination of the bottles.
8. If the urine specimen is cloudy or there is an obvious sediment, centrifuge the specimen and take the sample from the supernatant. This is because many urinary sediments are colored, and these can produce falsely positive tests. hCG is *not* centrifuged out, since it is in solution.

QUALITY CONTROL

Many of the currently available pregnancy tests have built-in positive and negative controls in the test system. In this way the controls are run each time a patient is tested. If such a test kit is used, then no other reagent quality control is required. We therefore recommend this type of pregnancy kit.

This position may be viewed by some traditional laboratorians as unreasonable, since past recommendations were to do daily runs with urines that were positive or negative for hCG. However, hCG is hCG! Therefore the positive control that is built into the filter paper accomplishes the same thing as a urine that has been spiked with hCG. In addition, a separate negative control is hard to justify, since the two-antibody reagent system and the shift from agglutination end points makes nonspecific false-positive reactions very unlikely. The most common reason for false-positive or false-negative tests are not reagent problems but rather are due to problems in test interpretation. The most common false-positive test is found in a person who has had a very early spontaneous abortion. The test correctly identifies the presence of hCG; however, the person is no longer pregnant. Likewise, the most common reason for a "false-negative" test is a patient who has conceived but in whom the pregnancy is too early for the hCG level to be detectable. This situation is much less common with the very sensitive pregnancy tests that are available today.

TEST SENSITIVITY

Pregnancy can be reliably detected by very sensitive methods as early as 6 to 9 days following conception. The hCG concentration then doubles every 2 to 3 days. Many of the current generation pregnancy tests can detect 20 mIU/mL of hCG. This corresponds to reliably detecting pregnancy nearly a week prior to the missed menstrual period. If pregnancy is suspected but the test is negative, then the patient should be asked to return in 1 to 2 days with a first-morning urine sample. The hCG concentration in the first-morning urine is about equal to that in the serum. If that test is also negative, then it is unlikely that the patient is pregnant. The sensitive urine pregnancy tests make a qualitative serum-pregnancy test unnecessary.

hCG is present in the urine of most women with ectopic pregnancies, although at a lower concentration than with normal intrauterine pregnancies. Pregnancy tests that are sensitive to 20 mIU of hCG are likely to indicate ectopic pregnancy in most women who have ectopic pregnancies. Studies have shown a 95% to 99% positive rate in those patients. If, however, the pregnancy test is negative and the clinical suspicion is still high for ectopic pregnancy, a diagnostic ultrasound is indicated.

Routine pregnancy tests are not useful in evaluating fetal death. The tests may, in fact, remain positive for up to 4 weeks after an abortion. Quantitative serum hCGs are the preferred tests to follow patients with suspected fetal death or threatened abortions. A normal intrauterine pregnancy should have a 66% increase in quantitative serum hCG over a 2-day period of time. The best way to handle this is to hold the first specimen 48 hours until both can be sent for testing. This prevents problems with between-run test variability.

TEST SPECIFICITY

The older latex agglutination pregnancy tests were often falsely positive because of proteinuria or patient medications (i.e., the phenothiazines). These problems are not found with the newer ELISA pregnancy tests. A false-positive test is occasionally seen because the patient has a tumor that secretes hCG. These tumors include choriocarcinomas, hydatidiform moles, lung cancers, or breast cancers. These tests are not really in error, since they are detecting hCG; however, this hCG is not due to pregnancy.

COMMENTS

- In a study of the proficiency testing program for the state of Idaho, half of all office laboratories submitted erroneous or unacceptable results for pregnancy testing on unknown samples. Much of this error was likely due to problems in identifying endpoints with the older latex agglutination tests. Most errors today probably are related to specimen handling and test interpretation.
- When examining the results of a large proficiency testing program (i.e., EXCEL from the College of American Pathologists) it is common to find that with the same sample, some manufacturers' tests report "positive", while other kits report "negative". This is usually not due to errors but rather the fact that different tests detect different levels of hCG. If your laboratory is using a less sensitive method, it should consider changing to one of the newer pregnancy tests.

REFERENCES

Bandi ZL, Schoen I, Delora M. Highly sensitive qualitative methods for serum choriogonadotropin (hCG): Clinical specificity studies. *Clin Chem.* 1987; 33:677-681.

Fletcher JL. Update on pregnancy testing. In: Fischer PM, Addison LA, eds. *Primary Care Clinics in Office Practice.* W. B. Saunders, 13(4):667-677, 1988.

SECTION 5: Testing for Rupture of the Amniotic Membranes

BACKGROUND

The nitrazine and ferning tests are used to detect the leakage of amniotic fluid from the membranes surrounding the fetus during pregnancy. Testing for rupture of the membranes is important because premature rupture of the membranes (i.e., the spontaneous rupture before the onset of labor) leads to a doubling in perinatal mortality after 24 hours and a second doubling within 48 hours. This increase in mortality is due to fetal infection and can be largely eliminated by the artificial induction of labor. The decision to proceed with the induction of labor is frequently dependent on a sensitive test that can detect small quantities of amniotic fluid in specimens taken from the vagina.

TEST INDICATIONS

The test is used for pregnant patients with a history of a watery discharge from the vagina. There is often a ''slow leak'' rather than the expected gush of fluid.

TEST MATERIALS

- Sterile vaginal speculum.
- Sterile cotton-tipped swab.
- Nitrazine pH strips or paper. These use a phenaphthazine indicator to cover a pH range from 4.5 to 7.5. They can be purchased as either individual strips or as rolls. The paper should be stored at room temperature and not used after its expiration date. A color chart is supplied with each package to indicate the pH corresponding to the various colors of the test paper.
- Microscope.
- Clean microscope slides, with frosted ends.

SPECIMEN COLLECTION

1. Write the patient's name on the frosted end of the glass slide.
2. Position the patient in the usual fashion for a pelvic examination.
3. Place a sterile speculum into the vagina. No lubricant or antiseptic should be used, as these chemicals may interefere with the testing procedure.
4. Using a sterile swab, obtain a sample of vaginal secretions from the posterior vaginal pool. Be sure not to touch the mucous plug in the cervix. Mucus can give a false-positive nitrazine test.
5. Touch the specimen to a strip of a nitrazine paper.
6. Immediately after touching the nitrazine paper, rub the swab against the glass slide, creating a very thin smear.
7. Set the slide aside to dry while the nitrazine test is evaluated.

NITRAZINE TEST PROCEDURE

Normal vaginal secretions have a pH of 4.5 to 5.5, whereas amniotic fluid has a pH of 7.0 to 7.5. With rupture of the membranes, the amniotic fluid leaks into the vagina and raises the pH of the vaginal secretions.

1. Examine the nitrazine paper as soon as possible after touching the specimen to it. The color that develops will fade after several minutes.
2. Compare the paper color to the color chart provided with the test paper.
3. A negative test (i.e., no amniotic fluid) will show a yellow-to-olive green color, corresponding to a pH of 4.5 to 6.0. A positive test (i.e., the presence of amniotic fluid) will appear blue green-to-deep blue, corresponding to a pH of 6.5 to 7.5.

SOURCES OF ERROR

The nitrazine test is highly sensitive but not very specific. Most studies have reported about a 5% false-positive rate and a 1% false-negative rate.

False-Positive Results

Specimen contamination from heavy vaginal discharge, blood, cervical mucus, semen, alkaline urine, and soap.

False-Negative Results

These can be produced by prolonged rupture of membranes (longer than 24 hours) or when only a small quantity of fluid has leaked.

FERNING (AMNIOTIC FLUID CRYSTALLIZATION) TEST

Unlike urine or vaginal secretions, amniotic fluid crystallizes to form a fernlike pattern when dried on a microscope slide. This is believed to be due to the relative concentrations of sodium chloride, proteins, and carbohydrates in the fluid.

PROCEDURE

1. Make sure that the specimen has completely dried on the slide. This may require up to 5 minutes. Do not heat the slide to dry.
2. Examine the slide under low power without a coverslip.

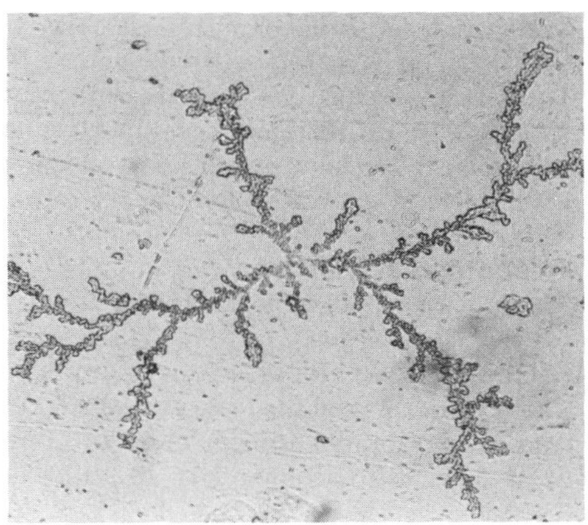

Figure 19–27. Positive ferning test. Viewed under low-power objective.

3. Ferning will usually be seen in most fields if the test is positive. Occasionally it may be difficult to locate, and you will need to thoroughly examine all fields on the slide (Figs. 19–27 and 19–28).

Figure 19–28. Positive ferning testing. Viewed under high-dry objective.

SOURCES OF ERROR

The ferning test is less sensitive but more specific than the nitrazine test. The test has been reported to have only a 1.5% false-positive rate and a 3.6% false-negative rate.

False-Positive Results

Presence of blood, urine, cervical mucus.

False-Negative Results

These can be produced by prolonged rupture of membranes (longer than 24 hours) or if only a small quantity of fluid has leaked.

COMMENTS

- Both the ferning and nitrazine tests should be performed whenever a patient is tested for rupture of the membranes. If the nitrazine test is negative but the fern test is positive, there is probably rupture of membranes because of the ferning test's greater specificity. If the nitrazine is positive and the fern test is negative, a second specimen should be obtained. Be sure to avoid collecting the sample near the cervical os, as cervical mucus may lead to false-positive results.
- The patient history has often been discredited in the diagnosis of rupture of membranes. This is especially true when intact membranes are seen on clinical examination. However, in one study of patients who had a history compatible with rupture of membranes, the clinical history alone resulted in only a 11.6% false-positive rate (i.e., 89.4% of these patients giving a compatible history were shown to have rupture of the membranes.) In comparison, only 31% of the patients in that study had obviously ruptured membranes by physical examination.
- Other testing for rupture of membranes has included Pap staining for fetal squamous cells, identification of fetal fat globules, and dipstick protein testing. These offer no advantage to the combined use of the nitrazine and ferning tests.

REFERENCES

Friedman ML, McElin TW. Diagnosis of ruptured fetal membranes: clinical study and review of the literature. *Am J Obstet Gynecol.* 1969; 104:544.

CHAPTER 20

Dermatology

SECTION 1: Introduction

Ten percent of all office visits to primary care clinicians are for disorders of the skin. For many of these common dermatologic diseases, a thorough history and examination of the rash are sufficient to make an accurate diagnosis. However, there are several simple office laboratory tests that can be used to confirm the suspected diagnosis. Described in this chapter are the preparation of KOH slides and Tzanck smears, examination for skin parasites, and skin biopsy procedures.

Specimen collection is the single most important factor in being able to demonstrate the causative organism in many dermatologic procedures. This often requires the clinician to be patient, determined, knowledgeable about the disease, and skillful with a scalpel blade. Be sure to pay attention to the specimen collection sections of the following procedures.

PROCEDURES NOT RECOMMENDED FOR OFFICE LABORATORIES

Some dermatologic testing is not recommended for the primary care office laboratory:

Darkfield Examination

Darkfield microscopy is the definitive method for identifying the lesions of primary syphilis. Most office laboratories are not equipped with a darkfield microscope and do not have a person adequately trained to identify the spirochetes. This testing should be referred to a reference laboratory. A telephone call to the state health laboratory can usually help in finding the nearest laboratory with this capability.

Histology

Office laboratories are not set up to perform histologic studies on biopsied lesions. This requires expensive equipment, special stains, and specially trained technicians. Skin biopsies can be easily done in the office, but the microscopic examination should be done by your reference laboratory.

Wound Cultures

Most skin infections can be adequately treated without any laboratory testing or with just the information obtained from a Gram's stain. However, if an unusual organism is suspected (i.e., anaerobes) or the infection is very serious, then a culture is indicated. These cultures should be done by your reference laboratory so that you can be certain of the organism identification and its antibiotic sensitivity.

Fungal Cultures

Many fungal infections of the skin, nail, or scalp can be identified by the physical examination and the KOH smear. Fungal cultures are therefore rarely needed. The occasional specimen that requires culturing can be sent for identification to the reference laboratory.

Biopsy of Suspected Melanoma
Any lesion that is strongly suggestive of melanoma should be biopsied by a specially trained surgeon. Such lesions are pigmented, with irregular borders, uneven density, varied coloration, or a recent change in appearance. Melanomas should be treated by wide excision and serial histology to properly grade the cancer. This is beyond the scope of most office practices.

SECTION 2: Skin Biopsies

BACKGROUND

Many of the small skin lesions that are seen in the primary care office can be biopsied for diagnosis, treatment, or cosmetic concerns. These include nevi, keratoacanthoma, skin tags, and seborrheic keratoses. Biopsies from the active borders of larger lesions or rashes can also be useful when the diagnosis is unclear.

The shave biopsy and punch biopsy techniques can be used in most office situations. A third technique, the excisional biopsy, requires more time and may result in a less satisfactory cosmetic result. It should be used primarily for large lesions and will not be described here. Any pigmented lesion that is strongly suggestive of malignant melanoma should be referred to a surgeon for wide excisional biopsy.

TEST INDICATIONS

- Pigmented lesions that change in shape or color.
- Lesions that recur or fail to heal.
- Lesions that bleed easily or ulcerate.
- Inflammatory rashes for which the diagnosis is uncertain.
- Areas of alopecia with an unclear etiology.

MATERIALS

- Disposable dermatologic punches can be purchased in various diameters from 2 mm to 6 mm.
- Surgical scalpel blades: A No. 15 scalpel is used for shave biopsies. They can be purchased complete with a plastic handle or used without a handle. A less expensive alternative is to break a double-edged razor blade into its two halves. Each half makes a convenient shave biopsy instrument.
- 1% lidocaine.
- Small, curved scissors.
- Monsel's solution or electrocautery to control bleeding.
- Syringes, 3 mL, 5 mL, with 27-gauge needles.
- Specimen containers with 10% formalin preservative.
- Mailing containers for specimen shipment to the reference laboratory.

PROCEDURE: PUNCH BIOPSY

1. Check that the patient is not allergic to lidocaine and does not easily form keloid scars.
2. Have ready a small biopsy container filled with 10% formalin.
3. Choose the lesion to be biopsied or the particular area of a large rash. Avoid areas that have been excoriated, as this tissue may be distorted and useless for the pathologic examination. If the rash is vesicular, choose a small, early, unbroken blister. If the lesion is large, choose an area at the outer advancing border, as this is where the biopsy is most commonly diagnostic.

4. Clean the area with an alcohol wipe and allow it to air dry.
5. Slowly infiltrate the skin around the lesion with 1 or 2 mL of 1% lidocaine using a 27-gauge needle. Most of the pain from the injection is caused by the infiltration of the anesthetic, not the needle puncture. The pain can be minimized by injecting the lidocaine very slowly. Explain to the patient that there will be a pin prick and then some burning as the anesthetic is injected.
6. Stretch the skin in the area to be biopsied, perpendicular to the wrinkle lines. This will minimize scarring and will bring the skin edges of the circular punch defect together when the biopsy specimen is removed.
7. Choose a punch size that will allow you to include a small amount of normal skin in the biopsied specimen.
8. Rotate the punch clockwise and then counterclockwise over the lesion while exerting gentle pressure against the skin. Continue until there is a slight decrease in skin resistance, which signifies that the punch has passed through the dermis and into the subcutaneous fat.
9. Apply gentle traction to the skin adjacent to the biopsy core, then clip the base of the specimen with a small curved scissors. Try to obtain a small amount of the subcutaneous fat with the skin biopsy. Handle the biopsy specimen carefully to avoid damaging the tissue structure. Do not crush the tissue!
10. Immediately place the biopsy specimen into a labeled container with 10% formalin.
11. If a 2- or 3-mm punch is used, no suture is required, and hemostasis can be achieved with pressure or the application of Monsel's solution. For larger punch biopsies, a single suture is required to stop the bleeding and to provide a better cosmetic result.

PROCEDURE: SHAVE BIOPSY

1. Check that the patient is not allergic to lidocaine and does not form keloid scars.
2. Have a specimen container with formalin ready to use.
3. Carefully examine the area to be biopsied. Note the height of the lesion so that the biopsy depth will bring the surface of the biopsied skin down to the level of the adjacent skin.
4. Clean the area with an alcohol wipe and allow it to air dry.
5. Slowly inject 1 to 2 mL of 1% lidocaine into the area under the lesion to be biopsied. This will raise the lesion up above the level of the surrounding skin.
6. Pinch the skin to further raise the lesion above the surrounding skin, and then "shave off" the lesion with a back-and-forth motion using a No. 15 scalpel blade. Alternately, the two ends of a half of a double-edged razor blade can be held between the thumb and index finger. Apply slight pressure so that the blade takes on a slight curve, and then shave off the lesion at the desired height. Feathering the edges of the remaining skin border will produce a better cosmetic result.
7. Immediately place the biopsy specimen in a labeled container with 10% formalin.
8. Use a drop of Monsel's solution or electrocautery to stop any bleeding in the biopsied area.

COMMENTS

- Many reference laboratories provide pathology services for skin biopsies. They are likely to supply the preservative, labels, containers, and mailing tubes. Typically results are available in 7 days, and charges range from $20 to $30 for each specimen submitted. If there is any questions about the pathology results, have the pathologist review the slides and discuss the findings with you.

- If more than one specimen from a single patient is submitted, they can be strung sequentially on a suture. Be sure to record for the pathologist the site of each of the lesions in the order that they will be removed from the suture.
- Small biopsies should be handled very carefully. This avoids crushing, which could otherwise cause tissue changes that make the pathologic examination difficult. Always hold the specimen by its edges.

SECTION 3: Identifying Herpes Infections With Tzanck Smears

BACKGROUND

Herpes viruses include two common pathogens, herpes simplex and herpes zoster, each of which causes a vesicular (blistering) disease in humans. Examination of the cells at the base of these blisters will often show multinucleated giant cells that are produced when the virus invades the squamous cells of the skin. The examination for these giant cells is called a Tzanck smear. This test is helpful in differentiating herpetic rashes from other vesicular diseases.

The five most common types of herpetic infections of the skin include:

Fever Blisters (herpes simplex)

This disease is transmitted through a break in the oral mucosa. There is a 2- to 12-day incubation period before the appearance of the vesicle. The initial infection usually occurs by the age of 2 years and is often subclinical. Ten percent of people with their first infection will develop herpes gingivostomatitis, which is characterized by many painful blisters on the gums and lips. After the primary disease, the virus remains in a latent phase, but clinical disease can recur. Recurrences are usually less severe than the primary infection.

Herpetic Witlow (herpes simplex)

These are blisters filled with a purulent fluid, occurring at the end of a finger. They are often recurrent and are due to oral-hand transmission or a herpes simplex infection.

Genital Herpes (herpes simplex)

This is a sexually transmitted disease that has reached epidemic proportions in the United States. It is now the most common cause of genital vesicles and ulcers. This disease follows the infection pattern of the oral herpetic disease (i.e., a severe primary infection followed by recurrent, less severe infections). Painful ulcers on the vulva and penis are the characteristic physical findings.

Chickenpox (herpes zoster)

This is the disease caused by the first infection with herpes zoster. It is spread by the respiratory route and has a 2-week incubation period. It is very contagious, with an 80% attack rate among household contacts who have not previously had the disease. The symptoms include fever and a pruritic rash. All stages of vesicles can be seen at one time (i.e., papules, blisters, and crusted lesions).

Shingles (herpes zoster)
This is the recurrent form of the zoster infection. The virus remains in the nerves after the chickenpox. It can then be reactivated and produces a painful, blistering rash. The distribution of the rash is along a sensory nerve root.

TEST INDICATIONS

Any blistering or ulcerating rash, especially those on the chest, genitalia, and mouth. The test should be done only if there are fresh, undisturbed vesicles. Excoriated or crusted lesions rarely show a positive test.

MATERIALS

- Alcohol wipes.
- No. 15 scalpel blade.
- Microscope slide.
- Hansel stain: Many people use a Wright's stain for the Tzanck preparation, but we have found the Hansel stain to be superior. It is available in 125-mL bottles and can be kept in an amber dropper bottle for convenient use. The dropper bottle should not be allowed to drop below two-thirds full, as this will cause the stain to become too concentrated. When it drops to this point, discard the remaining stain. Wash the bottle and dropper with stain-removing soap (Eradosol) and let them dry completely. Then refill the dropper bottle from the stock bottle. The stain is available from Lide Laboratories (15422 Cousteau Dr, Florissant, MO 63034).
- Distilled water: This is available in 1-gallon jugs from pharmacies or supermarkets. The water can be transferred to a 125-mL plastic squeeze bottle for easy use.
- 95% Methyl alcohol (methanol): This is available from scientific supply companies. It should be stored in a 125-mL squeeze wash bottle and kept near the laboratory sink. A Coplin jar should also be labeled and filled with methanol.
- Coplin jar: These jars are designed to hold microscope slides while they are being stained. A glass Coplin jar with a screw-on lid is required.
- Microscope.
- Immersion oil.

PROCEDURE

1. Choose a fresh vesicle that has not been traumatized. If no intact vesicle can be found, find an ulcer with a clean, denuded base. For genital herpes in women, a Pap smear specimen from the cervix can also be used. (Fig. 15–22)
2. Excise the roof of the vesicle with a No. 15 scalpel blade to expose the vesicle base. The multinucleated giant cells are found in the cells at the base, not in the blister fluid!
3. Blot the fluid from the vesicle. This is especially important if the fluid is purulent.
4. Scrape the base of the vesicle with the scalpel blade. It may be necessary to pinch the base of the lesion to prevent bleeding into the lesion. Avoid contaminating the specimen with blood.
5. Thinly spread the specimen onto a clean microscope slide.
6. Place the slide for a few seconds into a Coplin jar filled with 95% methanol, or flood the slide with methanol from the squeeze bottle. This will fix the specimen to the slide.
7. Allow the slide to air dry. This will take 1 or 2 minutes.
8. Place the slide on a slide rack over the sink and cover the slide with Hansel stain.
9. After 30 seconds add distilled water to the stain on the slide and allow this to stand for 30 seconds.

10. Pour off the stain and flood the slide with distilled water from a wash bottle. This will remove any excess stain.

11. Flood the slide with 95% methanol from a wash bottle; then drain and air dry.

12. Place the slide on the microscope stage. Check that the condenser is all the way up.

13. Examine the slide under low power to find an area with cells. You may be able to identify the multinucleated giant cells at this power.

14. Place a drop of immersion oil on the slide and examine it under the oil objective.

15. Look for multinucleated giant cells. They are several times larger than the normal epithelial cells and often have bizarre shapes. There will be two or more nuclei. (Color Plate IV, Fig. 2).

COMMENTS

- It is impossible to differentiate herpes simplex from herpes zoster by this technique. These two organisms may occasionally mimic each other clinically.
- If a pregnant woman is found to have a herpes simplex genital infection, she should be closely followed to prevent neonatal herpes. The mother's care should include more sensitive tests for herpes (i.e., viral cultures or antibody tests) on a regular basis during the pregnancy. Tzanck smears alone are inadequate for this clinical use.
- Tzanck smears have been reported to be positive in two out of three herpetic infections when a fresh, vesiculated lesion can be found. Our laboratory has not found such a high positive rate for the test.
- In the past herpes simplex was divided into types 1 and 2 by culture characteristics. This differentiation is not routinely done by many laboratories today. Both types of herpes simplex viruses can cause both oral and genital disease.
- If Hansel stain is not available, the slide should be stained with Wright's stain. Follow the procedure as described in Chapter 15, Section 9.

REFERENCE

Blank H, Burgoon CF, Baldridge GD, McCarthy PL, Urback F. Cytologic smears in diagnosis of herpes simplex, herpes zoster and varicella. *JAMA*. 1951;146:1410.

SECTION 4: Examination of Skin Scrapings for Fungi

BACKGROUND

Fungi cause a number of scaling lesions on the mouth, scalp, and skin. The diagnosis of a fungal infection can be rapidly made by identifying the fungus in a microscopic examination of a KOH slide. This examination should be considered a "superficial biopsy" of the stratum corneum, the skin layer in which the fungi live. The KOH reagent causes the cells of the stratum corneum to degenerate, leaving the hyphae and spores of the fungi to be more easily seen.

TEST INDICATIONS

- Areas of broken hair or baldness, especially with scaling of the scalp.
- Scaling lesion on any part of the body.
- Crumbling, scaling nails or nail beds.
- Excoriated papules.
- White plaques in the mouth.

TEST MATERIALS

1. Scraping instrument: a variety of common office instruments can be used, including a No. 15 or No. 10 scalpel blade, a skin curette, the edge of a microscope slide, or the blunt edge of a finger lancette.
2. Microscope slides.
3. Coverslips (22 × 30 mm and 22 × 50 mm).
4. Alcohol lamp.
5. Soap or alcohol wipes to remove cosmetics or topical medications.
6. Nail clippers or nail file.
7. Reagents: The function of these solutions is to "clear" the specimen of things that might confuse the examiner (i.e., the edges of epithelial cells when they occur in sheets).
   a. 10% KOH solution. This is available from pharmacies or scientific supply companies.
   b. Swartz-Lamkin fungal stain: This is a special mixture of KOH, Parker ink, and other chemicals. It is available from Dermatologic Lab and Supply Company, Council Bluffs, IA 51501. It can be used alone or with a 10% Rose bengal counterstain (available from scientific supply companies). The use of these stains highlights the fungi for easier location and recognition.
8. 6-inch wooden applicator sticks.

SPECIMEN COLLECTION

No matter what the specimen source, it is vital that the scrapings be as thin as possible to allow easy clearing of the skin cells.

Scalp

The scales should be vigorously scraped from the scalp in the area of alopecia. Any broken-off hairs can be plucked for examination.

Skin

Clean the skin with an alcohol wipe to remove cosmetics and oils. The skin should be vigorously scraped. The outer "active" border of the rash is the preferred site. If candidiasis is suspected, the satellite lesions apart from the main rash should also be scraped (Fig. 20–1).

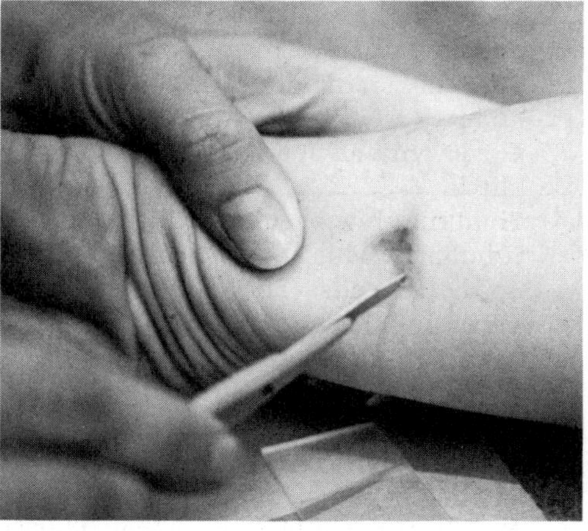

Figure 20–1. Skin scraping technique with a scalpel blade.

Nails

The scales on the surface of the nail or between the nail and the nail bed should be collected. It may be necessary to clip or file away a portion of the nail to collect these subungual scales.

Mouth

The white plaques of candidiasis are adherent to the mucosa of the mouth. A satisfactory specimen can be collected by scraping the plaque with a tongue blade. This will reveal a red, inflamed base to the lesion.

TEST PROCEDURE

1. Place the specimen in the center of the microscope slide.
2. Add two drops of 10% KOH to the specimen

   or

   Add two drops of Swartz-Lamkin stain

   or

   Add two drops of Swartz-Lamkin stain. Add the Rose bengal counter stain by taking a clean, wooden applicator stick and dipping it into the bottle of Rose bengal. Use the reddened end of the stick to stir the Swartz-Lamkin stain on the microscope slide.
3. Cover with an appropriate size coverslip.
4. Hold the slide over a flame, just until it bubbles. This allows clearing of the epithelial cells. The cell edges of a partially cleared specimen can be mistaken for fungi.
5. Examine the slide with the 10x objective, low intensity light, and a lowered condenser. The fungi can be missed if the light intensity is too high. If a suspicious area is seen under the 10x objective, re-examine this field with the 40x objective.

TEST RESULTS

Tinea Versicolor

The fungal spores and hyphae of the organism *Malassezia furfur* will give a "spaghetti-and-meat balls" appearance. The spores appear in clusters, are thick walled, and are round or slightly rectangular. The hyphae are long, branched, and twisted rods. They are about as wide as a spore (Fig. 20–2).

Tinea Capitis

The fungi that cause this scalp infection appear either as hyphae parallel to the hair cortex or as small BB shot-like spores just outside the hair shaft.

Tinea Corporis

A smear from an active ringworm will usually show some hyphae. If none are seen, reconsider your diagnosis.

Tinea Cruris

This infection spreads peripherally and produces a red, wormlike lesion at its borders. This active border is the best site for collecting a specimen for examination. The spreading rash leaves behind a reddish

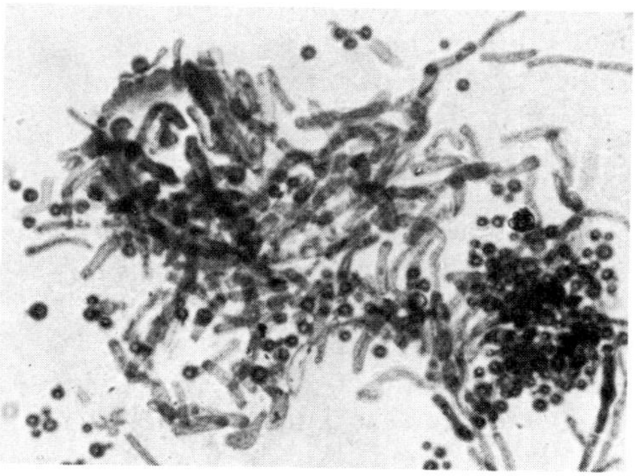

Figure 20–2. Tinea versicolor. Note the classic "spaghetti-and-meatball" configuration. Stain was added to enhance photographic contrast. Viewed under high-dry objective.

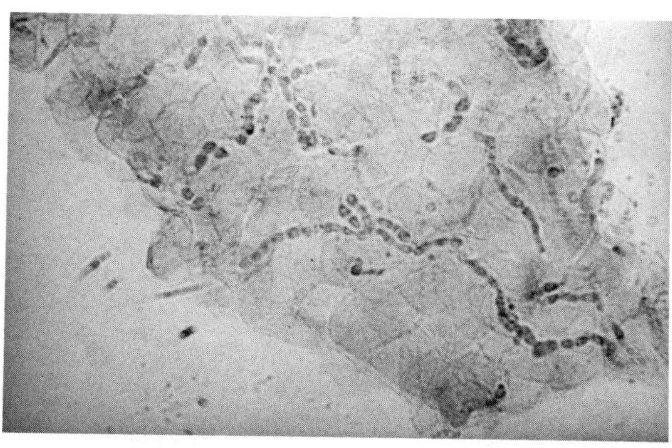

Figure 20–3. Macrocondria of T. Rubrium. Stained with lactophenol blue. Viewed under high-dry objective.

brown hyperpigmented area. It is important to prove that there is an actual infection (by microscopic examination of the scrapings) rather than just the residual hyperpigmentation from a past infection.

Tinea Pedis (athlete's foot)

There are few diseases that mimic tinea pedis, and a KOH smear is therefore not often needed. If the diagnosis is in question, the KOH smear is usually helpful.

Tinea Unguium

There are few diseases that mimic tinea pedis, and a KOH smear is therefore not often onychodystrophy. The Tinea infection must be treated for months with griseofulvin, so a fungal culture should be used to confirm the diagnosis. Consult your reference laboratory for the proper technique of specimen collection for a fungal culture.

Trichophyton Rubrum

This fungus frequently causes tinea corporis, tinea cruris, tinea pedis and tinea unguium in adults. The lesions have marked red margins and central clearings. Microscopically you can see large macrocondria (Fig. 20–3).

Candidiasis

Oral candidiasis (thrush) is a common infection in infants and is usually caused by *Candida albicans*. Intertriginous candidiasis can be found in the diaper area, axillae, and umbilicus. It is characterized by well-circumscribed, red, moist lesions and satellite pustules. In adults, oral candidiasis may be an indication of unsuspected diabetes or acquired immunodeficiency syndrome (AIDS). On microscopic examination, the fungi appear as long, branched organisms with a uniform diameter. These pseudohyphae are indistinguishable from the true hyphae of the dermatophytes. The filaments will frequently contain budding spores and oval yeast bodies. Figure 20-4 shows what the organism looks like Gram's stained. See Figures 19–24 and 19–25 for how it looks with a KOH preparation.

COMMENTS

- If the skin is not cleaned before scraping, the oil will get mixed in with the skin scrapings. The oil droplets can mimic fungal buds except that they will not have a uniform diameter.
- Lint or thread can be confused with hyphae, but they will not be segmented and will vary in diameter.

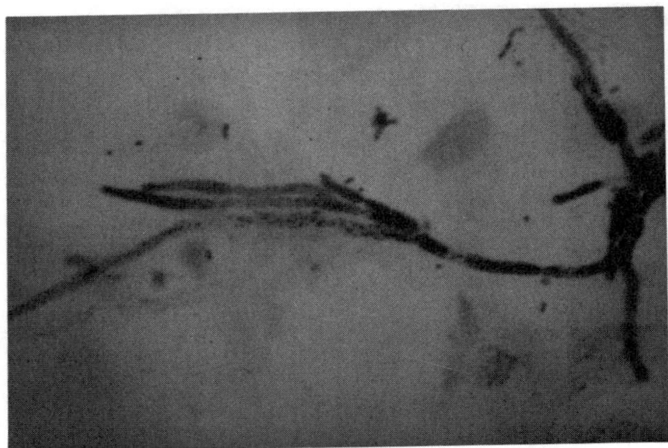

Figure 20–4. Gram's stain showing yeast. Viewed under oil-immersion objective.

SECTION 5: Examination for Scabies

BACKGROUND

Infection with the scabies mite (*Sarcoptes scabiei*) is spread by direct personal contact with an infected person, infected clothing, towels, or bed sheets. The diagnosis of scabies should be considered in any patient with a rash that itches and for which there is no other obvious etiology.

The diagnosis can often be established by the history and examination of the skin. For less typical lesions, identifying the mite, its eggs, or feces (scybala) can help to confirm the clinical suspicion. We have found, however, that laboratory confirmation often proves to be an unsuccessful undertaking.

MATERIALS

- No. 15 scalpel blade.
- Microscope slide.
- Mineral oil.
- Alcohol pads.
- Microscope.

SITE SELECTION

A new papule that has not been scratched must be found. The mite is very difficult to find in older excoriated lesions. The early papules are often the most pruritic, and the patient can therefore direct you to a likely site. The most common sites are between the fingers, wrist, elbow, axilla, scrotum, buttocks, popliteal spaces, ankles, and beneath the breasts. The papules will be 0.2 to 0.5 cm across and may be capped with a pinpoint crust. The mite's burrow may be seen as a faint zigzag line extending from the papule for 1 or 2 cm. This line will be slightly raised and white or gray in color. A helpful sign is that the burrow crosses the normal skin lines.

PROCEDURE: PAPULE SCRAPE TEST

1. Apply two to three drops of mineral oil to the papule.
2. Stretch the skin in the area of the papule between the thumb and the index finger of your left hand.

3. Hold a No. 15 scalpel blade in your right hand. Place it perpendicular to the skin and scrape back and forth several times. (By holding the blade perpendicular to the skin, you are less likely to produce an unwanted laceration.)
4. Collect the scrapings on the edge of the scalpel and transfer them to a microscope slide.
5. Place a coverslip over the scrapings and examine the slide under the 10x objective for mites, feces, and eggs. Suspicious areas should be more closely examined with the 40x objective. The female mite is about 0.4 mm long. It has four pairs of legs with visible spines and bristles on its back surface (Fig. 20–5). The male mite is rarely identified and is about half the size of the female. The eggs are transparent oval bodies about 0.1 to 0.15 mm long. They are uniform in size and shape and are usually found in multiples. They should not be confused with air bubbles , which are also round but will vary in size. Air bubbles will enlarge when presure is applied over the coverslip. The feces appear as clumps of small, round debris around the eggs. Without the presence of the eggs or the mite, it is difficult to differentiate the feces from other cellular debris.

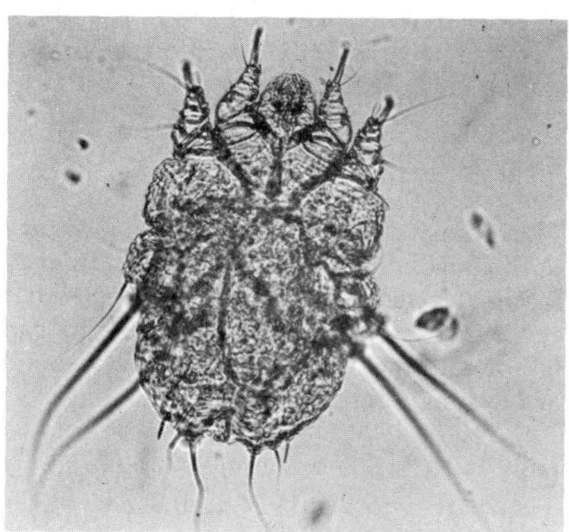

Figure 20–5. Female scabies mite. Viewed under low-power objective.

PROCEDURE: EPIDERMAL SHAVE BIOPSY

A shave biopsy can be useful when the previous technique fails to establish the diagnosis of scabies. There is an improved success rate with this technique because the biopsy preserves the mite eggs, which otherwise could be destroyed by the vigorous scraping of the papule.

1. Choose a new papule to biopsy.
2. Squeeze the skin in this area between the forefinger and the thumb to slightly raise the papule. For papules on the hand, it may be easier to raise the skin by gently squeezing with a curved hemostat.
3. Hold a No. 15 scalpel blade parallel to the surface of the skin. Using a back-and-forth motion, "saw off" the top of the papule and the burrow. The biopsy should be very superficial, and therefore no local anesthetic is needed.
4. Remove the biopsied tissue and place it on a microscope slide.
5. Add a drop of immersion oil on top of the biopsy specimen and cover it with a coverslip.
6. Scan the specimen under the microscope for mites, eggs, and feces, as was done for the scraped specimen.

COMMENTS

- If a young child is suspected of having scabies, ask the parents or an older sibling if they have a similar pruritic rash. The specimen will be easier to obtain from the parent than from the frightened young child.
- Each of these tests is highly specific for the diagnosis of scabies when properly

performed. Of the two tests, the shave biopsy technique is the most sensitive. It has been reported to be positive in 95% of patients with scabies. Our experience has not been nearly as rewarding.

- Resist the temptation of calling cellular debris mite feces. Identifying the feces without seeing the mite or its eggs is dubious.
- It is appropriate to treat someone for scabies even if the laboratory tests are negative. This can be a difficult organism to find.

REFERENCES

Martin WE, Wheeler CE. Diagnosis of human scabies by epidermal shave biopsy. *J Am Acad Deramtol.* 1979;1:335.

SECTION 6: Identification of Lice

BACKGROUND

Three species of lice infect humans. Each is a bloodsucking ectoparasite (i.e., a parasite that lives on the outside surface of the body). Humans are the only host for these parasites.

The diagnosis of a louse infection is suspected in patients with itching. If a louse is found on examination, there is little need for laboratory confirmation. We nevertheless find it educational for patients to examine the organism under the microscope. If no louse is found, the diagnosis can be confirmed by identifying a nit. These are louse eggs that adhere to hair shafts. Hairs can be examined under the microscope to differentiate nits from dandruff or other material found in the hair.

TYPES OF LICE

Pubic Lice (''Crab'' *Phthirus pubis*)

These organisms are primarily found in the pubic area but can be seen in other hairy areas, such as the axilla. They are 0.8 to 1.2 mm long. A casual examination of the pubic area might mistake them for a freckle, but with closer observation they may move. With the 10x objective on the microscope, the organism is seen to have three pairs of legs, two of which have large claws (Fig. 20–6). The nits are usually found close to the base of the pubic hair shaft.

Head Louse (*Pediculus humanis capitis*)

These parasites are found in the hair of the scalp. They are fairly large (3 to 4 mm). Patients may report finding them on their pillows. Under the microscope they cannot be fully seen with a single microscopic field (10x objective). The organism has three pairs of legs, all about the same length, and an elongated, hair-covered body (Fig. 20–7). Its nits appear as small, white specks that are adherent to the hair shaft.

Body Louse (*Pediculus humanis corporis*)

This parasite is very similar in appearance to the head louse (Fig. 20–7). It is primarily found attached to the fibers of clothes that come in contact with the skin. Its nits can be found attached to either the hair shafts or clothing fibers.

Nits

These are 1 to 2 mm long and encase the entire hair shaft (Fig. 20–8). It is important to pull the hair in question and examine it

Figure 20–6. Pubic louse (crab). Viewed under scanning (extra low power) objective.

Figure 20–7. Head louse. Viewed under scanning (extra low power) objective.

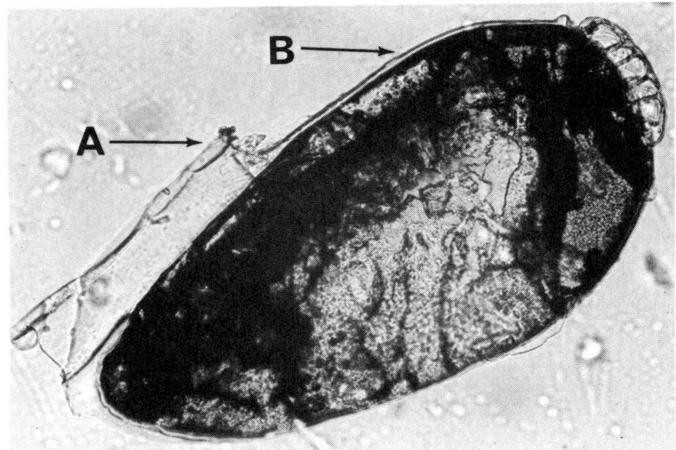

Figure 20–8. Head louse nit. **A.** Small hair shaft fragment. **B.** Double-wall shell around the nit. Viewed under low-power objective.

under the microscope. Many things may mimic nits to the naked eye (Fig. 20–9).

MATERIALS

- Forceps
- Collection cup: plastic urine collection cup with a lid can be used.
- Microscope.
- Microscope slides: nonfrosted slides are best.
- Coverslips.
- Immersion oil or mounting medium.

PROCEDURE

1. Carefully examine the area that itches. Look carefully for a moving louse or a nit on the hair shaft. Nits may look like dandruff but will be tightly adherent to the hair shaft. Examine any clothing the patient is wearing that comes in contact with the involved skin. The inside of the patient's underwear often reveals small, dark excrement particles in pubic-lice infestations.
2. If a louse if found, pick it up with the forceps and place it into a plastic specimen container. If a suspected nit is

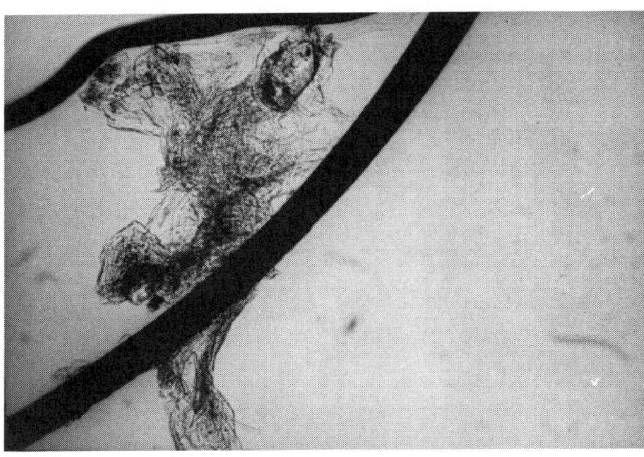

Figure 20–9. When viewed with the naked eye, this hair shaft looked like it had a nit on it. Under the microscope what grossly appeared to be a nit turned out to be debris.

found, pluck the hair shaft and place it into the collection cup.

3. Cover the collection cup and take it to the laboratory for examination.
4. Open the cup and place the louse or hair onto a microscope slide.
5. Place a drop of immersion oil or several drops of mounting medium over the specimen and cover it with a coverslip.
6. Examine the slide under the microscope with the 10x objective and with the condenser about halfway down. Nits can be distinguished from other debris because the nit has an intact shell casing totally around it.
7. Wash your hands carefully after handling the specimen. (Even when you have not become infected, you may notice that you start itching!)

REFERENCES

Brown H, Neva F. *Basic Clinical Parasitology*. 5th ed. New York, NY: Appleton; 1983:252-255.

Scott M Jr, Scott M Sr. Nits or not? *JAMA*. 1980;243:2325.

CHAPTER 21

Other Tests

SECTION 1: Introduction

This chapter describes a variety of tests that can be easily done in the office laboratory. Included are Hansel staining for eosinophils, examination for pinworms, stool examination for white blood cells (WBCs), testing for occult blood in the stool, the microscopic examination of synovial fluid, and laboratory studies for diagnosing infectious mononucleosis.

These tests do not easily fit into the standard laboratory disciplines highlighted in the previous chapters. It is likely that the next 10 years will lead to a rethinking of what constitutes a separate laboratory discipline. Already WBC differentials are being done by instruments that "count" cells based on entirely different characteristics from those seen under the light microscope. Cultures are being replaced by chemistry tests that detect either specific antigens or DNA sequences. A woman's risk for cervical cancer is being measured not by cytologic changes but rather by identifying infectious agents.

SECTION 2: Tests for Occult Blood in the Stool

BACKGROUND

The first use of guaiac-impregnated slides to detect asymptomatic colorectal cancer was in 1967. This testing became one of the standard adult health-maintenance screening procedures during the 1970s. The American Cancer Society currently recommends fecal occult blood screening once each year for persons over the age of 50 years.

Within the past few years there has been a growing debate in the literature about these recommendations. Research has shown that the guaiac slides are neither highly sensitive nor specific. Most individuals with positive tests do not have colon cancer, and many persons with colon cancer have negative tests. The cost-benefit ratio of this screening has come under serious questioning. Furthermore, no study to date has shown that occult blood screening reduces the mortality from colon cancer.

Colorectal cancer is second only to lung cancer as a cause of cancer-related deaths in the United States. It is estimated that in the population over the age of 40 years there is a prevalence of one to two colorectal cancers per 1,000 persons and a 5%

to 10% prevalence of colon adenomas. A small number of these adenomas are thought to progress to cancer.

The 5-year survival rate for colon cancer is 40% to 45%. This survival rate has changed little during the past decade, despite the emphasis on occult blood screening during this period of time. This fact has led to a re-examination of the literature. Most of the studies that have evaluated occult blood screening have shown that, in general, cancers were detected at an "earlier" stage. It was assumed that this would lead to a lower mortality figure; however, this assumption has not yet been proven. An alternative explanation is that screening of asymptomatic individuals detects cancer sooner but does not delay death. If this were true, mortality would not be affected. Currently there are long-term studies in progress that should settle this question.

Most of the available occult blood tests use paper impregnated with a guaiac reagent. Hemoglobin from occult blood acts like a peroxidase enzyme and develops a blue color when combined with guaiac. This reaction is facilitated by adding hydrogen peroxide, which is used as the developer in most test kits. Unfortunately guaiac slide testing is extremely problematic. False-positive test reactions can be found with diets that contain other peroxidases or with the use of a variety of medications. False-negative testing may be due to prolonged slide storage, low test sensitivity, medications, and dietary bulk. Most population-based studies have shown that 2% to 6% of screened individuals will have a positive test. However, only 5% to 10% of persons with a positive test will have colon cancer. In addition, only 50% of patients with proven colorectal cancer will have a positive guaiac test.

TEST INDICATIONS

- Colorectal cancer-screening examination for any patient over 50 years of age.
- Screening examination for patients with a history of colonic polyps or a family history of colon cancer.
- Testing of patients with suspected gastritis, esophageal varices, ulcers, gastric cancer, hiatal hernia, or other gastrointestinal lesions that produce bleeding.
- Evaluation of patients with abdominal pain.
- Evaluation of patients with a drop in hematocrit.
- Evaluation of patients taking aspirin, Coumadin (sodium warfarin), or other drugs likely to cause blood loss from the stomach or the bowel.

COMMENTS

- By the time this book is widely available, a new antibody-based test for occult blood in the stool should be in wide use. We expect that this test will replace guaiac testing because of its improved sensitivity and specificity. The new test is based on antibodies that are specific to human red blood cells (RBCs). The new test procedure, its interfering substances, and its performance will all be quite different from testing with guaiac cards. Because of this development, a detailed description of the guaiac test has not been included.

REFERENCES

Simon JB. Occult blood screening for colorectal carcinoma: A critical review. *Gastroenterology.* 1985;88:820-837.

Ahlquist DA. Fecal blood levels in health and disease. A studying using Hemoquant. *N Engl J Med.* 1985;312:1422-8.

Personal communication: Smith-Kline Diagnostics, 880 West Maude Avenue, PO Box 61947, Sunnyvale, CA 94086.

SECTION 3: Testing for Infectious Mononucleosis

BACKGROUND

Infectious mononucleosis is an incredibly varied disease that can range from asymptomatic (especially in young children), to very benign, to life threatening. It is caused by infection with the Epstein-Barr virus (EBV). It can affect the lymph nodes, spleen, liver, nervous system, bone marrow, kidney, heart, and adrenals. It is usually a more serious disease in older patients.

Although it is called the "kissing disease," which suggests a high transmissibility, studies have shown that it is rarely passed to spouses or other household members. The incubation period has been poorly documented but can range from 1 to 3 months. Patients can excrete the virus for up to 2 years after recovery from the clinical disease.

The classical patient is a teenager with a sore throat, fever, adenopathy, and fatigue. Patients can also have splenomegaly, jaundice (from EBV hepatitis), vomiting, and thrombocytopenia. Figures 21–1 and 21–2 show the frequency and duration of the signs and the major physical symptoms in young adults with infectious mononucleosis. Figure 21–3 shows the major laboratory findings.

Patients with infectious mononucleosis develop certain serologic markers of infection. The ones that are most commonly tested for are the heterophile antibodies. These are nonspecific IgM antibodies.

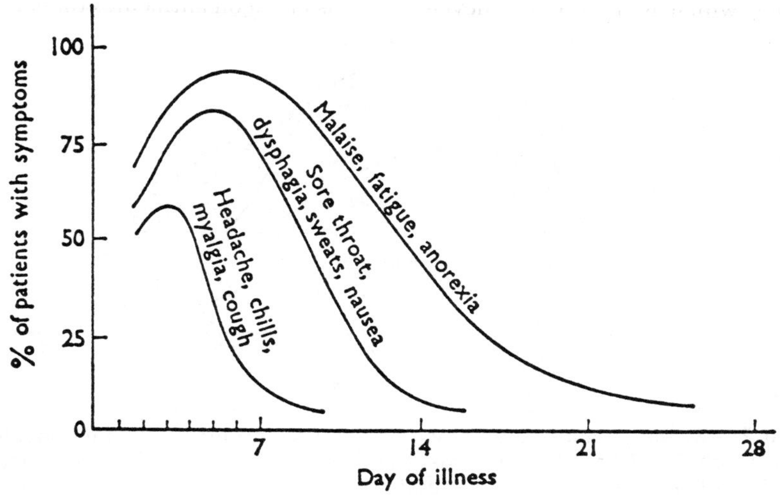

Figure 21–1. Usual frequency and duration of symptoms in adults with infectious mononucleosis. (From Finch SC. Clinical Symptoms and Signs of Infectious Mononucleosis. In *Infectious Mononucleosis,* RL Carter and HG Penman (eds), Oxford, Blackwell Scientific Publications 1969, with permission)

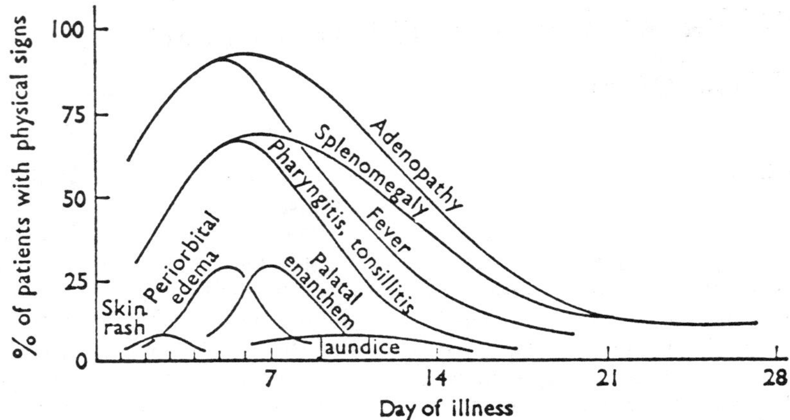

Figure 21–2. Usual frequency and duration of physical findings in adults with infectious mononucleosis. (From Finch SC. Clinical Symptoms and Signs of Infectious Mononucleosis. In *Infectious Mononucleosis,* RL Carter and HG Penman (eds), Oxford, Blackwell Scientific Publications 1969, with permission)

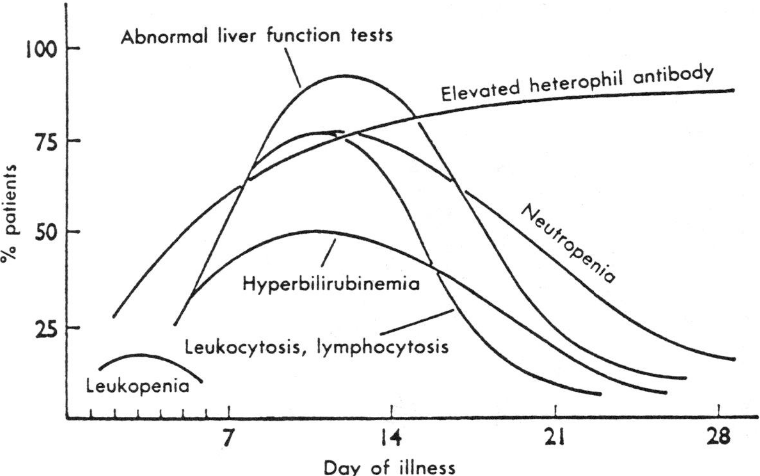

Figure 21–3. Major laboratory findings in adults with infectious mononucleosis. (From Finch SC. Clinical Symptoms and Signs of Infectious Mononucleosis. In *Infectious Mononucleosis,* RL Carter and HG Penman (eds), Oxford, Blackwell Scientific Publications, 1969, with permission)

There are also specific EBV serologic markers:

- IgM antiviral capsid antigen (IgM VCA)
- IgG antiviral capsid antigen (IgG VCA)
- antidiffuse component of the early antigen (D-EA or EA)
- anti-EBV nuclear antigen (EBNA)

These other tests are occasionally done in confusing cases (i.e., when the heterophile antibody is negative but there is a strong clinical suspicion of mononucleosis.)

Figure 21–4 shows the time course of the heterophile and EBV antibodies in cases of infectious mononucleosis.

In addition to developing antibodies, patients may also develop hematologic changes. There is an increase in circulating lymphocytes and in "atypical" or "reactive" lymphocytes. Figure 21–4 shows the time course of those hematologic changes.

It was once thought that a person could have mononucleosis only once in his or her life. In the past few years, however, it has been suggested that there is a recurrent, or chronic, form of the disease. The diagnosis of chronic mononucleosis has been suggested in patients who suffer disabling fatigue for over 6 months and who have a certain pattern of EBV serologic test results.

TEST INDICATIONS

Patients with fever, adenopathy, pharyngitis, splenomegaly, jaundice, liver tenderness, fatigue, vomiting, headache, or myalgias.

TESTS TO ORDER

An accurate diagnosis of infectious mononucleosis relies on laboratory confirmation.

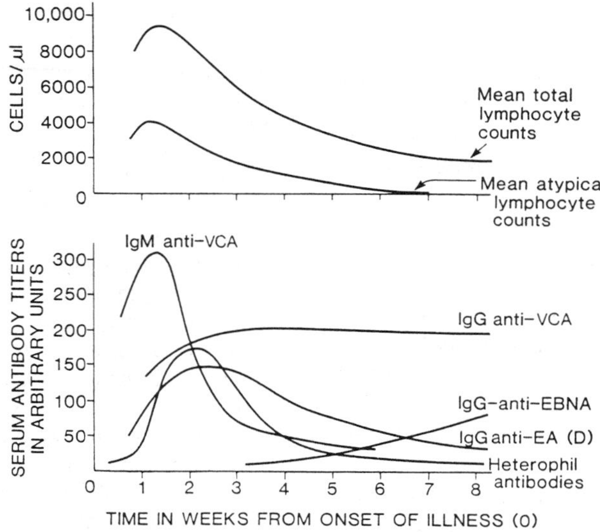

Figure 21–4. The relationship between heterophil antibodies and other Epstein-Barr antibodies and the mean total lymphocyte and atypical lymphocyte counts. VCA = Viral capsid antigen; EBNA = Epstein-Barr nuclear antigen; EA = Early Antigen; D = Diffuse component (From Wintrope M. *Clinical Hematology* 8th ed. Philadelphia, Lea and Febiger, 1981, with permission).

The *routine* laboratory evaluation of mononucleosis should include:

- A test for heterophile antibodies. This is the serology test that is done by most office or reference laboratories to detect infectious mononucleosis. It is not a specific EBV antibody test! The heterophile antibody test usually turns positive 2 to 3 weeks after the onset of symptoms; however, it may take up to 3 months to appear. It usually remains positive for 4 to 8 weeks, but this can be as short as 7 days or as long as 18 months. Children under 3 years of age rarely have a positive test. Only 50% of patients under 8 years will have a positive test.
- WBC Differential: Hematologic changes occur before the appearance of the heterophile antibodies. These usually appear between day 5 and 20 of the illness. They usually last for 1 month but can persist for up to 3 months. The WBC differential provides information about "relative" lymphocytosis and the presence of atypical or reactive lymphocytes. In addition, because the heterophile antibody test can be falsely positive in serious diseases such as leukemia and hepatitis, the peripheral WBC differential provides the clinician with an essential confirmatory test.
- WBC Count. This test is performed because it is necessary to calculate the "absolute lymphocyte count." This count is calculated by multiplying the percentage of lymphocytes (both regular and "atypical") by the total WBC count. Absolute lymphocytosis is an increase in lymphocytes to greater than 4.5×10^9/L (4,500/mm^3)
- Extra Red-Top Tube: Because the diagnosis is still suspected in many patients who have "negative" routine tests, we usually collect an extra red-top tube. This can then be used for EBV serologies, if necessary.

SPECIMEN COLLECTION

A purple-top and plain red-top tube should be collected.

The advantage of using plasma for the tests is that you do not have to wait the 5 to 10 minutes for the blood to clot. Prepare the specimen for the WBC count *before* you centrifuge the specimen to obtain the plasma.

To obtain serum, the blood must first clot. This takes about 5 to 10 minutes. Check for clotting by holding the red tube upside down. The tube can be centrifuged when the blood no longer flows. (See Chap. 13 for a detailed description of preparing plasma or serum.)

TEST PROCEDURES

Heterophile Antibody Test

There are a variety of tests available for office use. These usually can be run on serum, plasma, or fingerstick whole blood. The product insert will give you the specific test characteristics. Check with your scientific supply catalog for specific products. Two commonly used tests are Monospot (Ortho Diagnostics), and Mono-test (Wampole). These two kits use specially treated blood cells that are added to the patient sample to see if there is hemagglutination. These tests are quick and simple to perform. The biggest problem is in identifying subtle agglutination. While trained laboratory staff usually have little difficulty recognizing agglutination, untrained personnel may have difficulty with weak reactions.

Ventrex Laboratories has introduced a new type of heterophile antibody test (VENTRESCREEN), which is based on a solid-phase enzyme immunoassay (EIA) endpoint. The advantage of this test is that it has a colorometric endpoint that may be easier to recognize that agglutination. The disadvantages are that it requires more time

(7 to 10 minutes) and has many more steps (including five washings).

Peripheral Blood Smears

Our modified WBC differential (Chap. 15, Section 10) does not attempt to describe the differences between the various types of mononuclear cells. It is exceedingly difficult to make these distinctions without concentrated hematology training. If you have a technician who has had such concentrated training and is able to reliably recognize and distinguish normal lymphocytes, monocytes, atypical or reactive lymphocytes, and blast forms, then the differential can be done in the office. Otherwise, request a WBC differential from your reference laboratory. A reference laboratory usually requires a purple-top tube for this test.

WBC Count

See Chapter 15, Section 6.

TEST INTERPRETATION

Routine Tests

1. The patient with all of the following results has infectious mononucleosis:
   a. A positive heterophile serology.
   b. A relative lymphocytosis (i.e. > 50% lymphocytes).
   c. Greater than 12% reactive or "atypical" lymphocytes.
   d. An absolute lymphocytosis (i.e., > 4.5 × 10^9/L [4,500 lymphs/mL])
2. If the patient has a characteristic blood smear but a negative heterophile antibody test:
   a. It may be too early in the clinical disease to show a positive test. Repeat the heterophile antibody test several times during the next few months.
   b. The patient may be young and therefore will not develop a positive serology, despite having infectious mononucleosis.
   c. The patient may be infected with another disease that produces atypical lymphocytes (i.e., cytomegalovirus disease, rubella, rubeola, roseola, viral hepatitis, or toxoplasmosis) (Table 21-1).
3. If the patient has a positive Mono-Test but a normal peripheral smear, the patient may have had infectious mononucleosis but be beyond the time when the hematologic changes are seen, or the antibody test may be falsely positive.

EBV Serology

In clinically confusing cases, send a red-top tube to your reference laboratory for specific EBV testing.

1. Early Infection: EBV-specific antibodies to VCA or IgM and IgA classes
2. Recent or Acute Infection
   a. IgM anti-VCA: titer > or equal to 1:10
   b. IgG anti-VCA: titer > or equal to 1:320
   c. Anti-D-EA: titer > or equal to 1:10
   d. Absence of anti-EBNA: titer < or equal to 1:2
   e. Follow-up to have disappearance of IgM anti-VCA and anti-D-EA
   f. Follow-up to develop anti-EBNA
3. Past Infection (Lifelong)
   a. IgG anti-VCA: constant titer between 1:10 and 1:160 in acute and convalescent specimens
   b. Anti-EBNA: constant titer in acute and convalescent specimens
   c. Absent IgM anti-VCA: titer < or equal to 1:10
   d. Absent anti-D-EA: titer < or equal to 1:10
   e. Anti-R-EA: low titer
4. Re-emergence
   a. Increase in IgG anti-VCA
   b. Reappearance of anti-EA

5. Chronic Mononucleosis Syndrome: symptoms for at least 6 months, plus:
   a. IgG anti-VCA: titer > 160
   b. Anti-EA (R-EA and/or D-EA): titer >5
   c. Positive anti-EBNA
   d. Absent IgM anti-VCA
   e. Absent Paul-Bunnell heterophile antibody
6. No Infection (either recent or past): Absence of all EBV antibodies in convalescent and acute specimens

QUALITY CONTROL

- Run positive and negative controls with the first patient specimen of the day. Record the results on a form designed to accommodate your specific test (Fig. 8–8). Indicate the date, technician, kit reagents, expiration date, and control numbers. If the controls do not react properly, open another kit.
- The combination of the heterophile serology test and the WBC count and differential add important information—increasing the positive and negative predictive value of the tests.

SOURCES OF ERROR

Heterophile Antibody Test

The test is reported to have a sensitivity of 98.6% and a specificity of 99.3%. False-negative tests are seen in patients who have very low heterophile antibody titers and in the very early part of the disease. False-positive tests have been reported in patients with leukemia, viral hepatitis, myocardial infarctions, Hodgkin's disease, malaria, lymphoma, multiple myeloma, diabetes with atherosclerotic heart disease, pancreatic cancer, and rheumatoid arthritis.

Peripheral Blood Smear

Atypical lymphocytes can be seen in conditions other than infectious mononucleosis.

TABLE 21–1. CAUSES OF ATYPICAL LYMPHOCYTOSIS

I. Often more than 20% of the WBCs are atypical lymphocytes:
   (a) Infectious mononucleosis
   (b) Infectious hepatitis
   (c) The post-transfusion syndrome
   (d) Dilantin hypersensitivity

II. Less than 20% of the WBCs are atypical:
   Mumps
   Varicella
   Rubeola
   Rubella
   Herpes simplex
   Herpes zoster
   Roseola infantum
   Influenza
   Nonspecific upper respiratory infection
   Many other viral illnesses
   Tuberculosis
   Agranulocytosis
   Lead intoxication
   Stress

Adapted from Wood T, Frenkel E. The atypical lymphocyte. Am J Med. *1967;42:929.*

Table 21–1 lists the causes of atypical lymphocytes.

REFERENCES

Chang RS. *Infectious Mononucleosis.* Boston, Mass: Hall; 1970.

Henle W, Henle G. Serodiagnosis of infectious mononucleosis. *Res Staff Phys.* 1981;27:1.

Horowitz C. Practical approach to diagnosis of infectious mononucleosis. *Postgrad Med.* 1979;65:179.

McKenna R. Infectious mononucleosis: Part 1. Morphologic aspects. *Lab Med.* 1979;10:135.

Miale J. Laboratory Medicine, Hematology. 6th ed. St. Louis, Mo: Mosby; 1982.

Ophoren J. Infectious mononucleosis: Part 2. Serologic aspects. *Lab Med.* 1979; 10:203.

Scheupner C, Overall J. Infectious mononucleosis and Epstein-Barr virus. *Postgrad Med.* 1979;65:83.

Thompson M. The diagnosis of mononucleosis in the office laboratory. *Primary Care.* 1986;13(4):647-655.

Wintrobe M. Clinical Hematology. 8th ed. Philadelphia, Pa: Lea & Febiger; 1981.

Product Inserts:

MONO-TEST Wampole Laboratories, Division of Carter-Wallace, Cranbury NJ 08512

Ventrascreen: Mono Ventrex Laboratories. Portland Maine.

SECTION 4: Examination for Eosinophils/ Hansel Stain

BACKGROUND

Allergic diseases of the conjunctiva, nasal mucosa, and lung are frequently seen in primary care. These patients complain of itchy or red eyes, hay fever, a runny nose, sinus problems, recurrent cough, or wheezing. When the disease is seasonal or related to a particular environmental allergen, the diagnosis is easy and no laboratory testing is required; but at other times the disease may be difficult to distinguish from common infections. Allergic conjunctivitis may be confused with bacterial or viral conjunctivitis. Allergic rhinitis or sinusitis may be mistaken for common upper-respiratory infections or vasomotor rhinitis. Asthma may be thought to be bronchitis or bronchiolitis.

Allergic disease can be easily differentiated from the infectious etiologies by demonstrating the presence of eosinophils in stained specimens from the secretions of the affected areas. In contrast, infections will show many neutrophils but few eosinophils.

TEST INDICATIONS

- A patient with itching, injection, or tearing of the eye.
- A patient with sneezing and a watery nasal discharge.
- A patient with cough or wheezing who is able to produce a sputum for examination.
- To follow the treatment course of a known asthmatic patient.

TEST MATERIALS

- Hansel stain (Lide Laboratories, 15422 Cousteau Dr, Florissant, MO, 63034): This is a modified Wright's-Giemsa stain that colors eosinophils bright red while neutrophils and mucous secretions are stained blue. This makes it easy to identify eosinophils on a microscopic smear. The stain has a stable shelf life of several years. It is available in 125-mL bottles and can be kept in a smaller dropper bottle for convenient use. The dropper bottle should not be allowed to fall below two-thirds full, as the stain may then become too concentrated. When it drops to this point, discard the remaining stain. Wash the bottle with stain-removing soap (Eradosal) and let it dry completely. Then refill the dropper bottle from the stock bottle.

Regular Wright's stain can also be used for staining eosinophils, but the Hansel modified stain is vastly superior in its ease of use and its excellent eosinophil-staining characteristics.

- Microscope Slides: Calgiswab 1 (Inolex Biomedical Corp, Glenwood, IL): These synthetic fiber swabs have a flexible aluminum shaft. This allows safe sample collection from the eye or the nose (Fig. 18–30).
- Sputum Collection Cup: A covered plastic cup like those used for urine collection can be used.
- Cotton Swabs: A regular 6-inch sterile cotton swab can be used to streak a sample of sputum from the collection cup onto a microscope slide.
- Distilled Water: This is available in 1-gallon jugs from local pharmacies or super markets. The water should be transferred into a 125-mL squeeze wash bottle for easy use.
- 95% Methyl Alcohol (Methanol): This is available from scientific supply companies. It should be stored in a 125-mL squeeze wash bottle and kept near the laboratory sink. A Coplin jar should also be labeled and filled with methyl alcohol.
- Coplin jar: These jars are designed to hold microscope slides while they are being stained. A glass Coplin jar with a screw-on lid is required (Fig. 15–22).
- Grease pencil

SPECIMEN COLLECTION

Nasal Secretions

This specimen can best be obtained by having the patient blow his or her nose into a piece of cellophane or wax paper. A tissue can also be used, but the specimen will adhere to the paper and be difficult to use. If no sample can be obtained by this method, a Calgiswab can be inserted into each nares and rubbed against the inflamed nasal mucosa.

Eye Secretions

A Calgiswab should be used to gently wipe secretions from the conjunctival surface. Be sure not to touch the sensitive cornea.

Pulmonary Specimen

The patient should be instructed to cough up a specimen from deep in the chest (Chap. 18, Section 4). A satisfactory sample will be thick and mucousy. A specimen that is thin and foamy is likely to be from the mouth rather than the lung. In children a sputum sample is difficult to obtain, but a nasal smear will usually show the same findings.

TEST PROCEDURE

1. If a Calgiswab was used to obtain the specimen, wipe the fiber tip onto the center of a microscope slide. Try to spread the sample as thinly as possible onto the slide.
2. If the specimen was collected in a container or wax paper, examine it closely. Look for an area in the sample that has a thick, purulent consistency. Collect this portion onto a cotton swab and thinly streak it onto the center of a microscope slide.
3. Dip the slide into a Coplin jar containing methanol for a few seconds or use a squeeze bottle to flood the slide. This will fix the sample to the slide.
4. Allow the slide to air dry. This will take several minutes. The time can be shortened by placing the slide in the incubator.
5. With a grease pencil draw a circle on the slide around the specimen. This will act as a well to keep the stain from spilling off the slide.
6. Cover the entire specimen with Hansel stain and allow it to stand for 30 seconds (slightly longer for thick specimens).

7. Add an equal volume of distilled water to the stain on the slide and allow this to stand for an additional 30 seconds.
8. Pour off the stain and gently rinse the slide with distilled water to remove any excess stain.
9. Flood the slide with methanol from the squeeze bottle and then quickly drain this off. Do not use an excessive amount of methanol, as this will wash the blue stain from neutrophils and give them a pinkish appearance. Even though they will then appear pink, they still can be differentiated from eosinophils because they will not have large red granules in their cytoplasm.
10. Set the slide aside for several minutes.
11. Remove the blue filter from the condenser and then raise the condenser.
12. Scan the entire slide under low power (10x), looking for sheets of WBCs. Neutrophils will be stained blue and will have no cytoplasmic granules. Eosinophils will be red and will contain the characteristic large reddish granules. Areas that appear to contain eosinophils should be checked by examining with the oil objective (Color Plate IV, Fig. 3). More detailed instructions for using the microscope can be found in Chapter 10, Section 8.
13. If the specimen is thick or the staining is too dark, it can be slightly decolorized by rerinsing the slide with methanol.

TEST INTERPRETATION

In specimens from the eye, nose, or lung, the presence of eosinophils indicates an allergic component to the patient's symptoms. If 10 to 20 eosinophils are seen per low-power field and no neutrophils are present, the condition is likely to be totally allergic. The eosinophils are frequently in clumps or adhere to mucous threads. In strongly allergic conditions, such as asthma or hay fever, the smear may show sheets of eosinophils. A patient's response to treatment over time (especially with steroids) can be followed with serial smears. An exacerbation of asthma will show an increase in the eosinophil response, while a respiratory infection in an asthmatic will show an increased neutrophil response.

Some smears will show an even mixture of neutrophils and eosinophils. This indicates an underlying allergic condition with a superimposed infection.

A smear that shows a predominance of neutrophils and macrophages but with few eosinophils is likely to represent an infectious process rather than allergy. See the Hansel stain product insert for more detailed information and pictures .

SECTION 5: Examination for Pinworms

BACKGROUND

Pinworm (*Enterobius vermicularis*) infestation is one of the commonest intestinal parasitic infections in children. The adult female worm is yellowish white and 0.8 to 1.3 cm long. The worms attach to the mucosa of the cecum and at night migrate to the perianal area to lay their eggs. The eggs cause intense itching and vigorous scratching that may keep the child awake at night or alert the teacher at school. Laboratory confirmation of the diagnosis is made by collecting a perianal specimen from the child at night and examining it for worm eggs. It is uncommon to identify the adult worm in the specimen.

MATERIALS

- Pinworm Strip: This is a specially designed strip that includes a sticky end for collection of the specimen (Fig. 21–5). They are available free of charge from Janssen Pharmaceutica, 40 Kingsbridge Rd, Piscataway, NJ 08854 (201-524-9881). This company produces Vermox, which is the most commonly used medication to treat pinworm infections.

  A tongue depressor can be used for specimen collection if no pinworm strip is available. (See Comment 5 below.)
- Microscope: See Chapter 10 for a description of how to use the microscope.

SPECIMEN COLLECTION

1. Instruct the parent to collect the specimen just before the child rises in the morning. However, if the child is still in diapers, the specimen should be collected between 10 PM and midnight. Avoid collecting the specimen right after defecation, since the eggs may be wiped away. The test could then be falsely negative.
2. Give the parent three pinworm strips and advise him or her to collect specimens on 3 consecutive days. All three samples can be returned to the office after the third sample is collected. Write the patient's name on each strip with a grease pencil. Instruct the parent to:
   a. Peel the paper off the rounded, sticky end of the pinworm strip.
   b. Hold the strip in one hand with the sticky side down. With the other hand, separate the child's buttocks.
   c. Press the sticky end of the pinworm strip to the skin around the anus. Be sure to touch all of the skin in a circle around the anus. The pinworm eggs will not be visible to the naked eye.
   d. After the specimen has been collected, bend the sticky part back so that it sticks to the plastic shaft of the strip. Put the strips in an envelope and return them to the laboratory (Fig. 21–5).
   e. Wash hands, including under the nails, after collecting the specimen.

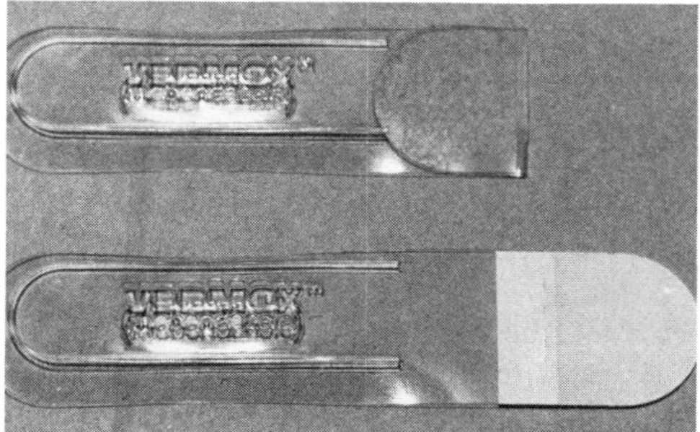

Figure 21–5. Pinworm strips. The top strip has been used and is bent back on itself. The bottom strip is unused.

PROCEDURE

1. Lay a frosted microscope slide on the table top and place a very small drop of xylene in the center. This is used to clear away the sticky material and thereby allows easier identification of the pinworm eggs.
2. Using a hemostat or forceps, peel back the used, sticky portion of the pinworm strip so that it can be pressed onto the microscope slide on top of the drop of xylene.
3. Using a piece of gauze, press the pinworm strip onto the slide, being careful not to catch air bubbles under the strip. The strip can be rubbed gently with a piece of gauze to remove any air bubbles.
4. Examine the specimen under the microscope with the low-power (10x) objective. Be sure that the condenser is in the

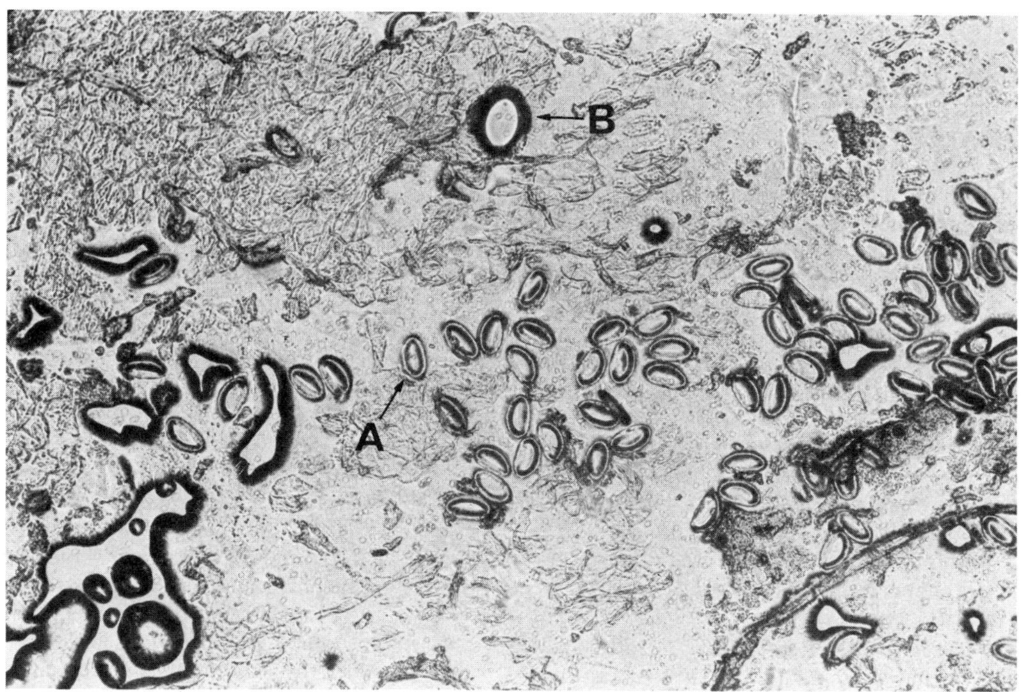

Figure 21–6. A. Pinworm eggs amid fecal debris. **B.** An air-bubble artifact at the top center of the photograph, 10x low-power objective. Viewed under low-power objective.

low position and that the light source is reduced.

5. Pinworm eggs are oval and asymmetrical, with one flattened side (Fig. 21–6). They are 50 to 60 μ long and 20 to 30 μ wide (i.e., about as long as the diameter of a squamous epithelial cell seen in a urine specimen but not as wide). The pinworm egg shell appears as a smooth, thin, double line around the egg. The inner portion of the egg is usually granular and colorless. Some eggs may show the curled-up embryo of the worm (Fig. 21–7). A fresh specimen may show the larva moving within the egg. Air bubbles can be confusing artifacts (Fig. 21–6B). They can usually be differentiated by examining the slide under the high-dry (40x) objective, or they can be pressed away with an applicator stick. In contrast to air bubbles, eggs will not change shape with gentle pressure. Air bubbles are usually encircled by a very dark outline.

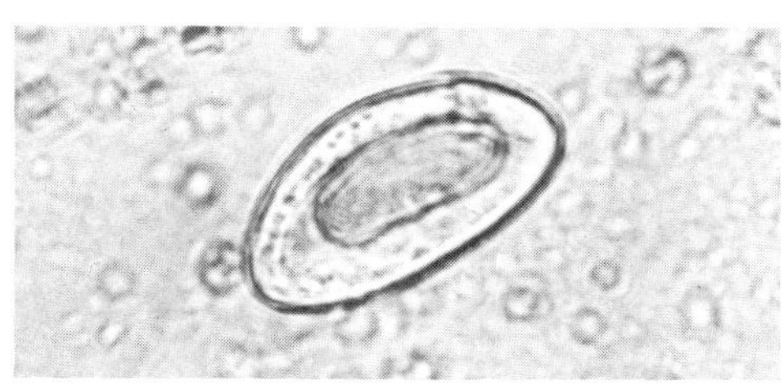

Figure 21–7. Pinworm egg, high-dry objective.

6. Discard used strips in the biologic hazard container. Wash your hands.

COMMENTS

1. If the patient washes or defecates before the specimen is collected, it will decrease the chances of finding the pinworm eggs.
2. A single specimen collection will reveal eggs in infected patients only 50% of the time. With three specimen collections this rate increases to 90%.
3. Threads and lint can mimic the microscopic appearance of the adult pinworm; however, it is unusual with the tape test to see the adult worm. Parents may report seeing worms in the perianal area when they collect the specimen.
4. Specimen collection in the office is possible by doing a digital rectal examination. Apply a small amount of lubricant to the tip of a gloved finger. Insert the finger 1 to 2 cm into the anus and rotate the finger. Remove the fecal material from the tip of the gloved finger and mix it on a microscope slide with one or two drops of xylene. Apply a coverslip and examine as above for the presence of the pinworm eggs.
5. If a pinworm collection strip is not available, an improvised collection system can be made from a tongue depressor, cellophane tape, and a microscope slide. Give these materials to the patient's parent.
   a. Tear off 4 inches of clear cellophane tape.
   b. Stick a 1.5-inch piece of this tape onto the tongue depressor so that the extra length of tape hangs over the end of the depressor (Fig. 21–8A).
   c. Bend the tape back over itself, covering the already stuck tape.
   d. Pull the tape tight and then bend the free end back on itself so that it sticks to the first layer of tape (Fig. 21–8B).
   e. This produces a sticky layer of tape that can then be pressed to the perianal skin for the specimen collection.
   f. After the specimen is collected, a microscope slide should be pressed onto the exposed tape surface. This permits easy transport. The extra tape is cut away and the slide is taken to the office for examination.
6. It is helpful to give the patient information about the pinworm infection. The pinworm eggs are usually transmitted by the fecal-oral route. Chil-

Figure 21–8. A. Preparing a homemade pinworm collection strip. **B.** Prepared collection strip, ready for use.

dren can therefore reinfect themselves by scratching and then putting their hands in their mouths. Both the parent who collects the sample and the laboratory worker who examines the slide should use gloves or wash the hands well. The eggs can also be spread in the air as dust particles. Ninety-two percent of dust samples from the homes of infected children will reveal pinworm eggs. Most of these eggs are not alive, since they can survive for only 2 days in the air. It takes 4 to 6 weeks from the time the egg is ingested until the perianal migration of the adult female worm. One gravid female can deposit about 11,000 eggs! Dogs and cats do not become infected with pinworms, but they may carry the eggs on their fur. Man is the only known host for the pinworm.

7. Keep positive slides for future reference or for teaching purposes. No special care is needed to make them into permanent specimens.

REFERENCES

Brown H, Neva F. Basic Clinical Parasitology. 5th ed. Norwalk, Conn: Appleton & Lange; 1983.

SECTION 6: Stool Examination for Leukocytes

BACKGROUND

Acute diarrhea is a frequently seen problem in office practice. The common form of the disease is a benign, self-limiting, viral enteritis. However, some patients will have more worrisome symptoms, such as bloody stools and abdominal pain. In this latter group of patients a diagnosis of parasitic, bacterial, or inflammatory colitis must be considered.

The laboratory workup for diarrhea includes cultures for pathogenic bacteria (*Salmonella, Shigella, Yersinia,* and *Campylobacter*) as well as a stool examination for parasites (most commonly *Giardia*). These tests are best referred to a reference laboratory. On the other hand, the stool examination for leukocytes is a rapid and convenient test that can be easily performed in the office laboratory and can be helpful in differentiating common viral enteritis from other more serious diseases. If there are no fecal leukocytes, it is usually safe to treat empirically and avoid culturing. Since *Giardia* does not lead to fecal leukocytes, a test for O & P should, however, be considered.

TEST INDICATIONS

- Patients with acute bloody diarrhea.
- Patients with new diarrhea that does not resolve in 2 to 3 days.
- Patients with diarrhea who have recently traveled to an area endemic for parasites.
- Patients with diarrhea who give a history of eating poorly refrigerated dairy or poultry products.

TEST MATERIALS

- Cotton swabs.
- Stool collection cup: the covered plastic

cups that are used for urine collections are useful for stool collections.

- Microscope slides and coverslips (22 × 50 mm) coverslips work best; however, 22 × 30 mm will do for a small specimen).
- Microscope (Chap. 10).
- Methylene blue stain. This is a common bacteriologic stain and has a long shelf life. It can be purchased in small bottles from scientific supply companies. Since only small quantities of the stain are needed in the office laboratory, it may be better to obtain a dropper bottle full from the local hospital laboratory. The stain should be stored in a brown dropper bottle. Write the expiration date, if available, on the bottle label. Do not use the stain beyond this date.

SPECIMEN COLLECTION

Have the patient collect a fresh stool specimen in a plastic collection cup. The sample can be tested for up to 24 hours after collection if it is kept at room temperature and is not allowed to dry out. (A rectal swab is an inadequate specimen, as it results in many falsely negative results.) Mucus, rather than stool, is the specimen of choice. WBCs are secreted by the bowel mucosa, so they are usually found in the mucus or on the outside of the stool.

TEST PROCEDURE

1. Using a cotton swab, separate out a fleck of mucus from the stool sample. If no mucus is found in the specimen, a small portion of liquid stool or a scraping from the outer portion of a formed stool can be used. Do not collect the specimen from the inner portion of a formed stool.
2. Place the small sample on the center of a clean microscope slide.
3. Add two or three drops (this amount varies with the size of the specimen) of methylene blue stain to the sample and mix thoroughly with a wooden applicator stick. Dispose of the stick in the biohazard container.
4. Place a coverslip over the specimen on the microscope slide.
5. Allow the specimen to sit for at least 2 minutes to provide time for good nuclear staining of the leukocytes.
6. Place the slide on the microscope stage and scan the low-power (10x) objective. Find an area that has a thin specimen layer.
7. Examine this area under the high-dry (40x) objective. All slides will show a heavy background of food particles and bacteria. Epithelial cells and RBCs are frequently seen. Positive smears will show many blue-staining leukocytes. In most positive smears the predominant cell type will be polymorphonuclear leukocytes; however, mononuclear leukocytes can also be seen (Color Plate IV, Fig. 4).

RESULTS

The presence of fecal leukocytes in the stool from patients with acute diarrhea indicates a breakdown in the integrity of the intestinal mucosa. This can be caused by both infections and inflammatory diseases. Illnesses that do not cause a break in the bowel mucosa will produce diarrhea without leukocytes in the stool specimen (Table 21–2).

COMMENTS

- Some of the organisms that cause diarrhea are difficult to culture. *Salmonella,* for example, may not be isolated by culture but will give a positive fecal leukocyte test.

TABLE 21–2. FECAL LEUKOCYTES AND CAUSES OF DIARRHEA

Disease	Findings
Viral enteritis	No leukocytes
Noninvasive *Escherichia coli*	No leukocytes
Irritable colon	No leukocytes
Cholera	No leukocytes
Giardia lamblia	No leukocytes
Invasive *E. coli*	More than 24 WBC/HPF
Shigella	More than 25 WBC/HPF
Ulcerative colitis	More than 25 WBC/HPF
Salmonella	1 to 3 WBC/HPF
Typhoid fever	1 to 3 WBC/HPF
Amebic dysentery	Variable
Campylobacter fetus, subspecies jejuni	3 or more WBC/HPF

- If ulcerative colitis or other inflammatory bowel diseases are suspected, a small specimen of fecal mucus can be spread onto a slide and stained with Hansel stain (Chap. 21, Section 4) or Wright's stain. Some of the leukocytes that are seen will be eosinophils. These are uncommon in other diseases that cause acute diarrhea.
- To help learn how to recognize white cells stained with methylene blue, take urine sediment that contains a large number of WBCs and add several drops of methylene blue. Mix using the transfer pipette and allow it to sit for at least 2 minutes. Examine under high dry and identify the stained white cells.

REFERENCES

Harris JC, Dupont HL, Hornick RB. Fecal leukocytes in diarrheal illnesses. *Ann Intern Med.* 1972;76:697.

Siegel D, Cohen PT, Neighbor M, et al. Predictive value of stool examination in acute diarrhea. *Arch Pathol Lab Med.* 1987;111:715-718.

SECTION 7: Microscopic Analysis of Synovial Fluid

BACKGROUND

The examination of synovial fluid for white cells, bacteria, and crystals can yield immediately helpful clinical information. While the number of white cells is not specific for any one type of arthritis, the combination of these three types of microscopic information can often help the clinician decide on an immediate course of treatment.

SPECIMEN COLLECTION

"Tapping" a joint is an important clinical skill to develop. Many joints are easy to tap, and the procedure can be as painless as a venipuncture. A good reference for this procedure is McCarthy DJ. A basic guide to arthrocentesis. *Hosp Med.* November 1968: 77-97.

SPECIMEN PROCESSING

If several milliliters of aspirate or more are available, we:

1. Put a small amount into a plain pediatric red-top tube to be sent for culture, if indicated.
2. Put a small amount in a pediatric purple-top tube. This can be sent to a reference laboratory for an accurate cell count.
3. Deposit a few drops of the unspun synovial fluid onto a microscope slide and coverslip it. This is used for the WBC estimate.
4. Put the remaining sample into pediatric red-top tube. Centrifuge this specimen, pipette off the supernatant, and use the sediment to look for crystals. An additional slide is made from the sediment for Gram's staining.

TEST PROCEDURE

White Cell Estimate

This is not a real WBC count, just a general impression of the specimen. The actual count, sent to a reference laboratory for testing, will often take a day to get back. However, it is clinically helpful to know if there are no white cells, a few white cells, a moderate number, or a packed field. (See Chap. 10, Section 7 for a more detailed description of how to use the microscope.)

1. Place the coverslipped slide of the unspun specimen on the microscope stage between the slide-holding clips.
2. Use the minimum light setting.
3. Scan the specimen under the low-power objective.
4. Switch to high dry and slightly increase the amount of light.
5. Examine at least ten different fields and average the number of WBCs seen/HPF: none, occasional, few (1 to 3/HPF), packed (Figs. 17–18 and 19–21).

Examination for Crystals

1. Polarize your microscope (Chap. 10, Section 10). *Do not put the full-wave compensator in at this time.*
2. Drop a few drops of the centrifuged sediment onto a microscope slide. Coverslip it.
3. Turn off the room lights in the laboratory and slowly examine the cover-slipped specimen. The background should appear dark. Look for *white* needles or small plates (Fig. 21–9). These crystals may be either intracellular or extracellular.
4. When you find a white needle or plate, switch to the high-dry objective. You may have to increase the light at this point.
5. Once you find a "birefringent," or polarized, needle or plate, slide the full-wave compensator in place. Rotate the compensator until the background is magenta or orange/pink.
6. With the compensator in place, see if the crystals are parallel or perpendicular to the long axis of the compensator.
7. Note the color of the crystal. Switch the compensator 90 degrees. Note the color change.

Examination for Bacteria

See Chapter 18 for a detailed description of how to prepare, Gram's stain, and examine a synovial fluid specimen for bacteria.

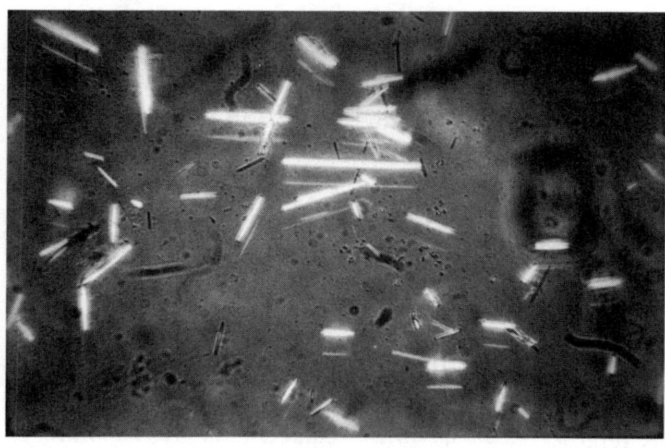

Figure 21–9. These are gout crystals viewed under polarized light. Viewed under high-dry objective.

TEST INTERPRETATION

White Cells

	per μL	Disease
Normal	0-200	None
Noninflammatory arthritis	0-5,000	Osteoarthritis, traumatic arthritis neurogenic joint disease, mild rheumatic fever, systemic lupus erythematosus (SLE), or early bacterial infection.
Mild inflammatory arthritis	0-10,000	Early rheumatic fever, SLE, septic arthritis, or the arthritis associated with ulcerative colitis, regional enteritis, or psoriasis.
Severe inflammatory arthritis	500-100,000	Rheumatoid arthritis, gout, pseudogout.
Septic arthritis	500-200,000	In viral arthritis the leukocyte count is usually less than 10,000/μL, while in bacterial arthritis it is usually greater than 10,000.

Crystals

- Gout: Monosodium urate (MSU) crystals are either needle or rod shaped. They are *yellow* when parallel to the long axis of the red compensator (yellow/parallel) and blue when perpendicular (Color Plate IV, Fig. 5).
- Pseudogout: Calcium pyrophosphate dihydrate (CPPD) crystals are rods, rectangles, or rhomboids. They are *blue* when parallel to the long axis of the red compensator (Color Plate IV, Fig. 6).

Bacteria

The Gram's stain is positive in only about 50% of the patients with septic arthritis. The culture is positive in 30% to 80% of these patients. The most common septic arthritis seen in primary care is staphylococcal (Color Plate II, Fig. 3). A septic arthritis can coexist with other nonseptic types of arthritis (i.e., gout, SLE).

COMMENTS

- Centrifuging the specimens before examining for crystals and bacteria enhances your chances of a positive result.
- If only extracellular gout crystals are found, it is unlikely these are responsible for the acute arthritis. In acute gout the majority of crystals are intracellular. In between acute attacks the crystals tend to be extracellular.
- Joint fluid should be *very* carefully examined before reporting it as negative.
- It is far easier to scan the specimen for crystals using regular polarized light instead of compensated polarization. This is because it is easier to see white crystals against a dark green background than it is to see yellow/blue crystals against magenta.

REFERENCES

Kjeldsberg C, Kreg A. Cerebrospinal fluid and other body fluids. In: Henry J, ed. *Todd Sanford Davidsohn Clinical Diagnosis and Management by Laboratory Methods.* 17th ed. Philadelphia, Pa: Saunders; 1984:475-483.

Appendix

SECTION 1: Laboratory Reference Books

The procedure sections of this book have described a variety of resource material that is available free or for a nominal cost. (See the Gram's Stain section, WBC Differential/RBC Morphology, and the Urinalysis Introduction.) In addition, every office laboratory should have several catalogs from which to order supplies and equipment:

- Fisher Scientific
  Corporate Headquarters
  711 Forbes Ave
  Pittsburgh, PA 15210
- Scientific Products
  Division, American Hospital Supply Corporation
  1430 Waukegan Rd
  McGaw Park, IL 60085
- Curtin Matheson Scientific
  PO Box 1546
  Houston, TX 77251

These companies will supply you with catalogs as well as the addresses of their representatives.

Several other books are handy references in the office laboratory:

- Gardner P, Provine HT. *Manual of Acute Bacterial Infections*. 2nd ed. Boston, Mass: Little, Brown; 1984. This is a spiral-bound book that has good pictures and information about the laboratory and clinical findings in the common bacterial infections.
- Haber MH. *Urinary Sediment: A Textbook Atlas*. Educational Products Division, American Society of Clinical Pathologists, 2100 W Harrison St, Chicago, IL 60612. This book, published in 1981, is one of the best collections of color photographs of the microscopic urine sediment. It also has a short but informative description of the pathophysiology of these sediment findings.
- Henry JB, ed. *Todd Sanford Davidsohn, Clinical Diagnosis and Management by Laboratory Methods*. 17th ed. Philadelphia, Pa: Saunders, 1984. This text is the bible for laboratory medicine. It's emphasis is on those tests performed in hospitals, but there are a few good sections that are relevant to office laboratories.
- Unditz E. *Sandoz Atlas of Haematology*. 2nd ed. Basle, Switzerland: Sandoz; 1973. This is an inexpensive atlas with hundreds of color plates showing both red blood cell (RBC) and white blood cell (WBC) morphology.

The authors have found one laboratory journal that occasionally covers areas of interest for the office laboratory worker:

- Laboratory Medicine
  American Society of Clinical Pathologists
  2100 West Harrison St, Chicago, IL 60612
  (312-738-1336)

This journal is available with membership to the American Society of Clinical Pathologists. Write the ASCP for information about the different types of membership.

The *Journal of the American Medical Association* has also had a series of articles since 1985 about office testing in its "Toward Optimal Laboratory Use" section.

SECTION 2: Exercise in Test Predictive Value

DEFINITIONS

True Positive (TP): A positive test in a diseased patient.
False Positive (FP): A positive test in a nondiseased patient.
True Negative (TN): A negative test in a nondiseased patient.
False Negative (FN): A negative test in a diseased patient.
Sensitivity: The rate of positive test results in a patient with a disease.

$$= \frac{\text{True positives}}{\text{Total diseased patients}}$$

$$= \frac{\text{TP}}{\text{TP} + \text{FN}} \times 100$$

Specificity: The rate of negative test results in patients without the disease.

$$= \frac{\text{True negatives}}{\text{Total nondiseased patients}}$$

$$= \frac{\text{TN}}{\text{TN} + \text{FP}} \times 100$$

Prevalence: The frequency of diseased patients in a defined population.
Positive Predictive Value (+PV): Percentage of patients with a positive test who are diseased.

$$= \frac{\text{True Positives}}{\text{Total Positives}}$$

$$= \frac{\text{TP}}{\text{TP} + \text{FP}} \times 100$$

Negative Predictive Value (−PV): Percentage of patients with a negative test who are free of the disease.

$$= \frac{\text{True Negatives}}{\text{Total Negatives}}$$

$$= \frac{\text{TN}}{\text{TN} + \text{FN}} \times 100$$

Baye's Formula:

$$+\text{PV} = \frac{(\text{Prevalence})\,(\text{Sensitivity})}{[(\text{Prevalence})\,(\text{Sensitivity})] + [(1 - \text{Prevalence})\,(1 - \text{Specificity})]}$$

Truth Table

	+ Test	− Test	Total
Diseased	TP	FN	TP + FN
Nondiseased	FP	TN	FP + TN
Total	TP + FP	FN + TN	TP + FN + FP + TN

PROBLEM 1

The laboratory examination for pinworm eggs is a useful test for children who complain of rectal itching. The test involves the collection of a perianal specimen and then examining it for the pinworm eggs.

1. If properly performed, what is this test's specificity?
2. A clinician is presented with a 6-year-old child who has rectal itching. The mother is given one pinworm strip to collect a specimen and is instructed to return it to the office the next day. The use of a single specimen results in a test sensitivity of 50%. From this data can you determine what the chances are of a child with a single, negative pinworm test being free of the infection?
3. Other causes of rectal itching in children include anal fissures or rectal inflammation. If 40% of children with rectal itching have a pinworm infection, what are the chances of a child with a single, negative pinworm strip being free of infection?
4. A different clinician is presented with another 6-year-old who has rectal itching. This clinician gives the mother three pinworm strips to collect three different specimens. This increases the test sensitivity to 90%. What are the chances of a child with three negative pinworm strips being free of a pinworm infection?

PROBLEM 2

The old agglutination urine pregnancy test was useful in detecting pregnancy. When it is performed at least 41 days after the start of the last menstrual period, it had a 98% specificity and a 95% sensitivity.

1. If 60% of women with a missed period are pregnant, what are the chances that a woman who has a positive pregnancy test is pregnant at 41 days after the last period?
2. What are the chances that a woman with a negative test at 41 days is not pregnant?

PROBLEM 3

A diagonal ear-lobe crease has been described as a clinical sign for the presence of coronary artery disease. It is named Frank's sign after the clinician who first described it. This ear-lobe crease is found in 47% of patients with coronary artery disease but also in 30% of patients without coronary artery disease. Five percent of the adult population has coronary artery disease.

1. What is this test's sensitivity?
2. What is this test's specificity?
3. How likely is a patient with a diagonal ear-lobe crease to have coronary artery disease?
4. What is the false-positive rate for this clinical sign?
5. How does this false-positive rate compare with the false-positive rate obtained if the flipping of a coin was used as a diagnostic test for coronary artery disease (i.e., heads means the presence of coronary artery disease, tails means free of disease)?

REFERENCES

Frank ST. Ear-crease sign of coronary disease. *N Engl J Med.* 1977;297:282.

Haines SJ. Nonspecificity of ear-crease sign of coronary-artery disease. *N Engl J Med.* 1977;297:1181.

PROBLEM 4

A clinician set out to compare a new test-tube turbidity screening test for sickle cell hemoglobin with the older metabisulfite test. (See the sickle cell section of the hematology chapter.) Test results were confirmed by hemoglobin electrophoresis. A total of 2,355 adult black patients were tested. Of these, 2,149 patients had negative testing by both methods and negative hemoglobin electrophoresis. A total of 155 patients had both a positive turbidity test and a positive metabisulfite test. Hemoglobin electrophoresis confirmed the presence of Hb-S in these patients. An additional 47 patients had positive turbidity tests and hemoglobin electrophoresis but falsely negative metabisulfite testing. Four patients had falsely positive turbidity tests, and they each had negative metabisulfite tests.

1. What is the sensitivity, specificity, and positive predictive value for the turbidity and metabisulfite tests?
2. Based on these findings, which of the two tests would be preferred as a "screening" test for the presence of Hb-S?

REFERENCE

Wei-Ping Loh. Evaluation of a rapid test tube turbidity test for the detection of sickle cell hemoglobin. *AJCP.* 1971; 55:55-57.

ANSWERS

Problem 1

1. Specificity is the rate of negative test results in patients who are free of the disease. The pinworm is not part of the normal bowel flora; therefore any patient who is found to have pinworm eggs is infected. Noninfected patients will never have positive pinworm examinations if the test is done properly. The test therefore has a 100% specificity.
2. There is no way of telling. This question asks for the predictive value of a negative test. To determine this, you need to know the disease prevalence (i.e., how many children with rectal itching have a pinworm infection as the cause?).
3. To answer this question, we set up a truth table:
   Given: Specificity = 100%
   Given: Sensitivity = 50%
   Given: Disease prevalence = 40%
   Assume: Population size of 100 children with rectal itching.
   Therefore: 40 of the children will have an infection. Half of these 40 will have a positive test and half will have a negative test. (Remember, the test sensitivity is 50%.)
   And: The remaining 60 children will not have a pinworm infection, and all of these children will have a negative test. (Remember, the test specificity is 100%.)

	+ Test	− Test	Total
Infected	20	20	40
Not infected	0	60	60
Total	20	80	100

The negative predictive value (i.e., the chance of a child with a negative test being free of the infection) is:

$$-\text{PV} = \frac{\text{TN}}{\text{TN} + \text{FN}} = \frac{60}{60 + 20} = 75\%$$

4. Sensitivity = 90%
   Specificity = 100%
   Prevalence = 40%
   Population = 100 children

	+ Test	− Test	Total
Infected	36	4	40
Not infected	0	60	60
Total	36	64	100

$$-\text{PV} = \frac{60}{64} = 94\%$$

Therefore 94% of patients who have three negative pinworm specimens will be free of the infection.

Problem 2

1. Sensitivity = 95%
   Specificity = 98%
   Prevalence = 60%
   Population = 1,000 women with a missed period.

	+ Test	− Test	Total
Pregnant	570	30	600
Not Pregnant	8	392	400
Total	578	422	1,000

2. $+PV = \frac{TP}{TP + FP} = \frac{570}{578} = 99\%$

   $-PV = \frac{TN}{TN + FN} = \frac{392}{422} = 93\%$

Problem 3

1. The test's sensitivity is the percentage of patients with the disease who have a positive test (i.e., 47%).
2. The test's specificity is the percentage of patients free of the disease who have a negative test (i.e., 70%; there is a 30% false-positive rate.)
3. Sensitivity = 47%
   Specificity = 70%
   Prevalence = 5%
   Patient population = 10,000

	+ Test	− Test	Total
Diseased	235	265	500
Not diseased	2,850	6,650	9,500
Total	3,085	6,915	10,000

$$+PV = \frac{235}{3,085} = 7.6\%$$

Therefore only 7.6% of patients with an ear crease will have coronary artery disease!

4. Since only 7.6% of patientswith the ear-lobe crease will have the disease, 92.4% of patients will be falsely classified by this clinical sign.
5. If you flipped a coin to diagnose coronary artery disease, half of the patients with the disease would be diagnosed correctly (i.e., sensitivity = 50%), and half of the patients free of the disease would be diagnosed correctly (i.e., Specificity = 50%).
   Sensitivity = 50%
   Specificity = 50%
   Prevalence = 5%
   Patient population = 10,000

	+ Test	− Test	Total
Diseased	250	250	500
Not diseased	4,750	4,750	9,500
Total	5,000	5,000	10,000

$$+PV = \frac{250}{5,000} = 5\%$$

Therefore 95% of patients with a positive coin-flip test would be incorrectly diagnosed as having coronary artery disease. This is not that much different than the 92.4% error rate with the ear-lobe test. (Remember, if the sum of the sensitivity and specificity is less than 100, the test is no better than chance.)

Problem 4

1.

	Metabisulfite Test		
	+ Test	− Test	Total
Diseased	155	47	202
Nondiseased	0	2153	2153
Total	155	2200	2355

$$\text{Sensitivity} = \frac{155}{202} = 77\%$$

$$\text{Specificity} = \frac{2,153}{2,153} = 100\%$$

$$+ \text{ Predictive Value} = \frac{155}{155} = 100\%$$

Test-Tube Turbidity Test

	+ Test	− Test	Total
Diseased	204	0	202
Nondiseased	0	2,149	2,153
Total	206	2,149	2,355

$$\text{Sensitivity} = \frac{202}{202} = 100\%$$

$$\text{Specificity} = \frac{2,149}{2,153} = 99.8\%$$

$$+ \text{ Predictive Value} = \frac{202}{206} = 98\%$$

2. A screening test should be used to identify the maximum number of cases of the disease, even if this means that the testing will result in some false positives. A "sensitive" test is therefore required rather than a highly "specific" test. The test-tube turbidity test is therefore a much better screening test than the metabisulfite test.

SECTION 3: Preparation of Urine Sediment Reference Slides

It has generally been difficult to preserve urine-sediment for future teaching or examination. This is because it is a wet specimen and its cellular, cast, and crystalline elements easily deteriorate. It is not uncommon, for example, that RBC casts will break apart and be unrecognizable by the time they are received in the laboratory for examination. Urine sediment deterioration is due to the fragility of its elements, changes in the urine pH as it is exposed to air, and the action of bacteria that grow in the specimen. An improved urine preservation technique has recently been described.

MATERIALS

- Several drops of a suitable urine sediment.
- Microscope slide.
- Large glass coverslips (22 × 30)
- Pasteur pipettes and bulb.
- 2-cc syringe with a 25-gauge needle.
- Mounting medium: This clear chemical is used to permanently mount a coverslip onto a microscope slide. It can be purchased in 450-mL bottles. One mounting medium is named Pro-Texx (Lerner Laboratories, Scientific Products, McGraw Park, IL). A single bottle should last an office laboratory for many years.
- 2% Glutaraldehyde-saline solution: Glutaraldehyde is a favorite electron-microscopy fixative. It can be purchased as a 25% solution from scientific supply companies and can then be diluted with saline (about 8 mL of the 25% solution in 92 mL of saline.) This diluted solution should be kept in a brown bottle.

PROCEDURE

1. Spin and decant the urine sediment in the usual fashion. (See the urinalysis chapter.) About 0.50 mL of urine sediment will remain after decanting.

2. Add 4 to 6 drops of the 2% glutaraldehyde-saline solution to the sediment in the centrifuge tube and gently mix.
3. Draw up mounting medium into a 2-cc syringe. Then place a 25-gauge needle onto the syringe. Use this syringe to "draw" a 1 cm-in-diameter circle in the center of a clean microscope slide. This will form a shallow-walled well of mounting medium in which the urine sediment will be placed.
4. Allow the mounting medium to fully dry. This will take about 10 minutes. Then put one to two drops of the preserved urine solution into the well on the slide. It will take some practice to get the right amount of sediment into the well. Too much sediment will flow out of the well when the coverslip is added. Too little sediment will result in air bubbles between the sediment and the coverslip.
5. Using the syringe, fill the area around the circular well with mounting medium. The covered area should approximate the size and shape of the rectangular coverslip.
6. Quickly place the coverslip over the sediment well and the mounting medium. Avoid getting air bubbles under the coverslip.
7. Allow the prepared slide to dry. This will take about one-half hour. It can then be stored with other reference slides for future teaching use.

RESULTS

A wide range of crystals, cells, and casts have been preserved with this technique. Even RBC casts have been nicely preserved for up to 2 years. These slides have been found to be a great improvement over the use of photomicrographs in teaching the urine sediment examination to medical students and residents.

REFERENCE

Fischer, PM, Addison LA. Preparation of urine teaching slides. *J Fam Pract.* 1980;11:1124-1126.

SECTION 4: Preparation of Permanent Reference Slides—Wright's, Gram's, and Hansel Stained

BACKGROUND

Many of the microscopic slides that are prepared each day can be saved for later reference. If the slide has unusual findings that would make it a valuable teaching slide, it can be prepared in a manner that will allow it to be repeatedly reused without damaging the specimen.

MATERIALS

- Mounting Media: Several manufacturers produce a chemical that can be used to make permanent reference slides. Pro-Texx (Scientific Products) is one such material. It is available in 450-mL bottles. A small amount of this liquid can be poured into a clean urine collection cup for easier use when slides are being prepared.

- Glass coverslips: 22 × 50-mm glass coverslips are needed to cover a large area of the microscope slide.
- Forceps.
- Soft cloth.
- Coplin jar: A Coplin jar with a lid should be filled with xylene. This should be a different Coplin jar than the one that is used to clean the immersion oil from microscope slides. Xylene is a respiratory irritant. Be sure that there is good ventilation when it is used.
- Glass stirring rod with a rounded tip or a wooden applicator stick.

PROCEDURE

1. Select a slide for permanent mounting.
2. Pour a small amount of mounting media into a small cup.
3. If the slide has any oil on it, clean it in the regular Coplin jar filled with xylene.
4. Dip the slide into the Coplin jar with the "clean" xylene (i.e., the Coplin jar used only for preparing permanent slides).
5. Remove the slide from the xylene and hold it upright. Touch the opposite long end of the slide to the cloth to draw off excess xylene. Do not blot or wipe it dry.
6. Lay the slide on the cloth with the specimen side facing up.
7. Dip the glass stirrer or wooden applicator stick into the container with the mounting media. Then touch the tip of the rod to the slide and deposit a medium-sized drop of mounting media onto the center of the slide. With experience you will learn the proper size of the drop of mounting media.
8. Take a large coverslip in one hand and the pair of forceps in the other hand. Place the coverslip on the specimen slide at a 30-degree angle. Slowly lower the coverslip until it is flat on the slide. As you are doing this, use the forceps to press any air bubbles from under the coverslip. Do not press too hard or the coverslip will break. If there are any bubbles, remove the coverslip and start over with a new coverslip and a recleaned slide.
9. After the slide is sealed, carefully wipe any excess mounting media from the edges of the slide with a piece of gauze.
10. Leave the slide on a flat surface for several hours to allow the mounting media to dry. If you stack the prepared slides together before they have fully dried, they will stick together. If this happens, put them in the cleaning Coplin jar with xylene and allow them to sit until they can be separated.

COMMENTS

- Pinworm-egg specimens do not need to be mounted with mounting media. The slides can be stored as they are when they are received from the patient.
- If you want to make a permanent slide of a louse or tick, merely place the specimen in the middle of a microscope slide and cover with a large drop of mounting media. Then carefully press a coverslip on top.

Index